Impact of Meat Consumption on Health and Environmental Sustainability

Talia Raphaely
Curtin University, Australia

Dora Marinova
Curtin University, Australia

A volume in the Practice, Progress, and
Proficiency in Sustainability (PPPS) Book Series

Published in the United States of America by
Information Science Reference (an imprint of IGI Global)
701 E. Chocolate Avenue
Hershey PA, USA 17033
Tel: 717-533-8845
Fax: 717-533-8661
E-mail: cust@igi-global.com
Web site: http://www.igi-global.com

Library of Congress Cataloging-in-Publication Data

Names: Raphaely, Talia, 1966- | Marinova, Dora.
Title: Impact of meat consumption on health and environmental sustainability
 / Talia Raphaely and Dora Marinova, editor.
Description: Hershey, PA : Information Science Reference, [2016] | Includes
 bibliographical references and index.
Identifiers: LCCN 2015032770| ISBN 9781466695535 (hardcover) | ISBN
 9781466695542 (ebook)
Subjects: LCSH: Meat industry and trade--Environmental aspects. | Meat
 industry and trade--Health aspects. | Food of animal origin--Health
 aspects.
Classification: LCC HD9410.5 .I47 2016 | DDC 333.95/414--dc23 LC record available at http://lccn.loc.gov/2015032770

This book is published in the IGI Global book series Practice, Progress, and Proficiency in Sustainability (PPPS) (ISSN: 2330-3271; eISSN: 2330-328X)

British Cataloguing in Publication Data
A Cataloguing in Publication record for this book is available from the British Library.

All work contributed to this book is new, previously-unpublished material. The views expressed in this book are those of the authors, but not necessarily of the publisher.

For electronic access to this publication, please contact: eresources@igi-global.com.

Practice, Progress, and Proficiency in Sustainability (PPPS) Book Series

Ayman Batisha
International Sustainability Institute, Cairo

ISSN: 2330-3271
EISSN: 2330-328X

MISSION

In a world where traditional business practices are reconsidered and economic activity is performed in a global context, new areas of economic developments are recognized as the key enablers of wealth and income production. This knowledge of information technologies provides infrastructures, systems, and services towards sustainable development.

The **Practices, Progress, and Proficiency in Sustainability (PPPS) Book Series** focuses on the local and global challenges, business opportunities, and societal needs surrounding international collaboration and sustainable development of technology. This series brings together academics, researchers, entrepreneurs, policy makers and government officers aiming to contribute to the progress and proficiency in sustainability.

COVERAGE

- Eco-Innovation
- Global Business
- E-Development
- Knowledge clusters
- Technological learning
- Socio-Economic
- Green Technology
- Outsourcing
- Innovation networks
- Environmental Informatics

IGI Global is currently accepting manuscripts for publication within this series. To submit a proposal for a volume in this series, please contact our Acquisition Editors at Acquisitions@igi-global.com or visit: http://www.igi-global.com/publish/.

Titles in this Series

For a list of additional titles in this series, please visit: www.igi-global.com

Handbook of Research on Pedagogical Innovations for Sustainable Development
Ken D. Thomas (Auburn University, USA) and Helen E. Muga (University of Mount Union, USA)
Information Science Reference • copyright 2014 • 800pp • H/C (ISBN: 9781466658561) • US $345.00 (our price)

Sustainability Science for Social, Economic, and Environmental Development
Nilanjan Ghosh (Multi Commodity Exchange of India Limited, India) and Anandajit Goswami (The Energy and Resources Institute, India)
Information Science Reference • copyright 2014 • 324pp • H/C (ISBN: 9781466649958) • US $195.00 (our price)

E-Innovation for Sustainable Development of Rural Resources During Global Economic Crisis
Zacharoula Andreopoulou (Aristotle University of Thessaloniki, Greece) Vagis Samathrakis (Alexander Technological Educational Institute of Thessaloniki, Greece) Soulla Louca (University of Nicosia, Cyprus) and Maro Vlachopoulou (University of Macedonia, Greece)
Business Science Reference • copyright 2014 • 317pp • H/C (ISBN: 9781466645509) • US $185.00 (our price)

Sustainability Integration for Effective Project Management
Gilbert Silvius (HU University of Applied Sciences Utrecht, The Netherlands & Van Aetsveld, The Netherlands) and Jennifer Tharp (Mastodon Consulting, USA)
Business Science Reference • copyright 2013 • 482pp • H/C (ISBN: 9781466641778) • US $185.00 (our price)

www.igi-global.com

701 E. Chocolate Ave., Hershey, PA 17033
Order online at www.igi-global.com or call 717-533-8845 x100
To place a standing order for titles released in this series, contact: cust@igi-global.com
Mon-Fri 8:00 am - 5:00 pm (est) or fax 24 hours a day 717-533-8661

To our children, their children and all future children. Thank you for lending us your world.

List of Reviewers

Diana Bogueva, *Curtin University, Australia*
Paula Brügger, *Federal University of Santa Catarina, Brazil*
Matthew Chrulew, Curtin University, Australia
Hans Dagevos, *Wageningen University, the Netherlands*
Xiumei Guo, *Curtin University, Australia*
Jonathan Hallett, *Curtin University, Australia*
Amzad Hossein, *Curtin University, Australia & Rajshahi University, Bangladesh*
Ian Phau, *Curtin University, Australia*
Geoff Russell, *Curtin University, Australia*
Angela Saunders, *Sanitarium, Australia*
Jane Scott, *Curtin University, Australia*
Kurt Seemann, *Swinburne University, Australia*
Vladislav Todorov, *University of Forestry, Bulgaria*
Myrna Tonkinson, *University of Western Australia, Australia*
Jiuchang Wei, *University of Science and Technology of China, China*

Table of Contents

Section 1
Climate and Water

Section 2
Diet and Wellbeing

Section 3
Society and Politics

Section 4
Conclusion

Detailed Table of Contents

Section 1
Climate and Water

Chapter 1

Celia Green, Australian National University, Australia
Andrew Joyce, Swinburne University of Technology, Australia
Jonathan Hallett, Curtin University, Australia
Toni Hannelly, Curtin University, Australia
Gemma Carey, Australian National University, Australia

This chapter examines the link between dietary choices and greenhouse gas (GHG) emissions and possible interventions to reduce this impact. The connections between climate change, food systems and public health are explored. It is shown that there is variance in the impact of different food types on GHG emissions, with animal products having the greatest impact. The role of food system activities in the production of GHG emissions is also explored. Dietary choices and GHG emissions are examined using case studies from a variety of countries. Results show that reduced animal food production has increased potential to reduce GHG emissions compared to technological mitigation or increased productivity measures. Finally, a systems science approach is used to explore possible interventions aimed at reducing consumption of animal products.

Chapter 2

Robert Goodland, World Bank Group, USA

The gravest environmental impact of food production is the impact of its greenhouse gas emissions; that's because of the uniquely diverse, unprecedented, and irreversible risks that it involves. According to the International Energy Agency, atmospheric carbon must be reduced significantly by 2020 or else the world may not be able to avert uncontrollable climate change. This chapter compares the most recent assessment of livestock and climate change by livestock specialists employed by the United Nations' Food and Agriculture Organisation with assessment of livestock and climate change by two World Bank Group environmental specialists, Robert Goodland and Jeff Anhang. It is explained how the only pragmatic way to reverse climate change before it is too late is to replace a substantial amount of today's livestock products with better alternatives.

Chapter 3

In Australia, the public got its first mass marketing about climate change and the measures that would be required to avoid it, by TV images of black balloons and Professor Tim Flannery turning off light switches. Journalistic coverage has been similarly dominated by household electricity. More technical literature is generally dominated by the concept of "carbon dioxide equivalence" (CO2eq) as spelled out in the Kyoto protocol. This concept isn't used in climate models because it makes no physical sense. The use of CO2eq and the focus on household electricity has lead to a profound mismatch between the causal factors as understood by climate scientists and causal factors as perceived by the public. "The public" here isn't just the general public, but people of many backgrounds with a strong interest in climate change but without the deep knowledge of professional climate scientists. We need images consistent with climate models, which accurately rank the causes of climate change and guide proposed actions. Such images point to meat as a key focal issue.

Chapter 4

The meat processing produces solid, liquid and gaseous wastes from stockyards, abattoirs and packing plants contain blood, fats, protein, gut contents, heavy metals, antibodies, hormones and other substances. Solid and liquid wastes can lead to an impairment or disruption of water eco-functionality and a preponderance of disease-causing organisms. In developing nations, many abattoirs dispose of their waste directly into streams or rivers and also use water from the same source to wash slaughtered meat. Most solid, liquid and gaseous wastes are released into the immediate environs of the abattoir. This chapter is a literature review of the impact of untreated slaughterhouse wastewater on water quality in selected areas of Southern Africa. The review shows that abattoir activities significantly reduced the physical, chemical and microbiological quality of water bodies. Although an improvement of the water quality was observed downstream, this was a result of self-purification and dilution effects.

Chapter 5

A global shift away from diets dominated by meat, dairy and eggs to mainly plant-based diets is as necessary in mitigating anthropogenic climate change as the shift away from fossil fuels. Yet a large awareness gap exists about animal agriculture's contribution to greenhouse gas emissions. Recent studies in Australia and the United States show this issue is represented in less than 1 percent of all newspaper articles about climate change. This chapter examines the opportunities and barriers in addressing the livestock sector's impact on climate change. Policy recommendations in the literature are compared with the responses of governments, industry and the NGO sector. Australia's unique socioeconomic and cultural ties to livestock production and the consumption of animal products represent a significant barrier to demand-side mitigation. An analysis of newspaper articles mentioning animal agriculture's link to climate change in The Sydney Morning Herald between 2006 and 2014 provides insights into the facilitation and shaping of public awareness on the issue to date. The findings can inform strategies to increase future media coverage and encourage a more engaged discourse on demand-side mitigation.

Chapter 6

Nick Pendergrast, Curtin University, Australia

Increasing awareness about the environmental consequences of consuming animal products together with recognition of animal rights and health benefits, has played a significant role in the rising interest in veganism. The chapter analyses recent trends in veganism in the Western world, focusing on the substantial environmental impact of animal agriculture being highlighted by respected agencies including the Food and Agriculture Organization of the United Nations and widely read publications such as the Guardian and the New York Times. Growing awareness of the environmental impact of animal products and greater visibility of veganism in the mainstream media have assisted in veganism emerging as an important response to the environmental crisis.

Section 2
Diet and Wellbeing

Chapter 7

Rosemary Stanton, University of New South Wales, Australia

This chapter aims to describe how meat fits into recommended dietary guidelines. In Australia, meat is included in one of the five food groups. However, this food group should not be described as the 'meat group' as it includes alternative choices. These include animal products such as seafood, poultry and eggs but also plant-based alternatives such as legumes, tofu, nuts and seeds. Choosing a range of foods from within this group contributes to a healthy dietary pattern with nuts, seeds and legumes providing extra benefits. Increasing plant-based choices also makes it easier for those who consume meat to keep to the weekly limit recommended to reduce the risk of health problems associated with a high consumption of red meat. Processed meats are not included in any of the five food groups and are now seen as 'discretionary' foods.

Chapter 8

Kate Marsh, Northside Nutrition and Dietetics, Australia
Angela Saunders, Sanitarium Health and Wellbeing, Australia
Carol Zeuschner, Sydney Adventist Hospital, Australia

Despite its nutritional benefits, there is an increasing body of evidence to suggest that regular consumption of red meat may negatively impact health and disease risk, including the risk of most common chronic diseases. This chapter reviews the current evidence linking red and processed meat intakes with chronic disease, obesity and mortality risks and discusses possible mechanisms to explain these associations. Research on the health benefits of diets low in red meat, including vegetarian, vegan, Mediterranean and other plant-based diets, is also reviewed.

This chapter discusses antibiotic use in the livestock industry and potential ramifications for human health. Antibiotics are routinely administered to food animals, primarily at sub-therapeutic levels. The extensive use of antibiotics in global animal husbandry in quantities greater than used for humans is creating antibiotic resistance. There is evidence that antibiotic resistant organisms emerging in food animals transfer to humans through the food chain, environmental contamination, direct association with animals or through mobile resistant genetic elements resulting in co-resistance to other antibiotics. No new classes of antibiotics have been developed since the 1980s. Intensifying use of existing antibiotics for meat production poses new challenges for treating humans, needs to be taken seriously and dealt with urgently. This chapter argues that reduced meat consumption is an under-considered but essential part in any suite of solutions aimed at preserving the use of antibiotics for human treatment.

Food consumption has both direct, immediate impacts and longer-lasting effects on human health and wellbeing. This chapter considers the evidence behind the impact of animal foods on human health. It also reviews the impact of intensive animal production on the health and wellbeing of the animals themselves, including their breeding and conditions in which they are reared. The potential for factory farms to contribute to the global threat of antibiotic resistance and to be the breeding ground for a viral pandemic is also considered. Intensive animal farming is further adversely affecting the earth's resources of cereals, soy and water. The conclusion is that there is overwhelming evidence for a reduction in consumption of animal products on the grounds of health, the use of the earth's precious resources and the wellbeing of the animals consumed for food.

Chapter 11

The chapter examines China's growing meat demand and its implications. Australia and China are currently set to expand trade in meat and livestock facilitated by a government negotiated Free Trade Agreement. China is already the world's largest meat consumer and with the increasing consumerism and wealth of its rapidly growing middle and upper class, the demand for animal products is likely to grow. This country's unprecedented appetite for animal proteins has stimulated the Australian livestock and related sectors, potentially enabling vast growth and profitability within these industries. Chinese customers have strong purchasing power and are eager to buy imported frozen and locally slaughtered Australian meat. While Australian farmers are capitalising on these economic opportunities, only the animal welfare sector voices any concern. This chapter highlights the ignored health and environmental costs.

Section 3
Society and Politics

Chapter 12

Hans Dagevos, Wageningen University, the Netherlands

Broad scholarly consensus exists nowadays that high meat consumption is particularly critical from an ecological perspective. Traditionally, technological progress and efficiency innovations in food supply processes are identified as key to solving food sustainability problems. However, it is increasingly recognised that technological innovation and efficiency gains alone are not enough to reduce the environmental impacts of growing meat production and consumption. Therefore, this chapter's point of view is consumption-oriented. Are consumers part of the solution by making transitions towards more sustainable consumption patterns in general and less meat-centric diets specifically? This chapter explores flexitarianism as a present-day food style that consists of different forms or levels, ranging from minor adjustments to regular meat consumption patterns to fundamental departure from habitual meat eating practices.

Chapter 13

Paul Shapiro, The Humane Society of the United States, USA

The animal agribusiness industries often proclaim a libertarian mantra when asked to accept rules for their conduct in regard to animal welfare, the environment, and food safety. However, in this chapter, the author explores how when these industries suffer from lack of demand, their clamor toward socialism is stark. They consistently come to the US Congress and the United States Department of Agriculture with outstretched arms and cupped palms, seeking to defy the normal laws of economics that other businesses must navigate. In fact, the meat, egg, and dairy industries are enormous beneficiaries of generous federal subsidies, research and development, and even surplus buy-ups of unwanted product. Such a reliance on federal handouts by animal agribusiness calls into question their proclamation of libertarianism and free market principles.

Chapter 14

Amzad Hossain, Curtin University, Australia & Rajshahi University, Bangladesh

It is difficult to separate western consumerism from excessive meat consumption and through globalization this culture is spreading through the planet to traditional places, such as Bangladesh and the Indian subcontinent. The chapter argues that the socio-economic and planetary cost of increasing meat consumption is clearly untenable and initiating a process that restores natural resources is imperative. A major objective of this chapter is to raise awareness about the consequences from unsustainable meat production and consumption and the negative implication from a Western type of diet. Drawing on the spiritual messages from the Baul philosophers, it makes the case that preserving traditional flexitarianism, defined here as meat in the absence of any other food options or rare ceremonial meat consumption, is essential for the health of the planet and its inhabitants.

This chapter explores how marketing uses the creation and perpetuation of myths to reinforce demand for meat amongst mainstream consumers. It explores advertising misinformation including with regards the place of meat in our culture, its nutritional value, its association with affluence, masculinity and the benefits of small-scale production. The power of marketing is within the context of whether marketing has a role to play in decreasing rather than perpetuating meat-consumption.

Chapter 16

This chapter brings together ethical, social and territorial implications of the meat industry in the regions of Brazil where animal husbandry is the main economic activity. This article thus articulates concepts elaborated in the field of environmental ethics applying them to sociological analysis of territorial development using the notion of 'regions of intensive speciesism'. The notion is elaborated as a conceptual tool to highlight territories whose socio-economic development pattern accepts the production of meat as a supposed "regional vocation". In such territories, non-human sentient beings are unquestionably put into the same category as things. The identification of this pattern enunciates interrelated implications of meat production that are not usually recognized. These implications include the high rates of health problems affecting workers in slaughterhouses; the symbolic and economic domination over territories and people by the agroindustry; and the drastic moral inconsideration of sentient beings. The article is based on the case study of Concórdia, a micro region located in the state of Santa Catarina in the south of Brazil.

Chapter 17

This chapter presents the results of studies that unveil how meat and other animal derived products are causing severe environmental impacts, social problems and ethical concerns regarding both human and non-human animals. Although there are many ways to tackle the issue a critical non-anthropocentric education that encompasses ethics as a dimension of sustainability, is proposed. Traditional non environmental education often legitimizes values that are averse to an ethic that could be described as correct regarding the relationship between humans and the other animal species and even many educational currents that call themselves "environmental" are guided by a shallow conservationist point of view. Although welfarist practices may in some contexts be of help, the authors propose the animal abolitionist perspective as the unique genuine foundation for education to build this new paradigm.

Section 4
Conclusion

Chapter 18

Jonathan Balcombe, Author, USA

This chapter describes a future scenario in which humankind abruptly stops eating animals. Its purpose is to explore the impacts of animal agriculture using a fictional device. Several ways, in which the end of meat might happen are suggested, including a virulent global human epidemic originating from the meat supply. The ensuing state of the world is examined along six axes: animals, the environment, human health, the economy, food, and society. Because of the enormous scale of animal agriculture today, a sudden end to meat would cause massive upheavals across all of these domains. However, it is predicted that positive gains would be significant, outweighing negative outcomes.

Preface

This book is a powerful collection of diverse research findings presenting the detrimental impacts of excessive meat consumption on human and environmental wellbeing. Despite increasing academic and scientific evidence, inherited beliefs together with wide-ranging motives, including powerful vested economic and political interests, ensure these mitigatable threats are typically neither transparent nor readily available to decision-makers or the general public.

Today, particularly in western countries, meat is easily accessible and is often cheaper than fruit and vegetables. It is also seen as nutritionally necessary, culturally inviolable and as a status symbol. Yet rapidly rising global meat production and consumption is one of the greatest threats and dangers to the short and longer-term sustainability of humankind and the planet.

There is increasing scientific evidence about meat's negative environmental and public health impacts. Dietary guidelines around the world recommend that the maximum safe intake of red meat is 500g per week per person (with a public health goal of 300g per week per person). However people in countries, such as Australia, US and UK, consume amounts much higher than these recommended limits. Developing countries are increasingly adopting western-style meat-rich diets, accelerating global meat production and consumption with worrying projections and trends. Meat's contribution to global greenhouse gas emissions is estimated to be between 13.5% and 51% with the latter more comprehensive assessment being from the World Bank (Robert Goodland, the World Bank's long-time environmental advisor has a post-humous chapter in this book). Increasingly there are calls internationally from scientists, researchers, academics and authors who are voicing concerns about the lack of awareness and understanding of the connections between meat, health and environmental sustainability.

Climate change continues to loom as the biggest threat of our times and the livestock industry is the largest contributor. This sector, together with related industries such as pharmaceuticals, is negatively impacting on the wellbeing of the planet and on human health. If left unchecked meat consumption is likely to continue to significantly exacerbate our current ecological and social problems. There is an urgent need to share the research outcomes in order to empower the broader community to participate in improving health and environmental sustainability through reduction in meat consumption.

Meat, despite the danger it represents, is excluded from most public policy debates, including international meetings aimed at climate change mitigation. The main objective of this book is to raise broad-based awareness of the risks and choices involved in high meat consumption. It positions meat firmly in the discourse about health and environmental sustainability. Reduction in meat consumption and production must clearly be part of any discussions and initiatives to arrest climate change. In the era of the anthropocene, failure to recognise this and appropriately respond will make life challenging for all species.

The international contributors to this book address the difficulties, challenges and opportunities in reducing excessive meat consumption to mitigate human and environmental damage. They discuss and present different aspects and considerations related to meat, including current trends, power and influence exerted by vested political and economic interests, public health impacts and dietary recommendations, impacts on climate change, biodiversity, water and land use as well as ethical issues and the increasing westernization of diets. Additionally, the book explores policy responses and strategies to assist dietary changes for urgent environmental and health benefits.

This book is deliberately addressed to a wide and diverse audience. Policy makers, academics, philosophers, researchers, advanced-level students, technology developers, public servants, politicians and government officials will find this text useful in furthering their research exposure to pertinent topics in sustainability and population health. Moreover, the book provides insights to support individuals and communities in understanding the destructive impacts of meat production and consumption. It empowers them to take responsibility for pursuing more sustainable personal choices and actions.

As global human population moves beyond 7 billion, more than 70 billion animals are raised and slaughtered each year for consumption. This causes unimpeded climate change, water pollution, land use and degradation, increasing antibiotic resistance, exploitation, western domination, speciesism, mismarketing and misinformation. All these result in global health and environmental crises in which no country will be left untouched. Rather than feeling overwhelmed or succumbing to negative outlooks related to socio-ecological destruction, we all have choices about how to look after ourselves and the planet. We hope this book reaches the individual – whatever your vocation, occupation, perspective, conviction or diet.

At the heart of the problem is that, despite calls from health and environmental authorities, meat consumption is on a steady increase (see Figures 1 and 2). For example, the World Cancer Research Fund and other national and international health organizations recommend curbing individual meat consumption to 26 kg per year as an individual goal. The Chair of the UN Intergovernmental Panel on Climate Change (IPCC) Rajendra Pachauri and other influential scientists have made calls for global meat reduction to combat greenhouse gas emissions and environmental deterioration. Yet, between 1961 – the year from when we have reliable data from the United Nations Food and Agriculture Organization – and 2011, the average annual per capita meat supply has increased globally from 26 to 46 kg per person (see Figure 1). In some countries it is even higher than 105 kg per person per year, including in New Zealand – 127 kg, Australia – 121 kg, United States – 118 kg and Austria – 106 kg. We cannot blame population numbers alone for humanity's mounting predilection for meat, nor can we attribute this solely to lifestyle improvements. Whilst indeed global population has been increasing, meat supply has been growing at a disproportionately faster rate (see Figure 2) triggered by economic opportunities and unsustainable dietary changes. As the chapters in this book show, this increasing meat consumption has an escalating health and ecological price tag.

The book contains the latest material which can motivate even experienced professionals in this field. Whilst livestock producers and related industries may be challenged by the content of this collection we invite them to read it and reassess their options. Decision-making, policy development and direction must be based on credible evidence and information, particularly in cases of intractable problems such as the question of meat consumption. One of the hoped-for outcomes of this academic collection is thus to contribute to informed policy processes, based on more readily available information, concerning the greater public good.

Figure 1. Global per capita meat consumption, 1961-2011 [kg]
Compiled from FAO (2015)

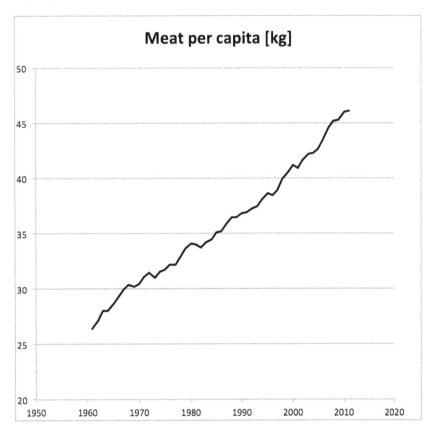

In a world of sometimes overwhelming messages about the bleakness of the future, this compilation highlights the value of individual and collective empowerment, choice and action as a way of countering environmental and health challenges. There is an urgent need for solutions that offer accessible opportunities for regaining, repairing and renewing human and biophysical wellbeing. Each one of us, independently and together, holds a key to a more sustainable world. It is as simple as what we chose to eat.

Our call for contributions was met by an overwhelming response and we received submissions from scientists, academics, poets, politicians, community members, professionals, essayists, industry and the informal sector from all continents. Such diverse and multifaceted interest, passion and participation shaped the direction of this collection. It is a book of hope and faith in the power and capacity of humanity to address and solve the environmental, social and animal welfare crisis now facing the world.

Internationally renowned Australian poet, novelist, critic, essayist, editor, Cambridge scholar and publisher of over 30 books John Kinsella begins this collection with *Sweeney the vegan*, a poem about becoming and being vegan. He shares a personal journey, parts of which may resonate with many readers. This establishes the book's direction illustrating that we all have choices.

The first section of the book *Climate Change and Water* begins with the chapter by the Australian interdisciplinary team of academics Celia Green, Andrew Joyce, Jonathan Hallett, Toni Hannelly and Gemma Carey which examines greenhouse gas (GHG) emissions associated with dietary choices. A systems science approach is used to explore possible interventions aimed at reducing the consumption of

Figure 2. Indices of population growth and meat supply, 1961-2011 [%]
Compiled from data from Unites States Census (2015) and FAO (2015)

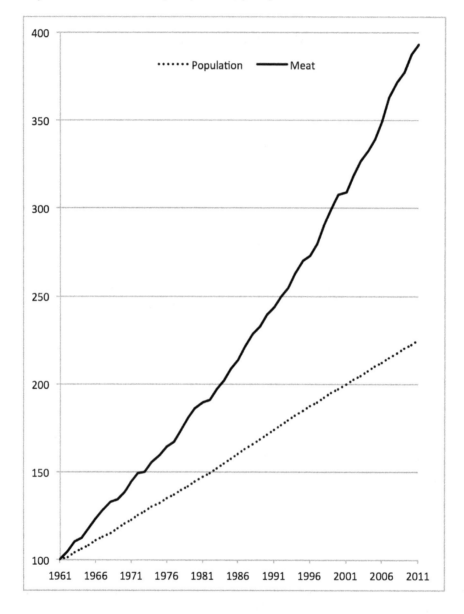

animal products. In *A fresh look at livestock greenhouse gas emissions and mitigation* two assessments of livestock's impacts on climate change – one by livestock specialists employed by the United Nations' Food and Agriculture Organisation and the other by World Bank Group environmental specialists, Robert Goodland and Jeff Anhang – are compared. Quantifying and describing the unprecedented environmental risks involved in meat production this chapter that replacing a substantial amount of today's livestock products with better, plant-based alternatives is the only pragmatic opportunity to reverse climate change before it is too late. Statistician, philosopher and author of CSIRO Perfidy, Geoff Russell in Chapter 3 reveals the poor public understanding of the factors driving climate instability. He points out the current profound mismatch in the way climate scientists and the public understand the causes of climate change.

It becomes abundantly clear that we urgently need accurate and accessible images that identify meat as a key focal issue. This contributes to achieving broader based understanding, and thus mitigation, of the factors driving climate change. The University of Zimbabwe lecturer Never Mujere brings the focus to developing countries, describing, in selected areas of Southern Africa, the detrimental impacts of untreated slaughterhouse waste on water quality and water bodies. The role of print journalism in shaping public awareness of animal agriculture's link to climate change is explored by Xavier Mayes who investigates the representation of the issue in Australia's daily newspaper The Sydney Morning Herald. His chapter will hopefully help those developing strategies to increase media coverage and encourage a more engaged discourse on demand-side mitigation. The first section of the book concludes with a chapter by the academic and activist Nick Pendergrast analysing recent trends in veganism in the western world as an important response to the growing global environmental crisis. This chapter conveys the hope that we are capable of, and on track for, addressing and mitigating the impending challenges.

The section on *Diet and Wellbeing* starts with the internationally recognized Australian nutritional expert Rosemary Stanton who describes how meat fits into recommended dietary guidelines. She highlights the opportunity to increase plant-based choices to reduce the risk of health problems associated with a high consumption of red meat. To further emphasise the value of increasing plant-based choices over animal-based protein options, Kate Marsh and the expert dietary team present and review the evidence that regular consumption of red meat negatively impacts health and disease risk, including the risk of most common chronic diseases. They also present research on the health benefits of diets low in red meat, including vegetarian, vegan, Mediterranean and other plant-based diets. The chapter on *The Future of Antibiotics and Meat* by Talia Raphaely and mother-daughter team Dora and Mira Marinova, draws attention to the controversial human health implications of antibiotic use in animal husbandry. Reduced meat consumption is an under-considered but essential part in any suite of solutions aimed at preserving the use of antibiotics for human treatment. Previous Chief Executive and current ambassador of Compassion in World Farming Joyce D'Silva considers the evidence behind the impact of animal foods on human, non-human and environmental health and wellbeing. China's unprecedented appetite for animal proteins is stimulating the Australian livestock and related sectors potentially enabling vast growth and profitability within these industries. Whilst there may be economic advantages, the unaccounted health and environmental costs need to be recognised as discussed in the chapter by Xiumei Guo and colleagues, concluding this section of the book.

The last section of this collection covers broader societal and policy issues. From Wageningen University in the Netherlands, Hans Dagevos discusses the adoption of flexitarianism as a way to reduce meat consumption in meat-based food cultures. This includes minor adjustments to regular meat consumption patterns as well as fundamental departures from habitual meat eating practices. Executive member of the Humane Society of the United States Paul Shapiro, in his frank and hard-hitting chapter *Feasting from the federal trough: how the meat, egg and dairy industries gorge on taxpayer dollars while fighting modest rules*, highlights the animal agribusiness' reliance on federal handouts. He calls into question their proclamation of libertarianism and free market principles. Dietary colonisation is brought to light in by the Bangladeshi academic Amzad Hossain. In *Mission impossible because of the West*, he explains that sustainable meat consumption is becoming increasingly unattainable in traditional places such as Bangladesh and the Indian subcontinent because of the globalisation of western culture and consumerism. Drawing on spiritual messages from the Baul philosophers, this chapter makes the case that preserving traditional flexitarianism, defined here as meat in the absence of any other food options or rare ceremonial meat consumption, is essential for the health of the planet and its inhabitants.

Marketing and media experts Ian Phau and Diana Bogueva illustrate how creation and perpetuation of myths reinforce demand for meat amongst mainstream consumers. They explore popularly used advertising misinformation including meat's cultural context, its nutritional value, its association with affluence, masculinity and the benefits of small-scale meat production. The opportunity for marketing to play a role in decreasing rather than perpetuating meat-consumption as a means for climate change mitigation is highlighted. The South American academics, Luciano Félix Florit and Cristiane Sbardelati correlate the ethical, social and territorial implications of the meat industry in the regions of Brazil where animal husbandry is the main economic activity. The philosophical questions and sociological implications include the high rates of health problems affecting workers in slaughterhouses; the symbolic and economic domination over territories and people by the agroindustry and the drastic moral inconsideration of sentient beings. This article is based on the case study of Concórdia, a micro region located in the state of Santa Catarina, in the south of Brazil and highlights the correlation between domination of animals and suffering of human beings. From the same major meat producing country, Paula Brügger and her co-authors explore and propose an ethical educational approach towards meat production and consumption. They argue that the animal abolitionist perspective is the unique foundation for education to build a new paradigm governing the relationship between humans and other animal species.

Jonathan Balcombe, author of numerous best-sellers, concludes this book with a future scenario. In *Beyond meat*, humankind abruptly stops eating animal proteins and this results in beneficial consequences for animals, the environment, human health, the economy, food and society. Ending on such a positive note, we hope this collection inspires and empowers you to participate in a sustainability transformation to make this scenario a reality by increasingly taking a personal lifestyle stand against current damaging dietary trends.

What you now hold in your hands is a representative starting point – a sample collection of the manifold, compounding impacts, perspectives and dimensions – of meat consumption's impact on human and environmental health. Each chapter has its own unique character and message and we deeply appreciate the cooperation and contribution of all authors. All chapters are equally important and valuable in the context of meeting this book's outreach objectives. Each of the contributions included in this selection has been double blind peer-reviewed (meaning the author was not known to the reviewer and vice versa).

Clearly the question of whether we are eating too much meat is not one where anyone is ambivalent. People are either strongly against or in favour of reduced meat consumption. Some readers may say that it seem we are prosecuting a reduced meat agenda. Indeed, they are right. In light of the many failed summits, meetings and conferences and the burgeoning planetary and human crisis, the time is urgent. Now, and ever more so, even small actions can have large consequences. Release latent forms of consciousness and political association, ushering in a united force that opens a door to world repair and renewal. Dietary intervention, particularly in countries which show excessive or growing quantities of individual meat consumption, hold enormous potential to mitigate environmental destruction.

We hope this book, through its diverse contributions, will help position the critical discussion about growing and excessive meat consumption where it belongs, at the forefront of the sustainability discourse and policy environment. Policy direction needs to address the role of meat consumption and its dangerous social and environmental consequences. More sustainable dietary choices, facilitated through personal awareness and appropriate public policy will have a beneficial outcome for health and environmental sustainability.

Irrespective of personal opinion, the consideration shouldn't be about whether curbing meat consumption will solve all human and ecological problems. Rather, it should be whether the existing evidence, as highlighted in the chapters of this book, suggests that reducing personal meat intake is an appropriate and reasonable part of the solution to safeguard human and planetary wellbeing. The health and environmental consequences of vast-scale and industrialised meat production and consumption un-questionably represent the greatest challenges humanity has ever faced. Instead of stopping this march of social and natural devastation, most governments, their agencies and instruments together with their corporate allies, continue to promote meat and ignore the numerous existential threats such increasing consumption is causing.

Together, the authors in this book are making an urgent call, based on the existing and growing scientific evidence, for global reductions in meat consumption both through policy direction and individual choices and actions. In the words of one of the contributors: "When the writing is on the wall, it only increases the pain and suffering to ignore it" (Russell, 2009, p. vi).

We are optimistic that this compilation of writings and research will firmly place reduction of meat consumption on the sustainability agenda. We also hope you find something that personally resonates for you.

SWEENEY THE VEGAN

They say I am mad, out of my tree, as I eat fruits and nibble leaves, harvest nuts and make tempeh.
As a teenager I played war games
and dreamed of being a general.
I collected guns and ammunition,
hunted foxes and parrots.
I grew sleepless and made night
my daylight, colouring the sky
with hallucinogens and narcotics,
wandering with agitation.
I watched bullets
fly unspent from the breech,
my brother unloading
as fast as I could load.
And watching over the farm,
I struck a ram in the ute,
and cradling its heavy, horned head,
its broken neck, decided to shoot it dead.
Something shifted, something disconnected,
and I went up to the wheatbins
with a damaged sense of self,
distressed as fellow workers shot cats.
And then backpacking from Bali
to Nepal, other possibilities
mocked and harried my predicament:
the hunger to score, the cliché of searching.
Part of me broke free

in the highest mountains
and settled in a temple tree,
though I didn't know it.
A bus accident and a litany
of death — a chopper in to take
away the wounded, corpses
left broken at the bottom of ravines.
To return without having really left,
to drink and drug to oblivion —
without art, without creativity —
just damage and loss and death.
To move into the squat
in Fremantle, where vegetables
were the only food when there was food,
to sign away from meat in self-disgust.
To retreat south with my brother
and girlfriend, to climb the flooded gum
on the roadside beside the dairy,
where tired cows dragged tonnes of hoodwink.
To wake one morning in asbestos walls,
the spring cold leaching in from irrigated fields,
swollen jerseys calling into the fog
to be relieved of their burdens.
To talk it out over breakfast,
dollops of cream the body and blood.
Haycarting we'd been told
the 'old girls' would make blood and bone.
To take wing and flit north
then south again, into the settler's chestnut tree,
looking down at the haters
poisoning water, driving us out.
Living in a field of cattle
earmarked for slaughter — young
'Molly B12' nuzzling our hands,
the screaming fox at night.
I was vegan and returned
to my girlfriend alone in the shack.
My brother had found his own way
and I sang my song of living flesh.
A decade of flying, staying
above ground while every fibre
yearned for obliteration, held together
by an ethics of Pythagoras.
In rehab, living in the Globe Hotel

or the Supreme Court Gardens,
hearing of friend after friend
dying of overdose, the song
kept me alive. Shivering
by the sea, colder inland
on a star-blown night I listened
to animal-sounds secure.
And when I joined forces
with the woman who knew
the same — one who had abandoned
habits of flesh-eating years before,
I vanished into the Indian Ocean,
went crazy on a coral speck eating
coconuts and rice, shreds of green and chilli
from the islands' greenhouses.
To break the cycle. Break free.
Rejoin the animal world, the kingdom
without hierarchy. But still sleepless,
some call me crazy, edgy.
I renounce all organised religion
and feel liberated spiritually,
no animals are slaughtered
to pave my way to plenty.
Plenty is sharing space,
Plenty is hearing another's breath,
Plenty is every atom of the biosphere,
Plenty is the weapon that cannot hurt me.
Returning home the other day,
we discovered a spatterwork of blood
by the front door, and only today
after noticing a lone doe with joey do I see.
The mob broken up in our absence
by gunfire, the doe sheltering by the house,
our shelter — this is more than Heidegger
could make of dwelling to Celan in the forest.
Refuge is the key. Refuge is where
no creature will be killed by us for flesh
but will make its own way — fences down
and passage no rite de passage condescended by us.
Almost three decades have passed.
I have learnt not to proselytise, and this
song is not a commandment.
My song is still a lament, and I perch high
in the old York gum that lost a limb

in the last storm — I hear the owl
homing in on its prey, and have nothing
to say against its way, knowing it's not my way.
High temperatures are shredding this environment
self-designed over millennia to take the heat —
failsafes have failed and a backup isn't in place.
Fire ran close to us just last night.
My fingers are not claws,
my teeth are not for tearing,
my legs are not for running down,
my feet not for trampling.
The music I hear is not all sweetness —
the abattoir fills my ears with blood,
the paddock with sheep conversations
firebreaks itself with burning flesh.
There's no denying the truth —
the 'sacrifice' of animals to human
addiction and thanatos. This omnicultural
worshipping of death to affirm life.
I breathe past the smoke, breathe
in clear blue sky. Though no watercress
to hand, I eat pulses and leafy greens —
the water deep below quivers under our weight.
They say I am mad,
out of my tree, as I eat
fruits and nibble leaves,
harvest nuts and make tempeh.
I listen to the peace
like static around a world at war,
I know the real clichés are
in the consuming of the living and the dead.
-- John Kinsella

Talia Raphaely
Curtin University, Australia

Dora Marinova
Curtin University, Australia

REFERENCES

Food and Agriculture Organization of the United Nations (FAO). (2015). *FAOSTAT.* Retrieved from http://faostat.fao.org/site/610/DesktopDefault.aspx?PageID=610

Russell, G. (2009). *CSIRO Perfidy.* Fremantle, Australia: Vivid Publishing.

Unites States Census. (2015). *International data base.* Retrieved from http://www.census.gov/population/international/data/idb/region.php?N=%20Results%20&T=13&A=aggregate&RT=0&Y=2001,2002,2003,2004,2005,2006,2007,2008,2009,2010,2011&R=1&C=

Acknowledgment

Without the 29 contributors this book would have not come to fruition. Their competent, knowledgeable and skillful writing and individual inputs about the wide impacts of meat consumption on health and environmental sustainability are invaluable. Our thanks also go to Phil Webster, a researcher at Curtin University Sustainability Policy (CUSP) Institute, whose many hours of editing and professional eye for detail were of enormous help.

We also acknowledge the patience and support of our loved ones who often missed our company while we were writing, organising and editing the book. Many thanks to Ariel, Mira, Ana, Daniel, Alexander, Roger, Erez and John for being the inspiration behind our efforts.

Section 1
Climate and Water

Chapter 1
The Greenhouse Gas Emissions of Various Dietary Practices and Intervention Possibilities to Reduce This Impact

Celia Green
Australian National University, Australia

Jonathan Hallett
Curtin University, Australia

Andrew Joyce
Swinburne University of Technology, Australia

Toni Hannelly
Curtin University, Australia

Gemma Carey
Australian National University, Australia

ABSTRACT

This chapter examines the link between dietary choices and greenhouse gas (GHG) emissions and possible interventions to reduce this impact. The connections between climate change, food systems and public health are explored. It is shown that there is variance in the impact of different food types on GHG emissions, with animal products having the greatest impact. The role of food system activities in the production of GHG emissions is also explored. Dietary choices and GHG emissions are examined using case studies from a variety of countries. Results show that reduced animal food production has increased potential to reduce GHG emissions compared to technological mitigation or increased productivity measures. Finally, a systems science approach is used to explore possible interventions aimed at reducing consumption of animal products.

INTRODUCTION

Climate change is now acknowledged as a significant public health issue and its impact on food security has become a major concern (AIWH, 2008; McMichael, Powles, Butler & Uauy, 2007). The links between climate change, food supply systems and dietary choices are complex (Sulda, Coveney & Bentley, 2010), and with a growing population, the global food system (comprising agriculture, food processing, distribution, retail and consumption) is predicted to experience an unparalleled number of converging pressures in coming decades (Pearson, Friel & Lawrence, 2014). In response to the global increase in

DOI: 10.4018/978-1-4666-9553-5.ch001

population growth, food production in the second half of the 20th century more than doubled (Kahn & Hanjra, 2009). This has resulted in changes to climate systems, loss of biodiversity, degradation of land, and increased resource inputs placing unprecedented demands on the food system (Ericksen, Ingram & Liverman, 2009). A large amount of natural resources (land, minerals, water, energy) are used for food production and consumption, which generates significant emissions of greenhouse gases (GHGs) (Elferink, Nonhebel & Moll, 2008). Projected variability in climate systems without sufficient mitigation strategies to address climate change, anticipated loss of agricultural productivity, pasture growth and livestock production, as well as increased costs for agricultural production will influence trade patterns and have substantial consequences for global agricultural production (Gunasekera, Tulloh, Ford & Heyhoe, 2008). Such effects on the food system foresee long-term impacts on both the environment and public health, with increasing difficulty in achieving food security (McMichael et al., 2007).

As climate change affects diet and nutritional status, so too do dietary choices and food systems effect levels of GHG emissions, and consequently climate change. The level of GHG emissions from food production has been shown to be on par with levels produced by the transport sector, with the latter frequently viewed as a major GHG contributor (Garnett, 2009). A regional analysis for Europe found that food accounts for 31% of the EU-25's total GHG impact (Tukker et al., 2006), and studies from developed countries show that food consumption contributes between 15% and 28% to overall national GHG emissions (Garnett, 2011). This chapter explores how dietary choices impact on climate change by reviewing case studies from countries with diverse food cultures and habits. It then examines possible intervention approaches to shift dietary practice, particularly reduced meat consumption, from a systems thinking perspective.

BACKGROUND

Food Type and GHG Emissions

The type of food produced, animal-based or plant-based, has a large influence on GHG emission levels. Raising livestock in particular, is GHG intensive and responsible for approximately half of all food-generated GHGs. Ripple et al. (2014) reported that the livestock sector accounts for approximately 14.5% of all anthropogenic GHG emissions worldwide, with ruminant production (mainly cattle and sheep) the largest source of anthropogenic methane, contributing more than monogastric livestock (pigs) (Friel, Barosh & Lawrence, 2013). In addition to the methane released by enteric fermentation and dung, other gases that contribute to the overall level of GHG emissions are carbon dioxide and nitrous oxide from agricultural soil, and fertiliser production and use (Risku-Norja, Kurppa & Helenius, 2009).

The dairy industry also contributes to GHG emissions with approximately 85% occurring at the primary production stage (mostly methane). The emission levels increase as the degree of processing increases, with yoghurt and cheese production yielding 1.4 and 5.7 times, respectively, the amount of GHGs produced by milk production (Bradbear & Friel, 2011). On a weight basis, less emissions are generated from the production of vegetables and grains than from animal-based foods, and these generally stem from the power used for irrigation, processing and packaging (Friel et al., 2013).

The global demand for food, including meat, is increasing due to rising incomes and urbanisation (Steinfeld et al., 2006). Steinfeld et al. (2006) have predicted demand for meat to double by 2050 from 1999/01 levels, resulting in considerable challenges for society given that livestock production affects almost all components of the environment (air, land, soil, water, biodiversity) (de Vries & de Boer, 2010). In particular, land use for livestock production and feed grains consumes significant natural resources and alters and reduces natural habitats (Garnett, 2009). Intensive livestock production not only contributes a high level of GHG emissions, but may lead to overuse of fertilisers and other inputs which create environmental problems such as contaminated soil and increased nutrient levels in waterways. Intensive livestock production may also lead to degradation and nutrient stripping of soil (Steinfeld et al., 2006). As well as the type of food produced, the systems used for producing food can also have significant impacts on GHG emissions. The next section will examine how these different food system activities can contribute to GHG emissions.

Food System Activities

Food system activities can be categorised as preproduction, production, and postproduction. Within each category, many activities produce GHGs, with emissions varying significantly between activities within the global food system (Vermeulen, Campbell & Ingram, 2012). There are key differences among countries, with postproduction usually having a larger role in high-income countries, while in other countries specific subsectors may be more important (e.g. high contribution from fertiliser manufacturing in China) (Vermeulen et al., 2012). Overall, the food system contributes 19% - 29% of total global anthropogenic GHG emissions, with agricultural production making up 80% - 86% (Vermeulen et al., 2012). Most GHG emissions associated with livestock production occur at the farm stage. The remainder result from preproduction activities such as fertiliser production and postproduction activities such as processing, packaging, refrigeration, retail, transport, catering, and waste disposal.

Preproduction Activities

Vermeulen et al. (2012) identify the main preproduction activities contributing to GHG emissions as the manufacturing of fertilisers and energy use in animal feed production, which can include such activities as production of pesticides and herbicides. While there may be other indirect inputs, for example antibiotic production and infrastructure (i.e. buildings and machinery), they are not discussed here as they have not been categorised as significant drivers of GHG emissions in relation to food systems.

Fertilisers

Production of plant food for human consumption or animal feed relies on inputs such as seeds, irrigation, pest control and fertilisers. Among these inputs, fertiliser production accounts for the majority of GHG emissions due to the energy intensity and emissions of nitrous oxide released when nitrate fertiliser is manufactured (Snyder, Bruulsema, Jensen & Fixen, 2009). Steinfeld et al. (2006) estimate that of the 80 million tonnes of nitrogen fertiliser produced yearly, approximately 20% is used in the production of livestock feed. In the United States where consumption of meat is high, around 50% of nitrogen fertiliser is used to fertilise crops for livestock feed (Eshel, Shepon, Makov & Milo, 2014).

Animal Feed

There are both direct and indirect factors contributing to GHG emissions through the production of animal feed. Producing feed for livestock and the aquaculture industry directly contributes to GHG emissions as a result of fossil fuel use in transport, plant cultivation and feed processing, with indirect inputs from land cover change (grazing and feed cultivation) (Vermeulen et al., 2012). While there is scant literature in this area it has been estimated that production of animal feed accounts for more than half of the total energy used within livestock production itself (Steinfeld et al., 2006). As Vermeulen et al. (2012) discuss, energy is used for production of animal feed via fertilizer production, manufacturing and distribution of herbicides and pesticides, fossil fuels (i.e. diesel) used in machinery, electricity used for irrigation and any heating, drying or processing which occurs.

Production Activities

Agricultural production is a significant driver of GHG emissions through current agricultural practices and increasing land area being used for agriculture (Vermeulen et al., 2012). Although there has been much research into the contribution of agriculture to climate change, including reviews conducted by the Intergovernmental Panel on Climate Change (IPCC), there remains ambiguity in many estimates (Smith et al., 2007). The IPCC estimates that direct GHG emissions from agriculture were 14% of total global anthropogenic emissions in 2004 (Metz, Davidson, Bosch, Dave & Meyer, 2007). Of all direct agricultural GHG emissions, 39% can be attributed to livestock production (enteric fermentation 32% and manure management 7%), with the remaining 61% to nitrous oxide from soil (38%), biomass burning (12%) and rice production (11%) (Vermeulen et al., 2012).

Indirect GHG emissions from agriculture are also a significant contributor to global GHG emissions with an estimated 12% (range 6% - 18%) of total anthropogenic emissions being a result of land cover change (van der Werf et al., 2009). In 2005 37% of the earth's terrestrial surface was used for agriculture (Smith et al., 2007) with around 80% of new agricultural land used for crops and pasture coming from replacement of forests (Gibbs et al., 2010). The livestock sector alone now accounts for 30% of global agricultural land use (Havlík et al., 2013). Such changes to livestock production practices and land use can thus be anticipated to have a significant effect on future GHG emissions.

Postproduction Activities

Postproduction activities relate to what happens to food once it has been produced. These activities include processing and packaging food, transport and refrigeration, food preparation and food waste. While some activities such as packaging may have minor impact on GHG emission, others such as refrigeration may have a more significant impact.

Processing and Packaging

Advances in food processing and packaging have played a key role in extending shelf life of food products and maintaining food safety for consumption (Marsh & Bugusu, 2007). This extended shelf life allows food to be stored for longer periods and to be transported safely over long distances before consumption, thereby improving food security (Marsh & Bugusu, 2007). Although packaging is an essential component of many food products, it is also a source of environmental burden and waste (Roy et al., 2009). For some food products, packaging is excessive and often unnecessary, resulting in release of additional emissions (Berners-Lee, Hoolohan, Cammack & Hewitt, 2012).

There is little information available on GHG emissions associated with food packaging and as Vermeulen et al. (2012) note the research is difficult to interpret as packaging can include manufacturing of packaging materials, the process of packaging and possibly refrigeration costs. Garnett (2011) attributes 7% of the GHG emissions associated with the food chain in the United Kingdom to packaging and a further 2% to waste disposal. One study by Jungbluth, Tietje and Scholz (2000) found that packaging for both animal and plant products is of only minor consequence in relation to total food system GHG emissions.

Transport

Food transport contributes to carbon dioxide emissions from agricultural products and is responsible for up to 12% of total food-chain emissions in the United Kingdom (Garnett, 2011; Meisterling, Samaras & Schweizer, 2009). The distance that food is transported is increasing as arable land is abandoned because of erosion and degradation and food crops are grown farther from urban areas. This change translates to increased GHG emissions from transport and greater cost to deliver produce to markets, thus raising the cost of food supplied to the consumer (Meisterling et al., 2009). Garnett (2009) and Berners-Lee et al. (2012) both noted that in addition to usual transport costs, 'out of season' produce that is freighted from another state or country significantly adds to the GHG emissions associated with a food product.

Refrigeration

Refrigeration is energy intensive and is reportedly a larger contributor to energy use in the food chain than transport or food miles (Pelletier et al., 2011). While there is limited data, it is estimated that approximately 1% of total global GHG emissions can be attributed to the food cold chain (James & James, 2010). This percentage varies by country with a higher percentage in high-income countries.

Retail Activities and Food Preparation

Vermeulen et al. (2012) discuss that modern food retail enterprises are a significant contributor to GHG emissions. In the UK an estimated $4MtCO_2e$ total annual emissions are related to major retail food outlets (Tassou, Ge, Hadawey & Marriott, 2011). Supermarkets can also contribute to GHG emissions via energy consumption, the main source being refrigeration. Food preparation contributes to GHG emissions through energy use from cooking and refrigeration. An estimated 9% of direct food chain GHG emissions in the UK can be attributed to home based cooking, storage and preparation of food (Garnett, 2011). Vermeulen et al. (2012) also notes that as household incomes increase there is likely to be an increase in emissions related to changing types of fuels used for domestic cooking. As an example, in China the shift towards using commercial fuels such as coal based electricity from more traditional biomass fuels has nearly doubled rural residential energy consumption.

Food Waste

In addition to the composition of the diet, actual consumption patterns should be considered when looking at the environmental impacts of food production. Berners-Lee et al. (2012) reported that less than two-thirds of available food globally is consumed and that reducing or eliminating both pre- and post-purchase food waste could prevent large quantities of GHGs emissions. Unused food is effectively 'wasted' emissions being generated throughout the entire length of the food chain: growing, processing, storing, retailing, transporting and cooking. This large proportion of food entering the waste stream is a considerable loss of energy and emissions to landfill (PHAA, 2009). Garnett (2011) suggested that approaches to reducing food waste, such as better coordination between primary producers and industry, public awareness campaigns, and innovative packaging developments, may help to reduce these unnecessary emissions. In addition, overconsumption also contributes to GHG emissions. With reference to the 2013 Australian Dietary Guidelines, Friel et al. (2013) commented that any food that is consumed above a person's energy requirements represents 'avoidable' GHG emissions in producing that foodstuff.

DIETARY CHOICES AND GHG EMISSIONS

Dietary Studies

There have been a number of recent studies examining the connections between dietary choices, food production systems and resultant impact on GHG emissions. Joyce, Hallett, Hannelly and Carey (2014) conducted a systematic review of research modelling different dietary patterns and GHG emissions with exploration of the effects on public health and policy making. They identified 21 primary studies that modelled GHG emissions and dietary patterns published since 1995. Due to considerable differences in study designs and assumptions a qualitative review was conducted which described the study methods and findings. Most studies examined were European, with one study from each of Australia, New Zealand, North America and India.

In all studies, an average diet based on food surveys or population data was used as a reference point for other additional diet scenarios. All diets used were based on reducing meat consumption with 11 studies including a diet without meat and 8 studies including a diet without any animal products (vegan). The majority of studies used a life cycle assessment (LCA). This is a method which allows for identification of inputs (i.e. water, electricity, other resources) and outputs (i.e. GHG emissions, waste) connected to production of products. It is a comprehensive method for calculating the global warming potential or GHG emissions emitted over the life cycle of a product. The following case studies represent the most recent research from the UK, Nordic region, continental Europe, Australia, New Zealand,

North America and India. In examining these case studies what is apparent is that despite significant differences in dietary types across these examples there are great similarities in the direction required to reduce GHG emissions.

Australia: Hendrie, Ridoutt, Wiedmann and Noakes (2014)

Hendrie, Ridoutt, Wiedmann and Noakes (2014) compared four different diets in the Australian context according to GHG emissions. The first diet was the average Australian diet taken from the 1995 National Nutrition Survey. This included data from males and females aged over 19 years and was weighted to provide an average intake per person per day of different food types. The second diet modelled was the average diet with minimal non-core foods, which were defined as those high in energy but low in nutrients and included processed meat, snack foods, confectionary, soft drinks, saturated fats and oils, and alcohol. Tea, coffee and offal meat were also classified as non-core foods but were included in this diet scenario. The third diet modelled was called the "Total Diet" and based on the Australian Dietary Guidelines. Average serves of food groups were modelled for Australian males of height 160cm and females 150cm aged 31 to 50 years assuming a very sedentary lifestyle. The fourth diet modelled was labelled a "Foundation Diet" which met all the minimum nutrient and energy requirements and only included core foods.

The GHG emissions of the different diet scenarios were modelled using input-output analysis based on food category. These estimates are larger than those produced with food product life cycle analysis as the approach takes a broader focus on whole of economy impacts and thus cannot be compared to other studies using the product-focussed LCA. As well as food category there was some differentiation of meat products given their larger GHG impact so that more accurate figures could be produced. The average Australian diet per person per day had a GHG emission of 14.5kg of carbon dioxide. The best performing diet was the Foundation Diet at 25% less than the average diet. The Total Diet and average diet without non-core foods were similar with an approximate 23% relative reduction each compared to the average diet.

The difference in GHG emissions between the diets in these models related to the presence or absence of non-core foods, which contributed 27% of GHG emissions in the average diet. While the Foundation Diet was optimal it was considered unlikely to be accepted by the Australian population. Despite the fact that red meat had the largest GHG emissions it was not recommended that this food group be reduced due to what the authors considered important nutrients provided and potential environmental impact of the "production of alternative protein sources – especially environmental impacts related to land and water use, and indirect land use change" (Hendrie et al., 2014, p. 298). Suggestions instead related to improved agricultural practices in this sector which could lower emissions. The study was 'funded and approved for publication by Meat & Livestock Australia' (Hendrie et al., 2014, p. 300). Alternatives posed for reducing GHG were adoption of the Total Diet which balanced nutrition and environmental impact or more modest reductions in food intake across the population such as a 10% reduction across all food groups (noting that this was not a nutritionally complete diet). It was also noted in the introduction, that while there are various policy mechanisms available to government to influence food choices, currently there is no evidence of population wide change to reducing food intake.

North America: Soret et al. (2014)

The objectives of the study by Soret et al. (2014) were to compare the GHG emissions of vegetarian, semi-vegetarian and non-vegetarian dietary patterns consumed by a large population in North America, and to independently assess mortality in consumers of each dietary pattern within the same population. This study is the first to assess the impact of reduced meat consumption diets on GHG emissions and health outcomes using non-simulated data from a large and diverse population. Data from the Adventist Health Study 2 (AHS-2) were used to characterise and measure the dietary patterns. The AHS-2 is a sizeable prospective cohort study of Seventh-Day Adventists in North America and Canada, who form a large ethnically, culturally and geographically diverse population. A self-administered food-frequency questionnaire of 210 food items including both plant and animal foods, was used to assess dietary intake. Dietary patterns were classed according to the stated combined intake of all meats (including fish), and were categorised as vegetarian (meat rarely or never consumed), semi vegetarian (meat consumed >1 time/month, but <1 time/week) and non-vegetarian (meat consumed at least 1 time/week).

A LCA approach was used to calculate GHG emissions associated with the foods consumed in the AHS-2 population. The carbon dioxide equivalents (CO_2e) within the 210 foods in the diet questionnaire were estimated by using LCAs and from published GHG emission data. Mean annual GHG emissions for each diet were calculated, with 1113 kg CO_2e for non-vegetarians, 872 kg CO_2e for semi vegetarians, and 788 kg CO_2e for vegetarians. Using the non-vegetarian diet as a reference point there was found to be a 29.2% (325 kg CO_2e) reduction in GHG emissions for the vegetarian diet and a 21.6% (241 kg CO_2e) reduction for the semi vegetarian diet. The difference in emissions reflected the amount of meat and plant foods consumed in the different diets. Mortality rates and mortality hazard ratios (HRs) were also examined for each diet, with the mortality rate for non-vegetarians found to be 20% higher than for vegetarians and semi vegetarians. The mortality HRs, were adjusted for a large range of lifestyle factors and also mirrored these results, showing a lowered risk for vegetarians and semi vegetarians compared with the non-vegetarians.

The results showed both public health and environmental sustainability benefits to diets with reduced meat consumption. By using observational data, a realistic assessment of dietary patterns over an extensive time frame as consumed by a large population was enabled.

Other studies examining the GHG impacts of different diets have used designed diets to model nutritional requirements and GHG emission reductions. However Soret et al. (2014) noted that the drawbacks of designed diets are that they may differ markedly from what populations actually consume on a daily basis, and the choice of foods to construct an 'ideal' diet can be influenced by researcher bias. Additionally, other research, which has assessed the impact of meat free diets has concentrated on single food exchanges (i.e. replacing meat with soy). In the real world people are unlikely to replace meat with soy only. Thus these approaches lead to results, which are theoretical in nature and do not necessarily reflect real life, potentially causing misrepresentative results and thus limits on their value in policy making.

Soret et al. (2014) concluded that relatively small changes in consuming plant foods rather than meat in populations whose dietary profiles are on a spectrum of no meat to low/moderate meat intakes can have a non-trivial impact on reducing GHG emissions. The meat consumption of the non-vegetarians in the AHS-2 study was 2 times less than that of average Americans and thus the difference in levels of GHG emissions could be greater if compared with typical non-vegetarian American diets. The public health benefits of reduced meat diets were also made clear with mortality rates being 16-17% lower in vegetarians and semi vegetarians. Soret et al. (2014) however, cautioned against making outright extrapolations to the general public as the association between GHG emissions and mortality with dietary patterns are not directly assessable from their study. Any implications made are only relevant to Western societies where populations have sufficient access to reduced meat in favour of plant based dietary choices. Within this context modification of dietary choices can be both a practicable and effective instrument for climate change mitigation and public health improvement.

New Zealand: Wilson et al. (2013)

Wilson et al. (2013) used linear programming to model 16 diets within the New Zealand context. Linear programming is a mathematical technique, which generates optimal solutions, for example identifying a mix of foods of the lowest cost, which fulfil minimum and maximum nutrient levels. The authors used a "bottom up" approach by acquiring data on a range of specific food items and optimizing diets meeting nutritional requirements from that point on. They used four groups of diet scenarios which all meet nutrient requirements. The groups identified were

1. Low-cost (scenarios C1 – C4),
2. Low in GHG emissions and low-cost (scenarios G1 – G4),
3. "Relatively healthy diets" utilising high vegetable intake – a Mediterranean style diet (scenario MED) and an Asian style diet (scenario ASIAN and ASIAN-G), and
4. "More familiar meals" designed to be more acceptable to most New Zealanders.

To select foods to be included in the models the study used a range of foods such as those covered by the New Zealand Food Price Index (includes 44 commonly purchased food items), previous research identifying low-cost protein sources in New Zealand, unprocessed food (i.e. lentils) and low-cost canned food, and lists of foods from a previous French optimization study. For the low-cost group, scenario C1, all energy and nutrient requirements modelled for NZ men were met via only 9 selected foods all of which were vegetarian (wholemeal flour, pasta, dried peas, eggs, sugar, milk powder, carrots, vegetable oil and kiwi fruit). Scenario C2 included slight changes to increase cooking suitability, C3 for ease of preparation and C4 had a relatively high vegetable intake with all other food amounts optimized. For the low-cost low GHG group all diets were lacto-ovo vegetarian with the exception of G4, which was fully vegan. The fully vegan diet resulted in slightly higher GHG emissions and cost than the other low GHG emission diets. While there was no direct explanation for this in the study it may have been due to the G4 scenario being a high-vegetable diet and the use of alternative foods containing micro-nutrients such as calcium in order to meet nutritional requirements. In contrast the other low GHG emission diets may have been based around a small number of foods with less vegetables but higher inclusion of lower GHG food items.

All modelled diets were markedly cheaper and produced less GHG emissions than the NZ average diet, which was modelled using national data from the New Zealand Adult Nutrition Survey. In addition the low-cost and low-GHG diets provided health benefits in comparison to the typical NZ diet such as less saturated fat from meat and lower sodium consumption. The high vegetable diets (C4, MED, ASIAN) were also identified as being healthier. All diet scenarios including meat and fish had larger GHG emissions than the vegetarian and vegan diets. The study identified a number of vegetarian dietary patterns, which were low-cost, low GHG and had health benefits while meeting all nutritional requirements for NZ adults. Including more dietary variability and acceptability of diets (more familiar meals which included meat) increased cost and GHG emissions, but were still lower cost and lower GHG emissions than the NZ average diet.

The authors noted as a study limitation that GHG emissions data are likely to be underestimates given the lack of consideration of factors such as land use required for food production which can significantly increase GHG values for meat and dairy foods, and the emissions associated with food transport, refrigeration and preparation. Policy implications from this study were identified, such as local government policies to ensure that optimized foods (low-cost, low GHG, and healthy) are exempt from food taxes imposed on less healthy foods and identification of foods to be promoted by the public health sector and environmental stakeholders, i.e. food labelling. These foods could then be incorporated into food being served in public institutions such as schools, retirement homes, prisons and hospitals.

India: Pathak, Jain, Bhatia, Patel and Aggarwal (2010)

Pathak et al. (2010) examined the carbon footprint of food commonly consumed in India and analysed differences in GHG emissions from vegetarian and non-vegetarian foods. As India is a diverse country in terms of both climate and culture, there are wide variances in the types of food consumed in different regions. For the purpose of the study the GHG emissions at various stages of the life cycle of the 24 most common Indian food products were analysed. A LCA was used which factored in four stages: production, processing, transportation and preparation. The GHG emissions from food item production varied substantially – GHG emissions from rice were 10.2 times greater than for wheat and 43.3 times greater than for vegetables. Higher emissions from rice were due to methane emissions from anaerobic soil conditions whereas wheat and vegetables are grown aerobically. Production of meat and milk also emitted large amounts of GHG emissions due to methane production from ruminant livestock. In terms of global warming potential (GWP) mutton contributed 36.5 times that of chapatti (wheat based product) over the food life cycle. Indians generally consume local, fresh food, so 87% of emissions were generated from production. This is in contrast to Europe, America, Australia and New Zealand where more processed food is consumed and thus the GWP contribution of processing and transport are greater.

The GHG emissions for a balanced diet per day for an adult man and woman at a moderate level of work were calculated based on dietary requirements (different foods in quantities and proportions that meet nutritional needs to withstand short duration of leanness). Five common diets, vegetarian, lacto-vegetarian (vegetarian with milk), ovo-vegetarian (vegetarian with egg), non-vegetarian with poultry and non-vegetarian with mutton were also used as a comparison of GHG emissions from each diet. Overall, GHG emissions for the non-vegetarian diet were 40% greater than for the vegetarian diet. In the lacto vegetarian diet rice and milk were the greatest contributors to GHG emissions. In the non-vegetarian meal mutton was the largest contributor.

The authors concluded that differences in food consumption could be used for mitigation of GHG emissions, for example, a shift from animal food products to crop foods. Within crop foods, wheat based products produced less GHGs than rice and thus either rice consumption needs to be reduced or grown in a way which does not emit methane. The authors also noted that in substituting food products it is important to pay attention to nutrient levels and achieving a balanced diet. They suggested that meat could be replaced by foods such as pulses and vegetables, but that replacing meat with milk products would be less effective in reducing GHG emissions.

United Kingdom: Scarborough et al. (2014)

Scarborough et al. (2014) estimated the variance in diet related GHG emissions of self-selected meat eaters, fish eaters, vegetarians and vegans in the UK. Data from participants in the EPIC-Oxford cohort were used for analysis. This cohort comprised 65,000 participants over the age of 20 recruited between 1993-1999 in the UK via collaboration with general practitioners, vegetarian and vegan societies and advertisements in health food and vegetarian magazines. Participants also recruited friends and relatives. A validated semi-quantitative food-frequency questionnaire was used whereby participants estimated their intake of 130 different food items consumed in the previous twelve months.

Six dietary groups were characterised: high meat eaters, medium meat eaters, low meat eaters, fish eaters, vegetarians and vegans. Participants were then classified as meat-eaters, fish-eaters, vegetarians and vegans in response to yes/no questions on dietary behaviours, with the final dataset comprising 29,589 meat eaters, 8,123 fish eaters, 15,751 vegetarians and 2,041 vegans. Nutritional analyses of the 130 food items were based on nutritional data from UK food composition tables for 289 food codes. The authors estimated GHG emissions for these 289 food codes using a method, which utilised a life cycle approach to estimate GHG emissions for food commodities from the earliest production stages to retail distribution. The GHG emissions for each participant were standardised to a 2,000 kcal daily diet to ensure that any difference in estimated energy consumption between different dietary groups did not affect results.

The study results showed that dietary GHG emissions were highest for high meat eating men and lowest for vegan women. Mean observed values for dietary GHG emissions for meat eaters were 52% higher than for vegetarians and 100% higher than for vegans. Statistical analysis (ANOVA) revealed highly statistically significant differences in dietary GHG emissions between the six dietary groups with increasing emissions in groups, which consumed more animal products. After adjusting for sex and age it was shown that on an average 2,000 kcal daily intake a high meat diet produced 2.5 times as many GHG emissions than an average vegan diet. The authors concluded that consuming less animal based products is a feasible method for reducing individual GHG emissions as well as providing health benefits. Further, the authors commented that any updates to national government dietary recommendations, which aim to define sustainable and healthy diets must include the recommendation to reduce consumption of animal based products.

Continental Europe: Vieux, Darmon, Touazi and Soler (2012)

Vieux et al. (2012) analysed the GHG emissions of the average French diet from a survey conducted in 2006-2007 where a nationally representative random sample of adults completed a 7-day diet recall. The final sample size used for these analyses was 1918 adults (776 men and 1142 women). All food items that were reported as consumed by the participants in the survey (1314 food items) were categorized into 16 food groups and 36 food categories. For each food the percentage of participants who consumed it was calculated and within the 36 food categories foods were ranked based on these percentages. At least one food item from among the most widely consumed foods was used to represent each category. This resulted in 73 representative foods, of which most were consumed by 10 or greater percent of participants.

Life cycle analysis up until the point of sale was used to calculate the GHG emissions of the 73 food types. The mean GHG emissions of the male diets were 4170g carbon dioxide/day and for women this figure was 3667g carbon dioxide/day. Simulated models were conducted whereby the energy levels of the diets were reduced to meet the energy expenditure of someone with low physical activity and another model to meet the energy needs of someone with moderate physical activity levels. Under the low physical activity scenario relative to the average diet the required reduction in calories resulted in a 10.7% reduction in GHG emissions. For the moderate physical activity simulation this reduction was 2.4%.

The contributions of the food groups to diet associated GHG emissions was analysed, showing that the meat and deli meat food group had the greatest impact with a mean contribution to GHG emissions of 27% - more than two times higher than emissions from other food groups. From these findings the authors devised two scenarios to analyse the impact of meat reduction on diet associated GHG emissions. In the first simulation meat/deli meat reduction was reduced by 20% without caloric compensation, which resulted in reducing mean GHG emissions by 4.1%. When the caloric loss was compensated for, there was a smaller decrease in the observed emissions. In the second simulation the meat content was held at 50g per day and deli meat was removed. This resulted in a 12% decrease in GHG emissions when there was no caloric compensation.

When fruits and vegetables were used for iso-caloric compensation there was a gain of 2.7% in GHG emissions owing to the large amounts of fruit and vegetables needing to be consumed (426g/day). However, the meat reduction analysis was conducted on the caloric content of the current diet situation (2118 kcal/day), which the authors had shown to be greater in energy than required to meet energy needs. Thus in both meat reduction scenarios without caloric compensation, the caloric content of the diet still exceeded that which is needed in the low activity level diet scenario. As Friel et al. (2013) showed, any food consumed above a person's energy requirements is an avoidable source of GHG emissions given the GHG emissions required to produce the extra food. Additionally while substituting fruit and vegetables for meat as an iso-caloric compensation resulted in higher GHG emissions (due to the amounts needed) this may not be a realistic real life scenario as other higher energy density foods with lower GHG emissions than meat, i.e. legumes and soy protein are typically consumed by vegetarians and those on low meat diets as substitutes for meat. For example in the study by Scarborough et al. (2014) a daily 2,000 kcal vegan diet produced less than half the GHG emissions than a daily 2,000 kcal high meat diet.

The authors concluded that simulating reduced meat diets to reduced GHG emissions using health measures such as caloric intake and energy density showed no systematic benefits from a health viewpoint or union between health and environmental impacts. This was because of assumptions related to caloric compensation for animal products; in that environmental effects are cancelled when meat is iso-calorically compensated for by only large amounts of fruit and vegetables. Thus, while this may beneficial for health, environmental and health effects do not converge. Additionally the findings in this paper are in contrast to other studies who found beneficial interactions between meat reduction, GHG emissions and public health (Hallström, Röös & Börjesson, 2014; Soret et al., 2014).

Nordic: Hallström et al. (2014)

Hallström et al. (2014) looked at the impact of reducing meat consumption on the dietary contribution of nutrients, GHG emissions and land requirements in Sweden. Three diet scenarios were modelled: a reference scenario (REF) based on current average Swedish meat consumption, and two nutritional scenarios (NUTR-1 and NUTR-2) of reduced meat intake. The REF scenario used data from 2009 national statistics on meat supply and consumption. For the NUTR-1 scenario the amount of meat was based on dietary guidelines with meat limited to 126g of uncooked meat per day and red meat restricted to 60g uncooked. In NUTR-2 the same amount of meat is used but beef is sourced entirely from production systems producing both meat and milk, which are more resource efficient.

The nutritional findings of the study were that Swedish per capita consumption of meat is double that recommended by public health guidelines. Reducing the average meat consumption to recommended levels according to dietary guidelines has only small effects on overall energy and protein contributions but a considerable reduction in fats, iron and zinc. From a nutrition point of view the decreased fat would have positive health benefits but more research is needed to examine the effect of replacing meat with different plant based foods to provide adequate levels of iron and zinc. Owing to overconsumption, it may however be possible to reduce meat intakes without any requirements for nutritional compensation. In environmental terms the average Swedish meat consumption accounts for an estimated 40% of the long-term per capita budget for sustainable GHG emissions and occupies half the available cropland. Reducing meat consumption to recommended levels (approximately 25% reduction in current intake) would reduce GHG emissions from meat production by around 15-25%. The authors conclude that more research is required in order to develop recommendations for dietary patterns with a lower environmental impact, which also meet nutritional requirements.

Summary

These case studies are representative of the 21 studies to date that have modelled the GHG impact of different dietary scenarios (Joyce et al., 2014). Nineteen of these 21 studies found that the greater the reduction in animal-based foods, the lower the GHG emission impact. The two studies that did differ in results produced divergent results due to differences in methodologies. As discussed, Vieux et al. (2012) used fruits and vegetables as an iso-caloric compensation for a 20% reduction in meat consumption rather than a broader and more realistic plant based substitute and caloric intake in the reduced meat diets was above what was required to meet energy needs. Wilson et al. (2013) found that while vegetarian diets had lower GHG compared to meat based diets, the vegan option had higher GHG compared to other vegetarian options due to the replacements for eggs and dairy which were more costly economically and had greater GHG emissions for the equivalent nutrient levels.

Hence, it can be concluded that reduced animal food production has increased potential to reduce GHG emissions compared to technological mitigation or increased productivity measures (Hedenus, Wirsenius & Johansson, 2014; Popp, Lotze-Campen & Bodirsky, 2010), with the greatest GHG reduction potential achieved by a combination (Popp et al., 2010). In respect of particular food types, beef has a much greater GHG impact relative to other types of animal products and vegetable products (Eshel et al., 2014). Modelling of future scenarios shows that current consumption patterns are unsustainable and that reduction in livestock will be an important element in achieving future food security (Bajželj et al., 2014).

The diet modelling studies have provided guidance on the areas requiring policy and program attention, particularly, reduced consumption of animal-based foods, but this research base does not provide much direction about the form of those interventions (Joyce et al., 2014). There have been conclusions made about the need to educate consumers on the health and/or environmental benefits of certain dietary types but these recommendations were based on the results of the simulated diet studies, not from the literature on social and behavioural models of change (Joyce, Dixon, Comfort & Hallett, 2012; Joyce et al., 2014). At present these recommendations on educating consumers seems naïve given the challenges involved in achieving behaviour change at a population level (Joyce et al., 2014). The remainder of this chapter will explore what research is required to better inform the development of policy and program interventions.

INTERVENTION POSSIBILITIES TO REDUCE THE GHG IMPACT OF DIETARY CHOICES

A Systems Science Approach to Reducing Meat Consumption

The term 'systems' is used in many different ways within public health and food industry. It is important to clarify our meaning of the word in the context of the 'systems thinking approach' discussed within this chapter. Here, 'systems' is used to refer to the interrelationship between parts of a system and draws upon many theoretical approaches including general systems theory, complexity science, chaos theory, and network theory (NCI, 2007). Pivotal to this approach is identifying system behaviours such as feedback loops, information flows, and organisational structure (Foster-Fishman, Nowell & Yang, 2007; NCI, 2007). This is in contrast to the food system as described in the first section of the paper (in relation to the mechanics of the different stages of pre-production, production, and post-production). Similarly, it differs from how the processes of each stage could be improved through system redesign and engineering with technological developments in the agricultural sector. However, it is worth noting that we are not proposing in this chapter that a social and behavioural perspective should be considered in isolation from other elements of the food system.

The remainder of this chapter explores how social and behavioural approaches to reducing meat consumption could be enhanced by applying a systems thinking perspective. To date a programmatic approach for shifting environmental behaviours is the predominant model used (McKenzie-Mohr & Schultz, 2014) and while there is good evidence for their effectiveness, there are significant challenges in scaling up such interventions and/or sustaining change. These include resource constraints of implementing evidence based approaches across multiple topics in real life conditions compared to the single issue focus of research (Bond & Butler, 2009) and the challenge of transferring the results of a project from one context to another context (McTaggart, 1991). Part of the impetus for a systems based approach is the challenge of using a programmatic approach when aiming for population level change (Hawe, Shiell & Riley, 2009).

System thinking approaches and tools have gained traction in a range of fields to help shed light on how to address complex problems such as obesity prevention (Johnston, Matteson & Finegood, 2014). In public health, new systems methodologies have emerged which have utility for shedding new light on how to think about intervening in complex settings for effective and sustained change. In particular, Johnston et al. (2014) recently developed a new systems framework, which has shown to have heuristic value for planning interventions and evaluating recommendations for change. Based on Meadows (1999) seminal paper, they outline a number of critical leverage points to intervene in a system. Their Intervention Level Framework includes five levels, ranging from weak to strong leverage points: structural elements, feedback and delays, system structure, goals, and paradigm. The remainder of this chapter explores potential approaches for reducing meat based consumption within each of these five 'Intervention Levels'.

Structural Elements

The most common approach for intervention is the structural elements which include interventions designed to influence policies, subsidies, and community attitudes and behaviours. While there is evidence that in combination these types of strategies are effective (Johnston et al., 2014), Meadows (1999) considers this to be the least effective type of intervention in its ability to leverage significant change – many actions at this level are required to create substantive change. Unfortunately, structural elements is the area that has attracted greatest research attention. While this research can help intervention design, a quick overview of this type of research will outline its limitations in providing a base for systems level interventions.

The research on attitudes and knowledge of diet-related behaviour can provide some useful background for intervention projects. The general public is more aware of the health consequences rather than environmental costs of different dietary choices (Joyce, Dixon, Comfort & Hallett, 2008). The adoption of plant-based diets is driven more by health and ethical concerns rather than for environmental reasons (Fox & Ward, 2008a, 2008b). After an individual adopts a vegetarian diet, their self-identity develops and motivations broaden to include environmental benefits and a combination of health and ethical concerns. Research has explored some of the barriers to adoption of plant-based diets with convenience, ease of preparation, and lack of information identified as key (Lea, Worsley & Crawford, 2005; Lea, Crawford & Worsley, 2006). Specific research on barriers to adopting vegetarian diets highlighted enjoying eating meat (78% agreed), not wanting to change eating habits and thinking that humans were meant to eat meat as barriers. It was concluded that targeted information needs to be provided on nutrition and preparation of plant-based meals. There has also been research applying behavioural models including the Theory of Planned Behaviour (Ajzen, 1991) and the Trans-theoretical Model (Prochaska & DiClemente, 1983) to predict intention to follow a plant-based diet (Wyker & Davison, 2010). This research discovered that the decision process to eat more fruit and vegetables differed from the decision process to adopt a plant-based diet. It was also found that attitudes towards a plant-based diet varied according to gender and particular stage of change. This research has some potential in guiding appropriate education campaigns, and information style strategies have an important place within an overall multi component approach (Howat, Sleet, Elder & Maycock, 2004).

The limitation of this research in guiding intervention design is that it views decision making as a rational process (Beverland, 2014) and the stages of change model in particular has been criticised for its lack of scientific rigour and evidence from intervention trials (van Sluijs, van Poppel & van Mechelen, 2004; West, 2005). Resnicow and Page (2008) offered a critique of the linear, reductionist approach to understanding health behaviour change using a systems perspective. They suggested that health behaviour change may be better construed as an unconscious and chaotic process that resembles a complex adaptive system. They listed factors that could influence change such as knowledge, attitudes, mood, social support, social norms, environmental factors and genetics and contended that given the possible permutations of these factors is infinite, predicting health behaviour change becomes extremely difficult. However, while information type approaches may have little utility in shifting attitudes or behaviours of the public, they may assist in creating a climate whereby other more effective approaches are supported – known as a tension for change in a system. Diffusion of innovations research (which theorises how new ideas spread through complex systems) has shown that creating a 'tension for change' within a social system is essential for the adoption of future innovations (Rogers, 1983).

Educational interventions on their own are unlikely to yield behavioural change but can support and augment other interventions (Howat et al., 2004). Thus while increasing education may not directly alter behaviour it may increase attitudes and knowledge and lead to increased support for economic, organisational and policy interventions that might be more effective in driving change (Stafford, Allsop & Daube, 2014). This is commonly accepted within health promotion and this multidisciplinary approach incorporating economic, organisational, policy and education interventions has produced a number of successes, most noticeably in the areas of reducing smoking rates, decreasing road fatalities, decreasing rates of cardiovascular disease and reducing rates of skin cancer (Moodie, 2004). The strength lies in a multidisciplinary approach incorporating economic, organisational, policy and education interventions (Howat et al., 2003).

Critical to this multidisciplinary approach and pursuing both policy and individual change is engaging a broad coalition of support. Policymaking and policy implementation take place within, and are influenced by, networks. These networks consist of individuals, coalitions and organisations (Kickert, Klijn & Koppenjan, 1997). From this perspective, policymaking is viewed as cooperation or non-cooperation between interdependent groups with different interests, ideologies and strategies (Kickert et al., 1997). Thus recommendations made for consumer information on the environmental impact of foods (Hoolohan, Berners-Lee, McKinstry-West & Hewitt, 2013) may have little impact on consumer behaviour. Yet, it may increase knowledge and subsequent acceptance for more significant policy reform such as the recommendations to reduce subsidies for animal based agriculture towards plant based agriculture. Thus even if consumer behaviour is rigid around meat consumption, a contention which itself has been challenged when examining consumer behaviour (Dagevos & Voordouw, 2013), promoting messages around the environmental impact of food choices may increase support for policy change which could be more effective in changing consumption patterns. Further discussion of partnerships required for policy reform is presented in the section on system structure.

Feedback and Delays

Information itself can be a very important leverage point as it can highlight the need to change and motivate action (Meadows, 1999). Research has found that information flows receive less attention relative to other intervention areas, and yet, inadequate or missing information flows are often the reasons for poorly performing systems (NCI, 2007). Information itself can be a very strong motivator to take action and begin a change process (de Savigny & Adam, 2009; Meadows, 1999; Mrazek, Biglan, Hawkins & Cody, 2007; NCI, 2007). It assists in making decisions about allocation of resources based on need to address issues of most concern (Mrazek et al., 2007). Information can also be used to monitor the effectiveness of interventions (Mrazek et al., 2007).

Viewed as an important leverage point, monitoring data can motivate improvements in performance and attainment of goals (Rowe, 2009). Improved surveillance of plant-based diets and levels of meat consumption along with related attitudinal trends could highlight areas of need for educational campaigns, those with significant public support to leverage legislative change and for policy advocacy and track population impacts of changing policies and social mores. Improved corporate social disclosure and increasing transparency in animal agricultural practices could also assist in providing information to consumers that may impact behaviour. Consumer pressure and increased regulation to improve animal welfare outcomes during farming and slaughter may increase costs of animal products thereby reducing demand. The potential of feedback loops to create this pressure for change needs to be tested and from such a research base, effective surveillance and monitoring systems would need to be established.

System Structure

Johnston et al. (2014) described this level as referring to the connections between system elements. Many disparate groups have an interest in shifting dietary practice, and the policies required to support this change, including health, environmental, and animal welfare and rights groups. Exploring means of connecting these groups will be critical in advancing policy and practice given the importance of cooperation between interdependent groups with different interests, ideologies and strategies in advancing a policy position (Kickert et al., 1997). It has been suggested that encouraging the adoption of healthy diets could improve both population health and reduce carbon emissions (Hendrie et al., 2014; Macdiarmid et al., 2012). However, there is little evidence that health related messages are effective for obesity prevention (Robinson, 2010). Robinson (2010) suggests that messages around environmental sustainability or animal rights might be more effective in producing health gains.

Writing from an environmental sustainability perspective, Beverland (2014) put forward that certain types of health messages might influence dietary shift to realise environmental gains. There have also been suggestions that promoting the cultural value of more traditional diets that include less meat consumption may have resonance with the public (Beverland, 2014; Pairottia et al., 2014). Thus there is acknowledgement of the need for cross-disciplinary work to address this issue (Johnston et al., 2014). Clearly there needs to be further research examining which messages will resonate with different sections of the public and linking together related organisations and groups that can advocate for policy reform (Beverland, 2014; Joyce et al., 2014). Understanding how to connect disparate groups interested in shifting dietary practice, and the policies required to support this change, is a critical research need. The challenge for future research is therefore to explore how to promote cooperation between health and environmental organizations to enable policy change.

Goals

Goals have a very important role in influencing the previous leverage points of feedback and delays, system structure and system elements as the goals of the system influence its direction and scope (Meadows, 1999). Meadows (1999) comments on how values can determine the selection of goals, which has dramatic system effects. Johnston et al. (2014) describe how unrealistic goals can raise expectations so that programs are destined to be viewed as failures and that setting process system goals may better direct effort, the example they gave was: "Make physical activity an integral and routine part of everyday life" rather than a goal on reducing obesity rates (Johnston et al., 2014, p. 1276).

Establishing goals for sustainable diets will be an interesting process given the competing interests and viewpoints, even among those directly concerned with the research, let alone professional organisations. Vieux et al. (2012) recommended, due to the rigidity of diets, that the focus should be on improving efficiency in the supply side rather than trying to shift consumer behaviour. This neglects the previously cited research on the importance of a multi-component strategy to enable policy change. This finding also is contrary to recent research suggesting some malleability in dietary practice. Dagevos and Voordouw (2013) found a reduction in frequency of eating meat from 2009 to 2011 from a population survey and reported that certain sectors of the population could be receptive to messages about reducing meat consumption for reasons of ethics, health, social norms or price. A study of Finnish university students found that the feasibility of adopting a vegetarian diet was high although they did not rank this behaviour as important thus highlighting the low level of knowledge about environmental impact of dietary choices (Salonen & Helne, 2012).

While there is general consensus on the need to reduce meat consumption to reduce GHG (Joyce et al., 2014), in the Australian study sponsored by the Meat & Livestock Australia, it was concluded that improved agricultural practices rather than reduction in meat consumption were appropriate goals (Hendrie et al., 2014). The position of this industry within Australia both at a broad economic and social level and also as a funder of nutrition related research, underscores policy formation. This issue will vary considerably across countries and regions depending on the formation of coalitions and message framing (Joyce et al., 2014). The process by which goals and then consequent policies and programs get developed is an area requiring further research.

Paradigm

Although the most difficult leverage point to shift, intervening in a system's paradigm can be very effective (Johnston et al., 2014). A paradigm describes a system's deepest held beliefs (Johnston et al., 2014; Meadows, 1999) and underpins the identification of goals as well as many of the other leverage points. By changing systems at the level of their paradigm or their underlying assumptions, transformation of the system can occur (Meadows, 1999). Indeed without changing the institutional structures of a culture, sustained behavioural change of individuals or organisations is unlikely (Kilbourne, Beckmann & Thelen, 2002).

Kilbourne et al. (2002) described the dominant social paradigm in relation to environmental attitudes as one which was engendered during the Enlightenment and collates values, beliefs and institutions into a social lens of how the social world is interpreted. Although a paradigm may be held by the majority it is not this that enables it to be dominant but it being "held by dominant groups who use it to legitimize and justify prevailing institutions" (Kilbourne et al., 2002, p. 194) and may function as an ideology. Relevant aspects of the dominant social paradigm include self-interested individualism, anthropocentrism, technological optimism and support for economic growth and limited government (Kilbourne et al., 2002; Kilbourne & Polonsky, 2005). Studies suggest that when individuals value self-enhancing and materialistic aims they are more likely to have negative environmental attitudes, less likely to engage in simple environmentally beneficial behaviours and choose behaviours that degrade the environment (Kasser, 2011).

Related to anthropocentric values is the relationship socially conceptualised between humans and other animals. The dominant paradigm vacillates around a utilitarian view of animals that as long as they are treated humanely (varying in definition) their use for human purposes including human consumption is acceptable. An alternative viewpoint comes from animal rights activists who want to not just improve the conditions under which animals live and are treated but attempt to change the way we conceptualise animals and abolish their use by humans completely (Payne, 2002). An ethically driven agenda based on revising the human relationship to animals and the resulting moral imperative may be more persuasive to shift dietary behaviours for some than messages related to health or sustainability. This would align with Kasser (2011, p. 94) suggestion that interventions to increase the importance that people place on "values such as benevolence and universalism and goals such as self-acceptance, affiliation, and community feeling" would assist in de-emphasising the prevailing materialist values associated with ecological degradation.

CONCLUSION

The diet modelling research has provided consistent findings that reduced consumption of animal based products and increase in consumption of plant-based foods would result in considerable reductions in GHG emissions. These findings are similar across countries and regions with diverse agricultural and dietary practices. At present the intervention recommendations stemming from this research are not grounded in the research on social, behavioural and policy change (Joyce et al., 2012; Joyce et al., 2014). This chapter has made apparent that what is available on these topics is focused somewhat narrowly on research leading to programmatic interventions rather than broader systems thinking approaches. Research and debates on the merits of promoting a paradigm shift in people's attitudes to animals compared to promoting a small reduction in meat consumption need to be explored. How to best implement surveillance and monitoring systems on food consumption and attitudes to plant-based diets need to be investigated and the formation of system structures and goals in this domain requires further research as well. While this research is still in its infancy, as it develops, possible strategy solutions should emerge that can be tested and refined and progress made towards reducing dietary related GHG emissions.

REFERENCES

Ajzen, I. (1991). The theory of planned behaviour. *Organizational Behavior and Human Decision Processes, 50*(2), 179–211. doi:10.1016/0749-5978(91)90020-T

Australian Institute of Health and Welfare (AIWH). (2008). Australia's health 2008. The eleventh biennial health report of the Australian Institute of Health and Welfare. Canberra, Australia: AIWH. Retrieved September 8, 2015, from http://www.aihw.gov.au/WorkArea/DownloadAsset.aspx?id=6442453674

Bajželj, B., Richards, K. S., Allwood, J. M., Smith, P., Dennis, J. S., Curmi, E., & Gilligan, C. A. (2014). Importance of food-demand management for climate mitigation. *Nature Climate Change, 4*(10), 924–929. doi:10.1038/nclimate2353

Berners-Lee, M., Hoolohan, C., Cammack, H., & Hewitt, C. N. (2012). The relative greenhouse gas impacts of realistic dietary choices. *Energy Policy, 43*, 184–190. doi:10.1016/j.enpol.2011.12.054

Beverland, M. B. (2014). Sustainable eating: Mainstreaming plant-based diets in developed economies. *Journal of Macromarketing, 34*(3), 369–382. doi:10.1177/0276146714526410

Bond, L., & Butler, H. (2009). The Gatehouse Project: a multi-level integrated approach to promoting wellbeing in schools. In A. Killoran & M. Kelley (Eds.), *Evidence-based public health: effectiveness and efficiency* (pp. 250–269). Oxford, UK: Oxford University Press. doi:10.1093/acprof:oso/9780199563623.003.016

Bradbear, C., & Friel, S. (2011). Food systems and environmental sustainability: a review of the Australian evidence. Working Paper October 2011. Canberra, Australia: Australian National University (ANU), College of Medicine, National Centre for Epidemiology and Population Health. Retrieved September 8, 2015, from http://nceph.anu.edu.au/files/Food%20systems%20and%20Environmental%20Sustainability%20A%20review%20of%20the%20Evidence%20NCEPHWorkingPaperOctober2011x%20(3).pdf

Dagevos, H., & Voordouw, J. (2013). Sustainability and meat consumption: is reduction realistic. Sustainability: Science, Practice, &. *Policy, 9*(2), 60–69.

de Savigny, D., & Adam, T. (2009). Systems thinking for health systems strengthening. Geneva, Switzerland: World Health Organization (WHO) & Alliance for Healthy Policy and Systems Research. Retrieved September 8, 2015, from http://www.who.int/alliance-hpsr/resources/9789241563895/en/

de Vries, M., & de Boer, I. J. M. (2010). Comparingenvironmental impacts for livestock products: A review of life cycle assessments. *Livestock Science, 128*(1), 1–11. doi:10.1016/j.livsci.2009.11.007

Elferink, E., Nonhebel, S., & Moll, H. (2008). Feeding livestock food residue and the consequences for the environmental impact of meat. *Journal of Cleaner Production, 16*(12), 1227–1233. doi:10.1016/j.jclepro.2007.06.008

Ericksen, P. J., Ingram, J. S. I., & Liverman, D. M. (2009). Food security and global environmental change: Emerging challenges. *Environmental Science & Policy, 12*(4), 372–377. doi:10.1016/j.envsci.2009.04.007

Eshel, G., Shepon, A., Makov, T., & Milo, R. (2014). Land, irrigation water, greenhouse gas and reactive nitrogen burdens of meat, eggs and dairy production in the United States. [PNAS]. *Proceedings of the National Academy of Sciences of the United States of America, 111*(33), 11996–12001. doi:10.1073/pnas.1402183111 PMID:25049416

Foster-Fishman, P. G., Nowell, B., & Yang, H. (2007). Putting the system back into systems change: A framework for understanding and changing organizational and community systems. *American Journal of Community Psychology, 39*(3-4), 197–215. doi:10.1007/s10464-007-9109-0 PMID:17510791

Fox, N., & Ward, K. J. (2008a). Health ethics and environment: A qualitative study of vegetarian motivations. *Appetite, 50*(2-3), 422–429. doi:10.1016/j.appet.2007.09.007 PMID:17980457

Fox, N., & Ward, K. J. (2008b). You are what you eat? Vegetarianism, health and identity. *Social Science & Medicine, 66*(12), 2585–2595. doi:10.1016/j.socscimed.2008.02.011 PMID:18378056

Friel, S., Barosh, L. J., & Lawrence, M. (2013). Towards healthy and sustainable food consumption: An Australian case study. *Public Health Nutrition, 17*(5), 1156–1166. doi:10.1017/S1368980013001523 PMID:23759140

Garnett, T. (2009). Livestock-related greenhouse gas emissions: Impacts and options for policy makers. *Environmental Science & Policy, 12*(4), 491–503. doi:10.1016/j.envsci.2009.01.006

Garnett, T. (2011). Where are the best opportunities for reducing greenhouse gas emissions in the food system (including the food chain)? *Food Policy, 36*, S23–S32. doi:10.1016/j.foodpol.2010.10.010

Gibbs, H. K., Ruesch, A., Achard, F., Clayton, M., Holmgren, P., Ramankutty, N., & Foley, J. (2010). Tropical forests were the primary sources of new agricultural land in the 1980s and 1990s. [PNAS]. *Proceedings of the National Academy of Sciences of the United States of America, 107*(38), 16732–16737. doi:10.1073/pnas.0910275107 PMID:20807750

Gunasekera, D., Tulloh, C., Ford, M., & Heyhoe, S. (2008, June 13). Climate change opportunities and challenges in Australian agriculture. Paper presented at the Faculty of Agriculture, Food and Natural Resources (FAFNR '08) Annual Symposium, Sydney, Australia. Retrieved September 8, 2015, from http://agricoop.nic.in/Climatechange/ccr/files/Climate%20Change-Australian%20Agriculture.pdf

Hallström, E., Röös, E., & Börjesson, P. (2014). Sustainable meat consumption: A quantitative analysis of nutritional intake, greenhouse gas emissions and land use from a swedish perspective. *Food Policy, 47*, 81–90. doi:10.1016/j.foodpol.2014.04.002

Havlík, P., Valin, H., Mosnier, A., Obersteiner, M., Baker, J. S., Herrero, M., & Schmid, E. et al. (2013). Crop productivity and the global livestock sector: Implications for land use changes and greenhouse gas emissions. *American Journal of Agricultural Economics*, *95*(2), 442–448. doi:10.1093/ajae/aas085

Hawe, P., Shiell, A., & Riley, T. (2009). Theorising interventions as events in systems. *American Journal of Community Psychology*, *43*(3-4), 267–276. doi:10.1007/s10464-009-9229-9 PMID:19390961

Hedenus, F., Wirsenius, S., & Johansson, D. J. A. (2014). The importance of reduced meat and dairy consumption for meeting stringent climate change targets. *Climatic Change*, *124*(1-2), 79–91. doi:10.1007/s10584-014-1104-5

Hendrie, G. A., Ridoutt, B. G., Wiedmann, T. O., & Noakes, M. (2014). Greenhouse gas emissions and the Australian diet - Comparing dietary recommendations with average intakes. *Nutrients*, *6*(1), 289–303. doi:10.3390/nu6010289 PMID:24406846

Hoolohan, C., Berners-Lee, M., McKinstry-West, J., & Hewitt, C. N. (2013). Mitigating the greenhouse gas emissions embodied in food through realistic consumer choices. *Energy Policy*, *63*, 1065–1074. doi:10.1016/j.enpol.2013.09.046

Howat, P., Maycock, B., Cross, D., Collins, J. D., Jackson, L., Burns, S., & James, R. (2003). Towards a more unified definition of health promotion. *Health Promotion Journal of Australia*, *14*(2), 82–85.

Howat, P., Sleet, D., Elder, R., & Maycock, B. (2004). Preventing alcohol-related traffic injury: A health promotion approach. *Traffic Injury Prevention*, *5*(3), 208–219. doi:10.1080/15389580490465238 PMID:15276921

James, S. J., & James, C. (2010). The food cold-chain and climate chain. *Food Research International*, *43*(7), 1944–1956. doi:10.1016/j.foodres.2010.02.001

Johnston, L. M., Matteson, C. L., & Finegood, D. T. (2014). Systems science and obesity policy: A novel framework for analyzing and rethinking population-level planning. *American Journal of Public Health*, *104*(7), 1270–1278. doi:10.2105/AJPH.2014.301884 PMID:24832406

Joyce, A., Dixon, S., Comfort, J., & Hallett, J. (2008). The cow in the room: Public knowledge of the links between dietary choices and health and environmental impacts. *Environmental Health Insights*, *1*, 31–34. PMID:21572845

Joyce, A., Dixon, S., Comfort, J., & Hallett, J. (2012). Reducing the environmental impact of dietary choice: Perspectives from a behavioural and social change approach. *Journal of Environmental and Public Health*, *2012*, 1–7. doi:10.1155/2012/978672 PMID:22754580

Joyce, A., Hallett, J., Hannelly, T., & Carey, G. (2014). The impact of nutritional choices on global warming and policy implications: Examining the link between dietary choices and greenhouse gas emissions. *Journal of Energy and Emission Control Technologies*, *2*, 33–43. doi:10.2147/EECT.S58518

Jungbluth, N., Tietje, O., & Scholz, R. W. (2000). Food purchases: Impacts from the consumers' point of view investigated with a modular LCA. *The International Journal of Life Cycle Assessment*, *5*(3), 134–142. doi:10.1007/BF02978609

Kahn, S., & Hanjra, M. A. (2009). Footprints of water and energy inputs in food production - global perspectives. *Food Policy, 34*(2), 140–150.

Kasser, T. (2011). Ecological Challenges, Materialistic Values and Social Change. In R. Biswas-Diener (Ed.), *Positive Psychology as Social Change* (pp. 89–108). New York, NY: Springer. doi:10.1007/978-90-481-9938-9_6

Kickert, W. J. M., Klijn, E., & Koppenjan, J. F. M. (1997). Introduction: Management perspective on policy networks. In J. M. Walter, E.-H. Kickert, & J. F. M. Koppenjan (Eds.), *Managing complex networks: strategies for the public sector* (pp. 1–11). London, UK: Sage. doi:10.4135/9781446217658.n1

Kilbourne, W. E., Beckmann, S. C., & Thelen, E. (2002). The role of the dominant social paradigm in environmental attitudes: A multinational examination. *Journal of Business Research, 55*(3), 193–204. doi:10.1016/S0148-2963(00)00141-7

Kilbourne, W. E., & Polonsky, M. J. (2005). Environmental attitutdes and their relation to the dominant social paradigm among university students in New Zealand and Australia. *Australasian Marketing Journal, 13*(2), 37–48. doi:10.1016/S1441-3582(05)70076-8

Lea, E. J., Crawford, D., & Worsley, A. (2006). Public views of the benefits and barriers to the consumption of a plant based diet. *European Journal of Clinical Nutrition, 60*(7), 828–837. doi:10.1038/sj.ejcn.1602387 PMID:16452915

Lea, E. J., Worsley, A., & Crawford, D. (2005). Australian adult consumers' beliefs about plant foods: A qualitative study. *Health Education & Behavior, 32*(6), 795–808. doi:10.1177/1090198105277323 PMID:16267149

Macdiarmid, J. I., Kyle, J., Horgan, G. W., Loe, J., Fyfe, C., Johnstone, A., & McNeill, G. (2012). Sustainable diets for the future: Can we contribute to reducing greenhouse gas emissions by eating a healthy diet? *The American Journal of Clinical Nutrition, 96*(3), 632–639. doi:10.3945/ajcn.112.038729 PMID:22854399

Marsh, K., & Bugusu, B. (2007). Food packaging – roles, materials and environmental issues. *Journal of Food Science, 72*(3), R39–R55. doi:10.1111/j.1750-3841.2007.00301.x PMID:17995809

McKenzie-Mohr, D., & Schultz, P. W. (2014). Choosing effective behaviour change tools. *Social Marketing Quarterly, 20*(1), 35–46. doi:10.1177/1524500413519257

McMichael, A. J., Powles, J. W., Butler, C. D., & Uauy, R. (2007). Food, livestock production, energy, climate change and health. *Lancet, 370*(9594), 1253–1263. doi:10.1016/S0140-6736(07)61256-2 PMID:17868818

McTaggart, R. (1991). Principles for participatory action research. *Adult Education Quarterly, 41*(3), 168–187. doi:10.1177/0001848191041003003

Meadows, D. (1999). *Leverage points: places to intervene in a system.* Hartland, VT: The Sustainability Institute.

Meisterling, K., Samaras, C., & Schweizer, V. (2009). Decisions to reduce greenhouse gases from agriculture and product transport: LCA case study of organic and conventional wheat. *Journal of Cleaner Production, 17*(2), 222–230. doi:10.1016/j.jclepro.2008.04.009

Metz, B., Davidson, O. R., Bosch, P., Dave, R., & Meyer, L. (2007). Climate change 2007: mitigation of climate change. Contribution of Working Group III to the Fourth Assessment Report of the Intergovernmental Panel on Climate Change (IPCC). New York, NY: IPCC, Working Group III. Retrieved September 8, 2015, from https://www.ipcc.ch/publications_and_data/publications_ipcc_fourth_assessment_report_wg3_report_mitigation_of_climate_change.htm

Moodie, R. (2004). Introduction: getting your hands on. In R. Moodie & A. Hulme (Eds.), *Hands-on Health promotion (i-xix)*. Melbourne, Australia: IP Communications.

Mrazek, P., Biglan, A., Hawkins, J. D., & Cody, C. (2007). *Community monitoring systems: tracking and improving the well-being of America's children and adolescents*. Falls Church, VA: Society for Preventive Research.

National Cancer Institute (NCI). (2007). Greater than the sum: systems thinking in tobacco control. Tobacco control monograph No. 18. Bethesda, MD: US Department of Health and Human Services, National Institute of Health, National Cancer Institute. Retrieved SEptember 8, 2015, from http://cancercontrol.cancer.gov/brp/tcrb/monographs/18/m18_complete.pdf

Pairottia, M. B., Ceruttib, A. K., Martinic, F., Vescea, E., Padovand, D., & Beltramoa, R. (2014). (in press). Energy consumption and GHG emission of the Mediterranean diet: A systemic assessment using a hybrid LCA-IO method. *Journal of Cleaner Production.*

Pathak, H., Jain, N., Bhatia, A., Patel, J., & Aggarwal, P. K. (2010). Carbon footprints of Indian food items. *Agriculture, Ecosystems & Environment, 139*(1-2), 66–73. doi:10.1016/j.agee.2010.07.002

Payne, R. (2002). Animal welfare, animal rights and the path to social reform: One movements struggle for coherency in the quest for change. Virginia Journal of Soial. *Policy & the Law, 9*(3), 587–633.

Pearson, D., Friel, S., & Lawrence, D. (2014). Building environmentally sustainable food systems on informed citizen choices: Evidence from Australia. *Biological Agriculture and Horticulture, 30*(3), 183–197. doi:10.1080/01448765.2014.890542

Pelletier, N., Audsley, E., Brodt, S., Garnett, T., Henriksson, P., Kendall, A., & Troell, M. et al. (2011). Energy intensity of agriculture and food systems. *Annual Review of Environment and Resources, 36*(1), 223–246. doi:10.1146/annurev-environ-081710-161014

Popp, A., Lotze-Campen, H., & Bodirsky, B. (2010). Food consumption, diet shifts and associated non-CO2 greenhouse gases from agricultural production. *Global Environmental Change, 20*(3), 451–462. doi:10.1016/j.gloenvcha.2010.02.001

Prochaska, J., & DiClemente, C. (1983). Stages and processes of self-change of smoking: Toward an integrative model of change. *Journal of Consulting and Clinical Psychology, 51*(3), 390–395. doi:10.1037/0022-006X.51.3.390 PMID:6863699

Public Health Association Australia (PHAA). (2009). A future for food: addressing public health, sustainability and equity from paddock to plate. Retrieved September 8, 2015, from https://www.phaa.net. au/documents/item/563

Resnicow, K., & Page, S. E. (2008). Embracing chaos and complexity: A quantum change for public health. *American Journal of Public Health*, *98*(8), 1382–1389. doi:10.2105/AJPH.2007.129460 PMID:18556599

Ripple, W. J., Smith, P., Haberl, H., Montzka, S. A., McAlpine, C., & Boucher, D. H. (2014). Commentary: Ruminants, climate change and climate policy. *Nature Climate Change*, *4*(1), 2–5. doi:10.1038/nclimate2081

Risku-Norja, H., Kurppa, S., & Helenius, J. (2009). Dietary choices and greenhouse gas emissions – assessment of impact of vegetarian and organic options at national scale. *Progress in Industrial Ecology*, *6*(4), 340–354. doi:10.1504/PIE.2009.032323

Robinson, T. N. (2010). Save the world, prevent obesity: Piggybacking on existing social and ideological movement. *Obesity (Silver Spring, Md.)*, *18*(1sSuppl 1), S17–S22. doi:10.1038/oby.2009.427 PMID:20107456

Rogers, E. M. (1983). *Diffusion of innovations*. New York, NY: Free Press.

Rowe, A. K. (2009). Potential of integrated continuous surveys and quality management to support monitoring, evaluation and the scale-up of health interventions in developing countries. *The American Journal of Tropical Medicine and Hygiene*, *80*(6), 971–979. PMID:19478260

Roy, P., Nei, D., Orikasa, T., Xu, Q. Y., Okadome, H., Nakamura, N., & Shiina, T. (2009). A review of life cycle assessment (LCA) on some food products. *Journal of Food Engineering*, *90*(1), 1–10. doi:10.1016/j.jfoodeng.2008.06.016

Salonen, A. O., & Helne, T. T. (2012). Vegetarian diets: A way towards a sustainable society. *Journal of Sustainable Development*, *5*(6), 10–24. doi:10.5539/jsd.v5n6p10

Scarborough, P., Appleby, P. N., Mizdrak, A., Briggs, A. D. M., Travis, R. C., Bradbury, K. E., & Key, T. J. (2014). Dietary greenhouse gas emissions of meat-eaters, fish-eaters, vegetarians and vegans in the UK. *Climatic Change*, *125*(2), 179–192. doi:10.1007/s10584-014-1169-1 PMID:25834298

Smith, P., Martino, D., Cai, Z., Gwary, D., Janzen, H., Kumar, P., & Towprayoon, S. et al. (2007). Policy and technological constraints to implementation of greenhouse gas mitigation options in agriculture. *Agriculture, Ecosystems & Environment*, *118*(1), 6–28. doi:10.1016/j.agee.2006.06.006

Snyder, C., Bruulsema, T., Jensen, T., & Fixen, P. (2009). Review of greenhouse gas emissions from crop production systems and fertilizer management effects. *Agriculture, Ecosystems & Environment*, *133*(3), 247–266. doi:10.1016/j.agee.2009.04.021

Soret, S., Mejia, A., Batech, M., Jaceldo-Siegel, K., Harwatt, H., & Sabaté, J. (2014). Climate change mitigation and health effects of varied dietary patterns in real-life settings throughout North America. *The American Journal of Clinical Nutrition*, *100*(Supplement_1), 490S–495S. doi:10.3945/ajcn.113.071589 PMID:24898230

Stafford, J., Allsop, S., & Daube, M. (2014). From evidence to action: Health promotion and alcohol. *Health Promotion Journal of Australia, 25*(1), 8–13. doi:10.1071/HE14001 PMID:24739773

Steinfeld, H., Gerber, P., Wassenaar, T., Castel, V., Rosales, M., & de Hann, C. (2006). Livestock's long shadow: environmental issues and options. Rome, Italy: Food Agriculture Organisation (FAO).

Sulda, H., Coveney, J., & Bentley, M. (2010). An investigation of the ways in which public health nutrition policy and practices can address climate change. *Public Health Nutrition, 13*(3), 304–313. doi:10.1017/S1368980009990334 PMID:19545472

Tassou, S. A., Ge, Y., Hadawey, A., & Marriott, D. (2011). Energy consumption and conservation in food retailing. *Applied Thermal Engineering, 31*(2-3), 147–156. doi:10.1016/j.applthermaleng.2010.08.023

Tukker, A., Huppes, G., Guinée, J. B., Heijungs, R., de Koning, A., & van Oers, L. … Nielsen, P. (2006). Environmental Impact of Products (EIPRO) Analysis of the life cycle environmental impacts related to the final consumption of the EU-25. Technical Report Series. Seville, Spain: European Commission, Joint Research Centre, Institute for Prospective Technological Studies. Retrieved September 8, 2015, from http://ec.europa.eu/environment/ipp/pdf/eipro_report.pdf

van der Werf, G. R., Morton, D. C., de Fries, R. S., Oliver, J. G., Kasibhatla, P. S., Jackson, R. B., & Randerson, J. T. et al. (2009). CO2 emissions from forest loss. *Nature Geoscience, 2*(11), 737–738. doi:10.1038/ngeo671

van Sluijs, E. M., van Poppel, M. N., & van Mechelen, W. (2004). Stage-based lifestyle interventions in primary care: Are they effective? *American Journal of Preventive Medicine, 26*(4), 330–343. doi:10.1016/j.amepre.2003.12.010 PMID:15110061

Vermeulen, S. J., Campbell, B. M., & Ingram, J. S. (2012). Climate change and food systems. *Annual Review of Environment and Resources, 37*(1), 195–222. doi:10.1146/annurev-environ-020411-130608

Vieux, F., Darmon, N., Touazi, D., & Soler, L. G. (2012). Greenhouse gas emissions of self-selected individual diets in France: Changing the diet structure or consuming less? *Ecological Economics, 75*, 91–101. doi:10.1016/j.ecolecon.2012.01.003

West, R. (2005). Time for a change: Putting the transtheoretical (stages of change) model to rest. *Addiction (Abingdon, England), 100*(8), 1036–1039. doi:10.1111/j.1360-0443.2005.01139.x PMID:16042624

Wilson, N., Nghiem, N., Ni Mhurchu, C., Eyles, H., Baker, M. G., & Blakely, T. (2013). Foods and dietary patterns that are healthy, low-cost and environmentally sustainable: A case study of optimization modelling for New Zealand. *PLoS ONE, 8*(3), 1–10. doi:10.1371/journal.pone.0059648

Wyker, B. A., & Davison, K. K. (2010). Behavioural change theories can inform the prediction of young adult's adoption of a plant-based diet. *Journal of Nutrition Education and Behavior, 42*(3), 168–167. doi:10.1016/j.jneb.2009.03.124 PMID:20138584

KEY TERMS AND DEFINITIONS

Climate Change: A change in the earth's climate, especially a change due to increased global average atmospheric temperature apparent from the mid to late 20th century onwards which is primarily recognised as being due to increased emissions of greenhouse gases into the atmosphere.

Dietary Choices: The types of foods chosen to be consumed by individuals.

Food System: The activities which relate to the production, processing, transport, storage and consumption of food.

Greenhouse Gases: Gases which absorb infrared (heat) radiation in the atmosphere, causing a greenhouse effect.

Greenhouse Gas Emissions: The emission of any of a variety of gases such as carbon dioxide and methane, into the earth's atmosphere which contribute to the greenhouse effect.

Life Cycle Assessment: A method which allows for identification of inputs (i.e. water, electricity, other resources) and outputs (i.e. GHG emissions, waste) connected to production of products.

Livestock Production: The raising of domesticated animals in an agricultural setting for the purpose of producing food.

Meat Consumption: The consumption of meat as part of an individual's diet.

Systems Thinking Science: An approach which looks at the interrelationship between parts of a system and identifies system behaviours such as feedback loops, information flows, and organisational structure.

Chapter 2
A Fresh Look at Livestock Greenhouse Gas Emissions and Mitigation

Robert Goodland
World Bank Group, USA

ABSTRACT

The gravest environmental impact of food production is the impact of its greenhouse gas emissions; that's because of the uniquely diverse, unprecedented, and irreversible risks that it involves. According to the International Energy Agency, atmospheric carbon must be reduced significantly by 2020 or else the world may not be able to avert uncontrollable climate change. This chapter compares the most recent assessment of livestock and climate change by livestock specialists employed by the United Nations' Food and Agriculture Organisation with assessment of livestock and climate change by two World Bank Group environmental specialists, Robert Goodland and Jeff Anhang. It is explained how the only pragmatic way to reverse climate change before it is too late is to replace a substantial amount of today's livestock products with better alternatives.

INTRODUCTION[1]

Jeff Anhang and I have estimated that the lifecycle and supply chain of livestock products are responsible for at least 51% of human-induced GHGs (Goodland & Anhang, 2009), which means that the only pragmatic way to reverse climate change by 2017 as needed is to replace at least 50% of today's livestock products with better alternatives. Our assessment began when we analysed some significant gaps that we found in a 388-page report by the UN Food and Agriculture Organization, or FAO. That FAO report was called *Livestock's Long Shadow* (Steinfeld et al., 2006), and it estimated that 18% of human-induced GHGs are attributable to livestock products – much less than our figure of 51%, but still a significant amount.

DOI: 10.4018/978-1-4666-9553-5.ch002

GREENHOUSE GAS EMISSIONS

We are approaching tipping points for climate change. It is important to have some understanding of food's contribution to greenhouse gas emissions and possible future scenarios.

Climate Tipping Points

Greenhouse gas (GHG) reductions have conventionally been sought in the replacement of fossil fuel infrastructure with renewable energy infrastructure. While that replacement is surely desirable, it has both long and complex product-development cycles and capital-intensive requirements. Indeed, sufficient renewable energy infrastructure to stop climate change is now projected to take at least 20 years and US$18 trillion to develop according to the International Energy Agency (IEA, 2011b). The IEA (2011a) and the Intergovernmental Panel on Climate Change (IPCC) (Spotts, 2012) have projected that climatic tipping points may be reached by 2017 or by 2020 at the latest, making later GHG reductions ineffective.

Climatic tipping points are fearsome. For example, the National Academy of Sciences published a new study that 1,700 American cities – including New York, Boston, and Miami – will be locked into some amount of submersion from rising sea levels as a result of climatic tipping points, unless expensive new dykes and levees can hold back the rising waters (Goldenberg, 2015).

Flooding is only one dire adverse impact expected from climate change. Another is drought – as a key aspect of climate change is that volatility will increase, so swings between one adverse impact and another will become increasingly wide. Increasing swings between flooding and drought can be expected to affect agriculture more than any other industry, as agriculture is worked on outdoors to an extent that's unique among all industries.

Another dire impact of climate change is global warming. For example, a one degree Celsius rise in temperature above optimum in a growing season causes a 10% decline in grain yields (Brown, 2011). This is already happening in some regions. Yet as conservative an organization as PricewaterhouseCoopers has warned of a possible six degree Celsius rise in global temperature this century (PwC, 2012) – that's a 10.8° Fahrenheit rise – which implies a 60% decline in agricultural outputs.

Unfortunately, climatic tipping points can no longer be avoided by constructing the amount of renewable energy infrastructure that would be required. The amount of renewable energy infrastructure needed to avert climatic tipping points projected to be reached by 2017 could have been constructed in time – if construction had started when the Kyoto Protocol was adopted in 1997. That was supposed to happen through the process that required each country that subscribed to the Kyoto Protocol to reduce GHG emissions in each year after the protocol came into force in each country. Instead, with only a few exceptions, GHG emissions ended up rising each year in each country that signed the Kyoto Protocol.

Food Future Scenarios

Now, there seems to be only one remaining pragmatic way to reverse climate change before it's too late – and that's by taking quick and large-scale actions in the realm of food, agriculture, and forestry. If the analysis by Jeff Anhang and me is correct in estimating that at least 51% of human-induced GHGs are attributable to livestock, then replacing 50 to 85% of today's livestock products with better non-animal based alternatives by 2017 could fully achieve the objective of the Kyoto Protocol and avert catastrophic climate change. According to our calculations in 2009, the amount of replacement needed would be more like 25%; but we had to revise that figure when we later found that anthropogenic greenhouse emissions worldwide rose by 61% from 1990 to 2013 (Radford, 2013).

Yet the FAO's *Livestock's Long Shadow* contemplates actions to manage a projected doubling in live-stock production to feed the projected 9 billion people who will be alive in 2050. Specifically, the report projected that "the production of meat will double between now and 2050" (Steinfeld et al., 2006, p. 388).

Let's examine a few parameters of the Food Agriculture Organization (FAO) projections to understand whether they can be justified. First, let's examine the FAO's assertion that we should expect livestock production worldwide to double by the year 2050. The Soil Association (2010) uses unusually harsh language to critique that projection, calling it a "big fat lie". It says that the FAO's projection assumes "high meat" growth in developing countries, which would generate "massive land use change" that would exacerbate climate change rather than slow it (Soil Association, 2010).

Conversely, a report by the International Food Policy Research Institute projects that consumption of livestock products could decline through at least 2030 (Msangi & Rosegrant, 2011). That institute co-published a report 14 years ago that kicked off what was called at the time the "Livestock Revolu-tion", which was said to be a recognition of a sort of inevitable increase in factory farming (Delgado, Rosegrant, Steinfield, Ehui, & Courbois 1999). So it's particularly striking to see the International Food Policy Research Institute now projecting such a different future.

Yet neither the International Food Policy Research Institute nor FAO has explicitly drawn any scenario by which disruptive climate events might force shrinkage of the livestock sector. The need to draw such a scenario seems apparent as large-scale die-offs of livestock, feed crops, and grasses for grazing started to be seen in every region of the world in 2009. In some regions, such die-offs reached unprecedented proportions recently – and climate change threatens to make them increase further.

Now, let's examine the FAO's projection that 9 billion people will need to be fed in 2050. That projection comes from what the UN Population Division calls its medium-variant projection for 2050. The UN Population Division also publishes a low-variant projection, which is for a human population in 2050 of 8.1 billion. That low-variant projection would yield 14% more people than today in 2050, while the medium-variant projection would yield a 28% increase. Neither one comes close to justifying a doubling in livestock production.

Moreover, none of the UN Population Division's projections factors in the possibility of a reduction in human population forced by climate change, which would mean that less food than is produced today might be required in the future. In any event, any switch from a livestock product to an alternative means that less crops need to be grown, which is equivalent to saying that less food may need to be grown in the future than is grown today.

How could the FAO go so wrong? Well, let's start with the FAO report that I've mentioned, namely *Livestock's Long Shadow*, published in 2006. Activists often use that FAO report to support advocacy for vegetarian foods. However *Livestock's Long Shadow* actually prescribes more factory farming: "The principal means of limiting livestock's impact on the environment must be... intensification" (Steinfeld et al., 2006, p. 236). That report's lead author and a co-author later wrote to confirm that prescription (Steinfeld & Gerber, 2010). Yet those authors are livestock specialists, not environmental specialists.

International good practice is to have any activity with major environmental impacts assessed by environmental specialists. My co-author Jeff Anhang and I are long-time environmental specialists employed by two UN specialized agencies, the World Bank and International Finance Corporation to conduct such assessments.

Why it matters who performs environmental assessment becomes clear when reviewing the report *Livestock's Long Shadow*, written by livestock specialists rather than environmental specialists. The report examined land degradation, climate change and air pollution, water shortage and water pollution, and loss of biodiversity. It asserted that those risks and impacts must be balanced with benefits available from raising livestock. Yet the report failed to measure livestock's various risks and impacts in any relative ranking, a basic element of environmental assessment. Moreover, the report failed to identify any climatic tipping points, even though they're actually the #1 risk of continuing to expand livestock production, rather than greenhouse gas in itself, or any other risk or impact. That failure enabled the fatal flaw of projecting a doubling in livestock production, and pairing that with only minor prescriptions for mitigating greenhouse gas, essentially asserting that benefits available from producing livestock products outweigh their risks. Also, the report assessed only livestock products and included no analysis of alternatives, another basic element of environmental assessment.

We know one of the authors of *Livestock's Long Shadow*, Cornelius De Haan, from when he worked at the World Bank. He was lead author of the World Bank's 2001 livestock strategy, which pegged livestock's adverse impacts at a lower level than in the 2006 *Livestock's Long Shadow* – yet the World Bank strategy correctly recommends that institutions should "avoid funding large-scale commercial, grain-fed feedlot systems and industrial milk, pork, and poultry production" (De Hann et al., 2001, p. 96).

Why would Cornelius De Haan move from being lead author of a report that prescribed avoiding factory farming to co-authoring a report that prescribed expanding factory farming? Maybe it's at least partly because the FAO, unlike the World Bank, has formed a formal partnership with:

1. The International Meat Secretariat,
2. The International Dairy Federation, and
3. The International Egg Association.

My assessment of that partnership was published by the New York Times (Goodland, 2012), and it's on our website, Chomping Climate Change (http://www.chompingclimatechange.org).

Our website includes links to many prominent citations of our analysis, including by UNESCO, NBC News, the Washington Post, Forbes Magazine, Canadian government climate scientists, and the Sierra Club. The website also has links to a number of universities' syllabi that refer to our analysis, including Harvard and the University of California.

In fact, worldwide interest in our analysis started to unfold the day it was first published, when Jeff Anhang and I got requests for interviews from all over the world, including from Reuters, the biggest news agency in the world. Worldwide interest in our analysis is important not for our sakes, but because climate change is one of a relatively small number of environmental issues that are trans-boundary. This means that GHG emissions and atmospheric carbon don't respect borders – so a molecule of carbon dioxide emitted in China can affect people anywhere in the United States or in Australia, for example, just as much as it will affect someone in Beijing. In other words, it's as important to be concerned about what happens with food and climate change elsewhere as it is to be concerned about what happens with food and climate change in California.

The FAO also projects that improvements in managing methane and nitrous oxide attributable to livestock will achieve most of the GHG reduction from livestock that is needed. Yet according to an article in Nature (Petherick, 2012), such improvements can achieve only 4% of global agricultural mitigation potential to 2030. That's much less GHG reduction than all other industries are being called upon to achieve, and indeed is a trivial amount relative to what the food industry should be called upon to achieve (as published in my response to Petherick, 2012 in Goodland, 2013a).

In any event, the FAO also projects that minimizing food waste will significantly help with GHG reduction from agriculture (Gustavsson, Cederberg, Sonesson, van Otterdijk & Meybeck, 2011). Minimizing food waste is surely desirable. Yet the FAO's best-case scenario for minimizing food waste would account for a small fraction of the GHG reduction that is needed. In fact, it might yield no GHG reduction at all. For example, it could mean that parents, instead of cooking less food for their children, will instead increase their insistence that children clean their plates, which some scientific studies have associated with increased obesity (Jacobsen, 2013). It could also mean that more animal fats will be incorporated in consumer products instead of being discarded, and this would have no beneficial effect on climate change.

SOLUTIONS AND RECOMMENDATIONS

Indeed, minimizing food waste has been an objective of humanity for centuries, during which food waste has only increased. It's been said that failing at the same thing over and over and expecting a different result constitutes insanity.

Yet virtually all efforts that have focused on reducing GHG emissions have failed over and over, and are nevertheless being tried again. What we really need is some sort of action that will yield a large-scale reduction in GHG emissions and at the same time will massively capture atmospheric carbon in a stable place, otherwise known as carbon sequestration. We need such a grand dual action because the average global concentration of atmospheric carbon continues to increase after it recently rose above 400 parts per million – far above the safe level of 350 parts per million; and once carbon is in the atmosphere, it normally stays up there for at least 100 years. Some of it stays up there for thousands of years.

Insanity vs. Better Alternatives

However there are ways to draw atmospheric carbon down to earth. The only known way to sequester atmospheric carbon on a large scale is by growing more trees. In recent decades, work to reforest land has been an overwhelming challenge, as policymakers and industry have both acted in various ways to do the opposite, namely to expand deforestation.

Simply proposing for the umpteenth time that we should grow more trees and expecting a different result could be considered insanity. Instead, Jeff Anhang and I propose replacing a substantial amount of today's livestock products with better alternatives which provide plant-based nutrients for human consumption. This will massively reduce GHG emissions and free up a vast amount of land to permit large-scale reforestation and GHG sequestration at the same time. The effect of this would be so enormous that it could actually be the only pragmatic way to reverse climate change. That's because livestock and feed production are estimated to occupy 45% percent of all land on earth, according to Thornton, Herrerro and Ericksen (2011)[2] – that's all land, both arable and non-arable, not excluding ice caps or mountaintops or anything else. Much of that 45% was once forested, and could be forested again.

In fact, there is documented potential for agricultural change to bring atmospheric carbon to pre-industrial revolution levels within five years. Some of the relevant literature deals with remineralisation (remineralize.org/research/the-potential-of-remineralisation), while other relevant literature uses terms such as afforestation, reforestation, forest regeneration, and rewilding.

While Jeff Anhang and I have identified the unique dual benefit of reducing GHG emissions and increasing GHG sequestration from the single action of replacing livestock products with better alternatives, the FAO strategy for GHG mitigation from livestock involves a relatively small amount of GHG reduction through methodological tweaks. So we weren't surprised when the lead author of FAO's *Livestock's Long Shadow* and a co-author participated on a team of livestock specialists who wrote a critique of our analysis. They were able to get their critique published only as a "commentary", not a regular article, even by the cattleman-friendly journal *Animal Feed Science and Technology* (Herrero et al., 2011).

Animal Feed Science and Technology then published our response, and provided the other team a chance to reply in turn, which they declined to take up (Hererro, 2011). This can be seen on our website (chompingclimatechange.org), and it can help people to decide whose analysis is more authoritative.

One of the mistakes that we've identified in *Livestock's Long Shadow* is that it doesn't count carbon dioxide in livestock respiration at all, as the FAO considers carbon dioxide to be balanced perfectly by photosynthesis. Indeed, the perfect balance of exhaled carbon dioxide with carbon captured by vegetation via photosynthesis is something that most children learn in school – namely, the carbon cycle.

However, reality no longer reflects the old model of the carbon cycle, in which photosynthesis balanced respiration. That model was valid as long as there were roughly constant levels of respiration and photosynthesis on Earth. In recent decades, respiration has increased exponentially with livestock production, which has risen from less than 10 billion land-based animals worldwide in 1945, up to more than 70 billion land-based animals worldwide this year[3] – while intensified livestock and feed production has been accompanied by large-scale deforestation and forest-burning, huge increases in volatilization of carbon, and a dramatic decline in both the Earth's photosynthetic capacity and in its GHG sequestration capacity (Goodland, 2013a).

Typically 200 tons of carbon per hectare are stored in forest, and are at risk of being released through deforestation. Yet more than 200 tons of carbon per hectare are typically stored in the soil beneath, and are at risk of being released when the forest above is degraded. The FAO estimates that each year between 2000 and 2010, an area larger than England was deforested (FAO, 2010).

As a result, either carbon dioxide in livestock respiration – or its reflection in carbon absorption forgone on land used for livestock and feed production – should be counted as emissions. When that is done, it seems clear that a priority for governments is to provide incentives for reforesting a significant amount of the 45% of all land on earth used today for livestock and feed production.

Jeff Anhang and I aren't the only environmental specialists to suggest that the carbon cycle has changed, and that carbon dioxide in livestock respiration should be counted. For example, our analysis references an estimate that carbon in livestock respiration exceeds 8 gigatonnes per year, about 8 times the volume of carbon dioxide from human respiration (Calverd, 2005). A lower estimate that relies on FAO data is available, but it similarly concludes that carbon dioxide in livestock respiration can no longer be considered offset by photosynthesis (Prairie & Duarte, 2007).

In contrast, some scenarios for future food production recommend that beef and dairy production should be prioritized on grassland. That's contrary to the FAO's recommendation for more factory farming. In fact, it seems clear that the FAO is correct in its assessment that beef and dairy cattle produced on grassland emit much more methane than intensively-produced ones and they take up much more land, leaving much less forest available to absorb atmospheric carbon (Eshel, 2010).

Scenarios involving more grass-fed livestock fail to account for the urgent need for reforestation. In any event, meat promoters can be seen spinning recommendations for more grass-fed livestock as if they apply equally to factory farmed meat (Henderson, 2013). Yet it's no new trick to promote factory-farmed meat as grass-fed. A grassland producer has himself noted that well over 90% of meat sold in the United States comes from feedlots, and most marketing of "grass-fed" beef is a hoax (www.americangrassfed-beef.com/grass-fed-hoax.asp). Beef marketed this way commands a 200-300% price premium – so the incentive for producers and sellers to cheat is overwhelming (Goodland, 2013b).

Yet there is an urgent need to halt climate change. In fact, the only real choice may be whether to look at replacing livestock products with better alternatives as a voluntary opportunity – or to deal with it involuntarily if large-scale die-offs of livestock, crops, and grasses continue to increase due to climate disruption.

As most livestock products worldwide are now produced on a large scale, all of the replacement of livestock products with better alternatives that's needed can be made to happen in ways that would allow livestock populations in poor rural communities to remain intact. If those poor pastoralists continue to suffer from livestock die-offs due to disruptive climate events, then rather than assist them to re-establish livestock, only to be imperilled in the next disruptive climate event, assistance should be provided to improve their livelihoods. Alternative livelihoods are now much more widely available than previously as a result of the dramatic growth in use of computers, mobile communications, mobile banking, microfinance and off-grid electricity (Sullivan, 2007).

In fact, the best development assistance for poor pastoralists would be to get them involved in reforestation. This would be more economical than building dykes and levees, as preventing adverse environmental impacts is almost always less expensive than adapting to adverse environmental impacts once they've occurred. More importantly, it has a chance to stop climate change – whereas large-scale dykes and levees would be practically a concession that there's no longer a chance to stop climate change. Indeed, if climate change would become unstoppable, then eventually sea levels could overwhelm any practical scale of dykes and levees.

Poor pastoralists who live in arid grasslands (as in Mongolia) would have to want to move to reforestable land, but such cases would be exceptional. Most poor pastoralists live in places that used to be forested and can be reforested, as in the Amazon rainforest. Reforestation may be easiest of all in parts of Sub-Saharan Africa, where forest must be constantly burned to maintain pasture.

Reforestation can be an enormous source of sustainable employment, with jobs in seed collection, planting, nurseries, fire prevention and conservation. There are already non-governmental organizations in place that specialize in reforestation, although their efforts haven't been nearly well-enough known or well-funded.

Indeed, in poor and well-off countries alike, there is a long history of passionate champions advancing the cause of reforestation. Yet it's been 200 years since the work of history's most notable champion of reforestation, Johnny Appleseed. The world could hugely benefit from the rise of a modern-day Johnny Appleseed championing the restoration of the carbon cycle to its former steady state. So if you know any young, ambitious person who wants to change the world, perhaps you could suggest to such a Jane Doe or John Doe that they change their name to Jane Appleseed or John Appleseed and become the change they want to see, to paraphrase a famous saying.

Technical GHG Assessment

So far, this chapter has mixed bits of technical assessment into more general analysis; but now it will focus directly on some technical GHG assessment of livestock products versus better alternatives. Doing so requires consideration of the GHG Protocol, the most widely-used methodology for GHG accounting (www.ghgprotocol.org). The GHG Protocol recommends counting GHGs as either Scope 1, Scope 2, or Scope 3 emissions. Scope 1 emissions are direct emissions from owned or controlled sources, while Scope 2 emissions are indirect emissions from purchased inputs, and Scope 3 emissions are indirect emissions other than those from purchased inputs, such as emissions attributable to disposal of food waste by consumers.

In food and agriculture, a first step is to pick which entities should be considered the key owners or controllers. It's commonly accepted that farmers are the key ones – rather than anyone else along the supply chain, such as processors, grocers, or consumers. Then, the key Scope 1 emissions are understood to be emissions from within a farm's boundaries – that is, emissions from plants and animals themselves, including emissions from their waste products. If a farmer owns livestock but not the plants that provide feed for those livestock, then livestock feed – and all the emissions attributable to that feed, including emissions attributable to deforestation – will be counted under Scope 2 or 3; otherwise, livestock feed is counted under Scope 1.

In the realm of Scope 1 emissions, a key question is whether or not to count carbon dioxide in animal respiration or its reflection in carbon absorption forgone in land used for livestock and feed production. The FAO says no, while analysis by Jeff Anhang and me says yes. When it comes to indirect emissions, they should be counted only when something can realistically be done to reduce them, according to guidance in a report by the World Resources Institute (Putt del Pino, Levinson & Larsen, 2006).

If we consider that there's no way to feed humans with a lower GHG footprint than by growing plants to feed humans directly, and we consider that farmers can't control emissions for transport, refrigeration, and packaging of foods once they leave the farmers' hands, then we can consider that the only emissions for transport, refrigeration, and packaging of foods that merit counting are any emissions over and above those required for plant-based foods. That's because farmers can control these emissions by choosing to produce higher-emission animal products or lower-emission alternatives.

Now, the GHG footprint of growing plants can be understood to depend principally on two factors:

1. The farmer's purchase of energy and fertilizer; and
2. Whether the farmer's planting methods either:
   a. Degrade soil and thereby cause soil carbon, or
   b. Improve soil and thereby yield net carbon absorption in the soil and the plants themselves.

Chemically-fertilized large-scale production of plants will generally degrade soil. In contrast, small-scale organic production will generally improve soil, so it can even yield more or less a zero-GHG operation for an organic farm producing alternatives to livestock products. The GHG footprint of a non-organic farm producing alternatives to livestock products is likely to be measurable – but it is likely to be small, since after all, it includes no carbon dioxide from livestock respiration, and no methane attributable to livestock, while the farm's plants are likely responsible for net absorption of GHGs from the atmosphere.

Ultimately, not only will every farm have a GHG footprint that differs from every other farm, but each different planting of the same plant on each farm is likely to have a unique GHG footprint. However, given all of what's explained above, it's possible to say that generally, the GHG footprint of alternatives to livestock products that are grown on appropriate lands (not recently deforested) is small, and barely merits counting. As an analogy, when renewable energy is promoted over fossil fuels, it's commonly accepted that the key GHGs to count are those relating to fossil fuels – as they're understood to drive catastrophic levels of atmospheric carbon – while renewable energy infrastructure is simply understood not to be such a driver.

Cultured Meat

There's a theoretical way to produce alternatives to traditional livestock products that's been in the news lately, variously called cultured meat or *in-vitro* meat or lab-grown meat (Agapakis, 2012). The base material consists of stem cells from livestock, and the growth medium is made with serum taken from the blood of calf foetuses. Still, some vegan activists had high hopes for this type of meat, especially if the foetal cow serum could be replaced with a vegan medium. However, that hadn't happened in time for a public test to see if the sample would taste like meat as required. In fact, the sample reportedly failed its taste test, and its proponents subsequently projected that commercializing a viable product would take at least 10-20 years, with no promise that the growth medium might ever become vegan.

Numerous analysts have questioned whether such meat can ever be commercialized (e.g. Timmer, 2011; McLeod, 2011). So this recent proof-of-concept test may turn out to be the high-water mark for cultured meat. While one article has asserted that cultured meat might eventually yield a 78 to 96 percent GHG reduction (Tuomisto & de Mattos, 2011), that seems optimistic. Cell culture in modern laboratories is incredibly energy-intensive, and therefore GHG-intensive, as it requires constant control of temperature and humidity, unlike outdoor livestock production. Also, it's said that cultured meat can gain the required consistency only with stretching machinery, essentially elaborate meat gyms (Agapakis, 2012). Again, that's energy-intensive, and therefore GHG-intensive.

Nevertheless, cultured meat has attracted a lot of media hype. Most media however have overlooked the fact that there are plant-based meats that long ago passed the same taste test that the cultured meat sample failed. The proponents of cultured meat by their own account are proposing no reduction in GHGs attributable to livestock for the next 10-20 years. Yet as I've suggested, climate change is likely to cause increasingly more large-scale die-offs of livestock and crops that feed livestock, and possibly a forced 60% reduction in crop production. That would force a large-scale replacement of livestock production with more efficient food production, without the cost and uncertainty of commercializing lab-grown meat. People can also voluntarily change their diet by buying great-tasting vegan products that have long been commercialized, and are available in abundance on grocers' shelves today.

Consumers commonly see processed vegan products as an important aid to transition from livestock products to better alternatives. Yet many nutritional experts would prescribe whole foods over those processed products, and it seems clear that whole foods are the most perfect foods. However, given the importance of large-scale replacement of livestock products by 2017 in order to avert catastrophic climate change, the best chance for such large-scale replacement to happen is likely to be fulfilled if the required dietary change is as imperceptible as possible. Once that critical transition is made, then health and environmental improvements could and should be pursued over time. For now, we should make sure that the perfect isn't the enemy of the good.

Back to Production

Now, let's move back from consumption to production. If privately-owned farms replace livestock and feed production with the production of better alternatives, they may need no subsidy or other incentives to do so, as they likely will profit more by switching production than by keeping on producing livestock or feed. On the other hand, they may need some sort of fiscal incentive to reforest any amount of their land.

Public lands, on the other hand, are often degraded by ranchers who do not pay the full cost of the adverse impacts caused by their ranching. So if governments would act to reduce or eliminate livestock grazing on any amount of public lands, then they will lose the relatively small amount of ranching fees paid on that land. That loss however would be more than offset by the elimination of adverse impacts on that land, and the huge value of carbon capture in forest and other vegetation that could be regenerated even passively, with little or no investment.

For example, eight researchers who studied the productivity of rangelands in the western United States concluded that climate change is causing additional stress to many western rangelands (Science Codex, 2012). As a result land managers should consider a significant reduction, or in some places elimination of livestock and other large animals from public lands (Science Codex, 2012).

While tropical forest is commonly thought of as being the world's most important carbon sink, boreal forest may absorb just as much carbon dioxide per hectare as tropical forest, according to Hance (2009), and it's perfectly possible for forest to regenerate on most sites on which boreal forest formerly grew. Similarly, burning of African forest and woodland-savannah to expand and maintain livestock grazing is even more intense than in the Amazon, and it must be stopped (Russell, 2010). Even if forest isn't regenerated, some plants, such as *Brachiaria,* may capture more carbon than can tropical forest, according to Tarré et al. (2001).

FUTURE POLICY DIRECTIONS: RAISING AWARENESS

Now let's consider some different ways to raise awareness about the need to replace a substantial amount of livestock products with better alternatives. Note that I don't simply say we should consider aspects of replacing "meat" and "milk". That's because vegan food producers often market their products as "meat" and "milk" (e.g., www.fieldroast.com and silk.com). While that may offend some activists, it seems more important that it offends livestock producers, some of whom have filed lawsuits to prevent vegan food producers from using the term "milk" (Robbins, 2010).

In fact, dictionaries define "meat" and "milk" as essential foods that include vegan versions (www. merriam-webster.com/dictionary/meat and www.merriam-webster.com/dictionary/milk). The various names for better alternatives to livestock products include soy milk, nut butters, vegan cheeses, and grain-based meats, and they can fully serve as essential foods. So it may not be the soundest of strategies to cede the terms "milk" and "meat" to livestock producers, and to press people to sacrifice those items.

Perhaps the most popular campaign that presses people to sacrifice meat is the Meatless Mondays campaign. If Meatless Mondays were to be fully adopted by everyone in the world, it could yield a 13 percent replacement of livestock products with better alternatives, which would be significant.

On the other hand, there are a number of ways in which Meatless Mondays could be improved upon. First, the Meatless Mondays campaign should recognize that by touting "meatless" eating in its name, it overlooks the fact that, as mentioned, dictionaries define meat as an essential food that includes vegan versions. Second, the Meatless Mondays campaign needs somehow to draw attention to the impacts of dairy and egg products – as at present, Meatless Mondays can lead people to switch from something like a salad with chicken to a cheese omelette or quiche, which could actually worsen both public health and climate change.

Moreover, Meatless Mondays' prescription for 13 percent of livestock products to be replaced is far less than the amount of replacement that's actually needed to reverse climate change, which is somewhere between 25 to 67 percent, depending on whose analysis is used. While the Meatless Mondays campaign publicizes a prescription from medical authorities which say that a 15% reduction in saturated fat is needed for good medical health. The campaign says this can be fully achieved by replacing livestock products one day per week, but it should similarly publicize how much livestock products need to be replaced for good climatic health.

Another issue for the Meatless Mondays campaign to consider is that if what's needed is framed as a meatless day, then people may perceive what's needed as a sacrifice – so the needed action may suffer the same fate as everything that promotes sacrifice; that is, people who will sacrifice meat one day may just crave it more the next day. Indeed, the Meatless Monday campaign anachronistically touts its basis in World War I rationing, after which meat consumption exploded (www.meatlessmonday.com/history).

Also, while the Meatless Mondays campaign presses people to consume something other than a livestock product one day per week, it seems that no consumer product is successfully sold by pressing people to use it just one day a week. For example, consumers might be wary of choosing Pepsi Cola at all if its marketing promoted it as a one-day-a-week drink, conceding that Coca Cola remains the drink of choice the rest of the week.

In fact, while Meatless Mondays has expanded around the world since 2003, the worldwide consumption of livestock products hasn't fallen, but rather has risen. This seems especially striking when considering the wave of economic downturns in one region after another since 2007 – as the consumption of livestock products meat-eating has historically dropped during economic downturns (The Afro American, 1974).

The actual trend of global meat consumption is surely more important than the trend of organizations to adopt Meatless Mondays. Perhaps the organizers of Meatless Mondays could achieve more success by framing alternatives to livestock products as better products, rather than a sacrifice. If alternatives to livestock products might seem hard to promote, well, then consider for example that bicycles may be hard to promote to car-lovers, but bike advocates normally don't frame bike-riding as a sacrifice.

In any event, the Meatless Monday campaign promotes sacrificing meat to reduce health risks, energy and water usage, and carbon emissions. Yet goals for such reductions are common in the production and consumption of many consumer products, and they usually don't motivate action to replace any products. In contrast, emergencies normally motivate major action – and both the UN Intergovernmental Panel on Climate Change and the UN International Energy Agency have warned that major action by 2017 may be the last real chance to reverse climate change before it's too late (IEA, 2011a; Spotts, 2012). That seems a more compelling motivation on which to dwell.

An alternative to the Meatless Mondays campaign is a campaign that has a similar but better name, namely Meat Free Mondays, operated by three members of a famous family, Mary, Stella, and Paul McCartney. By replacing "Meatless" with "Meat Free", they've moved from using a term that connotes taking something away – to a term that raises the idea of being free, or obtaining something for free.

The McCartney family is well-known for marketing a line of vegetarian foods named after Sir Paul's late wife Linda, and they recently published a vegetarian cookbook (Riggs, 2012). Stella McCartney runs a fashion company that is the largest of its type using no leather or fur. Paul McCartney has a well-known track record of promoting vegetarianism, veganism, and animal rights. The Meat Free Monday campaign organizes tabling at all of Paul McCartney's concerts around the world, whereby tables are set up and staffed by local volunteers who distribute literature and engage with concert-goers to promote the cause of meat-free eating.

CONCLUSION

Goals to produce low carbon products have long been promoted, yet they've achieved little success. In contrast, emergencies normally motivate major action – and as mentioned, both the Intergovernmental Panel on Climate Change and the International Energy Agency have warned that major action by 2017 may be the last real chance to reverse climate change before it's too late (IEA, 2011a; Spotts, 2012). Indeed, there's surely no more compelling motivation for action than that replacing livestock products with better alternatives may be the only pragmatic way to stop catastrophic climate change from imperiling much of life on earth.

REFERENCES

Agapakis, C. (2012). Steak of the art: the fatal flaws of in vitro meat. Retrieved from http://blogs.discovermagazine.com/crux/2012/04/24/steak-of-the-art-the-fatal-flaws-of-in-vitro-meat/UgAx51JQ0qt

Brown, L. (2011). *World on the edge: how to prevent environmental and economic collapse*. New York, NY: Norton.

Calverd, A. (2005, July). A radical approach to Kyoto. *Physics World*. Retrieved from http://physicsworldarchive.iop.org/pdf?site=pwa&bkdir=18/7&pdf=phwv18i7a46&pdfhash=77DB6720F54408CAD26A0A7B4DBB5D05&doctime=Fri%2C24 Apr 2015 03%3A35%3A32 GMT

De Hann, C., Schillhorn van Veen, T., Brandenburg, B., Gauthier, J., Le Gall, F., Mearns, R., & Simeon, M. (2001). *Livestock development: Implications for rural poverty, the environment and global food security*. Washington, DC: The World Bank. doi:10.1596/0-8213-4988-0

Delgado, C., Rosegrant, M., Steinfield, H., Ehui, S., & Courbois, C. (1999). Livestock to 2020: the next food revolution. Food, Agriculture and the Environment Discussion Paper 28. Washington, DC: International Food Policy Research Institute (IFPRI), Food Agriculture Organization (FAO) and International Livestock Research Institute (ILRI).

Eshel, G. (2010). Grass-fed beef packs a punch to the environment. Retrieved from http://blogs.reuters.com/environment/2010/04/08/grass-fed-beef-packs-a-punch-to-environment/

Food Agriculture Organization (FAO). (2010). *Global forest resource assessment*. Rome, Italy: FAO.

Goldenberg, S. (2013, July 29). Climate study predicts a watery future for New York, Boston and Miami. *The Guardian*. Retrieved from http://www.theguardian.com/environment/2013/jul/29/climate-new-york-boston-miami-sea-level

Goodland, R. (2012). FAO Yields to Meat Industry Pressure on Climate Change. Retrieved from bittman.blogs.nytimes.com/2012/07/11/fao-yields-to-meat-industry-pressure-on-climate-change/?_r=0

Goodland, R. (2013a). Lifting livestock's long shadow. *Nature Climate Change*, *3*(1), 2. doi:10.1038/nclimate1755

Goodland, R. (2013b). Meat, lies and videotape (A deeply flawed TED talk). Retrieved from http://www.freefromharm.org/agriculture-environment/meat-lies-videotape-a-deeply-flawed-ted-talk

Goodland, R., & Anhang, J. (2009, November/December). Livestock and climate change: What if the key actors in climate change are... cows, pigs and chickens? *World Watch Magazine*, *22*(6), 10–19.

Herrero, et al. (2011). Comment to the editor: Livestock and greenhouse gas emissions: the importance of getting the numbers right. In R. Goodland, & J. Anhang (Eds.), Animal Feed Science and Technology, 172(3-4), 166-167: 779-782. doi:10.1016/j.anifeedsci.2011.12.028

Gustavsson, J., Cederberg, C., Sonesson, U., Van Otterdijk, R., & Meybeck, A. (2011). Global food losses and food waste. Rome, Italy: Food Agriculture Organisation (FAO).

Hance, J. (2009). New report: boreal forests contain more carbon than tropical forests per hectare. *Mongabay*. Retrieved from http://news.mongabay.com/2009/1112-hance_boreal.html

Henderson, G. (2013). Commentary: Savory's solution will change views of livestock. *Drovers Cattlenetwork*. Retrieved from http://www.cattlenetwork.com/search/site/Savory%2527s%2520herding%2520instinct

Herrero, M., Gerber, P., Vellinga, T., Garnett, T., Leip, A., & Opio, C. et al. (2011). Livestock and greenhouse gas emissions: The importance of getting the numbers right. *Animal Feed Science and Technology*, *166–167*, 779–782. doi:10.1016/j.anifeedsci.2011.04.083

International Energy Agency (IEA). (2011a). IEA: time running out to limit earth's warming. *Inquirer.net*. Retrieved from http://newsinfo.inquirer.net/91179/iea-time-running-out-to-limit-earths-warming

International Energy Agency (IEA). (2011b). *World energy outlook 2011*. Paris, France: IEA.

Jacobsen, M. (2013). Saying good riddance to the clean-plate club. *The New York Times*. Retrieved from http://parenting.blogs.nytimes.com//2013/08/02/saying-goodbye-to-the-clean-plate-club/

Kruska, R. L., Reid, R. S., Thornton, P. K., Henninger, N., & Kristjanson, P. M. (2003). Mapping livestock orientated agricultural production systems for the developing world. *Agriculture Systems*, *77*(1), 39–63. doi:10.1016/S0308-521X(02)00085-9

McLeod, H. (2011). South Carolina scientist works to grow meat in the lab. *Reuters*. Retrieved from http://www.reuters.com/article/2011/01/30/us-food-meat-laboratory-feature-idUSTRE70T1WZ20110130

Msangi, S., & Rosegrant, R. W. (2011). Feeding the future's changing diets: implications for agriculture markets, nutrition, and policy. In *Proceedings of 2020 Conference: Leveraging Agriculture for Improving Nutrition and Health*. New Delhi, India. Retrieved from http://cdm15738.contentdm.oclc.org/utils/getfile/collection/p15738coll2/id/124834/filename/124835.pdf

Petherick, A. (2012). Light is cast on a long shadow. *Nature Climate Change, 2*(10), 705–706. doi:10.1038/nclimate1703

Prairie, Y. T., & Duarte, C. M. (2007). Direct and indirect CO_2 release by humanity. *Biogeosciences, 4*(2), 215–217. doi:10.5194/bg-4-215-2007

PricewaterhouseCoopers (PwC). (2012). Too late for two degrees? London, UK: PwC. Retrieved from https://www.thepmr.org/system/files/documents/Low%20Carbon%20Economy%20Index%202012.pdf

Putt del Pino, S., Levinson, R., & Larsen, J. (2006). Hot climate, cool commerce: a service sector guide to greenhouse gas management. Washington, DC: World Resource Institute (WRI). Retrieved from http://www.ghgprotocol.org/files/ghgp/tools/HotClimateCoolCommerce_lowrez.pdf

Radford, T. (2013). Carbon dioxide levels now 61% higher than 1990. *Responding to Climate Change*. Retrieved from http://www.rtcc.org/2013/12/31/carbon-dioxide-levels-now-61-higher-than-1990/

Riggs, D. (2012). *Meat free Monday cookbook*. London, UK: Kyle Books.

Robbins, J. (2010). What about Soy? *John Robbins: Tools, Resources, and Inspiration*. Retrieved from http://johnrobbins.info/blog/what-about-soy/

Russell, G. (2010). Burning the biosphere, boverty blues (part 1). *Brave New Climate*. Retrieved from http://bravenewclimate.com/2010/01/05/boverty-blues-p1/

Science Codex. (2012). Climate change increases stress, need for restoration on grazed public lands. Retrieved from http://www.sciencecodex.com/climate_change_increases_stress_need_for_restoration_on_grazed_public_lands-102039

Soil Association. (2010). Telling porkies: the big fat lie about doubling food production. Bristol, UK: Soil Association. Retrieved from https://www.soilassociation.org/LinkClick.aspx?fileticket=qbavgJQPY%2Fc%3D&tabid=390

Spotts, P. (2012, March 28). Climate change report: time to start preparing for the worst. *Christian Science Monitor*. Retrieved from http://www.csmonitor.com/Environment/2012/0328/Climate-change-report-time-to-start-preparing-for-the-worst

Steinfeld, H., & Gerber, P. (2010). Livestock production and the global environment: Consume less or produce better? *Proceedings of the National Academy of Sciences of the United States of America, 107*(43), 18237–18238. doi:10.1073/pnas.1012541107 PMID:20935253

Steinfeld, H., Gerber, P., Wassenaar, T., Castel, V., Rosales, M., & de Hann, C. (2006). Livestock's long shadow: environmental issues and options. Rome, Italy: Food Agriculture Organisation (FAO). Retrieved from ftp://ftp.fao.org/docrep/fao/010/a0701e/a0701e00.pdf

Sullivan, N. (2007). *You can hear me now: how microloans and cell phones are connecting the world's poor to the global economy.* San Francisco, CA: Jossey-Bass.

Tarré, R., Macedo, R., Cantarutti, R. B., de Rezende, C. P., Pereira, J. M., & Ferreira, E. et al. (2001). The effect of the presence of a forage legume on nitrogen and carbon levels in soils under Brachiaria pastures in the Atlantic forest region of the South of Bahia, Brazil. *Plant and Soil, 234*(1), 15–26. doi:10.1023/A:1010533721740

The Afro American. (1974, August 10). Tighten your belt: food prices to soar with world demand. *The Afro American.* Retrieved from https://news.google.com/newspapers?nid=UBnQDr5gPskC&dat=197 40810&b_mode=2&hl=en

Thornton, P., Herrero, M., & Ericksen, P. (2011). Livestock and climate change. Nairobi, Kenya: International Livestock Research Institute. Retrieved from https://cgspace.cgiar.org/bitstream/handle/10568/10601/IssueBrief3.pdf

Timmer, J. (2011). Lab-grown meats face long road to supermarket. *Wired.* Retrieved from http://www.wired.com/2011/11/artificial-meat-economics/

Tuomisto, H. L., & de Mattos, N. J. (2011). Life cycle assessment of cultured meat production. *Environmental Science & Technology, 45*(14), 6117–6123. doi:10.1021/es200130u PMID:21682287

KEY TERMS AND DEFINITIONS

Better Alternatives: Foods containing plant-based nutrients and no animal-based ingredients.

Greenhouse Gas (GHG): A gas released in the Earth's atmosphere which contributes to climate change through the greenhouse effect.

Intergovernmental Panel on Climate Change: United Nations panel with a mission to stabilize greenhouse gas concentrations in the atmosphere at a safe level.

International Energy Agency: Autonomous international organization focusing on energy security, economic development and environmental protection.

World Bank Group: A group of five affiliated organizations, consisting of the International Bank for Reconstruction and Development, the International Finance Corporation, the International Development Association, the International Centre for Settlement of Investment Disputes and the Multilateral Investment Guarantee Agency.

ENDNOTES

[1] Robert Goodland (1939-2013) served as lead environmental adviser at the World Bank Group for 23 years, and was the first ever winner of the World Conservation Union's Harold Coolidge medal for lifetime achievement in the conservation of nature and natural resources. This chapter was adapted from the circulated version of Dr Goodland's last presentation on livestock and climate change, available on the publications page at www.chompingclimatechange.org.

2 Before Thornton co-authored the 45% estimate, he co-wrote with other authors an article that contained a 55% estimate (Kruska, Reid, Thornton, Henninger & Kristjanson, 2003). According to that article's Table 2, livestock systems worldwide occupy 80.8 million km^2 of land, compared with the Earth's surface area of 149 million km^2, which means that livestock systems occupy 55% of all land on earth. Thornton is a livestock specialist employed by the International Livestock Research Institute, which generally promotes expansion of livestock and feed production.

3 The worldwide terrestrial livestock population in 2010 included 1 billion cattle, 1.8 billion sheep and goats, 1.5 billion pigs, and 68.8 billion poultry. These are not standing inventories (animals alive on a particular day) but the total number of animals alive in a year, which is particularly important for poultry where most commercial broiler chickens are slaughtered after 42 days. This data is for terrestrial livestock, so it excludes farmed fish and other farmed seafood. Half the fish consumed worldwide is projected to be farm-raised by the year 2025. As the definition of livestock is limited to animals raised by humans, data on livestock will always exclude non-farmed seafood, although there are major environmental impacts created in catching and processing such seafood, as indeed there are in farmed fish [After: ifahsec.org/wp-content/files_mf/ifahfactsheet0112fin.pdf].

Chapter 3
Improving the Understanding of Climate Change Factors with Images

Geoff Russell
Curtin University, Australia

ABSTRACT

In Australia, the public got its first mass marketing about climate change and the measures that would be required to avoid it, by TV images of black balloons and Professor Tim Flannery turning off light switches. Journalistic coverage has been similarly dominated by household electricity. More technical literature is generally dominated by the concept of "carbon dioxide equivalence" (CO2eq) as spelled out in the Kyoto protocol. This concept isn't used in climate models because it makes no physical sense. The use of CO2eq and the focus on household electricity has lead to a profound mismatch between the causal factors as understood by climate scientists and causal factors as perceived by the public. "The public" here isn't just the general public, but people of many backgrounds with a strong interest in climate change but without the deep knowledge of professional climate scientists. We need images consistent with climate models, which accurately rank the causes of climate change and guide proposed actions. Such images point to meat as a key focal issue.

INTRODUCTION

Nobody needs a poll to accurately rank the annual numbers of car driver, cyclist and skateboarder fatalities on our roads. Similarly, everybody understands that black pepper, vinegar and goji berries are insignificant parts of the global food supply; so yield increases aren't going to reduce global hunger. We all have passably accurate mental images and metaphors of such matters, which work.

However what about the global output of ocean fisheries? How does that compare to bananas? Or sausages? According to the Food and Agriculture Organisation of the United Nations (FAO, 2014a), the entire output of ocean fisheries supplies little more energy per person per day than a sugar cube (about half a teaspoon; 9 Calories). The entire meat output of the planet's 1.5 billion cattle is somewhat larger

DOI: 10.4018/978-1-4666-9553-5.ch003

but still quite small at just 1.4 percent of global food energy; about 4.5 sugar cubes per person per day (40 Calories). And sugar itself? About 22 cubes per person per day (197 Calories). Wheat and rice each come in at about 18 percent of global food energy; 57 sugar cubes (526 and 544 Calories respectively).

These kinds of facts surprise people because our knowledge of the relative importance of different foods is driven by our personal history coupled with our immersion in an ocean of advertising.

Similarly, when it comes to the factors contributing to climate change, few people understand even the rankings, let alone the relative size of causal factors. In Australia, the public got its first mass marketing about climate change and the measures that would be required to avoid it, by TV images of black balloons and Professor Tim Flannery turning off light switches. Journalistic coverage has been similarly dominated by household electricity. The degree to which food choices affect climate is largely unknown. Our mental maps and metaphors rarely bare any resemblance to the facts.

The use of CO2eq and the focus on household electricity has led to a profound mismatch between the causal factors as understood by climate scientists and causal factors as perceived by the public. "The public" here isn't just the general public, but people of many backgrounds with a strong interest in climate change but without the deep knowledge of professional climate scientists. This chapter begins with a survey of the extent of ignorance of the environmental impacts of food choices and then tackles what we think is a major impediment to clarity. Next we present some figures, which help people rank and quantify the environmental impacts of various activities.

First let's establish the necessity of global dietary reform in any plan to prevent further climate destabilisation. Figure 1 is adapted from (Pelletier & Tyedmers, 2010). The top quantity, 8.9 giga-tonnes, is measured as "carbon dioxide equivalent" (CO2eq) following the source image. It represents the Copenhagen Diagnosis Review (COP) (Allison et al., 2009) estimate of what global per person greenhouse gas emissions should be by 2050 on the path to climate stabilisation, given a population of 8.9 billion; it represents one tonne per person per year. Note that this emission budget covers all sources: food, electricity, transport, etc.

Figure 1. Diets and our greenhouse gas budget

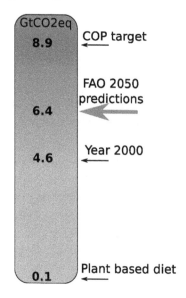

The word "equivalent" in "carbon dioxide equivalent" implies that it is the outcome of converting various greenhouse gases to commensurate units. If you want to know the climate impact of the combined emission of X tonnes of CO_2 and Y tonnes of methane (CH_4), most people who follow such matters assume that this is given by multiplying the CH_4 emissions by the Kyoto Protocol factor of 21 and then adding the CO_2 emissions. Thus the emission of 1 tonne of CO_2 and 1 tonne of CH_4 is said to be 22 tonnes of CO2eq. As will be explained shortly, this isn't true and is misleading, despite being used in all official national submissions to the international body responsible for such matters, the United National Framework Convention on Climate Change (UNFCCC). Let's proceed, on the assumption for now, that "carbon dioxide equivalent" means what it says.

The red region at the top of Figure 1, beyond 8.9 giga-tonnes of CO2eq, represents an unsustainable level. As of the year 2000, food production emissions were already 4.6 giga-tonnes of CO2eq and under United Nations FAO projections, given 2010 dietary trends, the emissions in 2050 from food production alone will constitute 6.4 giga-tonnes of the 8.9 budget; leaving just 2.5 giga-tonnes for electricity, transport and so on (for data references, see Pelletier & Tyedmers, 2010).

Note that none of this 6.4 giga-tonnes covers cooking, refrigeration or processing. It covers basic things such as methane from ruminants, paddy fields and nitrous oxide from fertilised soils.

With a plant based diet, the authors (Pelletier & Tyedmers, 2010) estimated that we can feed 8.9 billion people for just 0.1 giga-tonnes of CO2eq. This allows 8.8 giga-tonnes for all those other things such as electricity and the rest. However, on current trends, all we'll have left in the budget after food production is 2.5 giga-tonnes; this works out at 280 kg of CO2eq for each of the world's 8.9 billion people.

Building an electricity infrastructure to service people for just 1 tonne of CO2eq per year will be tough, and if food emissions cut the budget by more than two thirds, the problem gets that much harder. For example, Australia uses about 10,500 kilowatt hours (kWh) per person per year and we generate about 850 grams of CO_2 per kWh (IEA, 2013). This amounts to almost 9 tonnes of CO_2 per person and if we only have a budget of 280 kg, then we will need to be getting our electricity for about 26 grams per kilowatt hour. Estimates of the CO_2 produced per kilowatt hour for solar cells is between 40 and 200 g/kWh (NREL, 2012). So producing electricity for just 26 g/kWh will be tough. Moreover, we'll need far more electricity to substitute for as much as possible of the other three quarters of our fossil fuel (non-electrical) energy use. This implies a CO_2 g/kWh figure even lower than 25. There are some nuclear technologies that can meet this challenge, but it is still a huge task.

Clearly, dietary reform is an essential part of our climate change response. It is no less important than dealing with our energy infrastructure but so far has received very little attention.

Let's now turn to the complete disconnect between the story in this image and various views of the relative importance of food choices. It's not just the general public who is in the dark on the issue, but most of the people driving Australia's response to climate change.

CURRENT UNDERSTANDING OF CLIMATE CHANGE FACTORS

Little Awareness at Many Levels

One reason for the disconnect was found some years back in a US survey of news stories. This 2008 study found that only 0.5 percent of climate change articles mentioned farm animal greenhouse emissions, despite these being higher than emissions from transport (Neff, Chan & Smith, 2008). A 2008

Australian study (Lea & Worsley, 2008) is consistent with this finding in that people wrongly thought decreased packaging was the most effective way of reducing the environmental impact of food production. Later work implies that nothing much has changed. For example, a 2011 study (Tobler, Visschers & Siegrist, 2011) in Switzerland also found that people thought avoiding excess packaging was the most effective way of reducing the environmental impacts of their food choices while reducing meat was wrongly considered the least effective.

Often even experts whose job it is to understand the climate impact ranking of various activities don't appear to have any idea. For example, in Australia the Commonwealth Scientific and Industrial Research Organisation (CSIRO) has been surveying public attitudes to climate change annually since 2010 and the questions they ask, or don't ask, give clues about their understanding of the rank and strength of climate change causal factors. These questions influence survey participants and the media who cover the survey. In turn, the media reports influence the public who reads them.

The CSIRO polls (Leviston, Price, Malkin & McRea, 2014) explore people's personal involvement in climate change by asking them to list how many of 21 "pro-environment climate change relevant" behaviours they had engaged in and why. In the poll results report, the questions are sorted according to the number of people taking each action. They are presented below split into three groups. The first group describe well-defined actions. With a few exceptions, each can reasonably be expected to reduce a person's greenhouse footprint. Nevertheless nobody taking the survey would learn anything about the relative efficiency of these actions.

Clearly Defined Actions

- I have contacted a government member about climate change;
- I have installed a grey water recycling system on my property;
- I have taken part in a political campaign about an environmental issue;
- I buy carbon-offsets to reduce my carbon footprint;
- I have reduced my amount of air travel;
- I usually walk/cycle/carpool/take public transport;
- I have installed a solar hot water system, or solar panels, in my house;
- I am on Green Power electricity;
- I have installed insulation in my home;
- I have reduced the amount of petrol I use;
- I have reduced the amount of gas and/or electricity I use around the house;
- I switch lights off around the house whenever possible;
- I have reduced the amount of water I use around the house and garden;
- Most of my cleaning products are environmentally friendly; and
- I have switched to products that are more environmentally friendly.

The next four questions are somewhat different. We refer to them as "possibly counterproductive" because a person's greenhouse footprint might rise because of these actions instead of fall.

Possibly Counterproductive

- Where possible, I buy products that are made locally;

- I have installed a rain water tank on my property;
- I will usually try to fix things rather than replace them; and
- I recycle/compost as much household waste as possible.

None of these activities is certain to reduce emissions, and will sometimes increase them. For example, local goods might have a much higher greenhouse gas footprint than imported goods. Typically, Australian manufactured goods are made in coal-powered factories, but imported goods may be made using gas, hydro or nuclear based electricity. Similarly, is the energy and emissions embodied in a rainwater tank less than the energy required to distribute water to a property? The answer might be yes, but that will depend on quite a few factors. Similarly when does fixing an old car reduce emissions compared to buying a new more efficient car or going car free and using public transport? That's not simple to calculate either. Lastly, a home compost system may be far less efficient than a centralised system where the heap is professionally managed and perhaps the methane is captured and used.

Strange

- I have changed my gardening practices; and
- I have changed my diet

They assume people know how to change their diet or gardening habits to reduce their greenhouse emissions. How many people measure the methane emissions from their compost heap, if they have one? How many know the ranking of emissions from factory farmed chicken, grass fed beef and air freighted Japanese tofu? Whereas some previous questions were clear about what behaviours might be useful, these questions provide no information at all.

The poll didn't only ask people about the behaviours, but asked them why they engaged in them. Over 70 percent of people had reduced their petrol use in the past 12 months but 45 percent did it for non-environmental reasons. The reasons are certainly important in a poll aiming to measure attitudes, but does it matter why people reduce their climate footprint?

Of the whole 21 items, one related to diet and it was quite non-specific. Given polls like this, it is no surprise that people don't think changing their diet is effective.

Indeed, dietary change was among the least likely pro-environment actions taken by poll responders. Recycling/composting was the most popular, with 52 percent of people doing it and doing it for environmental reasons. While 37 percent of people had changed their diet, only 9 percent did it for environmental reasons and we have absolutely no idea if *any* of the changed diets resulted in a lower or higher greenhouse footprint.

The question was simply too vague. Perhaps some people switched from grain fed beef to grass fed beef under the mistaken belief that this would reduce their carbon footprint, when it would in fact do the opposite (Harper, Denmead, Freney & Byers, 1999). Perhaps some switched from imported pasta to local pork. Again, this would have an impact opposite to that intended, but such is the level of misinformation that such a mistaken belief is entirely understandable.

Perhaps the poll designers themselves weren't aware of the relative importance of their "pro-environment" activities. How else can we account for a question asking people whether they have reduced their air travel but no question asking whether they have reduced their beef consumption? There is a clear conflict of interest with CSIRO also doing work for the beef industry (Russell, 2009), but is it plausible to think that poll designers simply don't know about climate impact relativities?

One thing is certain; nobody doing the poll would be able to guess the relative value of the various activities. Lastly, for the past decade, CSIRO has been promoting a high red meat diet; the CSIRO "Total Wellbeing Diet" (Noakes, 2013; Noakes & Clifton, 2005, 2006). Not only is this a high red meat diet, it is a high *lean* red meat diet. The environmental cost of meat is inversely proportional to its fat content (Peters, Wilkins & Fick, 2007). These high profile (over a million books sold in Australia) CSIRO diets present the public with a powerful message that food isn't a climate issue.

Highly Infectious Expert Ignorance

If stories about lung cancer rarely mention of smoking but always refer to alcohol, then it would be natural for people to think the latter was a much bigger problem than the former. Similarly, if methane leaking from coal seam gas (CSG) wells is the only time the word methane gets a mention in relation to climate change, then that's going to focus people's minds and they won't give food a second thought.

Consider the *Australian Greens*. A search of their website (greens.org.au) on the 25th October 2014 finds just 12 mentions of 'methane', one in connection with waste and 11 associated with CSG leakage. While it may be that our current inventories don't accurately capture all CSG leakage, they currently record CSG leaks as being around 8 percent of the level of enteric fermentation from livestock.

A search of the *Friends of the Earth* website (www.foe.org.au) found links to 40 pages mentioning methane with just one devoted to food choices and their impact on climate change, with methane from livestock making its appearance near the bottom after a list beginning with buy local and buy organic. No attempt was made to rank or quantify the impacts of the various suggestions.

The *Australian Conservation Foundation* (ACF) is Australia's largest environmental organisation and has gone backwards over the past decade in terms of informing members about the ranking of factors affecting the climate or the environment more generally. Back in 2009, the ACF eco-footprint calculator was the best in Australia. It was a graphical tool built on the *Integrated Sustainability Analysis* calculator developed by Sydney University (www.isa.org.usyd.edu.au/index.html). It converted expenditure on various items such as meat, electricity, air travel and petrol into carbon dioxide equivalent emissions.

However, in 2014, all that remains of that web-based information is the *Consumption Atlas* (http://www.acfonline.org.au/sites/default/files/resource/index67.swf) and you won't find that without knowing it exists and searching for it.

There is no link to it on either the ACF front page (www.acfonline.org.au) or its "Sustainable Living" page (www.acfonline.org.au/be-informed/sustainable-living). For all their shortcomings (Blomqvist, Brook, Ellis, Kereiva & Nordhaus, 2013), eco-footprint concepts are an attempt to rank and quantify. The ACF has obviously backed away from such undertakings. It's not clear why. Its current "Be Informed" website (www.acfonline.org.au/be-informed) menu contains the following items: "Climate change and energy", "Land and forests", "New economics", "Northern Australia and nuclear free", "Oceans and rivers" and "Sustainable living".

What are the relative contributions of dietary choices, electricity choices, the size of your house? What's the ratio of the greenhouse gas emissions generated by building a car compared to those generated by typical usage over its lifetime? What are the relative climate impacts of producing meat compared to transporting it from farm to plate? How do various meats compare?

Few people know such things and neither the big environment groups nor our political parties seem interested in providing clear information. A search on "climate change" on the Australian Liberal party website (www.liberal.org.au) finds 486 hits, but a search on "methane" found just 9 with 8 relating to landfill with just one to livestock. That was a 2010 press release objecting to taxing farmers for livestock methane emissions (www.liberal.org.au/latest-news/2010/11/2CO2labor-tax-farmers-carbon-0). A search of the Climate Council website (www.climatecouncil.org.au) found no mention of methane.

Only the Climate Institute (www.climateinstitute.org.au) had anything at all detailed about methane and livestock (Watts, 2012). We'll discuss it later in the chapter. A search of the Climate Spectator (www.climatespectator.com.au) site, a prominent website devoted to climate change news, found 1,060 hits on methane but only 54 on "methane cattle" with 243 on "methane CSG".

This kind of misleading information bias helps explain, for example, the response of a researcher with a Sydney based climate research and communications group to an email from me mentioning the role of cattle in methane production. She responded that: "where methane is more of an issue [is] fugitive emissions from coal mines". This is simply not true. In Australia's 2014 UNFCCC National Inventory Report (2014) (NIR), methane emissions from coal mining and handling are just over half the size of methane production from cattle. So clearly it isn't *more* of an issue but *less* of an issue, even if still of considerable importance.

The spread of such mistaken ranking attributions isn't surprising in the light of the above simple searches on environmental and political organisational websites. Hence, the people driving our response to climate change aren't acting rationally because they don't have an accurate mental map of the relative contribution of different factors. If all our major environmental and political groups don't understand the relativities, then what hope is there for well-meaning people who, of necessity, rely on them for information and guidance?

Factors Distorting Perception

Figure 1 makes it explicitly clear that food choices are a major climate change issue. There are however two problems with this figure. First, as we suggested, it uses the false and misleading concept of carbon dioxide equivalence. Second, the figure sheds no light on where meat eating sits in relation to other activities. The article from which the image was taken (Pelletier & Tyedmers, 2010) explains why meat has such a big climate impact, but not where it sits in relation to other activities.

Let's first clarify the CO2eq issue. Is there some amount of sugar you can eat which is equivalent to eating a date? The short answer is no. You can weigh the date and eat the equivalent weight of sugar, but dates have some protein and sugar has none. Or you can measure the energy value of a date and match it with sugar, but what about the fibre? And the tanins? The list of differences is long. Similarly, suppose you reduce emissions of CO2 by some amount and plot the resulting temperature change. The fact is that there is no alternative reduction in CH4 that will reproduce that temperature curve. Have a look at Figure 2. The image follows (Shindell et al., 2012) and technical readers can consult the equations in Lauder et al. (2013) for more detail.

Note the different shape of the methane reduction curve compared to the CO2 reduction curve. A focus on methane (CH4) to the detriment of CO2 reductions would produce higher temperature peaks in the long run, but a CO2 obsession would see higher temperature peaks in the short term and could put us at risk of crossing dangerous climate tipping points.

Figure 2. Multi-gas reduction pathways
(Figure based on (Shindell et al., 2012)).

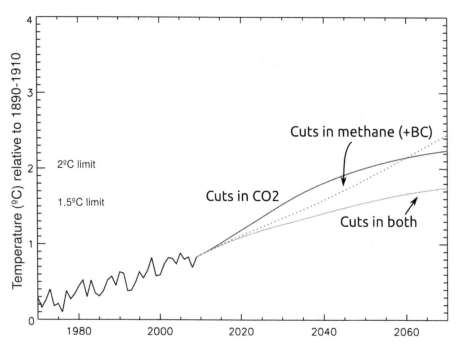

Just as you need to consider multiple nutrients when planning a diet, you need to consider multiple gases when planning the optimal emissions reduction strategy. We've assumed, as seems reasonable, that the principal focus of policy should be to minimise the peak temperature. If the peak is too high, there is a risk that positive feedbacks could take control of future climate out of our hands. For example, if planetary ice areas are reduced, then less heat is reflected back into space. This is a positive feedback. The critical issue is that you can get a lower peak by reducing both methane and CO2 than either on its own.

The factor currently used to convert a tonne of methane into a tonne of CO2eq was decided by the Kyoto Protocol and stands at 21. This is behind the CO2eq figures in newspaper articles about emissions since 1997 when the protocol was agreed. It is implicit when people say that, for example, agriculture is 14 percent of global emissions. It is also behind Figure 1.

However, is this factor of 21 used inside climate models? These model the physical impact of atmospheric concentrations of methane on the ratio of energy arriving at and leaving the planet, and the short answer is no, they don't use a figure of 21. In a climate model (Shindell et al., 2009) the whole ensemble of gases and aerosols in the atmosphere is used to determine the instantaneous effect. The use of a simple factor is certainly an attractive alternative to a complex truth, but its misleading impacts have been and continue to be profound.

Apply a blow torch to your leg for 10 seconds and measure the temperature as your flesh burns. Now pretend you applied the same amount of heat energy, but at a tiny level over 12 months. This would falsely predict no burn. That's precisely what has been done with methane in the Kyoto Protocol and UNFCCC submissions. They pretend that methane has a significant effect for a hundred years and are then calculating an average over that whole period. That's simply not true.

Add a tonne of CO_2 to the atmosphere and it has a small effect for a very long time; add a tonne of methane and it has a huge effect for a short time but almost 90 percent is oxidised (to carbon) within 20 years. Why would you average its impact over 100 years? It's nonsense, which is why physicists don't use anything remotely resembling the Kyoto factor in climate models. The reason is that it is wrong and far more seriously, it is misleading. It's like averaging the impact of consuming a bottle of whisky over a year when you drink it in an hour.

To summarise, the normal definition of CO2eq underestimates the warming impact of methane by a factor of about 5 (5x21=105). This figure 105 is the impact of a tonne of methane relative to a tonne of CO_2 over a twenty year period (Shindell et al., 2009). It's a reasonable approximation to the physical reality. If you redo all the climate change statistical documents with this figure, they'll provide a much better understanding of the relative impacts of different activities.

An increase in either gas is pernicious, but for very different reasons — CO_2 because it hangs around for a long time and methane because of its capacity for short term warming. Short term warming may be particularly nasty if it drives positive feedbacks like the ice-albedo effect mentioned earlier.

It should also be understood that 105 is *not* an *equivalence* factor. It is simply not true that you can trade emissions of the two gases by using this or any other factor. There is no factor that allows this. Our 105 factor is more like a Calorie measure in nutrition. It has a physical basis and we can meaningfully use it to compare foods in some circumstances; but definitely not all. Thinking that 1000 Calories of sugar is equivalent to 1000 Calories of vegetables could get you into serious trouble.

Carbon dioxide equivalence is a flawed concept which long blinded people to the requirement for multifaceted action to stop the further destabilisation of the climate. Figure 1 needs to be replaced by at least two targets ... just as we have separate nutrient targets. We can however use the 105 methane factor much like a Calorie measurement. That is, we can use it to better understand the relative importance of various activities, just so long as we understand it isn't the whole story. Let's now see what Australia's greenhouse gas emissions look like with this more realistic methane factor as an example of a greenhouse intensive economy.

A NEW UNDERSTANDING OF CLIMATE CHANGE FACTORS WITH IMAGES

The New Picture

Figure 3 is a "treemap" — a way of displaying hierarchical data by using nested rectangles with areas proportional to the specific dimensions of the respective data (Shneiderman & Plaisant, 2014). The interpretation is that if the size of one part of the map is double the size of another, then the climate impact of the larger part is double that of the smaller. Some parts of the map (for example, "Electricity") have been subdivided to represent the different levels of the data (for this example — "Other electricity", "Household electricity", "Commercial services" and "Aluminium"). Again the interpretation is similar with ratios of areas corresponding to ratios of climate impact. The image was built using data from Australia's latest submission to the UNFCCC (2014), but instead of using the Kyoto factor of 21 to relate methane to CO_2, we've used the 105 figure discussed above. The mismatch between the climate impact of animal and plant foods as shown in the treemap (see Figure 3) and the bland and uninformative use of the word "diet" in the CSIRO poll discussed is stark.

Figure 3. Australian climate forcings by sector

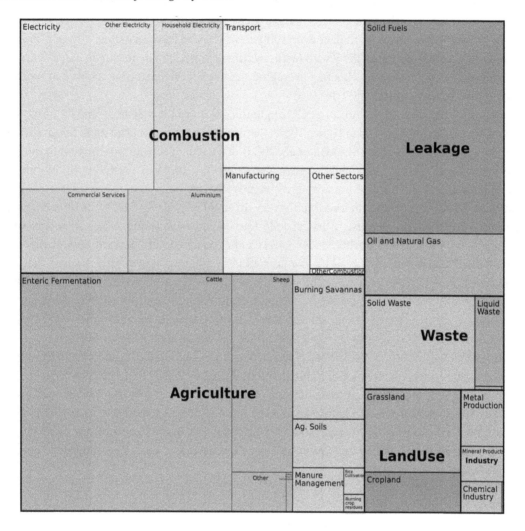

Within agriculture, enteric fermentation, the production of methane by animals, including us, dominates. This basic biological function makes greenhouse gas emissions from factories, steel mills, aluminium plants, planes and trucks seem small. Air travel is about 8 percent of the transport block in Figure 3, but the CSIRO poll gives it a nice clear question. Red meat has many times the impact but doesn't get a mention. Australia has a vast fleet of 81,000 articulated vehicles and but they can't compete with 29.3 million cattle, each with multiple stomachs digesting feed and expelling methane. The sectoral breakdown, as presented in Figure 3, is almost identical to those in the source data, with the exception that we've used Australian Bureau of Statistics (ABS) data (ABS, 2013) to split electricity into its four key components. The sub-division of enteric fermentation is as presented in Australia's UNFCCC data, except that any category responsible for less than 1 percent of enteric fermentation emissions has been ignored to simplify the image.

Recall Figure 1 and the very low level of emissions it attributed to a plant based diet. We can understand this now by looking at the breakdown within agriculture. Any emissions from "Manure management" are directly related to diets, which include meat and animal products. Similarly, "Burning savannahs" is predominantly associated with clearing land for grazing (Raphaely & Marinova, 2014). Emissions from "Agricultural soils" used to grow crops are the biggest of a few small components. Of course, Australia's livestock can eat far more of the cereal crop than people do. The ratios vary year to year depending on weather, but is generally between 4:1 and 5:1 with any extra cereal either stored or exported (FAO, 2014a). It's not straightforward to track the ratio of human to animal use of Australia's cereal exports. In 2011, for example, livestock in Australia ate about 9 million tonnes of cereals while Australia's human population ate about 2 million (FAO, 2014a). The amount exported can range from a few million to 25 million tonnes (FAO, 2014a).

One aspect of the image on Figure 3 can mislead. Since typically half of Australia's ruminant meat is exported, any per person calculations need to be cognisant of the export quantities. Nevertheless, Australian livestock is a major contributor to the country's greenhouse gas emissions and this fact needs to be recognised and reflected in trade and industry policies.

Interim Conclusion

The obvious conclusion from Figures 1 and 3 is that Australia should be phasing out both its fossil fuel use and ruminant populations with considerable urgency. In the case of ruminant meat, there is no nutritional need (Stanton, 2012) which can't be filled by healthier alternatives.

Treemaps of emissions do vary considerably from country to country in their structure. Brazil with roughly as many cattle as people looks similar to Australia, but places like Bangladesh or Nigeria have relatively few cattle per person.

Globally, the implication is also clear that deep cuts in ruminant numbers are necessary to avoid the worst that climate change will otherwise deliver. This follows with a fair degree of certainty from both Figure 1 and Figure 2. Action is also implied with regard to other major sources of methane, but ruminants happen to be the biggest (Ripple et al., 2014). Since their meat is a major cause of bowel cancer (WCRF, 2007) and other health problems, then we can tackle two problems with one action.

A Preemptive Reply to Critics

Such a call for ruminant reductions isn't new; with the most powerful recent call being in a 2014 Nature Climate Change paper (Ripple et al., 2014). The response to such calls is exemplified by a 2013 paper (Lauder et al., 2013), and an earlier paper in a special edition of Science (Godfay et al., 2010). Both papers respond defensively to calls for dietary change by claiming that livestock deliver many benefits, particularly to developing countries, and while both papers are rigorous academic papers replete with references on other matters, neither feel obliged to support this particular claim with evidence. Lauder et al. (2013) state that livestock reductions are incompatible with both 'sustainable land management [and] food security goals'; but provide no evidence for either claim. Taking a slightly different line, an Australian team, writing in the Lancet (McMichael, Powles, Butler & Uauy, 2007), thought that rich countries should reduce their meat intake to allow an increase in meat consumption in poorer countries. The target they proposed would have provided a 10 percent reduction in global meat production and about 50 grams per day of ruminant meat for everybody. As part of their argument they asserted, without evidence that this would lead to a dramatic decrease in stunting in developing countries.

There is a pattern here. Academic journals and research organisations seem happy for people to defend meat using all manner of claims without evidence. The article by Watts (2012), mentioned earlier,

Figure 4. The global food supply

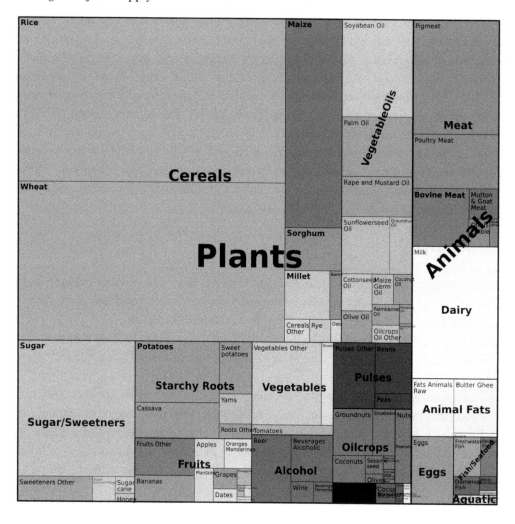

follows the pattern and warns " ... losing the benefits [of beef] — especially to those most in need." Again, no evidence is provided.

Figure 4 is a treemap where we can see the scale of meat in the global food supply. It makes clear how small a part of the world's food supply ruminants are. There is however considerable variation between countries. The UN FAO statistical data base (FAO, 2014a) allows us to calculate the range. Ruminants provide 6.1 percent of food energy in Australia and about 1.6 percent in the least developed countries. Ruminant meat is 1.1 percent of food energy for the 4.2 billion people in Asia. It may be objected that these figures would rise if protein rather than energy were used. This is another misconception, as we'll see later that it's a lack of energy rather than protein that's a problem in the least developed countries.

China has far more meat per person than Cuba but also more stunted children: 9 percent compared to 7 O, (WHO, 2014b). China also has more overweight children. Nigeria has a 35 percent rate of stunting in children (WHO, 2014b), but has it anything to do with meat? To see if the problem might be too little meat, it pays to think about the 65 percent of children who *aren't* stunted. Divide all of Nigeria's meat up among 65 percent of the population and they each only get 38 grams of meat a day. That can reasonably be considered an upper bound on what the 65 percent of children who aren't stunted can be eating. It seems obvious that this isn't enough explain the lack of stunting in these children.

The obvious has been experimentally verified in a study (Neumann et al., 2003) which tried adding meat to the diet of malnourished children and no, it didn't reduce stunting levels. Also, the children in the study received far more than this Nigerian maximum of 38 grams. We'll discuss it below.

Let's for now, stick to cattle meat in particular. Is there any general evidence to suggest that levels of beef in a country's food supply helps the nutritional status of the population? In 2014 the World Health Organisation (WHO) released updated figures on Disability Adjusted Life Years (DALY) lost due to ill-health for most countries on the planet (WHO, 2014a). This data has a category for "Nutritional Deficiencies".

What happens when we plot DALYs against bovine meat consumption in the countries considered by WHO to be the "Least Developed"? Figure 5 shows the result. Many countries with less than half the beef intake of Mali or Niger have far lower rates of nutritional deficiency.

We've drawn a regression line in Figure 5, but it shouldn't be taken seriously, given the obvious lack of any meaningful causal relationship between beef and DALYs. Note also the presence of lowish

Figure 5. Beef isn't a nutritional magic bullet

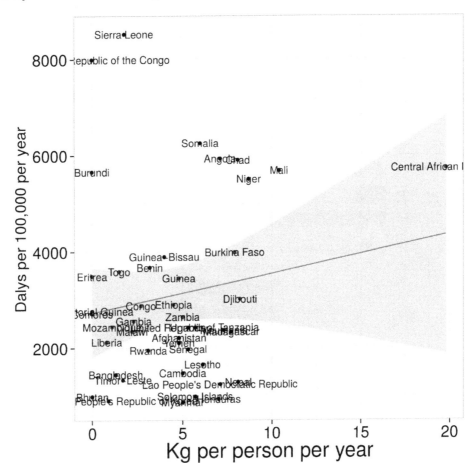

DALYs rates in many countries with very low bovine meat intake. The regression line is fairly clearly pulled up by the outlier; Central African Republic of the Congo. This graph however, doesn't capture other relevant characteristics of a country's food supply. Let's look in a little more detail at some selected countries. We've picked a few of the Least Developed Countries and added Sri Lanka and Nigeria. They aren't in the "Least Developed" list, but have a marginal food supply and serious nutritional problems. Now consider the charts in Figure 6.

Figure 6. Is beef a nutritional saviour for poor countries?

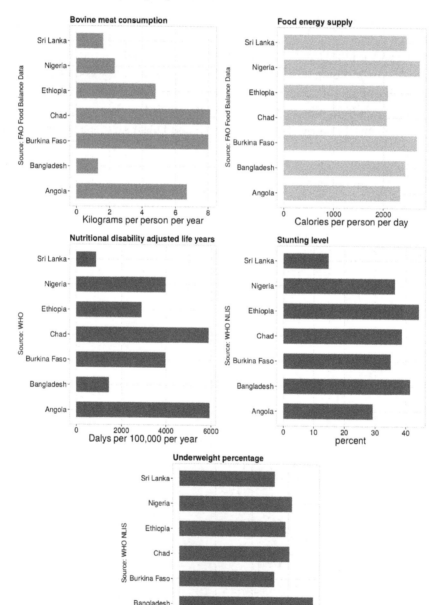

The countries Sri Lanka and Bangladesh have the lowest levels of bovine meat availability and also the lowest levels of nutritional DALYs. Bangladesh however, has a high stunting level. Clearly the food supply in Chad and Ethiopia is dismal with production of around 2,000 Calories per person, but Chad's relatively high levels of bovine meat are apparently useless at compensating for this. Burkina Faso has more food available than Sri Lanka and far more bovine meat but it still has higher level of DALYs.

One response to such data is to propose some ad-hoc theory to save the image of beef so vigorously advertised for decades in Australia and elsewhere as a wonder food. Perhaps the distribution system isn't working and the right people aren't getting the meat, or they aren't getting it at the right time in their life, etc., etc. That would be a predictable response. However, as mentioned above, a large, well-executed intervention study, funded by the US National Cattlemen's Beef Association, in Kenya in 1999-2000 showed exactly what plotting beef against DALYs showed. Nothing. The study (Neumann et al., 2003) was published as a series of papers in a special supplement of the *Journal of Nutrition*. It involved giving extra beef to 550 children (median age 7.4 years) for two years. They received twice as much red meat as Australian children of that age were getting at that time (Baghurst, Record & Leppard, 2000), but when compared to a similar amount of plant calories, it made no difference. More food made some difference, but it didn't matter what kind of food. There were 4 supplemental feeding groups given extra food. The supplement was added to the standard staple stew that the children normally ate. One group received 200ml of milk, another — 60 grams of beef (but this increased during the study to 85 grams), the third got more stew and the fourth got nothing. Note that this is more red meat than the *Lancet* study recommended (McMichael et al., 2007). It's almost as much red meat as the total amount of all meat that the study claimed would dramatically decrease stunting rates, it didn't (Grillenberger et al., 2003).

Lastly, this null result is consistent with a major study by the International Food Policy Research Institute (Smith & Haddad, 2000) into the causes of child malnutrition in developing countries. Their top four causal factors were: calories, water, mother's education and mother's status relative to her husband. Malnutrition is largely prevented by enough food and clean water coupled with adequate maternal education and power. You won't find any mention of beef, meat or protein in their report. Medical people made what was famously called "The big protein fiasco" back in the 1960s and 70s (Carpenter, 1986; McLaren, 2000). They attributed wonderful properties to protein and it took a couple of decades to realise that malnutrition wasn't a protein problem; it was a food and water problem. Apparently it's a fiasco that isn't famous enough and its lessons haven't spread to the wider scientific community and especially to non-medical scientists working outside this rather specialised area.

In any event, objections to calls for meat reductions by appealing to meat's role in developing countries is rather like objecting to calls for restrictions on high powered motor vehicles by pointing to the advantages of V8s in ambulances. Most of the world's cattle meat isn't consumed by the poor in developing countries. If there were evidence ruminant or other meat was useful to compensate for poor diets in food deficient countries, then that still leaves ample room for deep cuts in ruminant numbers while allowing for exceptions in developing countries.

Currently, countries in the FAO's "Least Developed" group have about 281 million of the world's 1.5 billion cattle and almost a billion people. This ratio is misleading. The average cattle carcass weight of these animals is 117 kilograms compared to 283 kilograms in the European Union, 351 kilograms in the US with the global average being 214 kilograms (FAO, 2014b). These poor countries may have 19 percent of the world's bovines, but closer to 10 percent of the planet's bovine biomass. This isn't to deny that cattle may have value in these poorer countries …but it certainly isn't as food.

Having defended meat with reference to developing countries, Lauder et al. (2013) go on to give an example of a methane-offsetting metric using Australia; which is neither poor nor undernourished. If the motivation for the spirited defence of ruminants were really the world's hungry, then why not use them as the case study for the proposed methane-offsetting metric?

An Aside on Vitamin B12

Vitamin B12 has assumed a special and unwarranted position in nutritional debates because it cannot be obtained from plant foods. It is essential for good health and can either be obtained from animal products or supplements.

The Kenyan study found that B12 was the only micronutrient whose status was improved by the meat or dairy supplementation (Siekmann et al., 2003). Unfortunately, plasma folate and zinc levels *decreased* in these groups, particularly the meat group. Iron levels rose most in the children receiving extra plant food. All in all, the micronutrient changes refused to behave, as the meat industry funders of the study would have liked. Dirty water and infections make for a complex multi-factorial problem. Just adding meat didn't work. Clearly, any suggestion that meat is the cheapest, most reliable, or most environmentally benign method of solving B12 or other nutrition deficiencies needs to be demonstrated rather than assumed.

For most of the world's cattle consumers, the claim is clearly false with supplementation being far cheaper (I pay about 10 cents per week for mine), than ruminant meat with none of the animal welfare, climate, or other environmental side effects. Vitamin B12 is cheap to make and distribute, doesn't need refrigeration, and doesn't interfere with cropping or pollute waterways. Cattle have been generating novel rotaviruses (Martella, Bányai, Matthijnssens, Bounavogolia & Ciarlet, 2010) and polluting water for thousands of years. Diarrhoea has been killing children throughout the period and still is, despite modern medicine. Vaccines are, of course, much better than nature's method of responding to such infections. Similarly, B12 supplements are much better than the historically natural source of the vitamin.

The grazing of cattle on crop residues also has major adverse impacts. Writing in *Science* in 2009, Lal and Pimentel, two of the world's top soil scientists summed up the problem of food production in sub-Saharan Africa as follows:

The agrarian stagnation and perpetual food deficit in sub-Saharan Africa is attributed to severe soil degradation..., caused by extractive farming practices that involve continuous removal of crop residues for use as traditional biofuels and cattle feed. This has created a negative nutrient budget, (2009, p. 1345).

The phrases "food deficit" and "negative nutrient budget" don't begin to capture the horrors of what could better be described as livestock driven starvation.

CONCLUSION

We need images consistent with climate models, which accurately rank the causes of climate change and guide proposed actions. The notion of CO2eq is a simplifying fiction with no physical basis. It obscures the fact that the prevention of climate destabilisation is a multi-faceted problem with meat as a key issue. Many climate scientists recognise this but few have articulated it as clearly as James Hansen (Hansen, 2009; Hansen et al., 2008) from the US National Aeronautics and Space Agency (NASA). He divides the necessary actions into 4 areas: the phase out of coal, sharp cuts in non-CO2 gases and aerosols, the phase out of remaining fossil fuels (oil, gas, shale oils and the like), and lastly, the rollback of 200 years of deforestation. Action on all four fronts is necessary, with no one being sufficient to bring the planet back into energy balance. The elimination of ruminant meat and the movement towards a plant based diet contributes in a major way on two fronts by reducing methane and allowing reforestation.

We did not cover reforestation in this chapter. It is obvious however that if you wish to reforest land that has been cleared for grazing or to raise crops for factory farmed animals, then you need to remove the animals and stop growing the crops.

Many people have talked about climate change as a moral issue and it is. It's about us robbing not just our human descendants of a healthy environment, but also depriving the remaining wildlife populations of the planet. It's no accident that Peter Singer, writing in the second edition of "Animal Liberation" over 20 years ago recognised existential nature of the problem and the fact that agribusiness will be a major opponent:

Forests and meat animals compete for the same land. The prodigious appetite of the affluent nations for meat means that agribusiness can pay more than those who want to preserve or restore forests. We are, quite literally, gambling with the future of our planet — for the sake of hamburgers, (1991, p. 169)

It is clear from the resistance to any kind of ruminant reduction that it isn't just agribusiness that puts today's profits before the planet's future. There are also plenty of climate change experts willing to use arguments to defend meat that they wouldn't dare use for any other purpose. How is the value of meat to stunted children in Africa relevant to the vast majority of meat consumption? Not only isn't it relevant, but the data clearly shows that meat is no magic bullet. It's time journals started to demand that arguments defending the dietary status quo obey the same standards of evidence that apply to other claims. Equally, it's time that politicians and climate lobby organisations started to present the full range of necessary actions required to prevent further disruption of the climate. It's time climate change surveys and education campaigns actually reflected the quantitative realities of the causes of the problem.

We've deliberately ignored moral arguments in this chapter but assumed that readers agree that protecting future generations from a destabilised climate is essential. Nobody without this basic concern for future generations will be reading. On the other hand, many have little or no concern for animals and it is these people that need to be convinced to change their diets. They are in the overwhelming majority and without them changing, avoiding climate destabilisation isn't possible. Agribusiness scientists are, of course working to reduce methane from ruminants, but they've been doing that for 40 years and achieved little outside of research papers and grants. It's a tough problem and livestock producers won't accept anything that isn't also profitable. Even if such projects were successful, how would that address the reforestation requirement? It wouldn't. No, the time is ripe to recognise that scientific consistency and intellectual integrity demand that people stop making excuses and learn how to eat differently.

REFERENCES

Allison, I., Bindoff, N. L., Bindschadler, R. A., Cox, P. M., de Noblet, N., & England, M. H. … Weaver, A. J. (2009). The Copenhagen Diagnosis: updating the world on the latest climate science. Retrieved June 10, 2015, from http://www.ccrc.unsw.edu.au/sites/default/files/Copenhagen_Diagnosis_HIGH.pdf

Australian Bureau of Statistics (ABS). (2013). Energy Accounts Australia 2011-12 (No. 4604.0). Retrieved June 10, 2015, from http://www.abs.gov.au/ausstats/abs@.nsf/mf/4604.0

Baghurst, K., Record, S., & Leppard, P. (2000). Red meat consumption in Australia: Intakes, nutrient contribution and changes over time. *Australian Journal of Nutrition and Dietetics*, *57*(4), S3–S37.

Blomqvist, L., Brook, B. W., Ellis, E., Kereiva, P., & Nordhaus, T. (2013). Does the shoe fit? Real versus imagined ecological footprints. *Public Library of Science (PLoS). Biology*, *11*(11), e1001700. doi:10.1371/journal.pbio.1001700

Carpenter, K. J. (1986). The history of enthusiasm for protein. *Nutrition (Burbank, Los Angeles County, Calif.)*, *116*(7), 1364–1370.

Food and Agriculture Organization of the United Nations (FAO). (2014a). 2013 food balance sheets. Retrieved June 10, 2015, from http://faostat3.fao.org/download/FB/FBS/E

Food and Agriculture Organization of the United Nations (FAO). (2014b). Production/ livestock, primary. Retrieved June 10, 2015, from http://faostat3.fao.org/download/Q/QL/E

Godfay, H. C. J., Beddington, J. R., Crute, I. R., Haddad, L., Lawrence, D., Muir, J. F., & Toulmin, C. et al. (2010). Food security: The challenge of feeding 9 billion people. *Science*, *327*(5967), 812–818.

Grillenberger, M., Newman, C. G., Murphy, S. P., Bwibo, N. O., van't Veer, P., Hautvast, J. G., & West, C. E. (2003). Food supplements have a positive impact on weight gain and the addition of animal source foods increases lean body mass of Kenyan schoolchildren. *The Journal of Nutrition*, *133*(11), 3957S–3964S.

Hansen, J. (2009). *Storms of my grandchildren: the truth about the coming climate catastrophe and our last chance to save humanity*. New York, NY: Bloomsbury.

Hansen, J., Sato, M., Kharecha, P., Beerling, D., Mason-Delmotte, V., Pagani, M., & Zachos, J. (2008). Target atmospheric CO_2: Where should humanity aim? *Open Atmospheric Science Journal*, *2*, 217–231. doi:10.2174/1874282300802010217

Harper, L. A., Denmead, O. T., Freney, J. R., & Byers, F. M. (1999). Direct measurements of methane emissions from grazing and feedlot cattle. *Animal Science (Penicuik, Scotland)*, *77*(6), 1392–1401.

International Energy Agency (IEA). (2013). Key world energy statistics. Retrieved June 10, 2015, from http://www.iea.org/publications/freepublications/publication/KeyWorld2014.pdf

Lal, R., & Pimentel, D. (2009). Biofuels: Beware crop residues. *Science*, *326*(5958), 1345–1346.

Lauder, A., Enting, I., Carter, J., Clisby, N., Cowie, A., Henry, B., & Raupach, M. (2013). Offsetting methane emissions - an alternative to emission equivalance metrics. *International Journal of Greenhouse Gas Control*, *12*(0), 419–429. doi:10.1016/j.ijggc.2012.11.028

Lea, E., & Worsley, A. (2008). Australian consumers' food-related environmental beliefs and behaviours. *Appetitie*, *50*(2-3), 207–214.

Leviston, Z., Price, J., Malkin, S., & McCrea, R. (2014). Fourth annual survey of Australian attitutes to climate change. Interim report. Perth, Australia: Commonwealth Scientific and Industry Research Organisation (CSIRO).

Martella, V., Bányai, V., Matthijnssens, J., Bounavogolia, C., & Ciarlet, M. (2010). Zoonotic aspects of rotaviruses. *Veterinary Microbiology*, *140*(3-4), 246–255.

McLaren, D. S. (2000). The great protein fiasco revisited. *Nutrition (Burbank, Los Angeles County, Calif.)*, *16*(6), 464–465.

McMichael, A. J., Powles, J. W., Butler, C. D., & Uauy, R. (2007). Food, livestock production, energy, climate change, and health. *Lancet*, *370*(9594), 1253–1263.

National Renewable Energy Laboratory (NREL). (2012). Life cycle greenhouse gas emissions from solar photovoltaics. Retrieved June 10, 2015, from http://www.nrel.gov/docs/fy13osti/56487.pdf

Neff, R. A., Chan, I. L., & Smith, K. C. (2008). Yesterday's dinner, tommorrow's weather, today's news? US newspaper coverage of food system contributions to climate health. *Public Health Nutrition*, *12*(7), 1006–1014.

Neumann, C. G., Bwibo, N. O., Murphy, S. P., Sigman, M., Whaley, S., Allen, L. H., & Demment, M. W. et al. (2003). Animal source foods improve dietry quality, micronutrient status, growth and cognitive function in Kenyan school children: Background, study, design and baseline findings. *The Journal of Nutrition*, *133*(11), 3941S–3949S.

Noakes, M. (2013). *CSIRO total wellbeing diet recipes on a budget*. Melbourne, Australia: Penguin Books.

Noakes, M., & Clifton, P. (2005). *CSIRO total wellbeing diet*. Melbourne, Australia: Penguin Books.

Noakes, M., & Clifton, P. (2006). *CSIRO total wellbeing diet. Book 2*. Melbourne, Australia: Penguin Books.

Pelletier, N., & Tyedmers, P. (2010). Forecasting potential global environmental costs of livestock production 2000-2050. [PNAS]. *Proceedings of the National Academy of Sciences of the United States of America*, *104*(12), 4814–4819.

Peters, C. J., Wilkins, J. L., & Fick, G. W. (2007). Testing a complete-diet model for estimating the land resource requirements of food consumption and agricultural carrying capacity: The New York State example. *Renewable Agriculture and Food Systems*, *22*(2), 145–153. doi:10.1017/S1742170507001767

Raphaely, T., & Marinova, D. (2014). Flexitaririanism: Decarbonising through flexible vegetarianism. *Renewable Energy*, *67*, 90–96.

Ripple, W. J., Smith, P., Harberl, H., Montzka, S. A., McAlpine, C., & Boucher, D. H. (2014). Ruminants, climate change and climate policy. *Nature Climate Change*, *4*(1), 2–5.

Russell, G. (2009). *CSIRO perfidy*. Fremantle, Australia: Vivid Publishing.

Shindell, D. T., Faluvegi, G., Koch, D. M., Schmidt, G. A., Unger, N., & Bauer, S. E. (2009). Improved attribution of climate forcing to emissions. *Science, 326*(5953), 716–718.

Shindell, D. T., Kuylenstierna, J. C. I., Vignati, E., van Dingenen, R., Amann, M., Klimont, Z., & Fowler, D. et al. (2012). Simultaneously mitigating near-term climate change and improving human health and food security. *Science, 335*(6065), 183–189.

Shneiderman, B., & Plaisant, C. (2014). Treemaps for space-constrained visualization of hierarchies: including the history of treemap research at the University of Maryland. Retrieved June 10, 2015, from http://www.cs.umd.edu/hcil/treemap-history/index.shtml

Siekmann, J. H., Allen, L. H., Bwibo, N. O., Demment, M. W., Murphy, S. P., & Neumann, C. G. (2003). Kenyan school children have multiple micronutrient deficiencies, but increased plasma vitamin B-12 is the only detectable micronutrient response to meat of milk supplementation. *The Journal of Nutrition, 133*(11), 3972S–3980S.

Singer, P. (1991). *Animal liberation* (2nd ed.). London, UK: Random House.

Smith, L. C., & Haddad, L. J. (2000). Explaining child malnutrition in developing countries: a cross-country analysis. Research report. Retrieved June 10, 2015, from http://www.ifpri.org/sites/default/files/pubs/pubs/abstract/111/rr111.pdf

Stanton, R. (2012). A plant based diet - good for us and for the planet. [MJA]. *The Medical Journal of Australia, 1*(Suppl2), 5–6.

Tobler, C., Visschers, V. H., & Siegrist, M. (2011). Eating green: Consumer's willingness to adopt ecological food consumption behaviours. *Appetite, 57*(3), 674–682.

United Nations Framework Convention on Climate Change (UNFCC). (2014). National inventory submissions: Australia. Retrieved June 10, 2015, from http://unfccc.int/national_reports/annex_i_ghg_inventories/national_inventories_submissions/items/8108.php

Watts, C. (2012). Methane and livestock: factoids help farmers least of all. Retrieved June 10, 2015, from http://www.climateinstitute.org.au/articles/opinion-pieces/methane-and-livestock-factoids-help-farmers-least-of-all.html

World Cancer Research Fund (WCRF). (2007). Food, nutrition and the prevention of cancer: a global perspective. Retrieved June 10, 2015, from http://www.wcrf.org/sites/default/files/Second-Expert-Report.pdf

World Health Organization (WHO). (2014a). Disability-adjusted life years (DALYs). Retrieved June 10, 2015, from http://www.who.int/gho/mortality_burden_disease/daly_rates/en/

World Health Organization (WHO). (2014b(). Nutritional landscape information system. Retrieved June 10, 2015, from http://apps.who.int/nutrition/landscape/report.aspx?iso=CHN&rid=1620&goButton=Go

KEY TERMS AND DEFINITIONS

Carbon Dioxide Equivalence: A concept by which different gases are multiplied by a factor in an attempt to make them equal to that amount of CO2.

Climate Forcing: This can be anything, which changes the ratio of energy arriving and leaving the planet. Examples: the global area of ice because it changes reflectivity, the level of CO2 in the atmosphere, the area of clouds and even the colour of roofs.

Enteric Fermentation: The process of digesting food. It is of concern to climate scientists because significant quantities of methane are produced by some species.

Treemap: A graphical way of displaying data, which highlights hierarchical relationships.

Chapter 4
Water Quality Impacts of Abattoir Activities in Southern Africa

Never Mujere
University of Zimbabwe, Zimbabwe

ABSTRACT

The meat processing produces solid, liquid and gaseous wastes from stockyards, abattoirs and packing plants contain blood, fats, protein, gut contents, heavy metals, antibodies, hormones and other substances. Solid and liquid wastes can lead to an impairment or disruption of water eco-functionality and a preponderance of disease-causing organisms. In developing nations, many abattoirs dispose of their waste directly into streams or rivers and also use water from the same source to wash slaughtered meat. Most solid, liquid and gaseous wastes are released into the immediate environs of the abattoir. This chapter is a literature review of the impact of untreated slaughterhouse wastewater on water quality in selected areas of Southern Africa. The review shows that abattoir activities significantly reduced the physical, chemical and microbiological quality of water bodies. Although an improvement of the water quality was observed downstream, this was a result of self-purification and dilution effects.

INTRODUCTION

The killing of cattle, sheep, pigs, goats, poultry and other equine animals to supply meat for human consumption is a common practice in Southern Africa. Abattoir waste contains high levels of organic matter due to presence of manure, blood, fats, grease, hair, grit or sediments and undigested feeds. Blood and fats contribute mostly to organic load; blood is also a major contributor of nitrogen content in the effluent. It can also contain high level of salts, phosphate, nitrates, trace or heavy metals, bacteria, viruses and other microorganisms (Muhirwa et al., 2010; Kosamu, Mawenda & Mapoma, 2011). Abattoir effluent affects water, land or air quality if proper disposal practices are not followed. Improper waste disposal reduces oxygen in water, thereby endangering aquatic life. It can lead to animal diseases being transmitted to humans through contact with animal faeces. Pollution of the environment in the vicinity of the abattoir as

DOI: 10.4018/978-1-4666-9553-5.ch004

well as reduced quality of health of residents in the local area result in elevation of excessive coughing, typhoid fever, diarrhoea, and malaria and muscle pains (Benker-Coker & Ojior, 1995). Noise pollution is also associated with abattoir activities (WHO, 2011). Respiratory and mucous membrane effects are common in areas close to intensive swine operation (Itodo, Awulu & Harrison, 1999). Thus, surface and ground water quality in vicinity of the abattoirs can be adversely affected by effluents.

BACKGROUND

Meat processing, slaughtering and processing remain critical in ensuring that consumers receive a hygienically safe product. Production processes for meat products have been shown to have a significant impact on the environment, accounting for between 15% and 24% of greenhouse gas emissions (Mittal, 2006). Abattoirs are an important industry in Africa's domestic meat supply and export industry, as well as employment opportunities to many communities. Abattoir effluent is known to degrade the quality of receiving water bodies. Waste from abattoirs changes the microbiological and physico-chemical water quality parameters such as pH, temperature, electrical conductivity, salinity, turbidity, total dissolved solids (TDS), total suspended solids (TSS), dissolved oxygen (DO) and coliforms (GDARD, 2009). The effluent characteristics of the abattoir are nutrients and biologically active constituents which pose a considerable challenge for water resources management particularly the delivery of essential water, sanitation services and environmental protection. Nitrates are very high in the abattoir wastewater especially at the evisceration and slaughter step where the urine and undigested stomach content concentrated in nitrate was mixed in wastewater streams (Mittal, 2004). This is because wastewater streams for these processes consist of mixed intestinal contents and blood with a high content of nitrates. This chapter focuses on the environmental effects of meat processing in general. In particular, it highlights how meat processing contributes to changes in water quality.

Water Quality

Water quality describes the physical, chemical, biological and aesthetic properties of water which determine its fitness for a variety of uses and for protecting the health and integrity of aquatic ecosystems. The quality of any body of surface or ground water is a function of either or both natural influences and human activities. Levels of water quality are assessed in terms of: physical characteristics e.g. color, temperature, taste, turbidity, hardness and smell or odor which are determined by senses of touch, sight, smell and taste; chemical characteristics such as conductivity, dissolved salts and oxygen demand; and biological e.g. amount of micro-organisms present such as bacteria, protozoa and algae parameters (WHO, 2011).The definition of water quality takes into account the fact that even when water is polluted say with sewage, depending on the level of the contaminant, it can be used for other purposes such as agriculture.

Environmental Pollutants from Abattoirs

An abattoir is loosely defined as any registered facility that is responsible for the conversion of animals to meat via a slaughtering process. This includes livestock, poultry and special classes of animals, e.g. crocodiles and game (GDARD, 2009). The slaughtering process remains critical in ensuring that consumers receive a hygienically safe product.

Effluent Salinity

Skin preservation by dry salting is a common procedure at small abattoirs that are remote from tanning operations and often export their hides and skins for tanning. After salting, often in converted cement truck mixers, the hides are hung to dry for a minimum of 5 days. During this period, the salt draws the moisture out of the hide, together with the protein-filled fluids contained in the attached flesh. The effluent from drying sheds is therefore highly saline and has a very high biochemical oxygen demand (BOD). It also contains high levels of fluoride, since the salt used contains up to 1 percent sodium fluoride as a bactericide.

Wastewater

Wastewater produced in animal slaughter areas typically has a high BOD. It is also very saline and has high levels of nutrients, suspended solids and bacterial contamination.

Solid Waste

Sources of solid wastes generated at abattoirs include: manure generated in animal holding areas; slaughterhouse and processing areas; sludge generated in waste treatment ponds; and unwanted carcasses, hide or skins, paper, cardboard, plastics, feathers and pieces, and unwanted carcasses and carcass parts (Mittal, 2004).

Odors

Potential sources of odors in abattoir operations include stale materials and fugitive emissions from cookers. Odors in animal holding pens are produced by manure and urine. Abattoirs odors come from solid wastes such as paunch contents and blood residues. Anaerobic waste treatment ponds may produce gases such as methane, ammonia and hydrogen sulphide, which give rise to objectionable odors (GDARD, 2009). Odor control may be a significant issue, particularly when the abattoir is located near residential areas or in a hot environment.

Dust

Potential sources of dust emissions at an abattoir are: paddocks, sale-yards and holding pens; and materials burned at an abattoir (GDARD, 2009).

Greenhouse Gases

Animals produce methane gas (a greenhouse gas) during the process of digestion which is released to the atmosphere during processes such as defecation. Little can be done to prevent or reduce the amount of methane produced. However, by varying the food types and quality, methane generation may be reduced. Chlorofluorohydrocarbons (CFCs) may be used in refrigeration and freezer plants. These are ozone depleting gases and their production and use are subject to national and international regulation (McLaughlin & Mineau, 1995). Ozone depleting gases used in refrigeration units should be replaced in existing abattoirs and new abattoirs must install refrigeration units that do not utilize such substances (Quinn & McFarlane, 1989).

Noise

In abattoirs noise can be generated by several sources, including: animals, especially when in concentrated groups; processing activities within the slaughterhouse; plant machinery; and plant and service vehicles (Muhirwa et al., 2010).

EFFECTS OF ABATTOIRS ON WATER QUALITY

The people and environment in Southern Africa are highly vulnerable to pollution from abattoirs. Availability of safe water in the region is worsened by poor sanitation and improper disposal of animal waste. The majority of the people living along and down river courses depend on untreated stream water supply for domestic purposes and sometimes for drinking. While rivers and streams serve as major sources of reliable domestic water supply, most abattoirs are located close to such water sources. Nevertheless, abattoirs discharge untreated wastewater into water bodies because they normally face difficulties in meeting costs of treating wastes such as fats, oils, greases and suspended solids (Chukwu, 2008). Thus, there are high chances of people drinking unclean water contaminated with bacteria or pathogens that could impact on public health. The following are case studies findings on the physico-chemical and microbiological impacts of abattoir activities on water quality in selected countries. Table 1 shows findings of a study conducted to assess the water quality of abattoirs in Ghana.

Generally the influent met the recommended standard for treated water quality except for fecal and total coliform counts of 36 and 84 CFU, respectively, instead of 0 CFU recommended by the World Health Organization (WHO, 2011). The effluent was highly polluted and did not meet the set standards for effluent discharges into the environment. Its extremely high levels of BOD (3300 mg/L) and low dissolved oxygen (0 mg/L) are strong indications of high organic pollution in the effluent. Effluent fecal coliforms level of 450,000/100 ml and total coliforms of 550,000/100 ml were recorded. Excessive turbidity, or cloudiness, in drinking water is aesthetically unappealing, and may also represent a health concern. Turbidity can provide food and shelter for pathogens. If not removed, turbidity can promote regrowth of pathogens in the distribution system, leading to waterborne disease outbreaks, which have

Table 1. Effects of Tamale abattoirs on water quality in Ghana

Parameter	Influent	Effluent	WHO Standard
DO (mg/l)	4.6	0	5
BOD (mg/l)	2.2	3,300	20
TSS (mg/l)	4	2,567	30
Turbidity (NTU)	4.8	>1,000	5
Fecal coliforms (CFU/100 ml)	36	450,000	0
Total coliforms (CFU/100 ml)	84	550,000	0

(Weobong & Adinyira, 2011, p. 388).

Table 2. Effects of meat processing on water quality on Mchesa stream in Malawi and Ogbogoro stream in Nigeria

Parameter	Upstream	Discharge Point	Downstream	WHO Standards
BOD (mg/l)	257.3	612.3	487	20
COD (mg/l)	42.87	42.08	42.57	1000
DO (mg/l)	3.8	1.8	2.4	5
SS (mg/l)	286.0	477.3	272.0	30
EC	99.3	128.3	177.7	400
pH	7.4	7.7	7.5	6.5-9.5
Total alkalinity (mg/l)	2.55	2.37	2.25	NA
Nitrate (mg/l)	0.39	0.83	0.62	10
Phosphate (mg/l)	0.05	0.07	0.06	NA
Turbidity (NTU)	0.46	0.77	0.56	5
Odor	Odorless	Offensive	Offensive	Inoffensive

(Kosamu et al., 2011).

caused significant cases of gastroenteritis throughout the United States and the world. Although turbidity is not a direct indicator of health risk, numerous studies show a strong relationship between removal of turbidity and removal of protozoa. Residents of the community where the abattoir is located complained about bad odor, pollution of their water source (dam) from the effluents and the outbreak of maggots, flies and diseases such as malaria, typhoid, dysentery and diarrhea (Weobong & Adinyira, 2011).

Table 2 shows the results of a study conducted in Malawi and Nigeria. The first sampling point was located 10 m upstream with respect to abattoir discharge point. It served as reference point. A second sampling point was located at abattoir effluent discharge point. The third sampling point located 10 m downstream from effluent discharge. The physico-chemical parameters tested were: pH, biochemical oxygen demand (BOD), dissolved oxygen (DO), suspended solids (SS) and electrical conductivity (EC).

From Table 2, it is clear that at a point upstream, the BOD value was 257.3 mg/l, and the concentration went higher to 612.3 mg/l at discharge point. This was because of the effluent discharged by the abattoir on the stream at this point and therefore not completely mixed. At the sampling point downstream, the mean value of BOD was reduced to 487.0 mg/l. This can partially be attributed to dilution due to mixing and partially as a result of settling along the stream course and dilution.

Because oxygen from the atmosphere is dissolved in the water, (Verheignjen, Weiersna, Hulshoff & de Wit, 1996). It is one of the most significant tests for measuring the quality of water. The standard for sustaining aquatic life is stipulated to be 5 mg/l. Concentration below 2 mg/l adversely affects aquatic and biological life hence lead to death of fish. The low DO concentration at discharge point was be due to high organic load as shown by BOD and suspended solid values (Table 2).

The concentration of suspended solids show the highest value of 477.3 mg/l at discharge point hence exceeded the WHO (2011) recommended maximum limit of 30 mg/l. This was due to lack of proper sedimentation facility to separate the solid waste from liquid waste before the effluent is discharged. High concentrations of suspended solids can also cause problems to aquatic life such as by reducing water clarity and clogging fish gills (GDARD, 2009). The changes downstream can be attributed partially to dilution and partially to settling.

Table 3. Mean wasterwater values of bacteriological characteristics upstream and downstream of Kuje, Gwagwalada, Kubwa, Deidei, Karu abattoirs in Nigeria

Parameter	Upstream Station	Downstream Station	Discharge Limit into Surface Water
TC $(x10^2)$	38	6100	4.0
FC $(x10^2)$	2.5	230	0
F. streptococcus $(x10^2)$	3.3	7	120
E. coli $(x10^2)$	2.7	24	0

(Chukwu, Adeoye & Chidiebere, 2011, p. 104; Agwa, Sito & Ogugbue, 2013, p. 882).

Electrical conductivity (EC) is used to indicate the dissolved solids in water because the concentration of ionic species determines the conduction of current in an electrolyte. The high values of electrical conductivity downstream therefore suggest that has a considerable loading of dissolved salts (WHO, 2011).

The mean values of pH obtained ranged from 7.4 to 7.7. These pH values were normal to unpolluted freshwater. Water having a pH range of 6.5 -8.5 will generally support a good number of aquatic species. Thus, the values obtained were also within the recommended ranges (6.5-9.5) as stipulated by WHO (2011).

Table 3 shows findings of a study conducted to determine the bacteriological characteristics of abattoir effluents from 6 abattoirs discharged into water bodies in Nigeria. The concentrations of total coliform (TC), fecal coliform (FC), F. streptococcus, and E. coli were determined from upstream water servicing the abattoir and water bodies receiving abattoir effluents downstream. Total coliform bacteria concentration in the abattoir wastewater discharged exceeded the recommended limit for the discharge of effluents into water bodies.

The receiving water bodies were used for drinking, bathing, washing, watering of animals, watering of crops, and other domestic purpose downstream. Total coliforms and fecal coliforms or E. coli are generally not harmful themselves. While total coliforms are common in the environment, fecal coliforms or E. coli are usually associated with sewage or animal wastes. The presence of these bacteria in drinking water is a result of a problem with water treatment or the pipes which distribute the water, and indicates that the water may be contaminated with organisms that can cause disease. Symptoms may include diarrhea, cramps, nausea, and possibly jaundice, and any associated headaches and fatigue (Benker-Coker & Ojior, 1995).

F. Streptococci are common bacteria that live in the human body, including the nose, skin, and genital tract. These bacteria can damage, and destroy red blood cells. Drinking water which meets the WHO standard is usually not associated with a health risk from disease-causing bacteria and is considered safe.

Twelve wells were sampled and assessed of their physico-chemical and bacteriological parameters in Nigeria (Adeyemo, Ayodeji & Aiki-Raji, 2002). Four wells were sited within the ground of the Bodija abattoir, four were about 50 m – 80 m upstream of the abattoir and four were 500 m – 100 m downstream of the abattoir in a residential area (Table 4).

From Table 4 it is clear that most parameters at discharge points and downstream stations met the WHO standards. However, microbiological, total solids, manganese, lead and total chloride were outside the recommended limits.

Table 4. Mean values of physical, chemical and bacteriological characteristics of the well water used by butchers in Bodija Abattoir Ibadan, Nigeria

Parameter	Upstream Stations	At Discharge Point	Downstream Stations	WHO Limits
Coliform count (x10⁴)	5.6	6.9	0	0
F. streptococcus (x10⁴)	4.3	8.8	0	0
Bacteria count (x10⁴)	11	15.4	2	0
pH	6.57	6.41	6.75	7.0-8.5
Temperature (°C)	27.8	26.1	25.8	----
Sodium (mg/l)	35	40	46.30	---
Potassium (mg/l)	73.80	127.01	125.00	---
Calcium (mg/l)	81.30	143.00	131.30	75-200
Magnesium(mg/l)	13.70	27.00	12.30	30
Zinc (mg/l)	0.002	0.06	0.007	5-15
Iron (mg/l)	0.44	3.00	0.82	0.10-1.00
Copper (mg/l)	0.30	0.39	0.32	0.05-1.50
Manganese (mg/l)	1.18	2.19	2.19	0.05-0.50
Lead (mg/l)	0.02	0.09	0.01	0.01
Total dissolved solids	950.00	2,100.05	750.00	500-1500
Total hardness (mg/l)	133.25	371.25	151.00	100-500
Total chloride (mg/l)	134.75	312.05	94.85	250
Clarity	Clear	Clear	Clear	Clear
Odor	Odorless	Odorless	Odorless	Odorless
Taste	Tasteless	Tasteless	Tasteless	Tasteless

(Adeyemo et al., 2002, p. 52).

Total dissolved solids in water supplies originate from natural sources, sewage, urban and agricultural runoff, and industrial wastewater. Salts used for treating animal hide can also contribute to the TDS loading of water supplies. Ingestion of TDS in drinking water appears to be associated with various health effects such as incidence of cancer, coronary heart disease and arteriosclerotic heart disease.

Manganese occurs naturally in air, soils, surface water and groundwater sources. However, human activities are also responsible for much of the manganese contamination in water in some areas. Manganese is an essential nutrient for the proper functioning of humans and animals, as it is required for the functioning of many cellular enzymes. Although manganese is an essential nutrient at low doses, chronic exposure to high doses may be harmful. The health effects from over-exposure of manganese are dependent on the route of exposure, the chemical form, the age at exposure, and an individual's nutritional status. The nervous system is the primary target organ with neurological effects generally observed (WHO, 2011).

Chlorine is generally found only in industrial settings. It enters the body breathed in with contaminated air or when consumed with contaminated food or water. It does not remain in the body, due to its reactivity. Most direct releases of chlorine to the environment are to air and to surface water. Exposure to chlorine effect on human health depends on the amount of chlorine that is present, the length and frequency of exposure. Breathing small amounts of chlorine for short periods of time adversely affects the human respiratory system. Effects differ from coughing and chest pain, to water retention in the lungs. Chlorine irritates the skin, the eyes, and the respiratory system. These effects are not likely to occur at levels of chlorine that are normally found in the environment.

Lead is a useful and common metal but can be a very dangerous poison, particularly for children, when it is accidentally inhaled or ingested. In children, lead is most damaging when they are six years and younger. High levels of lead are life threatening and can cause seizures, unconsciousness, and death. Even at low levels, lead can be harmful and be associated with learning disabilities resulting in a decreased intelligence, attention deficit disorder, behavior issues, nervous system damage, speech and language impairment, decreased muscle growth, decreased bone growth and kidney damage. High levels of lead in adults can cause increased chance of illness during pregnancy, harm to a fetus, including brain damage or death, fertility problems in both men and women, high blood pressure, digestive issues, nerve disorders, memory and concentration problems and muscle and joint pain (WHO, 2011).

Thus, the discharge of abattoir waste negatively imparts on the chemical and microbiological quality of the stream especially at discharge points and downstream stations. The spatial distribution of the physico-chemical parameters along the stream showed relatively higher values at the effluent discharge point with increased attenuation in the levels of the parameters downstream indicating the capacity of the stream for self-purification. Nevertheless, the organoleptic assessment showed that the samples from all the wells were clear, without odor and taste. It is vital to have adequate treatment of waste before discharge and waste management practices that ensure waste reduction, re-use and recycling, should be encouraged by environmental regulatory bodies in order to protect the water resources from negative impacts of abattoir wastes. Therefore, the discharge of untreated abattoir waste into rivers and streams can have adverse environmental consequences on water quality (Agwa et al., 2013).

CONCLUSION

While the slaughtering of animals results in meat supply and useful by-products like leather and skin, livestock waste spills can introduce enteric pathogens and excess nutrients into surface waters and can also contaminate ground waters. These leachates consist largely of solids, microbial organisms and in special situations chemicals and shallow wells like hand-dug wells are more dangerously polluted. Abattoir operations produce a characteristic highly organic waste with relatively high levels of suspended solid, liquid and fat. The solid waste includes condemned meat, undigested ingesta, bones, horns, hairs and aborted fetuses. The liquid waste is usually composed of dissolved solids, blood, gut contents, urine and water. As population grows and urbanization increases, more water is required and greater demand is made on ground and surface sources and an even greater amount of organic and inorganic wastes are speared back into water sources so that less potable water becomes available. Water is regarded as being polluted when it is unfit for its intended use.

In developing countries abattoirs are known to pollute the environment either directly or indirectly from their various processes. Wastewater from an abattoir is a particularly concentrated source of oxygen consuming waste. In most countries, abattoir operations are generally unregulated. Moreover, abattoirs are usually located near water bodies in order to gain unhampered access to water for processing. Abattoirs generally use large quantity of water for washing meat and cleaning process areas. The disposal of effluent into rivers and streams is a common practice.

This study has shown that the operations of abattoirs have some negative impacts on water quality in Southern Africa. The water used by the abattoir for its operations (influent) is generally of good physical, chemical and microbiological quality. If proper abattoir management practices are not followed, animal waste can be washed into unprotected water bodies. Abattoirs however pose the following contamination risks to water resources:

- Nutrients from animal manure and process wastewater leading to eutrophication in wetlands and waterways, and toxic concentrations of nitrogen compounds;
- Depletion of oxygen levels in surface waters due to the breakdown of organic matter;
- Pathogens such as salmonella bacteria from any diseased stock held at the site;
- Increased salinity derived from waste stabilization pond effluent evaporation;
- Increased turbidity due to solids transfer to waterways and wetlands; and
- Alkalinity and surfactants derived from equipment cleaning.

Even though there are a number of state of the art abattoirs around the world, waste management practices in such facilities remain incompetent in resolving the environmental and human contamination risks and impacts. In the context of Southern Africa, emergency response and remediation are not always readily available to address contamination and pollution considerations, which renders humans and the environment more vulnerable and mean the risk of environmental and human adverse impact is potentially more serious.

Based on current knowledge, no South African abattoirs operate on a closed water circuit, as in general, it is prohibitively costly to treat waste water to a water quality standard fit for recycling and/or re-use. Thus, most abattoirs discharge (after appropriate pre-treatment) to municipal sewers.

Many of these small abattoirs relied on denaturing as a sole option for the disposal of condemned products and blood. With the increase in slaughtering and structural changes to these abattoirs over the past ten years, by-product treatment, in any form, was not included as a prerequisite.

Discharge costs due to the high organic loads in the untreated abattoir waste water are relatively high. Abattoirs normally also have difficulty in meeting municipal by- law quality standards for fats, oils, greases and suspended solids. A degree of on-site pre-treatment is thus necessary. However, to minimize waste volumes, water conservation and optimum water housekeeping are essential. In addition to good abattoir housekeeping, abattoir waste management should be progressively implemented commencing with low-cost, low-technology practices and thereafter progressing to more sophisticated technologies.

Therefore, the importance of adopting appropriate abattoir wastewater treatment measures to prevent the chances of contaminating water bodies and ground water is recommended. Determination of specific pathogenic microorganisms in abattoir wastewater and their health impacts is also recommended. It is therefore instructive to treat the abattoir wastes appropriately and properly dispose them to avert environmental disasters. Furthermore, legislation on abattoir operations and management can reduce the negative impacts and enhance sustainability in the industry.

REFERENCES

Adeyemo, O. K., Ayodeji, I. O., & Aiki-Raji, C. O. (2002). The water quality and sanitary conditions in a major abattoir (Bodija) in Ibadan, Nigeria. *African Journal of Biomedical Research*, *5*, 51–55.

Agwa, O. K., Sito, E., & Ogugbue, C. J. (2013). A spatial assessment of the microbiological and physicochemical quality of a stream receiving raw abattoir waste. *Middle-East Journal of Scientific Research*, *14*(7), 879–886.

Benker-Coker, M. O., & Ojior, O. O. (1995). Effect of slaughterhouse wastes on the water quality of Ikpoba River, Nigeria. *Bioresource Technology*, *52*(1), 5–12.

Chukwu, O. (2008). Analysis of groundwater pollution from abattoir waste in Minna, Nigeria. *Research Journal of Dairy Sciences*, *2*(4), 74–77.

Chukwu, O., Adeoye, P. A., & Chidiebere, I. (2011). Abattoir wastes generation, management and the environment: A case of Minna, North Central Nigeria. *International Journal of Biosciences*, *1*(6), 100–109.

Department of Agriculture and Rural Development (DARD). (2009). Guideline manual for the management of abattoirs and other waste of animal origin. Gauteng, Southern Africa: DARD, Gauteng Provincial Government. Retrieved September 8, 2015, from http://www.gdard.gpg.gov.za/Services1/Guideline%20Manual%20for%20the%20Management%20of%20Abatooirs.pdf

Itodo, I. N., Awulu, J. O., & Harrison, R. M. (1999). *Understanding our environment: An introduction to environmental chemistry and pollution* (3rd ed.). London, UK: The Royal Society of Chemistry.

Kosamu, B. M., Mawenda, J., & Mapoma, H. W. T. (2011). Water quality changes due to abattoir effluent: A case on Mchesa Stream in Blantyre, Malawi. *African Journal of Environmental Science and Technology*, *5*(8), 589–594.

McLaughlin, A., & Mineau, P. (1995). The impact of agricultural practices on biodiversity. *Agriculture, Ecosystems & Environment*, *55*(3), 201–212.

Mittal, G. S. (2004). Characterization of effluent wastewater from abattoir for land application. *Food Reviews International*, *20*(3), 229–256.

Mittal, G. S. (2006). Treatment of wastewater from abattoirs before land application: A review. *Bioresource Technology*, *97*(9), 1119–1135.

Muhirwa, D., Nhapi, I., Wali, U., Banadda, N., Kashaigili, J., & Kimwaga, R. (2010). Characterization of wastewater from an abattoir in Rwanda and the impact on downstream water quality. *International Journal of Ecology and Development*, *16*(S10), 30–46.

Quinn, J. M., & McFarlane, P. N. (1989). Effect of slaughterhouse and dairy factory effluent on epilinium. *Water Resources*, *23*, 1267–1273.

Verheignjen, L. A. H. M., Weiersna, D., Hulshoff, L. W., & de Wit, J. (1996). Livestock and the environment. Finding a balance: management of waste for animal processing, Wageningen, the Netherlands: International Agriculture Centre. Retreived September 8, 2015, from http://www.fao.org/ag/againfo/programmes/en/lead/toolbox/Refer/IACwaste.PDF

Weobong, C. A.-A., & Adinyira, E. Y. (2011). Operational impacts of the tamale abattoir on the environment. *Journal of Public Health and Epidemiology*, *3*(9), 386–393.

World Health Organization (WHO). (2011). Guidelines for drinking-water quality (4th ed.). Geneva, Switzerland: WHO. Retrieved September 8, 2015, from http://www.who.int/water_sanitation_health/publications/2011/dwq_guidelines/en/

KEY TERMS AND DEFINITIONS

Abattoir: An abattoir or slaughterhouse is a place where animals are slaughtered for meat.

Dissolved Oxygen: The measure of the degree of pollution by organic matter, the destruction of organic substance as well as self-purification of the water bodies.

PH: Refers to hydrogen ion concentration, and is the indicator of acidity or alkalinity of water. It is a measure of the effective concentration (activity) of hydrogen ions in water.

Pollution: The change in the physical, chemical, radiological or biological quality of the resource (air, land or water) caused by people or due to human activities that is liable to cause hazards to human health, harm to ecological systems, damage to structures or amenity and is injurious to existing, intended or potential uses of the resource.

Water Quality: Describes the physical, chemical, biological and aesthetic properties of water, which determine its fitness for a variety of uses and for protecting the health and integrity of aquatic ecosystems. Water quality is the degree of usefulness of water for a particular purpose. It is a measure of the amount and types of substances contained in water with respect to what it is intended to be used for.

Chapter 5
Livestock and Climate Change:
An Analysis of Media Coverage in the Sydney Morning Herald

Xavier Mayes
Curtin University, Australia

ABSTRACT

A global shift away from diets dominated by meat, dairy and eggs to mainly plant-based diets is as necessary in mitigating anthropogenic climate change as the shift away from fossil fuels. Yet a large awareness gap exists about animal agriculture's contribution to greenhouse gas emissions. Recent studies in Australia and the United States show this issue is represented in less than 1 percent of all newspaper articles about climate change. This chapter examines the opportunities and barriers in addressing the livestock sector's impact on climate change. Policy recommendations in the literature are compared with the responses of governments, industry and the NGO sector. Australia's unique socioeconomic and cultural ties to livestock production and the consumption of animal products represent a significant barrier to demand-side mitigation. An analysis of newspaper articles mentioning animal agriculture's link to climate change in The Sydney Morning Herald between 2006 and 2014 provides insights into the facilitation and shaping of public awareness on the issue to date. The findings can inform strategies to increase future media coverage and encourage a more engaged discourse on demand-side mitigation.

INTRODUCTION

Humans are "eating away at our own life support systems" at a rate unseen in the past 10,000 years (Milman, 2015). Population growth, rising incomes and continuing urbanisation are changing the nature of our food and agricultural systems (Hoffman, 2013). The human population on this planet will continue growing to an estimated 9.6 billion in 2050 (UN, 2013). World meat production has doubled in the past three decades. In that same time it has tripled in the global South, pointing to the rapid changes in the diets of populations in developing countries (Cudworth, 2011, p. 106). Demand for meat and milk in particular is projected to increase by 73 percent and 58 percent respectively by 2050 (Gerber et al., 2013). Based on recent estimates, meeting the demand will require increases of between 60 and 100 percent in

DOI: 10.4018/978-1-4666-9553-5.ch005

cropland and pasture-based food production (Alexandratos & Bruinsma, 2012; Porcher, 2006), placing further strain on land, water resources, biodiversity and native forests (Steinfeld et al., 2006).

The world-wide production of meat and other animal-based products is a significant driver of climate change (Steinfeld et al., 2006). Current estimates of global livestock emissions along the supply chain range from 14.5 to 51 percent (Gerber et al., 2013; Goodland & Anhang, 2009). In 2010, an international agreement was reached stating greenhouse gas emissions need to be reduced so that average global temperatures increase no more than two degrees Celsius above pre-industrial times (UNFCCC, 2014). To reach this goal the growth in carbon dioxide emissions, the most prevalent of anthropogenic greenhouse gases, must be quickly contained. Yet immediate and drastic measures to decarbonise world economies will be prohibitively costly (Stehfest et al., 2009) and any response within the climate system will not occur for decades unless the more powerful and short-lived greenhouse gas methane is given equal consideration (Shindell, 2013). Achieving the two-degree target will be impossible if world economies only focus on energy and transportation. Serious measures to address agricultural emissions, specifically from livestock, will also be essential (Hedenus, Wirsenius & Johansson, 2014). A growing body of evidence suggests a global shift away from diets dominated by meat, dairy and eggs to mainly plant-based diets is as necessary in mitigating anthropogenic climate change as the shift away from fossil fuels (Bailey, Froggatt & Wellesley, 2014; Popp, Lotze-Campen & Bodirsky, 2010; Stehfest et al., 2009). Turning around the rising global consumption of animal products through significant changes to dietary practices will have profoundly positive consequences for the planet's climate, biodiversity and natural resources (Raphaely & Marinova, 2014b). Human health and equity, as well as the welfare of farmed animals, will also dramatically improve.

Despite these benefits, efforts by governments, industry and advocacy groups in developed nations to address the impact of animal agriculture on climate change have been limited, particularly in encouraging less consumption of animal-derived food and other products. There is a wide awareness gap concerning animal agriculture's significant contribution to climate change. A global survey found that twice as many people think transport is the bigger contributor (Bailey et al., 2014). Closing this awareness gap is a crucial prerequisite to behavioural change. As a wealthy, developed country with a high level of ingrained meat consumption and a powerful livestock industry, Australia faces significant barriers to the wide acceptance of effective solutions. Despite the news media's fundamental role in both public communications and agenda-setting (Schmidt, Ivanova & Schäfer, 2013), reportage on this issue in the country's major newspapers has been disproportionately low in recent years when compared with other climate change topics (Friedlander, Riedy & Bonfiglioli, 2014).

BACKGROUND

Below is a short discussion of the major impacts of livestock production. This sector is a key driver of climate change.

Major Impacts

Livestock production is one of the most destructive human activities on the planet. It directly and indirectly uses 30 percent of the entire ice-free terrestrial surface of the planet and 70 percent of all arable land (Gerber et al., 2013). One fifth of pastures and rangelands have been at least partly degraded by soil erosion, compaction and overgrazing because of livestock activity (Steinfeld et al., 2006). The shift to more intensive crop production has resulted in less soil fertility, more soil salinity and toxicity, waterlogging and an increase in pests (Sreinfeld et al., 2006). Phosphorous, an essential nutrient for plant growth once found naturally in organic soils, is now mined from phosphate rock, a finite resource that is likely to reach peak production this century due to intensive agriculture (Cordell, Drangert & White, 2009). Fertilisers used on feedcrops and on-farm animal waste are the main sources of agricultural pollutants in waterways that cause widespread anoxia, or dead zones, in coastal marine areas (Cordell et al., p135-138). The primary cause of species loss on the planet is due to overgrazing and habitat loss for livestock production on land, and overfishing the oceans (Clucas, 1997; Margulis, 2004). Globalisation and an increasing reliance on technological advances has led to less subsistence farming and more industrial and corporatized livestock practices (Idel, Fehlenberg & Reichert, 2013). Competition for scarce land, water and energy resources due to more intensive livestock practices in developing countries is a direct threat to global food security (Erb, Mayer, Kastner, Sallet & Haberl, 2012). The livestock sector worldwide uses up to seven times the amount of grain and other crops than for human use due to the inefficient conversion of plant protein to meat protein (Pimentel & Pimentel, 1997). One kilogram of beef requires up to 43 times the amount of water used to produce the same weight of cereal grain (Pimental et al., 2004). The consumption of animal products, particularly red meat, is linked to cardiovascular disease, Type II diabetes, various forms of cancer, hypertension and obesity (Bazzano et al., 2002; Friel et al., 2009; Friel, 2010; McMichael, Powles, Butler & Uauy, 2007; Tilman & Clark, 2014). The requirements of a more intensive and mechanised livestock sector is also responsible for significant effects on the mental health of abattoir workers that can lead to an increased rate of violence and abuse against their families, communities and especially the animals sent to slaughter (Porcher, 2006, p. 65; Richards, Signal & Taylor, 2013, p. 400). The wellbeing and treatment of livestock animals before their slaughter has been called into question, with claims they are routinely and legally subjected to physical, mental and emotional abuse while being denied the ability to fulfil many of their natural behavioural needs (HSUS, 2009).

A Key Driver of Climate Change

Livestock production is both directly and indirectly a significant source of anthropogenic greenhouse gas (GHG) emissions. It is the largest anthropogenic emitter of methane (CH_4), comprising 44 percent of all sources. This is primarily due to enteric fermentation, the methane-producing digestive process in ruminant animals such as cows, sheep and goats (Gerber et al., 2013, p. 15). Land-use changes, particularly deforestation, for the expansion of pasture and arable land to grow feed crops make up the main share of livestock's carbon dioxide (CO_2) emissions. The production of fertilized feed crops for livestock and the resulting manure is also responsible for 53 percent of all anthropogenic nitrous oxide (N_2O) emissions, a very potent and long-lived GHG (Gerber et al., 2013, p. 15). Beef and dairy are the most emissions intensive livestock products and account for 65 percent of the sector's emissions (Gerber et al., 2013, p. 15). Pig meat, chicken meat and eggs are much less intensive, comprising just under 10 percent of the sector's emissions, however rising global demand for these products mean they will have an increasing burden on the climate (Gerber et al., 2013, p. 16).

Various lifecycle estimates put global livestock emissions at 14.5 percent to 51 percent of all anthropogenic GHGs. The UN Food and Agriculture Organisation's (FAO) 2006 report *Livestock's Long Shadow* was the first look at global supply chain emissions for the sector, and attributed to it 18 percent of total GHGs (Steinfeld et al., 2006). Gerber et al. (2013) FAO report on the sectors emissions revised that overall figure to 14.5 percent. Many consider these figures overly conservative including Goodland and Anhang (2009), two former World Bank environmental assessors who, in a 2009 *World Watch* report, argued the sector's emissions are closer to 51 percent of total GHGs. Two of their reasons for such a significant attribution are important to note. Firstly, they argue CO_2 emissions from livestock respiration and soil oxidation (heavy animals compacting the soil) are no longer part of a natural closed loop in the carbon cycle. Steinfeld et al. (2006, p. 95) chose not to consider respiration as a net source of CO_2 because livestock animals consume plant matter that drew an equal amount of CO_2 from the atmosphere via photosynthesis. Goodland and Anhang (2009, pp. 11-12) assert that the natural carbon cycle is now out of balance to the tune of up two billion tons of CO_2 per year because of the sheer number of livestock animals reared by humans and the planet's decline in photosynthetic ability due to deforestation and degradation of other carbon sinks. Their second reason is concerned with how methane's effect on the climate is calculated relative to carbon dioxide. The practice of placing various GHGs on the same footing as CO_2, called 'carbon dioxide equivalent' (CO_2e), is done by calculating the amount of CO_2 that would have the same global warming ability as a unit of another GHG over a given time-frame (UNEP, 2014). The Intergovernmental Panel on Climate Change (IPCC) and the rules governing the Kyoto Protocol calculate the impact of CO_2e gases, such as methane, over 100 years (Shindell et al., 2009). Authors of the IPCC's latest climate change assessment admit that "there is no scientific argument for selecting 100 years compared with other choices" and that the "choice of time horizon has a strong effect on the GWP values" of a chosen CO_2e gas (Myhre et al., 2013, p. 711). Over 100 years, methane's global warming potential (GWP) is estimated to be as much as 34 times that of CO_2 (IPCC, 2013). However, as methane is a more short-lived GHG its GWP over this arbitrary time scale is muted (Shindell et al., 2009). A time period of 20 years which more closely reflects methane's lifespan in the atmosphere results in a GWP ratio of 86 times that of CO_2 (IPCC, 2013). Goodland and Anhang (2009) therefore argue that methane's GWP should be measured on a 20-year time scale to reflect its dominant role in driving short-term climate change within the critical period of the next few decades.

Global leaders agreed to limit atmospheric warming to two degrees Celsius relative to the global mean pre-industrial temperature at the 2010 Cancun Climate Conference, also known as the 16th UN Conference of the Parties (COP16). The two-degrees limit is understood to allow a one-in-two chance that current global warming won't set off so called 'tipping points', leading to dangerous and irreversible changes to the climate (IPCC, 2014). Examples of known tipping points include abrupt changes to global ocean currents that regulate heat and weather patterns, or the irreversible melting of the Greenland ice sheet that would guarantee several metres of global sea level rise (Cleugh, Smith, Battaglia & Graham, 2011). However, analysis of emissions reduction pledges by countries at COP16 shows that they would fail the two-degrees limit and instead lead to over three degrees of warming (Chen et al., 2011). Despite the number of climate mitigation policies around the world growing since the Cancun conference, anthropogenic GHGs continue to increase at an alarming rate (IPCC, 2013). Measures to stringently mitigate CO_2 emissions are of high importance over the long-term due to their primary ability to affect radiative forcing over a very long lifespan in the atmosphere (Rogelj et al., 2014). It is widely accepted that a certain degree of further rises in temperature and other observed effects on the climate are already locked in due to historical emissions (UNEP, 2014) and that a dramatic cut in current global GHG levels is required to avoid more than two degrees of warming. To what extent that cut should be and how it should be brought about is still the subject of both scientific and political debate. Recent analysis published in the journal *Nature* suggests that 82 percent of coal, half of gas and a third of all oil reserves must remain unused if global temperature increases are to be kept within this two-degrees limit

(Jakob & Hilaire, 2015). The United Nations Environment Programme (UNEP, 2014, p. pxv) advises the world must eliminate anthropogenic carbon dioxide emissions by 2070 and all other greenhouse gases, including methane and nitrous oxide, entirely by the end of the century. Yet other research concludes that phasing out potent but short-lived GHGs as an equal priority to CO_2 will likely result in a quicker response in the climate system and achieve significant short-term relief while the world identifies ways to cooperatively implement more difficult structural changes to realise major reductions in CO_2 (Goodland, 2010; Shindell, 2013; Smith, Desaia, Rogersa & Houghton, 2013). This is especially important as even a dramatic decline in CO_2 emissions is not expected to provide much relief from climate impacts until mid century at the earliest due to the long atmospheric life of historic emissions (Shoemaker, Schrag, Molina & Ramanathan, 2013).

MITIGATING LIVESTOCK EMISSIONS

There are both demand-side and supply-side solutions for mitigating livestock emissions. They are discussed below.

Demand-Side Solutions

If agriculture and land-use change emissions continue in a business-as-usual fashion they alone will almost use up the global emissions budget for two degrees of warming by 2050 (Bajželj et al., 2014). Even an ambitious scenario, requiring the roll-out of best-practice efficiency all over the world and the halving of all food waste, uses up over half of the planet's emissions budget – leaving little space for energy production and all other emissions-intensive processes (Bajželj et al., 2014). One of the most effective ways to cut methane emissions is to reduce demand for global livestock production, particularly ruminant animal populations such as cattle and sheep responsible for the majority of livestock emissions (Ripple et al., 2014). Demand-side solutions involving dietary change are therefore crucial in reducing dangerous climate change (Raphaely & Marinova, 2014b). Cutting European meat and dairy consumption in half would lead to as much as a 45 percent reduction in EU greenhouse emissions (Westhoek et al., 2014). A vegan diet leads to even more impressive results. It has been estimated that a vegan diet is up to seven times less emissions-intensive than a locally-grown omnivorous Western diet (Weber & Matthews, 2008). Heller and Keoleian (2014) found that switching the average meat-heavy U.S. diet to one that didn't contain animal products would lead to the country's biggest reductions in diet-related greenhouse emissions. Eshel and Martin (2006) tested the environmental impact of five diets and found that a vegan diet has the least negative effect on global warming – so much so that changing from an average meat-eating American diet to a vegan one was more effective than trading in an average American car for an electric hybrid Toyota Prius. Oxford University research that analysed food consumption data from around 55,000 UK residents was also consistent with the findings above (Scarborough et al., 2014).

A large-scale dietary shift to veganism in Western countries is an optimal solution but it requires fundamental behavioural changes that will be difficult for most people in countries with high levels of per capita meat consumption, such as Australia. Raphaely, Marinova, Crisp and Panayotov (2013) suggest 'flexitarianism', or part-time vegetarianism, as an alternative approach to dietary change that "allows for almost instantly beneficial yet gradual, incremental and progressive advances" in the replacement of meat with plant-based foods (Raphaely et al., 2013, p. 46). The key advantage in all demand-driven approaches is that they don't require expensive, unproven or time-consuming technological fixes. Yet such a significant shift in the production and consumption of animal products would require a comprehensive suite of policies and behaviour-change programs. Government interventions to reduce meat consumption can be implemented at various levels, ranging from consumer information campaigns to an increasing depth of regulatory intervention into market processes such as taxes or prohibitions (Cordts, Nitzko & Spiller, 2014, p. 88). Livestock businesses could change business models that are currently geared to supply overconsumption, be involved in consumer education, influence purchasing decisions, and change the retail environment in order to address emissions (Cameron, 2014, p. 12). An effective yet contentious way of incentivising reduced consumption of animal products is through a tax or emissions trading scheme on livestock production to modify consumer prices (Ripple et al., 2014). Such a demand-side measure will have more environmental and social co-benefits than supply-side measures (Ripple et al., 2014, p. 3). McMichael et al. (2007) recommend an international contraction and convergence strategy to keep within the two-degrees limit, whereby developed nations reduce their average meat intake to allow developing nations to modestly increase their intake. While global average meat consumption is around 100g per person per day, there is a ten-fold variation between high-consumption countries, such as Australia and the U.S., and low-consuming populations (McMichael et al., 2007, p. 1). McMichael et al. (2007) propose a more evenly shared global average of 90g per day as a working target, adding that no more than half that target per day come from red meat such as beef and lamb (McMichael et al., 2007, p. 8). Goodland (2010, 2014) asserts that while international negotiations on developing renewable technologies and energy efficiency measures continue, better alternatives to animal products can be scaled up and have a significantly fast and positive effect on the climate. He recommends replacing 25 percent of livestock production with plant-based alternatives to reduce emissions 13 percent by 2017, a target in line with recent international climate treaty negotiations (Goodland, 2014, p. 3). This would occur through joint action by consumers, government, industry and investors. Reducing food waste by addressing overconsumption is another complimentary measure to dietary change that would slow climate change and have positive effects on food security and equity (Bellarby et al., 2013).

Supply-Side Solutions

While policy-driven dietary changes are necessary they are also contentious. Authors of a recent study published in the journal *Climatic Change* say they are pessimistic such policies will be implemented in time, speculating they would "almost certainly emerge only after productivity improvement and technical measures largely have been exhausted". Havlik et al. (2013) argue that mitigation policies should target supply and not demand to minimise economic and social costs. Gerber et al. (2013) assert that wider use of existing best practices and technologies could cut global livestock GHG emissions by as much as 30 percent. Other research points to all known technological options only achieving at best a 20 percent reduction in GHG emissions if adopted (Weidema et al., 2008; Wirsenius & Hedenus, 2010). However, given Australia's proud record as one the most advanced agricultural sectors in the world in terms of productivity and technology (Cornwall, Collie & Ashton, 2000), Australian livestock operations have arguably already adopted many best practice opportunities yet to be taken up in other parts of the world.

In Australia to date, government responses to the impact of livestock on climate change have been minimal and narrowly focused on productivity and technology development. Before the repeal of Australia's carbon pricing legislation by the current Federal Government, consideration was given to the agriculture sector's participation in the planned emissions trading scheme starting in 2015. This would have placed a price on GHG emissions to factor in costs currently externalised by the market (Garnaut, 2008). The scheme would have seen the livestock industry liable for direct emissions from enteric fermentation and manure management. They also would have been affected by indirect costs that would apply to fossil fuel-intensive farm inputs like feed crops, electricity and petrol. Economic modelling produced to help explain the scale of the industry-wide impacts estimated that a $25 price per ton of CO_2e would result in a fall in beef output of four percent, increasing average beef farm operating costs by around 17 percent, which would cause up to one in seven beef farms to shut down – especially marginal small operations (Jiang, Hanslow & Pearce, 2009, pp. 41-42). The report's authors suggested agricultural operations could best avoid the impact of a carbon price by shifting from livestock farming to other agricultural goods with lower emissions, such as grain or horticulture crops, as well as investing in reforestation on the farm's property to offset operating emissions (Jiang et al., 2009, p. 43). However there are no current plans to implement a price signal under the current Australian government.

A number of research programs and trials have been designed and implemented with the agricultural industry that focuses on supply-side approaches to mitigation. The Carbon Farming Initiative, a program under the current Federal Government's emissions reduction policy, provides financial incentives for farmers and land managers to reduce GHGs through activities such as soil carbon sequestration, reforestation and fertiliser efficiency (DOE, 2014). Between 2008 and 2012 the Australian Government funded the Climate Change Research Program, a series of 50 research projects and on-farm demonstrations that focused on emissions reduction, climate adaptation and soil management in the agricultural sector. Selective breeding research under the program found that the progeny of certain cattle and sheep sires in a given herd produced between 11 and 24 percent less methane than the progeny of other sires (DOA, 2013, p. 6). Laboratory and field research found that a range of tropical legumes, unique forage plants such as chicory and turnip, and supplements such as by-products of winemaking might reduce methane emissions through modifications to livestock feed and forage. A demonstration project to capture methane emissions from covered manure ponds in intensive livestock operations was successful and could be burned for electricity generation (DOA, 2013, p. 7). Industry body Meat and Livestock Australia (MLA) has been handed the coordination of further research into methane reduction by the Australian Government since the Climate Change Research Program was de-funded in 2012. The National Livestock Methane Program continues to explore technological fixes such as rumen modification, feed supplements, forage changes and genetics (MLA, 2014c).

Many expert recommendations for supply-side responses are contradictory and would exacerbate other environmental, social and animal welfare impacts. Steinfeld et al. (2006, p. 236) recommend greater production efficiencies should be achieved by pursuing intensive livestock operations. Animals grown on grain in such intensive operations reach slaughter weight quicker when compared with extensive grazing and therefore emit less methane emissions from enteric fermentation, however there are numerous issues with this approach. Intensive livestock production results in more GHG emissions from waste management and crop production. Land previously occupied by extensive grazing operations could be used for reforestation, but would most likely need to accommodate the rapid increase in feed crop production. In any case, projected global demand for meat and dairy means any efficiencies gained through this measure would be negated by increased stock density. Some experts argue continued extensive livestock production is feasible if practices return to the regional feed base and shift from chemical nitrogen fertilisers back to the use of animal waste as a natural fertiliser (Idel et al., 2013, p. 139). In Europe, Bellarby et al. (2013) recommend the prioritisation of beef and dairy farming on grassland to limit emissions. While some have controversially pointed to optimised farming methods such as intensive rotational cattle grazing as an opportunity to not only use dedicated livestock land as a carbon sink but reverse desertification (Monbiot, 2014), peer-reviewed evidence points to all methods of extensive grazing as ecologically destructive (Briske, Ash, Dernerc & Huntsinger, 2014; Carter, Jones, O'Brien, Ratner & Wuerthner, 2014). It should be noted that the livestock industry's pursuit of greater productivity through technological measures often leads to further systemic and detrimental animal abuse (Moss, 2015). The urgent need to develop responses to livestock emissions should be locally appropriate and approached in a comprehensively holistic manner – that is they should not focus on economic development to the detriment of environmental sustainability and animal wellbeing (Thornton & Herrero, 2010).

BARRIERS

A recent survey in nine countries found that, though public awareness of anthropogenic climate change is very high, recognition of the livestock sector as a significant source of GHGs was the lowest compared to all emissions sectors (Bailey et al., 2014). This awareness gap contributes to indifference; only 9 percent of survey respondents would consider reducing their meat consumption to limit climate change. The socio-economic, political and cultural entrenchment of livestock production in Australia further compounding this indifference is illustrative of the considerable barriers facing dietary change in other developed countries.

Socio-Economic

Since European settlement in the late eighteenth century, livestock production has dominated farming in Australia (Henzell, 2007). From the time when cattle and sheep were brought on the long journey from England on the First Fleet with false hopes of a lush and fertile continent, the livestock industry has adapted to Australia's harsh, often unforgiving climate (Cornwall et al., 2000, p. 5). Beyond the seaboard, Australia is one of the driest continents on the planet, forcing farmers to pioneer dryland farming (Cornwall et al., 2000, p. 5). Animal agriculture has been by far the major cause of historic land clearing in Australia. From the 1870s through to the late twentieth century the tropical far north savannahs of the country were cleared for extensive cattle grazing (Cornwall et al., 2000, p. 18). Broad acre cattle and sheep farms still account for between 50 and 64 percent of Australia's landmass (Target 100, 2014a; Longmire, Taylor & Wedderburn-Bisshop, 2014). While the economic dominance of livestock production at a national level has waned, the industry remains a significant part of the country's economy, particularly in the export market. Australia's beef industry is a major employer in rural areas and still the nation's largest agricultural enterprise, comprising half of all farms with agricultural activity (NFF, 2015). Although Australia produces just 3.9 percent of beef world-wide, its export market accounts for 60 percent of all cows slaughtered or exported live to overseas markets, making Australia the third largest beef exporter in the world behind India and Brazil (MLA, 2014b).

Political

The ingrained consumption of meat, dairy and eggs in most western countries like Australia and the United States is in part due to the highly politicised nature of livestock production. A decreasing number of vertically integrated corporations hold political and economic power across most of the supply chain (Cudworth, 2011, pp. 109-119). A case in point is the involvement of the livestock industry in the development of national dietary guidelines. In the U.S., dietary guidelines are set by the United States Department of Agriculture (USDA), a body also established to promote the interests of the agricultural sector (Raphaely et al., 2013, p. 43). If Americans altered their current eating habits to match the USDA dietary guidelines without reducing the overall amount of calories they currently consume, their diet-related GHGs would actually increase (Heller & Keoleian, 2014). This is because the guidelines recommend an increase in dairy consumption. Interestingly, the USDA guidelines don't even explicitly state that a reduction in meat consumption is necessary (Erickson, 2014). The conflict of interest in the U.S. is less obvious in Australia, although Meat and Livestock Australia provides funding for government research organisations (Russell, 2009) and Australia's dietary guidelines are very similar to those drawn up by the USDA (Raphaely et al., 2013).

Cultural

With a national past so dominated by farming and a politically powerful livestock industry, it's no wonder that Australians have always been enthusiastic meat-eaters. Meat consumption per person in Australia is only second to the United States (FAOSTAT, 2014). An average Australian eats 304g of meat per day, of which at least 126g is from beef and lamb, according to Australian Bureau of Statistics (ABS) figures from 1998-1999. Australian dietary guidelines advise limiting the consumption of red meat to 455g per week, an average of 65g per day (NHMRC, 2013). An Australian study of 2011 data found that 40 percent of food expenditure is typically spent on meat; 111kg consumed per capita a year (Wong, Selvanathan & Selvanathan, 2013). While overall meat consumption has increased in the past decade, the types of meat being consumed are changing. Australians are eating less beef and lamb but consuming more chicken and pork (Marshall, 2014), reflecting a similar trend around the world. Traditionally seen as the centrepiece of every Australian dinner, some argue that this flip in demand for red meat in Australia today is due in part to an increasingly ethnic population (Ting, 2013). Global diets are changing and becoming more homogenous as average incomes rise and populations continue to become increasingly urbanised (Cameron, 2014).

Eating meat, dairy and eggs in Australia and other Western countries remains a culturally entrenched and routine dietary practice (Taylor, 2012). That livestock production has not been a focus in public campaigns to mitigate climate change is not an oversight but an indication that meat, dairy and eggs are 'natural' forms of consumption embedded in Western, middle-class socio-cultural practices (Heinz & Leeb, 1998). To alter individual practices of consuming animal products therefore requires an understanding of the symbolic nature of food consumption (Doyle, 2011). Discourse – in this context, to frame and dominate the conversation in order to create and control the knowledge and power about an issue – and language are used daily to reinforce and normalise the consumption of animal products. The invisible process of transforming livestock animals from individuals into consumable objects occurs both physically (outside our towns and cities) and culturally through language (Adams, 2013, p. 97). The fetishism of meat, milk, cheese and eggs as cuisine commodities, through the removal of the circumstances of their production, disposes their original meaning. The void is then filled with culturally positive meanings in the marketplace (Heinz & Leeb, 1998). The negation of any agency and individual identity allows us to see animals as "walking larders", further supporting another ingrained cultural belief that humans have dominion over animals and nature (Taylor, 2012). Recent research reinforces the notion that meat and other animal products are culturally entrenched through discourse. Taste, unwillingness to alter personal or family habits, the belief that humans are 'meant' to eat meat, and health concerns were the main barriers listed by respondents in a 2003 study when asked if they would consider making the switch to a vegetarian or vegan diet (Lea & Worsley, 2003). A follow-up study in 2006 that instead asked respondents if they would consider a plant-based diet that included small amounts of meat and other animal products elicited much less resistance (Lea, Crawford & Worsley, 2006). Western discourse normalises the consumption of meat by portraying alternative diets like veganism as deviant behaviour (Nath, 2010). A study assessing the portrayal of veganism in the British media in 2007 found that only 5.5 percent of 400 relevant articles were positive. A majority of the articles, 74.3 percent, were negative and broadly characterised veganism as a fad, impossible to sustain, hostile or similar to asceticism (Cole & Morgan, 2011). The trend towards health as a motivating factor in reduced meat consumption does not necessarily mean more of the population share the same underlying values of the minority who have adopted a vegetarian or vegan diet. Baker, Thompson and Palmer-Barnes (2002) found that people who have reduced their intake of meat products still identify meat with desires of pleasure, happiness and freedom. Producers within the livestock industry affected by these changes in meat consumption would not need to challenge underlying values, just shift the beliefs of these consumers about the healthiness of consuming meat (Baker et al., 2002, pp. 28-29).

When meat consumption began to drop in the 1970s and livestock producers could no longer rely on the Australian working man to "enjoy his meat three times a day – for breakfast lunch and tea", the industry responded with 'lean' cuts and advertising campaigns proclaiming "Feed the Man Meat" in an effort to retain that essential demographic (Cornwall et al., 2000, pp. 11-13). Advertising over the decades has consistently established meat as essential to a meal, to the point where editorials about meatless meals must first overcome why they are defying the norm (Nestle, 2002, pp. 15-16). Just as the hot-dog is an American ritual, a summer sausage barbeque or roast lamb on Australia Day are reinforced as rituals through ever more sophisticated retail and industry advertising (Nestle, 2002, pp. 20-21). Meat & Livestock Australia (MLA), a prominent industry body engaged in supporting cow and sheep producers, utilise marketing and public relations through the media to shape the perception of livestock products such as beef and lamb, and the industry as a whole. Through sponsorship and participation in major food events in capital cities deliberately frame Australian beef and lamb production as "sustainable" and keep alive the iconic image of the honest, hard-working farmer (MLA, 2014a). Increasing environmental concerns in the production of food have led to MLA establishing a new campaign using case studies of farmers to demonstrate the industry's sustainability credentials (Target 100, 2014a). Some have argued the campaign has relied on outdated or incorrect facts (Mahony, 2014). Recently, the campaign has been used by the industry to challenge a popular documentary examining the environmental impacts of global animal agriculture and reinforce industry messages about its supply-side measures to limit emissions (Target 100, 2014b).

ANIMAL AGRICULTURE AND CLIMATE CHANGE IN THE MEDIA

The news media are vital to a free, democratic society. As arbiters of knowledge they wield power through their choices of framing and sources, so much so that the issue priorities of the news become the issue priorities of the public (Bennett, 1996; Friedlander et al., 2014). News media are crucial actors in societal uptake of wicked problems such as climate change through their role as the "interpretative systems" of modern society (Peters & Heinrichs, 2004). News media, such as *The Sydney Morning Herald (SMH)*, are central to raising awareness and disseminating knowledge via public communication due to their mass appeal and high circulation (Schmidt et al., 2013). They also act as a central and public forum for debates on the legitimacy of policy and governance, allowing various societal actors such as experts, environment groups, industry leaders, and party and government officials to participate in the formation of political opinion (Schneider, Nullmeier & Hurrelmann, 2007, p. 136; Steffek, 2009, p. 315). It is this latter function that is most important in this context, as policymakers consider news media to be a reliable indicator in gauging the public importance of a topic that they may need to respond to (van Aelst & Walgrave, 2011, p. 298). In a developed country with a high level of meat consumption, such as Australia, media coverage of the link between diet and climate change is essential in raising public awareness and can support or obstruct structural changes and wider behavioural change.

With a growing awareness in the scientific community of the importance of livestock emissions, several recent studies have examined the issue's coverage in news media. Friedlander et al. (2014) recently studied the first seven news pages of eight Australian metropolitan newspapers between 1 July 2008 and 30 June 2013 for mentions of livestock and climate change. They found that just 1 percent of articles about climate change were related to meat or livestock (Friedlander et al., 2014, p. 33). This finding matched a similar study of climate change-related articles in 16 newspapers in the United States which found only 0.5 percent mentioned animal agriculture (Neff, Chan & Smith, 2008). Kiesel (2010) found that *The New York Times* and the UK's *The Guardian* published just three and six articles respectively on the topic of livestock emissions in a two-year time period following the release of the FAO's *Livestock's Long Shadow*. Almiron and Zoppeddu (2014) analysed the top five newspapers in both Spain and Italy between November 2006 and September 2013 and found that 1.5 percent (Spain) and 3.6 percent (Italy) of climate change-related media coverage mentioned the contributions of animal agriculture emissions. Friedlander et al. (2014) also analysed the phenomenon of media 'framing' – what details are focused on or omitted from the reporting of an issue or news event – and found that animal welfare and the economy were the most dominant frames in Australian reportage of meat in general.

Methodology

A content analysis of reportage in the *SMH* from 2006-2014 was employed to gain a deeper understanding of how public awareness in Australia on the issue of livestock's contribution to climate change is being shaped in the media. The analysis examined the proportion of content relevant to animal agriculture's impact on climate change, how articles were framed, what sources journalists relied on, and considered what internal and external factors may have influenced the reportage. The *SMH* was chosen because it is one of Australia's most prominent and well-read newspapers. It is also the most established, as the oldest continually published newspaper in the country. As well as national and international coverage, *SMH*'s news focus is on Sydney and to a lesser extend regional New South Wales (NSW). Meat and dairy industries in NSW are a significant part of the country's agricultural and economic activity – cattle and sheep in NSW account for one-fifth and one-third of the national herd respectively (Government of NSW, 2008) – and are well represented in news coverage. The newspaper has also run environmental campaigns, partnering with World Wildlife Fund for Nature Australia (WWF) to organise the launch of Earth Hour in 2007 (Salter, 2012). All of these aspects suggest the SMH would have appropriate reasons to report on various angles of the livestock emissions issue over the specified time period.

Articles published in the SMH between 1 January, 2006 and 31 December, 2014 were collected via Factiva using the search '(climate change OR global warming OR greenhouse gas OR emissions)

AND (livestock OR animal agriculture OR meat OR dairy)'. Eggs were not included in the search as it was determined that any relevant articles would be captured by the other search terms. The FAO's *Livestock's Long Shadow* published in 2006 was the first major report to widely communicate the many interconnected environmental impacts of livestock production and therefore served as a suitable base year. Irrelevant sections of the newspaper such as personal announcements, letters to the editor, reviews, weather news, arts and entertainment, sports and recreation stories, and market reports were filtered. A secondary manual filter was then applied to the total search result of 339 articles to discount irrelevant news stories, for example where the 'climate change' and 'livestock' keywords in an article were not in any way contextually linked.

Quantitative methods were used to analyse all articles over the eight-year period. Articles were coded to identify between one and three main themes or narratives. They were

1. 'Cause', an explanation of livestock emissions;
2. 'Solutions', discussion of mitigation opportunities; and
3. 'Consequences', discussion of adaptation to climate impacts by the sector.

If mitigation solutions (2) were mentioned, a secondary code was employed to record whether it was related to 2.1) 'Supply-side', discussion of industry measures; or 2.2) 'Demand-side', a discussion of consumer-led measures. Qualitative analysis was then used to examine the discourse and framing of a proportion of articles that substantively featured the research topic. 'Substantive' articles were determined as having at least two paragraphs that included an explanation of livestock emissions (1) and some discussion of mitigation (2). Articles that only referred to the consequences of climate change for the livestock sector (3), for example adaptation measures for farmers, were only analysed quantitatively. A lack of multiple coders poses a limitation to this research.

While the content analysis was approached in an objective and systematic manner in line with the core principles of the discipline (Lombard, Synder-Duch & Bracken, 2002), testing inter-coder reliability with one or more other coders would give both the methodology and findings a higher-level of validity (Neuendorf, 2002).

Findings

A search of the designated terms for climate change and livestock found 149 relevant articles published in the *SMH* between 1 January, 2006 and 31 December, 2014. Of those relevant articles, 19 or 12.8 percent were coded as 'Causes'; that is, they contained content explaining one or more causes of livestock emissions, but contained no content relating to mitigation. Articles coded as 'Solutions' totalled 86, or 57.7 percent; they mentioned direct or indirect mitigation opportunities for livestock emissions. Articles coded as 'Consequences' totalled 44, or 29.5 percent; they were only concerned with likely climate change impacts on the animal agriculture sector and how farmers could adapt.

As shown in Figure 1, stories coded as 'Consequences' dominated coverage in 2006 and were a significant part of the discourse through to 2008. From 2011 until 2014 they had dropped to near zero in the articles studied. Stories that contained content about the causes of animal agriculture emissions (but nothing else) were a minor proportion of all articles throughout the years studied. Content that mentioned solutions to mitigate livestock emissions spiked in 2007, remaining prominent but gradually declining

Figure 1. Coded articles by year
Sydney Morning Herald 2006-2014.

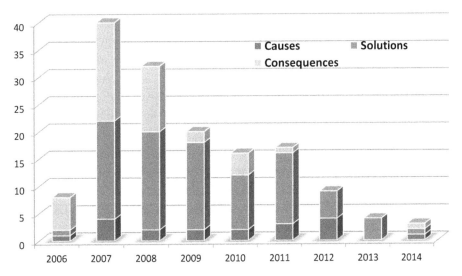

in frequency per year. While articles in 2006 were among the lowest of all the years before a big spike in 2007, a worrying decline in reportage can be seen from 2012 to 2014, with the latter recording the lowest of all years – just three articles in twelve months.

Of the total 149 articles studied, 34 or 22.8 percent contained substantial content regarding animal agriculture emission contributions to climate change. 61 articles, or 57.7 percent of the total analysed, were coded as mentioning the role of demand-side measures, such as diet change, in the mitigation of livestock emissions. However, only 26 (42.6 percent) of these demand-side articles were deemed to be substantially about the topic. Coverage of climate change and animal agriculture was most frequent in 2008 and 2009 – due in part to the *Garnaut Climate Change Review*, a wide-ranging economic analysis commissioned by the Australian Government. Friedlander et al. (2014, p. 36) also observed this spike in articles. News stories between 2006-2007 were primarily focused on the worsening drought and high world grain prices (with biofuels a scapegoat). The *SMH*'s food and lifestyle supplement 'Good Living' was responsible for 23.3 percent of all articles coded as solutions-focused. A string of feature stories in 2007-08 on sustainable eating espoused the 100-mile 'locavore' diet, even raising and killing your own animals for meat, but with only fleeting or in-direct mentions of livestock emissions and other impacts on the environment.

Sources and Journalistic Balance

The sources a journalist chooses in reporting a story is a crucial factor in the process of how news media shape the construction of problems for society and the subsequent legitimisation of policy responses. Journalistic balance is fundamental to the perceived objectivity of the news media and is a key determinant of how sources are used in a story. The concept of balance dates back to the Fairness Doctrine, established in the 1940s when public broadcasting licenses were few and considered a public trust. It was a guideline for U.S. public broadcasters to dedicate airtime to controversial issues in a balanced manner. Newspaper journalists and editors started giving equal weight to both sides of an issue in a great debate, a practice that was exploited by tobacco companies in the mid Twentieth Century to obfuscate the link between smoking and cancer, even though the scientific evidence was clearly one-sided (Oreskes & Conway, 2010). The exploitation of journalistic balance can be detrimental to news media's role as a public forum platform. Other studies have found evidence that the media sometimes convey inaccurate information in climate change reportage to balance two sides of a story. Antilla (2005) concluded that the media would sometimes cite climate sceptics to appear journalistically balanced, often giving undue weight to sensational, less scientific approaches and exaggerating the extent to which issues were debated (Bennett, 1996; Goodman & Goodman, 2005).

An average number of 1.8 sources were quoted in the 61 articles containing substantive content about livestock emissions. As seen in Figure 2, a significant majority of sources (41 percent) were experts from various academic disciplines including ecologists, economists, ethicists, nutritionists, animal welfare professors, anthropologists and agricultural researchers. Representatives of six NGOs (Australian Conservation Foundation, Climate Institute, Voiceless, Friends of the Earth (Australia), Australian Vegetarian Society and Australian Marine Conservation Society) were quoted eight times. Though academics and NGOs comprised over half the share of these source groups, not all were commenting on animal agriculture emissions – often they were commenting on wider agriculture or climate change issues. Interestingly, all of the industry sources (Meat and Livestock Australia, Kangaroo Industry Association, Cattle Council of Australia, Australian Chicken Meat Federation) were commenting specifically on the issue of livestock emissions. This is important to note as it indicates that public engagement by academics and NGOs on this issue is inflated.

'Hold the red, pass the white - meat that is' was the article to most substantively examine the issue of livestock emissions and also contained the most sources. Six academics from various disciplines were quoted by journalist Inga Ting to canvas cultural, social, nutritional aspects to meat eating as well as climate impacts (Ting, 2013).

Figure 2. Sources by type, substantive articles
Sydney Morning Herald 2006-2014.

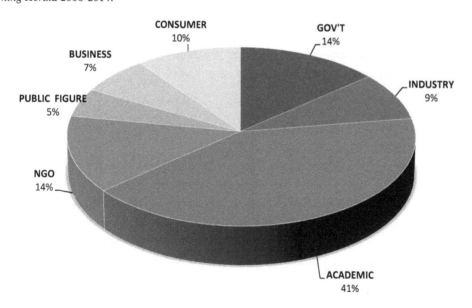

The analysis of substantive articles also reveals that certain journalists play an important role in sustaining coverage on this issue. Journalist Nick Galvin was responsible for five articles that substantively mentioned livestock emissions in the *SMH*'s food and lifestyle supplement 'Good Living'. He used no contrarian sources but relied on a range of experts and people in all levels of the food industry. Galvin is one of many journalists and editors around the world who are what Neff et al. (2008) describe as "part of an issue's idiosyncratic history" (2008, p. 1011). The only clear case where unscientific arguments were used to balance a story was in Miranda Devine's op-ed 'Act hastily, roo the scare tactics' (2008b). In a diatribe about the *Garnaut Review*'s recommendations that 30 percent of livestock be replaced with lower-emitting kangaroos for meat, Devine quoted geologist and renowned climate sceptic Bob Carter to question the merits of taking action on climate change at all.

Framing

An examination of the way the *SMH* framed supply-side and demand-side measures for mitigation gives an insight into systemic social and political barriers to their wider acceptance. While demand-side reductions in livestock emissions such as dietary change were mentioned more times overall than supply-side changes by industry, a significant proportion of the latter were superficial inclusions in lists related to the frame of sustainable living. For example, in a 60-page feature supplement for the first Earth Hour in 2007, a guide outlining 18 ways to a smaller ecological footprint listed dietary change last, recommending that one serving of meat less a week would suffice (Woodford, 2007). The fact that this guide accompanied a larger piece on many more changes that one could make in every room around the modern home yet neglected to mention diet further reinforced this superficiality. In comparison, supply-side measures were reported in only a third of mitigation-related stories but were more likely to be featured in the front news section, framing them in a much stronger fashion. In 'Gas-friendly cows an emission possible to cut global warming', news of the Department of Primary Industries researching ways to breed more methane-efficient beef cattle was featured on page five of the newspaper with an accompanying photo and cartoon (Macey, 2006). Even in articles that primarily reported on demand measures, journalists frequently balanced their story with an industry source quick to point to a technological solution. In 'Into the meat of the issue', the first in-depth examination of livestock emissions in the articles collected, the head of marketing at Meat and Livestock Australia implied their research into more efficient feed and animal breeds was a sufficient exercise in reducing livestock emissions (Dow, 2007, July 3). As a result, it can be argued that the framing of a smaller number of supply-side stories emphasised the legitimacy of technological solutions as the preferred method of climate change mitigation by governments and industry (Lewis & Boyce, 2009, p. 13).

Several studies examining climate change media coverage found it reflects and influences short-term public concern about the issue (Carvalho & Burgess, 2005; Sampei & Aoyagi-Usui, 2009; Trumbo, 1996). To a larger extent however, media's interest in climate change, both in Australia and internationally, is largely reflected in how much governments take the issue seriously or not. That is, climate change coverage is largely governed by national political agendas, so much so that the lack of conviction of some governments can be seen in the uncertainty in news reporting (Lewis & Boyce, 2009, p. 10). The spike in news articles reporting on the issue of livestock emissions is strongly linked to the Australian Government-commissioned *Garnaut Climate Change Review*. The frequency of articles reporting Garnaut's supply-side policy recommendation of replacing 30 percent of meat livestock with commercial kangaroo production far out-shadowed expert recommendations advocating for demand-side reductions in meat consumption. Of all supply-side articles, 56.3 percent were published as news or op-ed commentary compared to 44.2 percent of all demand-side articles. That the latter were featured most frequently in lifestyle supplements such as 'Good Living' shows that news media such as the *SMH* are more likely to frame dietary change and other demand-side policy recommendations as less newsworthy – a similar finding to that of Neff et al. (2008) in their analysis of U.S. newspapers.

Despite evidence that points to a vegan diet being the easiest and most effective way for individuals to lower their greenhouse gas footprint, veganism and even vegetarianism on the whole suffered from portrayals as too alternative, on the fringe, even radical. In an otherwise comprehensive article examining the various ways consumers can be conscious of their diet's impact on the planet, journalist Nick Galvin keeps things light when suggesting cutting back on meat: "That doesn't mean connecting with your inner vegan or discovering a passion for tie-died clothing…" (Galvin, 2008). In 2008 hard-right columnist Miranda Devine took exception to two pieces of news that called for changes to meat consumption. She derided the chair of the IPCC, Rajendra Pachauri, for saying meat should be taxed. Devine argued that green groups had now joined animal rights groups in being "no-meat killjoys" before quoting an anthropology paper that described how meat-eating was essential for the evolution of the human brain (Devine, 2008a).

Conflict is a type of framing frequently used by media in the production of news. Explicit conflict frames were present in some of the *SMH*'s coverage, for example where farming for food was pitted against biofuel crops, further reinforcing the dominance of food production over new and disruptive climate mitigation approaches. A study of two influential specialist farm media publications in Sweden also found that reporting of agriculture's contribution to national emissions was downplayed in a similar way through the use of conflict framing (Asplund, Hjerpe & Wibeck, 2013). Articles often portrayed other sectors such as transport as equally if not more responsible, for example by presenting mitigation policies as a choice between "the motorist or the farmer", thus legitimising agriculture's emissions (Asplund et al., 2013, p. 202). Conflict frames are more relevant in climate change communications where concerted campaigns by environmental NGOs have undermined the social operating license of resource companies, electricity generators, financial institutions, governments and others involved in supporting the development of fossil fuels. Findings from this content analysis point to a proactive and savvy approach by the livestock industry to a similar threat. Strategic approaches to messaging and policy-making are evident in articles where industry sources are quoted or referenced. Again, in 'Into the meat of the issue', the head of marketing at Meat and Livestock Australia gave examples of farmers planting native grasses and trees to protect soils and waterways, and described the process of farming livestock as relying on "resources which are naturally available, such as rain and grass and sunshine" (Dow, 2007, July 3). The president of the Cattle Council of Australia, approached by columnist Miranda Devine to provide his thoughts on Rajendra Pachauri's support of a meat tax, was happy to "acknowledge we are part of the problem", but argued farmers were environmental stewards of the land producing food for the world (Devine, 2008a). In 'Farmers say $100m policy must go further' the *SMH* reported that the National Farmers Federation wanted a more ambitious global warming policy from the Federal Government, including new initiatives such as an emissions scheme that farmers could participate in (Peatling, 2007). That the carefully crafted image of pasture fields in harmony with the earth does not face up entirely to scrutiny is beside the point. Being on the front foot with messaging and public policy suggestions is yet another example of the industry crafting a 'stewards of the land' image.

Framing a reduction in an individual's meat consumption was popularly (yet not at all prominently) included in several lists designed to give the reader agency to lighten their 'carbon footprint'. Articles such as 'How to live in a more sustainable way' and 'Test driving a green life' were published in 2007 and 2008 following a wave of 'green' awareness after the release of Al Gore's *An Inconvenient Truth*.

Images play an active and important role in the cultural politics of climate change. In terms of saliency, the image of a steam stack on a coal-fired power plant, made famous on the poster for *An Inconvenient Truth*, is perhaps one of the most iconic climate change images. Salient images that effectively capture the scale of livestock production were found wanting in the articles studied. Most images associated with news stories in the study featured scenes of close-to idyllic pastures, dotted with cows. Just one image of an intensive feedlot arguably qualified as a salient illustration of the demand for meat. Interestingly, while images related to self-efficacy make up a very small proportion of imagery used in news media (O'Neill, 2013), a study in the U.S., UK and Australia of climate change-related imagery that promoted self-efficacy found that people felt empowered most by images of energy futures – such as solar panels and electric cars – compared to images associated with other lifestyle choices such as red meat (O'Neill, Boykoff, Niemeyer & Day, 2013).

Other Factors Affecting Reportage

Any topic warranting coverage in a major daily newspaper such as the *SMH* is subject to multiple other factors driving its agenda. The pursuit of objective newsworthiness is arguably at the top any newspaper's agenda, if only for marketing purposes. Other factors include the judgement of the journalist and editor, presenting a 'good story' for their readership, advertiser and sponsor pressures, issue promotion by interest groups, and coverage in other media outlets (Clayman & Reisner, 1998; Smith, 2005). The newsworthiness of an issue is often governed by framing that favours systemic responsibilities – those of government and business – as more important than individual responsibilities (Stone, 1989). Several 'internal' factors influence the reporting of climate change issues. Routine pressures such as publication deadlines and available space can affect the number of sources and depth of an article (Lewis & Boyce, 2009). The restructuring of Australian newspapers in the wake of falling readership and advertising revenues is exacerbating these pressures as a smaller workforce is expected to do more with less. However, routine pressures are also being compounded by more widespread changes in the media landscape. The means to publish content digitally, whether through a website, blog or on social media sharing platforms, is leading to more organisations and corporations publishing their own news rather than relying on independent media outlets (Simons, 2012). The blurring of lines between journalism and public relations is evident in the partnership between *SMH*, WWF and advertising agency Leo Burnett in the creation of the first Earth Hour event in 2007. A special, well-timed supplement in the newspaper was produced with stories related to the campaign in 2008. Though it provided an opportunity for various climate change issues, including livestock's impact, to be explored in the lead up to the event it also breached journalistic independence. According to disgruntled staff at Melbourne newspaper *The Age*, editors were told to avoid negative coverage of the event and a story schedule for the event's coverage was chosen by Earth Hour organisers (Hanan & Davis, 2008). Recent and arguably more savvy climate-focused campaigns from other organisations, such as not-for-profit The Climate Council, rely on accessible and timely research to tap into current issues that news media will want to cover, while also producing their own multimedia content to share via digital platforms.

DISCUSSION

The disproportionately low level of media attention on the issue of livestock emissions is deeply concerning. Frequency, framing, sources and other journalistic decisions have all played a role but factors out of the *SMH*'s control can also be barriers to better coverage. There is strong correlation between considerable national media coverage of climate change and countries with robust Kyoto Protocol emission targets (Schmidt et al., 2013). Yet international policy-setting on livestock emissions is in its infancy. Negotiations under the United Nations Framework Convention on Climate Change (UNFCCC) – the body who governs the Kyoto Protocol – have to date overlooked livestock, and efforts to establish a specific work stream on agriculture have failed to gain support. Accordingly, agricultural mitigation projects account for a disproportionately small slice of international finance under the Kyoto Protocol's Clean Development Mechanism (Bailey et al., 2014). While other emission sectors such as fossil fuels and deforestation receive the majority of attention in international climate negotiations and subsequent media coverage, other means of driving media attention are clearly required in order to place livestock emissions higher on the international agenda.

The lack of nationally-relevant information on emissions across the livestock sector's supply chain is also hampering media coverage. A study of the discourses of a variety of North American stakeholders concluded that "while the problematisation of animal agriculture's contribution to global climate change has been acknowledged by a number of stakeholder groups, for the most part it has not increased since the release of *Livestock's Long Shadow*" (Bristow & Fitzgerald, 2011). Unlike in Europe, but similar to the United States (Neff et al., 2008), Australia does not have any life-cycle assessment data to properly quantify animal agriculture's contribution to national emissions. The Australian Government's National Greenhouse Gas Inventory accounts for direct emissions under the strict and conservative reporting guidelines for the Kyoto Protocol. Spikes in media interest related to the 2008 *Garnaut Review*, and to a lesser extent high profile comments from the IPCC Chair and other senior UN figures, demonstrate an appetite to report on this issue at a national and international scale. This current and significant data gap is not only reflected in policy and advocacy drivers that are far too disparate, it may also mean journalists find it difficult to report on the issue. Though academic experts were the most quoted sources in articles found in the case study, there is still a clear need for targeted research on this topic and subsequent public engagement that is nationally relevant. In particular, governments responsible for funding agricultural research into supply-side mitigation could equally match sponsorship into the development and take up of plant-based foods, in particular meat and dairy alternatives derived from soy. The public health and nutrition community can also play an important role in communicating the co-benefits of replacing animal products with plant-based analogues in both individual and social food choices (Neff et al., 2008).

A lack of funds within the plant-based food industry continues to be a barrier to heavier promotion of plant-based alternatives to meat, especially in comparison to the significant marketing budget of the livestock industry (Nestle, 2002). The market for meat and dairy analogs that look and taste similar to livestock products is growing and, with the involvement of well-known food companies and substantial marketing, could claim an equal share of the market (Goodland & Anhang, 2009). A broad-based communications approach with support from government, food retailers and opinion leaders could be a necessary step to achieve stronger awareness – particularly in the media – of the environmental, health and animal welfare benefits to entirely plant-based meals. Excessive government subsidies to the livestock industry that artificially lower the cost of meat and dairy (Raphaely et al., 2013) should be realigned to support such a campaign. Commercial financing can also be achieved through a push for ethical investment in plant food companies (Goodland & Anhang, 2009), similar to the fossil fuel divestment campaign currently backed by many environmental NGOs.

Perhaps the most significant barrier is the reluctance of environmental non-government organisations (NGOs) to campaign on this issue. When it comes to the behaviour change of its citizens, governments in western countries have been more inclined to focus on limiting fossil fuel emissions through household energy consumption than "tackle questions of personal choice and consumption" related to diet (Robins & Roberts, 2006, p. 39). The same can be said for most climate action advocacy groups. Former U.S. vice president and influential environmental campaigner Al Gore's Nobel Peace Prize winning documentary *An Inconvenient Truth* overlooked the role the average American's diet had on their climate footprint (Freudberg, 2013), instead focusing on changes to people's behaviours that would limit fossil fuel use, such as switching to energy efficient lighting or riding a bike to work (Gore, 2006). When asked why he didn't focus on animal agriculture in his film, Gore answered that for most people, the role of animal agriculture in climate change is too inconvenient of a truth (Moby, 2014).

Environmental NGOs play a powerful role in influencing climate policy and encouraging behaviour change (Hall & Taplin, 2008). Australian NGOs campaigning on fossil fuels and climate change have been effective at advancing public debate and influencing governmental policies despite restricted political access (Hall & Taplin, 2008, pp. 361-363). However, despite a growing scientific understanding of the importance in reducing emissions from livestock none of these NGOs have established a prescribed campaign to encourage policy reform that will reduce meat consumption and mitigate climate change. A similar experience occurs in other western countries with high meat consumption. A study of NGOs in the U.S., Canada and Sweden found few sought to promote national-level polices to reduce meat consumption, with environmental NGOs far less interested in active public outreach than animal protection groups (Laestadius, Neff, Barry & Frattaroli, 2013).

With the exception of WWF's involvement in Earth Hour, which featured one substantive article about livestock emissions, no other environmental NGOs were the source of reportage in this study. Though NGOs have played an important and prominent role as sources of climate change media coverage, specifically in relation to energy and transportation (Hall & Taplin, 2008), they have been much slower than other advocacy groups, such as animal rights organisations, to raise awareness of livestock emissions in the same way. This is perhaps due to a wariness or fear of alienating their supporter base and the wider community with messages about changing eating habits (Neff et al., 2008). Unsurprisingly then, studies have demonstrated that newspaper articles relating to climate change consequences of meat production elicit lower levels of concern in consumers than articles about animal welfare aspects, potentially because consumers are mostly unaware of the more abstract and less intuitive climate impacts (Cordts et al., 2014; Tobler, Visschers & Siegrist, 2011). Total avoidance of animal products is strongly linked to ethical and moral considerations of the subjugation and killing of animals for food. For example, the dairy industry in both Australia and NZ is keenly aware of the effect on their operations should public sensitivities be awoken by a broader awareness regarding bobby calves and systemic bovine health problems such as mastitis and lameness (Campbell, 2009). Increasing media attention on animal welfare and animal food production methods has been demonstrated to have a direct effect on demand, with one study finding that media coverage of animal wellbeing and welfare reduced pork and poultry demand in the U.S. (Tonsor & Olynk, 2011). Furthermore, expenditure was not reallocated to other meats such as chicken but spent on non-meat food. Research suggests however that often over time welfare considerations are augmented with concern about environmental impacts (Fox & Ward, 2008). Taylor (2012) proposes a more concerted push to highlight animal welfare and wellbeing concerns in the media as an effective strategy in efforts to reverse the culture of meat eating. An important point to note however is that half of respondents to a recent Australian study of attitudes to a plant-based diet were unsure of any environmental benefits (Lea et al., 2006). This correlates with previous research (Friedlander et al., 2014) that shows media coverage of the environmental impacts of animal agriculture are even less frequent than that associated with animal welfare. At the same time this paper's findings demonstrate that animal welfare messages are rarely associated with livestock emissions reportage. An increased awareness of both the environmental impacts of livestock production and the environmental gains of changing to a plant-based diet may be as strong a motivator as welfare concerns. The development of a strong link between animal welfare and environmental campaigns regarding animal agriculture could be a much more effective public engagement strategy.

CONCLUSION

When all the impacts of livestock production on the planet, human health and animals are weighed up, the overconsumption of meat, dairy and other animal products is morally indefensible (Raphaely & Marinova, 2014a). The livestock sector's contribution to climate change in particular is an urgent priority from an ecological perspective as it could singlehandedly tip the world's climatic systems into a dangerous and unpredictable state (Goodland & Anhang, 2009). Complicated and entrenched barriers continue to hamper public awareness of the need to reduce demand for animal products in order to avoid the worst consequences of future climate change.

The *SMH* case study findings partly validate the pessimism of Hedenus et al. (2014) in saying necessary policy-driven dietary changes will only be taken seriously once all supply-side measures have been exhausted. Though dietary changes and other demand-side measures were mentioned more often in the articles studied than supply-side measures, analysis revealed that the latter was framed as more newsworthy. Often there was an inability for reportage to move beyond highlighting dietary change measures as too contentious or radical, despite the scientific literature proposing them as legitimate policy ideas.

For more widespread dietary changes to take place and reverse the dominance of animal-derived foods, the benefits of changing to a plant-based diet need to outweigh the perceived barriers (Lea et al., 2006). That vegetarianism and especially veganism are still framed in the news as radical and unworthy of close attention demonstrates these challenges rather acutely. The sharp drop in *SMH* articles over the last three years is worrying and could be put down to a recent disinterest in the 'sustainable living' narrative. There is a clear need for environmental NGOs to replicate their sophisticated campaigning efforts on fossil fuels and bring animal agriculture to the heart of both national and international climate agendas as well. If Australia's leading environmental and climate action organisations can advocate for a future powered wholly on renewable energy, why not a 100 percent plant-based future too? Animal agriculture and fossil fuels share plenty in common. They are key drivers of climate change. They have adverse effects on human health. They are often heralded by industry as essential to lifting the world's poor out of poverty. Their markets for production and consumption are responsible for gross inequality. They are used every day around the world, yet their production is more often than not hidden from view. There are alternatives for both that will dramatically improve these impacts. Avoiding climate catastrophe may in reality only require 80 percent of fossil fuels to be displaced by energy sources that don't drive climate change. Nevertheless, a world powered with renewable resources has many other benefits to people and the environment. In the same way, a world that doesn't involve the raising and slaughter of billions of animals each year would reap many benefits over and above avoiding dangerous climate change. Awareness-raising strategies by other actors, including the plant food industry and public health community, can also help pressure governments to implement policies that address imbalanced subsidies and other incentives flowing to the livestock industry.

The more that is known about climate change news coverage, the more effective these interventions will be in improving message accuracy and framing. This study, and others in the literature, demonstrates that livestock emissions have not been reported proportionately to their share of total emissions. Future research on this topic could analyse factors in the reporting of renewable energy to learn of yet more successful strategies that increase media coverage. The path to demand-side livestock mitigation policies that are seen as legitimate in the eyes of government, media and broader society lies in creating a bigger and more nuanced public discourse regarding dietary changes that encourage reduced consumption of animal-based products. This must be an urgent priority of all actors concerned about the growing anthropogenic destruction of the planet's life support systems and the vital task of limiting climate change to two degrees.

REFERENCES

Adams, C. J. (2013). *The sexual politics of meat*. New York, NY: Bloomsbury Academic.

Alexandratos, N., & Bruinsma, J. (2012). World agriculture towards 2030/50 (no. 12-03). Food and Agriculture Organization of the United Nations (FAO). Retrieved from http://www.fao.org/fileadmin/templates/esa/Global_persepctives/world_ag_2030_50_2012_rev.pdf

Almiron, N., & Zoppeddu, M. (2014). Eating meat and climate change: The media blind spot - a study of Spanish and Italian press coverage. *Environmental Communication*, 9(3), 307–325. doi:10.1080/17524032.2014.953968

Antilla, L. (2005). Climate of skepticism: US newspaper coverage of the science of climate change. *Global Environmental Change*, 15(4), 338–352. doi:10.1016/j.gloenvcha.2005.08.003

Asplund, T., Hjerpe, M., & Wibeck, V. (2013). Framings and coverage of climate change in Swedish specialized farming magazines. *Climatic Change*, 117(1), 197–209. doi:10.1007/s10584-012-0535-0

Bailey, R., Froggatt, A., & Wellesley, L. (2014). Livestock - climate change's forgotten sector: global public opinion on meat and dairy consumption. London, UK: Chatham House, The Royal Institute of International Affairs. Retrieved from http://www.chathamhouse.org/sites/files/chathamhouse/field/field_document/20141203LivestockClimateChangeBaileyFroggattWellesley.pdf

Bajželj, B., Richards, K. S., Allwood, J. M., Smith, P., Dennis, J. S., Curmi, E., & Gilligan, C. A. (2014). Importance of food-demand management for climate mitigation. *Nature Climate Change*, 4(10), 924–929. doi:10.1038/nclimate2353

Baker, S., Thompson, K. E., & Palmer-Barnes, D. (2002). Crisis in the meat industry: A values-based approach to communications strategy. *Marketing Communications*, 8(1), 19–30. doi:10.1080/13527260110108319

Bazzano, L. A., He, J., Ogden, L. G., Loria, C. M., Vupputuri, S., Myers, L., & Whelton, P. K. (2002). Fruit and vegetable intake and risk of cardiovascular disease in US adults. *The American Journal of Clinical Nutrition, 76*, 93–99. PMID:12081821

Bellarby, J., Tirado, R., Leip, A., Weiss, F., Lesschen, J. P., & Smith, P. (2013). Livestock greenhouse gas emissions and mitigation potential in Europe. *Global Change Biology, 19*(1), 3–18. doi:10.1111/j.1365-2486.2012.02786.x PMID:23504717

Bennett, W. L. (1996). *News: the politics of illusion* (3rd ed.). White Plains, NY: Longman.

Briske, D. D., Ash, A. J., Dernerc, J. J., & Huntsinger, L. (2014). Commentary: A critical assessment of the policy endorsement for holistic management. *Agricultural Systems, 125*, 50–53. doi:10.1016/j.agsy.2013.12.001

Bristow, E., & Fitzgerald, A. J. (2011). Global climate change and the industrial animal agriculture link: The construction of risk. *Society & Animals, 19*(3), 205–224. doi:10.1163/156853011X578893

Cameron, E. (2014). *Climate change: implications for agriculture: key findings from the Intergovernmental Panel on Climate Change Fifth Assessment Report*. Cambridge, UK: University of Cambridge, Institute for Sustainability Leadership.

Campbell, D. (2009). Animal welfare and market access: a dairy industry perspective. Paper presented at the Proceedings of the Society of Dairy Cattle Veterinarians of the NZVA Annual Conference, Rotorua, NZ.

Carter, J., Jones, A., O'Brien, M., Ratner, J., & Wuerthner, G. (2014). Holistic management: Misinformation on the science of grazed ecosystems. International Journal of Biodiversity, *2014. Article ID, 163431*, 1–10. doi:10.1155/2014/163431

Carvalho, A., & Burgess, J. (2005). Cultural circuits of climate change in U.K. broadsheet newspapers, 1985-2003. *Risk Analysis, 25*(6), 1457–1469. doi:10.1111/j.1539-6924.2005.00692.x PMID:16506975

Chen, C., Hare, B., Hagemann, M., Höhne, N., Moltmann, S., & Schaeffer, M. (2011). Cancun climate talks: keeping options open to close the gap. Climate Action Briefing Paper. New York, NY: Climate Analytics, PIK and ECOFYS. Retrieved from http://climateactiontracker.org/assets/publications/briefing_papers/briefing_paper_cancun.pdf

Clayman, S. E., & Reisner, A. (1998). Gatekeeping in action: Editorial conferences and assessments of newsworthiness. *American Sociological Review, 63*(2), 178–199. doi:10.2307/2657322

Cleugh, H., Smith, M. S., Battaglia, M., & Graham, P. (Eds.). (2011). Climate change: science and solutions for Australia. Collingwood, Australia: Commonwealth Scientific and Industrial Research Organisation (CSIRO).

Clucas, I. (1997). A study of the options for utilization of bycatch and discards from marine capture fisheries. Fisheries Circular No. 928. Rome: Italy: Food and Agriculture Organization of the United Nations (FAO), Fishery Industries Division, Fisheries Department. Retrieved from http://www.fao.org/docrep/w6602e/w6602e00.HTM

Cole, M., & Morgan, K. (2011). Vegaphobia: Derogatory discourses of veganism and the reproduction of speciesism in the UK national newspapers. *The British Journal of Sociology, 62*(1), 134–153. doi:10.1111/j.1468-4446.2010.01348.x PMID:21361905

Cordell, D., Drangert, J., & White, S. (2009). The story of phosphorus: Global food security and food for thought. *Global Environmental Change, 19*(2), 292–305. doi:10.1016/j.gloenvcha.2008.10.009

Cordts, A., Nitzko, S., & Spiller, A. (2014). Consumer response to negative information on meat consumption in Germany. *International Food and Agribusiness Management Review, 17*(A), 83-106.

Cornwall, J., Collie, G., & Ashton, P. (2000). *Sustaining a nation, celebrating 100 years of agriculture in Australia*. Sydney, Australia: Focus Publishing.

Cudworth, E. (2011). *Social lives with other animals*. Basingstroke, UK: Palgrave Macmillan. doi:10.1057/9780230302488

Department of Agriculture (DOA). (2013). Australian agriculture: reducing emissions and adapting to a changing climate. Key findings of the Climate Change Research Program. Canberra, Australia: DOA. Retrieved from http://www.agriculture.gov.au/ag-farm-food/climatechange/australias-farming-future/climate-change-and-productivity-research

Department of the Environment (DOE). (2014). About the Carbon Farming Initiative. Retrieved December 13, 2014, from http://www.environment.gov.au/climate-change/emissions-reduction-fund/cfi/about

Devine, M. (2008a, September 11). Our dietary freedom is at steak. *The Sydney Morning Herald*. Retrieved from http://www.smh.com.au/news/miranda-devine/our-dietary-freedoms-at-steak/2008/09/10/1220857635063.html

Devine, M. (2008b, October 2). Act hastily, roo the scare tactics. *The Sydney Morning Herald*. Retrieved from http://www.smh.com.au/news/opinion/miranda-devine/act-hastily-roo-the-scare-tactics/2008/10/01/1222651169023.html?page=fullpage

Dow, S. (2007, July 3). Into the meat of the issue. *The Sydney Morning Herald*. Retrieved from http://www.smh.com.au/news/environment/into-the-meat-of-the-issue/2007/07/02/1183351125939.html

Doyle, J. (2011). Sustainable consumption? Reframing meat and dairy consumption in the politics of climate change. In J. Doyle (Ed.), *Mediating climate change* (pp. 123–144). Burlington, VT: Ashgate.

Erb, K., Mayer, A., Kastner, T., Sallet, K., & Haberl, H. (2012). The impact of industrial grain fed livestock production on food security: an extended literature review. Vienna, Austria: Alpen-Adria University Klagenfurt, Institute of Social Ecology. Retrieved from http://www.fao.org/fileadmin/user_upload/animalwelfare/the_impact_of_industrial_grain_fed_livestock_production_on_food_security_2012.pdf

Erickson, J. (2014, September 5). Dietary recommendations may be tied to increased greenhouse gas emissions. *Michigan News*. Retrieved from http://ns.umich.edu/new/releases/22359-dietary-recommendations-may-be-tied-to-increased-greenhouse-gas-emissions

Eshel, G., & Martin, P. (2006). Diet, energy and global warming. *Earth Interactions, 10*(9), 1–17. doi:10.1175/EI167.1

Food and Agriculture Organization of the United Nations. Statistics Division (FAOSTAT). (2014). Food Security Indicators. Retrieved from http://faostat3.fao.org/download/D/FS/E

Fox, N., & Ward, K. J. (2008). Health ethics and environment: A qualitative study of vegetarian motivations. *Appetite*, *50*(2-3), 422–429. doi:10.1016/j.appet.2007.09.007 PMID:17980457

Freudberg, D. (2013, October 1). The diet-climate connection. The Huffington Post. Retrieved from http://www.huffingtonpost.com/david-freudberg/the-dietclimate-connectio_b_2446929.html

Friedlander, J., Riedy, C., & Bonfiglioli, C. (2014). A meaty discourse: What makes meat news? *Food Studies*, *3*(3), 27–43.

Friel, S. (2010). Climate change, food insecurity and chronic disease: Sustainable and healthy policy opportunities for Australia. *New South Wales Public Health Bulletin*, *21*(6), 129–133. doi:10.1071/NB10019 PMID:20637169

Friel, S., Dangour, A. D., Garnett, T., Lock, K., Chalabi, Z., Roberts, I., & Haines, A. et al. (2009). Public health benefits of strategies to reduce greenhouse-gas emissions: Food and agriculture. *Lancet*, *374*(9706), 2016–2025. doi:10.1016/S0140-6736(09)61753-0 PMID:19942280

Galvin, N. (2008, December 16). Eat your greens, good living. *The Sydney Morning Herald*.

Garnaut, R. (2008). *The garnaut climate change review: final report*. New York, NY: Cambridge University Press.

Gerber, P. J., Steinfield, H., Henderson, B., Mottet, A., Opio, C., & Dijkman, J. … Tempio, G. (2013). Tackling climate change through livestock - a global assessment of emissions and mitigation opportunities. Rome: Italy: Food and Agriculture Organisation of the United Nations (FAO). Retrieved from http://www.fao.org/docrep/018/i3437e/i3437e.pdf

Goodland, R. (2010). Livestock and climate change: Critical comments and responses. *World Watch Magazine*, *23*(2), 7–9.

Goodland, R. (2014). A fresh look at livestock greenhouse gas emissions and mitigation potential in Europe. *Global Change Biology*, *20*(7), 2042–2044. doi:10.1111/gcb.12454 PMID:24166774

Goodland, R., & Anhang, J. (2009, November/December). Livestock and climate change: What if the key actors in climate change are cows, pigs and chickens? *World Watch Magazine*, *22*(6), 10–19.

Goodman, A., & Goodman, D. (2005). *The exception to the rulers: exposing oily politicians, war profiteers and the media that love them*. New York, NY: Hyperion.

Gore, A. (Writer) & D. Guggenheim (Director). (2006). An inconvenient truth. Hollywood, CA: Paramount Pictures.

Government of New South Wales (NSW). (2008). Atlas of NSW: economy, agriculture-livestock. Land and property information. Retrieved from http://atlas.nsw.gov.au/public/nsw/home/topic/article/agriculture-livestock.html

Hall, N. L., & Taplin, R. (2008). Room for climate advocates in a coal-focused economy? NGO influence on the Australian climate policy. *The Australian Journal of Social Issues*, *43*(3), 359–379.

Hanan, E., & Davis, M. (2008, April 11). Age blighted by bias, selling its soul: so say the staff. *The Australian*. Retrieved from http://www.highbeam.com/doc/1G1-177695483.html

Havlik, P., Valin, H., Herrero, M., Obersteiner, M., Schmid, E., & Rufino, M. C., ..., Notenbaert, A. (2013). Climate change mitigation through livestock system transitions. *Proceedings of the National Academy of Sciences (PNAS)*, *111*(10), 3709-3714.

Hedenus, F., Wirsenius, S., & Johansson, D. J. A. (2014). The importance of reduced meat and dairy consumption for meeting stringent climate change targets. *Climatic Change*, *124*(1-2), 79–91. doi:10.1007/s10584-014-1104-5

Heinz, B., & Leeb, R. (1998). Getting down to the meat: The symbolic construction of meat consumption. *Communication Studies*, *49*(1), 86–99. doi:10.1080/10510979809368520

Heller, M. C., & Keoleian, G. A. (2014). Greenhouse gas emission estimates of U.S. dietary choices and food loss. *Journal of Industrial Ecology*, *19*(3), 391–401. doi:10.1111/jiec.12174

Henzell, T. (2007). *Australian agriculture: its history and challenges*. Collingwood, Australia: CSIRO Publishing.

Hoffman, U. (2013). Agriculture at the crossroads: assuring food security in developing countries. In U. Hoffman (Ed.), *Trade and environment review 2013: Wake up before it is too late.* (pp. 1-8). Geneva, Switzerland: United Nations Conference on Trade and Development. Retrieved from http://unctad.org/en/PublicationsLibrary/ditcted2012d3_en.pdf

Humane Society of the United States (HSUS). (2009). *The welfare of animals in the meat, egg and dairy industries. HSUS Reports: Farm industry impacts on animals Paper 2* (Vol. 370, pp. 1253–1263). Washington, DC: HSUS.

Idel, A., Fehlenberg, V., & Reichert, T. (2013). *Livestock production and food security in a context of climate change and environmental and health challenges. Background Paper*. Berlin, Germany: Germanwatch.

Intergovermental Panel on Climate Change (IPCC). (2013). Summary for policy makers. In T. F. Stocker, D. Qin, G.-K. Plattner, M. Tignor, S. K. Allen, J. Boschung, & P. M Midgley et al. (Eds.), *Climate change 2013: the physical science basis. Contribution of Working Group 1 to the Fifth Assement Report of the Intergovernmental Panel on Climate Change (IPCC)* (pp. 3–30). Cambridge, UK and New York, NY: Cambridge University Press.

Intergovernmental Panel on Climate Change (IPCC). (2014). History. Retrieved from http://ipcc.ch/organization/organization_history.shtml

Jakob, M., & Hilaire, J. (2015). Climate science: Unburnable fossil-fuel reserves. *Nature*, *517*(7533), 150–152. doi:10.1038/517150a PMID:25567276

Jiang, T., Hanslow, K., & Pearce, D. (2009). On-farm impacts of Australian emissions trading scheme - an economic analysis. Barton, Australia: Australian Government (RIRDC).

Kiesel, L. (2010, June 27-30). A comparative rhetorical analysis of US and UK newspaper coverage of the correlation between livestock production and climate change. *Proceedings of the 10th Biennial Conference on Communication and the Environment*, Portland, OR. Retrieved from https://theieca.org/contents-2009-conference-communication-and-environment-proceedings

Laestadius, L. I., Neff, R. A., Barry, C. L., & Frattaroli, S. (2013). Meat consumption and climate change: The role of non-governmental organizations. *Climatic Change*, *120*(1-2), 25–38. doi:10.1007/s10584-013-0807-3

Lea, E., & Worsley, A. (2003). Benefits and barriers to the consumption of a vegetarian diet in Australia. *Public Health Nutrition*, *6*(5), 505–511. doi:10.1079/PHN2002452 PMID:12943567

Lea, E. J., Crawford, D., & Worsley, A. (2006). Public views of the benefits and barriers to the consumption of a plant based diet. *European Journal of Clinical Nutrition*, *60*(7), 828–837. doi:10.1038/sj.ejcn.1602387 PMID:16452915

Lewis, J., & Boyce, T. (2009). Climate change and the media: the scale of the challenge. In T. Boyce & J. Lewis (Eds.), *Climate change and the media* (pp. 3–16). New York, NY: Peter Lang.

Lombard, M., Synder-Duch, J., & Bracken, C. C. (2002). Content analysis in mass communication: Assessment and reporting of intercoder reliability. *Human Communication Research*, *28*(4), 587–604. doi:10.1111/j.1468-2958.2002.tb00826.x

Longmire, A., Taylor, C., & Wedderburn-Bisshop, G. (2014). Land use: agriculture and forestry. Discussion Paper, Zero Carbon Australia Land Use Report. Melbourne, Australia: The University of Melbourne, Melbourne Sustainable Society Institute.

Macey, R. (2006, June 22). Gas-friendly cows an emission possible to cut global warming. *The Sydney Morning Herald*. Retrieved from http://www.smh.com.au/news/national/emission-possible-gasfriendly-cows/2006/06/21/1150845247870.html

Mahony, P. (2014). Cowspiracy and the Australian red meaty industry. Retrieved from http://terrastendo.net/2014/11/09/cowspiracy-and-the-australian-red-meat-industry/

Margulis, S. (2004). Causes of deforestation of the Brazilian Amazon. World Bank Working Paper No. 22. Washington, DC: World Bank (WB). Retrieved from https://openknowledge.worldbank.org/bitstream/handle/10986/15060/277150PAPER0wbwp0no1022.pdf?sequence=1

Marshall, A. (2014, August 18). White meat still Australia's choice. *The Land*. Retrieved from http://www.theland.com.au/news/agriculture/agribusiness/general-news/white-meat-still-australias-choice/2708428.aspx

McMichael, A. J., Powles, J. W., Butler, C. D., & Uauy, R. (2007). Food, livestock production, energy, climate change and health. *Lancet*, *370*(9594), 1253–1263. doi:10.1016/S0140-6736(07)61256-2 PMID:17868818

Meat and Livestock Australia (MLA). (2014a). Community engagement. Retrieved from http://www.mla.com.au/Marketing-beef-and-lamb/Community-engagement

Meat and Livestock Australia (MLA). (2014b). Fast facts 2014: Australia's beef industry. Retrieved from http://www.mla.com.au/Cattle-sheep-and-goat-industries/Industry-overview/Cattle

Meat and Livestock Australia (MLA). (2014c). National livestock methane program. Retrieved from http://www.mla.com.au/Research-and-development/Environment-research/National-livestock-methane-program

Milman, O. (2015, January 16). Rate of environmental degredation puts life at risk, says scientist. *The Guardian*. Retrieved from http://www.theguardian.com/environment/2015/jan/15/rate-of-environmental-degradation-puts-life-on-earth-at-risk-say-scientists

Moby. (2014). Save the humans. *The Huffington Post*. Retrieved from http://www.huffingtonpost.com/moby/moby-meat_b_5889850.html

Monbiot, G. (2014, August 4). Eat more meat and save the world: the latest implausible farming miracle. *The Guardian*. Retrieved from http://www.theguardian.com/environment/georgemonbiot/2014/aug/04/eat-more-meat-and-save-the-world-the-latest-implausible-farming-miracle

Moss, M. (2015, January 19). U.S. research lab lets livestock suffer in quest for profit. *The New York Times*. Retrieved from https://www.organicconsumers.org/news/us-research-lab-lets-livestock-suffer-quest-profit

Myhre, G. D., Shindell, D., Bréon, F.-M., Collins, W., Fuglestvedt, J., & Huang, J. ... Zhang, H. (2013). Anthropogenic and natural radiative forcing. In T. F. Stocker, D. Qin, G.-K. Plattner, M. Tignor, S.K. Allen, J. Boschung, ..., P. M. Midgley (Eds), Climate change 2013: the physical science basis. Contribution of Working Group 1 to the Fifth Assessment Report of the Intergovernmental Panel on Climate Change (IPCC) (pp. 659-740). Cambridge, UK and New York, NY: Cambridge University Press.

Nath, J. (2010). God is a vegetarian: The food, health and bio-spirituality of Hare Krishna, Buddhist and Seventh-Day Adventist devotees. *Health Sociology Review*, *19*(3), 356–368. doi:10.5172/hesr.2010.19.3.356

National Farmers' Federation (NFF). (2015). Major commodities: beef. Retrieved from http://www.nff.org.au/commodities-beef-cattle.html

National Health and Medical Research Council (NHMRC). (2013). Australian dietary guidelines. Canberra, Australia: Australian Government, NHMRC, Department of Health and Ageing. Retrieved from https://www.nhmrc.gov.au/guidelines-publications/n55

Neff, R. A., Chan, I. L., & Smith, K. C. (2008). Yesterday's dinner, tomorrow's weather, today's news? US newspaper coverage of food system contributions to climate health. *Public Health Nutrition*, *12*(7), 1006–1014. doi:10.1017/S1368980008003480 PMID:18702838

Nestle, M. (2002). *Food politics: how the food industry influences nutrition and health*. London, UK: University of California Press.

Neuendorf, K. A. (2002). *The content analysis guidebook*. Thousand Oaks, CA: Sage.

O'Neill, S. J. (2013). Image matters: Climate change imagery in the US, UK and Australian mass media. *Geoforum*, *49*, 10–19. doi:10.1016/j.geoforum.2013.04.030

O'Neill, S. J., Boykoff, M., Niemeyer, S., & Day, S. A. (2013). On the use of imagery for climate change engagement. *Global Environmental Change, 23*(2), 413–421. doi:10.1016/j.gloenvcha.2012.11.006

Oreskes, N., & Conway, E. M. (2010). *Merchants of doubt*. New York, NY: Bloomsbury.

Peatling, S. (2007). Farmers say $100m policy must go further. *The Sydney Morning Herald*. Retrieved from http://www.smh.com.au/news/federal-election-2007-news/farmers-say-100m-policy-must-go-further/2007/11/13/1194766675352.html

Peters, H. P., & Heinrichs, H. (2004, June). Expertise for the public: the science-journalism interface in German discourse on global climate change. Paper presented at the 8th International Conference, Barcelona, Spain. Retrieved from http://juser.fz-juelich.de/record/42732

Pimental, D., Berger, B., Filiberto, D., Newton, M., Wolfe, B., Karabinakis, E., & Nandagopal, S. et al. (2004). Water resources: Agricultural and environmental issues. *Bioscience, 54*(10), 909–918. doi:10.1641/0006-3568(2004)054[0909:WRAAEI]2.0.CO;2

Pimentel, D., & Pimentel, M. (1997). Sustainability of meat-based and plant-based diets and the environment. *Clinical Nutrition (Edinburgh, Lothian), 78*(3), 660S–663S. PMID:12936963

Popp, A., Lotze-Campen, H., & Bodirsky, B. (2010). Food consumption, diet shifts and associated non-CO_2 greenhouse gases from agricultural production. *Global Environmental Change, 20*(3), 451–462. doi:10.1016/j.gloenvcha.2010.02.001

Porcher, J. (2006). Well-being and suffering in livestock farming: Living conditions at work for people and animals. *Sociologie du Travail, 48*, 56–70. doi:10.1016/j.soctra.2006.02.001

Raphaely, T., & Marinova, D. (2014a). Flexitarianism: A more moral dietary option. *International Journal of Sustainable Society, 6*(1-2), 189–211. doi:10.1504/IJSSOC.2014.057846

Raphaely, T., & Marinova, D. (2014b). Flexitarianism: Decarbonising through flexible vegetarianism. *Renewable Energy, 67*, 90–96. doi:10.1016/j.renene.2013.11.030

Raphaely, T., Marinova, D., Crisp, G., & Panayotov, J. (2013). Flexitarianism (flexible or part-time vegetarianism): A user-based dietary choice for improving personal, population and planetary wellbeing. *International Journal of User-Driven Healthcare, 3*(3), 34–58. doi:10.4018/ijudh.2013070104

Richards, E., Signal, T., & Taylor, N. (2013). A different cut? Comparing attitudes toward animals and propensity for aggression within two primary industry cohorts-farmers and meatworkers. *Society & Animals, 21*(4), 395–413. doi:10.1163/15685306-12341284

Ripple, W. J., Smith, P., Harberl, H., Montzka, S. A., McAlpine, C., & Boucher, D. H. (2014). Ruminants, climate change and climate policy. *Nature Climate Change, 4*(1), 2–5. doi:10.1038/nclimate2081

Robins, N., & Roberts, S. (2006). Making sense of sustainable consumption. In T. Jackson (Ed.), *The earthscan reader on sustainable consumption* (pp. 37–47). London, UK: Earthscan.

Rogelj, J., Schaeffer, M., Meinshausen, M., Shindell, D. T., Hare, W., & Kilmont, Z., … Schellnhuber, H. J. (2014). Disentangeling the effects of CO2 and short-lived climate forcer mitigation. *Proceedings of the National Academy of Sciences (PNAS), 111*(46), 16325-16330.

Russell, G. (2009). *CSIRO perfidy*. Fremantle, Australia: Vivid Publishing.

Salter, D. (2012, March 29). An hour to ponder why Fairfax bothers turning off the lights. *Crikey*. Retrieved from http://www.crikey.com.au/2012/03/29/history-of-earth-hour-at-fairax/?wpmp_switcher=mobile

Sampei, Y., & Aoyagi-Usui, M. (2009). Mass-media coverage, its influence on public awareness of climate-change issues and implications for Japan's national campaign to reduce greenhouse gas emissions. *Global Environmental Change, 19*(2), 203–212. doi:10.1016/j.gloenvcha.2008.10.005

Scarborough, P., Appleby, P. N., Mizdrak, A., Briggs, A. D. M., Travis, R. C., Bradbury, K. E., & Key, T. J. (2014). Dietary greenhouse gas emissions of meat-eaters, fish-eaters, vegetarians and vegans in the UK. *Climatic Change, 125*(2), 179–192. doi:10.1007/s10584-014-1169-1 PMID:25834298

Schmidt, A., Ivanova, A., & Schäfer, M. S. (2013). Media attention for climate change around the world: A comparative analysis of newspaper coverage in 27 countries. *Global Environmental Change, 23*(5), 1233–1248. doi:10.1016/j.gloenvcha.2013.07.020

Schneider, S., Nullmeier, F., & Hurrelmann, A. (2007). Exploring the communicative dimension of legitamacy: text analytical approaches. In A. Hurrelmann, S. Schneider, & J. Steffek (Eds.), *Legitamacy in an age of global politics* (pp. 126–155). Hampshire, UK: Palgrave Macmillan.

Shindell, D. T. (2013). Climate change: breaking the stalemate. *Milken Institute Review: A Journal of Economic Policy, 15*(1), 35-45.

Shindell, D. T., Faluvegi, G., Koch, D. M., Schmidt, G. A., Unger, N., & Bauer, S. E. (2009). Improved attribution of climate forcing to emissions. *Science, 326*(5953), 716–718. doi:10.1126/science.1174760 PMID:19900930

Shoemaker, J. K., Schrag, D. P., Molina, M. J., & Ramanathan, V. (2013). What role for short-lived climate pollutants in mitigation policy? *Science, 342*(6164), 1323–1324. doi:10.1126/science.1240162 PMID:24337280

Simons, M. (2012). *Journalism at the crossroads: crisis and opportunity for the press*. Brunswick, Australia: Scribe.

Smith, J. (2005). Dangerous news: Media decision-making about climate change risk. *Risk Analysis, 25*(6), 1471–1482. doi:10.1111/j.1539-6924.2005.00693.x PMID:16506976

Smith, K. R., Desaia, M. A., Rogersa, J. V., & Houghton, R. A. (2013). Joint CO_2 and CH_4 accountability for global warming. [PNAS]. *Proceedings of the National Academy of Sciences of the United States of America, 110*(31), E2865–E2874. doi:10.1073/pnas.1308004110 PMID:23847202

Steffek, J. (2009). Discursive legitimation in environmental governance: Discourse and expertise in forest and environmental governance. *Forest Policy and Economics, 11*(5), 313–318. doi:10.1016/j.forpol.2009.04.003

Stehfest, E., Bouwman, L., van Vuuren, D., den Elzen, M., Eickhout, B., & Kabat, P. (2009). Climate benefits of changing diet. *Climatic Change, 95*(1), 83–102. doi:10.1007/s10584-008-9534-6

Steinfeld, H., Gerber, P., Wassenaar, T., Castel, V., Rosales, M., & de Hann, C. (2006). Livestock's long shadow: environmental issues and options. Rome, Italy: Food Agriculture Organisation (FAO). Retrieved from ftp://ftp.fao.org/docrep/fao/010/a0701e/a0701e00.pdf

Stone, D. A. (1989). Casual stories and the formation of policy agendas. *Political Science Quarterly*, *104*(2), 281–300. doi:10.2307/2151585

Target 100. (2014a). About. Retrieved from http://www.target100.com.au/About

Target 100. (2014b). Cowspiracy. Retrieved from http://www.target100.com.au/Hungry-for-Info/Target-100-Responds/Cowspiracy

Taylor, N. (2012). Reversing meat-eating culture to combat climate change. Haslemere, UK: World Preservation Foundation (WPF).

Thornton, P., & Herrero, M. (2010). The inter-linkages between rapid growth in livestock production, climate change and the impacts on water resources, land use and deforestation. Policy Research Working Paper 5178. Nairobi, Kenya: International Livestock Research Institute. Retrieved from https://openknowledge.worldbank.org/handle/10986/9223

Tilman, D., & Clark, M. (2014). Global diets link environmental sustainability and human health. *Nature*, *515*(7528), 518–522. doi:10.1038/nature13959 PMID:25383533

Ting, I. (2013, April 16). Hold the red, pass the white - meat that is. *The Sydney Morning Herald*. Retrieved from http://www.smh.com.au/national/hold-the-red-pass-the-white--meat-that-is-20130415-2hvv6.html?skin=text-only

Tobler, C., Visschers, V. H., & Siegrist, M. (2011). Eating green: Consumers' willingness to adopt ecological food consumption behaviours. *Appetite*, *57*(3), 674–682. doi:10.1016/j.appet.2011.08.010 PMID:21896294

Tonsor, G. T., & Olynk, N. J. (2011). U.S. meat demand: The influence of animal welfare on media coverage. *Journal of Agricultural Economics*, *62*(1), 59–72. doi:10.1111/j.1477-9552.2010.00266.x

Trumbo, C. (1996). Constructing climate change: Claims and frames in US news coverage of an environmental issue. *Public Understanding of Science (Bristol, England)*, *5*(3), 269–283. doi:10.1088/0963-6625/5/3/006

United Nations. (2013). World population prospects: the 2012 revision. New York, NY: United Nations. Retrieved from http://esa.un.org/unpd/wpp/Publications/Files/WPP2012_HIGHLIGHTS.pdf

United Nations Environment Programme (UNEP). (2014). The emissions gap report 2014. A UNEP synthesis report. Nairobi, Kenya: UNEP. Retrieved from http://www.unep.org/publications/ebooks/emissionsgapreport2014/portals/50268/pdf/EGR2014_LOWRES.pdf

United Nations Framework Convention on Climate Change (UNFCCC). (2014). Background on the UNFCCC: The international response to climate change. Retrieved from http://unfccc.int/essential_background/items/6031.php

van Aelst, P., & Walgrave, S. (2011). Minimal or massive? the political agenda-setting power of the mass media according to different methods. *The International Journal of Press/Politics, 16*(3), 295–313. doi:10.1177/1940161211406727

Weber, C. L., & Matthews, H. S. (2008). Food-miles and the relative climate impacts of food choices in the United States. *Environmental Science & Technology, 42*(10), 3508–3513. doi:10.1021/es702969f PMID:18546681

Weidema, B. M., Wesnes, M., Hermansen, J., Kristensen, T., Halberg, N., Eder, P., & Delgado, L. (2008). Environmental improvement potentials of meat and dairy products: JRC Scientific and Technical Reports, No. EUR 23491. Seville, Spain: Institute for Prospective Technological Studies. Retrieved from http://ftp.jrc.es/EURdoc/JRC46650.pdf

Westhoek, H., Lesschen, J. P., Rood, T., Wagner, S., DeMarco, A., Murphy-Bokern, D., & Oenema, O. et al. (2014). Food choices, health and environment: Effects of cutting Europe's meat and dairy intake. *Global Environmental Change, 26*, 196–205. doi:10.1016/j.gloenvcha.2014.02.004

Wirsenius, S., & Hedenus, F. (2010). Policy strategies for a sustainable food system: options for protecting the climate. In J. D'Silva & J. Webster (Eds.), *The meat crisis: developing more sustainable production and consumption* (pp. 237–253). London, UK: Earthscan.

Wong, L., Selvanathan, E. A., & Selvanathan, S. (2013, July 7-10). Changing patterns of meat consumption in Australia 2013. Paper presented at the Australian Conference of Economists, Murdoch University, Australia. Retrieved from http://www.murdoch.edu.au/School-of-Management-and-Governance/_document/Australian-Conference-of-Economists/Changing-pattern-of-meat-consumption-in-Australia.pdf

Woodford, J. (2007, March 23). How to live in a more sustainable way. *The Sydney Morning Herald*.

KEY TERMS AND DEFINITIONS

Animal Agriculture: The raising and slaughtering of animals for food and other commodities.

Anthropogenic Climate Change: Observed human-caused shifts in long-term average weather patterns.

Behavior Change: An individual decision to alter their actions.

Discourse: The way knowledge and power is created and controlled through dominant ways of talking about an issue.

Framing: Presenting facts or other information in such a way that encourages certain interpretations and discourages others (in the context of mass media communication).

Greenhouse Gases: A gas in the atmosphere that absorbs and re-emits radiation from the sun.

Plant-Based Diet: The elimination of animal products from an individual's regular eating habits.

Chapter 6
Environmental Concerns and the Mainstreaming of Veganism

Nick Pendergrast
Curtin University, Australia

ABSTRACT

Increasing awareness about the environmental consequences of consuming animal products together with recognition of animal rights and health benefits, has played a significant role in the rising interest in veganism. The chapter analyses recent trends in veganism in the Western world, focusing on the substantial environmental impact of animal agriculture being highlighted by respected agencies including the Food and Agriculture Organization of the United Nations and widely read publications such as the Guardian and the New York Times. Growing awareness of the environmental impact of animal products and greater visibility of veganism in the mainstream media have assisted in veganism emerging as an important response to the environmental crisis.

There has been a huge change regarding veganism. When Animal Liberation first appeared [in 1975], you couldn't use the word 'vegan' without an explanation. The Vegan Society in Britain had only 300 members. There was a US vegan group that was small. There has been an enormous difference, an interesting difference. I was surprised – if you had asked me in 1975, I would have said I think vegetarianism would spread, and it would have been at a higher level in the public than now. I would have been surprised veganism caught on to the extent that it has, (Singer, 2012).

INTRODUCTION

The damage being done to the earth's environment as a result of rapidly increasing meat production and consumption is a central theme of this publication. Veganism emerges as one strategy that individuals can take to assist in combatting this alarming situation. The term "veganism" initially referred to a diet

DOI: 10.4018/978-1-4666-9553-5.ch006

free of animal products when it was coined by Donald Watson in 1944 (The Vegan Society, 2012). In 1951, The Vegan Society redefined the concept to mean seeking 'to end the use of animals by man for food, commodities, work, hunting, vivisection and all other uses involving exploitation of animal life by man' (Cross, 1951). Both meanings are still often used and confusion over the use of the term still prevails in the media and within the animal advocacy movement. Nonetheless, veganism as a term and concept is increasingly evident in the media and in public consciousness. Articles in mainstream publications promoting veganism have assisted in spreading awareness and providing a much more viable environment for vegan advocacy (Munro, 2012, pp. 169, 173; Sneijder & te Molder, 2009, pp. 626, 628).

Growing awareness of the environmental consequences of consuming animal products, in addition to mainstream recognition of animal rights and health benefits, has played a significant role in the rising interest in veganism. Major reports by respected agencies, including the Food and Agriculture Organisation of the United Nations (Steinfeld et al., 2006, p. 20) and the Worldwatch Institute, have warned about the enormous contribution that animal agriculture makes to worldwide greenhouse gas emissions (Goodland & Anhang, 2009). This chapter will examine the growth of veganism in the Western world, and the way in which its increasing profile has enabled wider recognition of the environmental and ethical benefits that it brings, with particular reference to the situation in Australia and the United States.

Snow (2004, p. 383) explains that social movements are 'embroiled in conflict over competing claims about aspects of reality'. An important sociological concept in understanding social movements and social change is "claims-making". 'Claims makers are people who articulate and promote claims' and their cause is benefited 'in some way if the targeted audience accepts their claims as true' (Ferrante, 2011, p. 167). Individuals and organisations encouraging people to become vegans are examples of claims makers. Carrie Freeman (2013, pp. 94, 109) notes that common claims made by vegan advocates are that it has benefits in terms of assisting non-human animals, the environment and human health (see, for example, ALV, 2012; PETA, 2011b).

Attracting media attention for such claims is important to successful claims-making (Ferrante, 2011, p. 169). Vegan advocates are now having more success in getting their claims heard, partly due to mainstream media being more receptive to a vegan message. This success has been strengthened by claims about environmental benefits of veganism emanating from sources outside the animal advocacy movement. When highly regarded institutions, like the United Nations, endorse claims commonly made by animal advocates, the perceived legitimacy of these claims increases (Williams, 2012, pp. 9-10). As a result of these supportive reports from respected organisations and their coverage in mainstream media, the impact of the consumption of animal products is getting increasing traction as an important environmental concern, not only within the animal advocacy movement, but in the wider public sphere.

This chapter will examine some articles published by widely read publications like the *Guardian* and the *New York Times*. Articles in these publications have been chosen due to their large reach, which extends well beyond animal advocates to a much more general audience (Boyd-Barrett, 2006, pp. 170-171). The fact that they are being included is significant: it demonstrates a belief within mainstream media that audiences have sufficient familiarity and interest in veganism to engage with the subject. Due to time and space constraints, media rarely present material with which their audience is unfamiliar (Flew, 2008, p. 11).

While veganism is increasingly entering the mainstream media and consciousness, attention has mainly focussed on a diet free of animal products, rather than the broader philosophy and lifestyle (Hill, 2011). This lifestyle involves an individual commitment to not only avoid eating animal products, but also to not participate in other instances of animal exploitation, such as the use of non-human animals for entertainment and clothing. This is closely tied to an animal rights ideology that contests the concept of other animals being here for humans to use and slaughter; regardless of how "humanely" or otherwise this is done (Williams, 2012, pp. 13-14). The reason for the dietary focus of much of the mainstream recognition of veganism is because it is particularly the environmental and health arguments that have attracted the most attention. Adopting a vegan diet for these reasons does not necessarily involve an embracing of animal rights ideology (Waters, 2014). This demonstrates a limitation of the mainstreaming of the vegan diet in terms of addressing the interests of the non-human animals who are harmed for other purposes besides food, however, this process also has the potential to lead to significant gains for the environment and the other animals we share it with.

THE MAINSTREAMING OF VEGANISM

Veganism is increasingly becoming a mainstream dietary choice. Below I review developments in Australia and the United States as well as in the online space.

Vegans in the United States and Australia

These two countries were chosen because the initial aim was to focus on Australia but through research it was found that media and animal advocacy in the United States exerts a strong influence on animal advocacy in Australia. The limited data on the number of vegans in the United States and Australia points to only a small proportion in both countries.

Elizabeth DeCoux argues that if the empirical data is explored, it is clear that the percentage of the population in the United States who are vegans is not increasing, and may actually be decreasing. Her conclusion draws on a variety of surveys in the United States between 1997 and 2009 commissioned by vegetarian organisations or publications and a *Time* magazine poll in 2002. The poll in *Time* showed the lowest proportion of vegans in America as compared to the other surveys (0.2 percent). DeCoux maintains that this is likely to be the most accurate, as it had the largest sample size and was not commissioned by pro-vegetarian sources. The highest figure cited was 1.8 percent in 2003 and the most recent survey she drew on, put the figure at only 0.8 percent. All the polls involved around 10,000 respondents or less (2009, pp. 18, 25-27). The very limited sample makes it difficult to make strong conclusions about trends or proportions of vegans in the population, but they do show that numbers were clearly very small. More recent surveys do show results of two to three percent, however, once again, the small sample sizes make it difficult to draw reliable conclusions (Newport, 2012; Stahler, 2011).[1]

Unfortunately, there is very little data available on the number of vegans in Australia (Lawson, 2008, October 3; Vegan Society NSW, 2012). More research would be valuable. In a 2009 Australian survey conducted by Newspoll for the Vegan/Vegetarian Society of Queensland, 1 percent of respondents claimed to be vegan. However, once these people were questioned about what they ate, most ate animal products, at least some of the time. In fact, only one of the 1 202 people surveyed was actually vegan – meaning that the proportion was only 0.06 percent (VVSQ, 2010, pp. 3-4). These Australian examples highlight the difficulties of getting accurate information from survey respondents about dietary preferences. The Vegan Society NSW (2012) plans to 'commission a poll to determine [the] number of vegans in Australia and number of people who consume vegan meals'. This is likely to provide more substantial Australian information.

While there is little reliable data to make conclusions about the number of vegans in Australia and the United States, this chapter will demonstrate that there is a growing awareness of, and interest in, veganism in both countries and beyond. Lyle Munro (2012, p. 169) proposes that the 'normalization of previously exotic issues' is crucial to the success of social movements. Petra Sneijder and Hedwig te Molder (2009, pp. 626, 628) analyse the ways in which vegan advocates commonly use the Internet to participate in normalising 'vegan food choice and practices' in order to 'construct and protect veganism as an ideology'. Carol Glasser (2011, p. 192) pointed to some success being achieved through such advocacy when she noted that 'there is clearly a cultural and corporate recognition that refusing to use animals for their flesh, skins, or excretions is a legitimate perspective'.

Veganism Online and into the Mainstream

While veganism has been discussed and promoted through alternative media sources for some time, it is now being increasingly taken up by more mainstream sources as well. Alternative media sources are those that are more likely than mainstream media to: originate from small, ideologically committed groups; have cheap and accessible distribution; rely on funding from users; exhibit non-commercial behaviour; contain a diversity of sources and perspectives; be politically oppositional; promote activism; and present ideologies not well represented in the mainstream media (Boyd-Barrett, 2006, pp. 206-207). Websites from individual advocates or organisations promoting veganism meet many of these criteria.

Activists promoting veganism have used the Internet to create blogs, podcasts, and websites where they can present their views, which have often been excluded in the mainstream media. The Internet is decentralised in comparison to other forms of media: it has put media tools in the hands of "ordinary people". New media practices and forms, new ways of consuming and using media, and alternative spaces for a variety of communities and interests not well catered for or represented in traditional media forms are emerging (Goggin, 2006, pp. 259-276; Petray, 2011, pp. 924-925).

The Internet has created the opportunity for arguments promoting veganism to be supported in detail through books, slideshows, documentaries, articles and podcasts. This development has enabled veganism to be explained without being reliant on traditional media, with its time and space constraints, which were referred to above. Paul McLaughlin and Marwan Khawaja (2000, p. 425) argue that resources are required to promote a social movement's frame, however, the media landscape is in a rapid period of change and the animal rights frame is being promoted very cheaply online.

Vegan activists are able to cheaply distribute their own leaflets, videos, blogs, articles, podcasts and much more using the Internet (Wrenn, 2012, p. 451; Yates, 2015). Pamphlets promoting veganism are readily available on a range of animal rights websites. While the creation of pamphlets and other information with this message were certainly created in the past, the Internet has allowed such materials to more easily spread beyond the location in which they are produced and reach more people.

Arguments for veganism are also gaining more currency in broader society through coverage in more mainstream media. These sources have many characteristics that alternative media are less likely to have, including: carrying advertising; conforming to professional conventional standards in terms of job specialisation and content; and addressing members as a mass, white or middle-class collectivity (Boyd-Barrett, 2006, pp. 206-207). It is important to avoid analysing alternative and mainstream media in the form of opposing binaries and totally isolated entities, as is sometimes done in media research. There is actually a crossover of journalists, ideas, content, and style between these two supposedly oppositional media forms (Harcup, 2005; Bolton, 2006). It is clear, however, that in recent times the vegan

Figure 1. The growing coverage of veganism in Australian newspapers
Compiled from data from Factiva (2014).

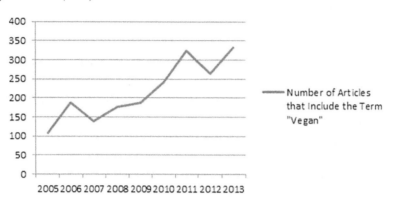

message has gone from being promoted in sources that score highly as far as Boyd-Barrett's (2006, pp. 206-207) criteria for alternative media, to sources that fit many of the characteristics he identifies for mainstream media.

This is new territory for vegan messages: it is difficult for alternative ideas to break into mainstream media coverage, which can assist in explaining why veganism was marginalised in the past. Scarcity of space (in print media) and time (in broadcast media) serve to marginalise different, new, and critical voices from the media (Flew, 2008, p. 11). 'In a three-minute stretch between commercials, or in seven hundred words, it is impossible to present unfamiliar thoughts or surprising conclusions with the argument and evidence required to afford them some credibility' (Chomsky, 1989, p. 10). Repeating familiar information that does not depart too far from dominant understandings does not face these problems (Chomsky, 1989, p. 10). Such understandings are accepted as "common sense" and therefore require little or no explanation.

The growing number of articles on veganism in more mainstream sources indicates that the public now has sufficient familiarity with the concept for media to be more receptive to the topic. It has reached a point where it has overcome some of the problems for attracting media coverage that Chomsky defined. Figure 1 demonstrates the increasing number of times the word "vegan" is used in Australian newspapers. The data was obtained through searching the *Factiva* database (2014) for the term "vegan" in all Australian newspapers.

Further evidence that veganism is entering mainstream consciousness can be found from *Google* data. In 2011, for the first time ever, "vegan" was regularly getting a higher "search volume index" (number of searches) than "vegetarian", as well as sometimes achieving a higher "news reference volume" (number of news stories) (Google, 2013). These statistics, displayed in Figure 2, clearly demonstrate a growing number of Internet searches for "vegan" (shown in blue/the darker colour) compared to "vegetarian" (shown in red/the lighter colour) over time.

As Singer's quote at the beginning of this chapter indicated, he has been surprised at how much interest in veganism has increased over the last few decades, particularly in relation to vegetarianism. As noted above, veganism is a relatively new concept, whereas the history of vegetarianism goes much further back. The idea of avoiding eating the flesh of other animals for health and ethical reasons has been around for millennia, in many different cultures all around the world, often as part of religious convictions and practice. In Western countries, the first formal efforts by adherents to organise began in

Figure 2. Google trends on veganism and vegetarianism (Google, 2013).

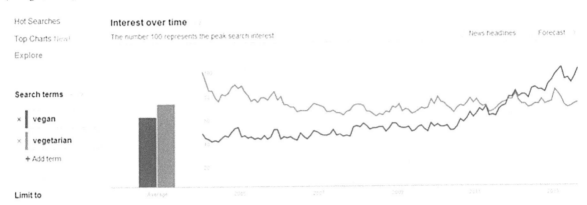

Britain. A national Vegetarian Society was set up in Britain in 1847 to promote a meat-free diet (Leneman, 1999, p. 219). The growing awareness of the less historically recognised concept of veganism is demonstrated in discussions about the environmental impact of animal agriculture.

ENVIRONMENTAL CONCERNS AND VEGANISM

Environmental concerns have played a significant role in the growing interest in veganism. Mainstream media have regularly raised the issue of links between environmental degradation and the consumption of animal products in recent times. This link was highlighted in the Food and Agriculture Organisation of the United Nations report 'Livestock's Long Shadow' in 2006, which revealed that 'the livestock sector emerges as one of the top two or three most significant contributors to the most serious environmental problems, at every scale from local to global' (Steinfeld et al., 2006, p. 20). This report had a powerful impact at the time due to the prestige and legitimacy of the United Nations. One particular fact from the report that has been repeated numerous times by media outlets and animal advocacy organisations alike, is that livestock are responsible for 18 percent of worldwide greenhouse gas emissions – more than all forms of transport combined (FAO, 2006).

The findings of this report were highlighted by a range of animal advocacy organisations. The Humane Society of the United States (HSUS) promoted them in an advertising campaign (see Figure 3) and People for the Ethical Treatment of Animals (PETA) attempted to convince environmentalist Al Gore to become a vegetarian based on the results of this report (see Figure 4) (PETA, 2007). Both of these campaigns were covered in the article 'Trying to Connect the Dinner Plate to Climate Change,' which was published in the *New York Times* (Deutsch, 2007). The *New York Times* newspaper has a weekday circulation of over 900,000 and a Sunday circulation of 1.4 million (Perez-Peña, 2008) and many more than this view the content of the paper online (Bianco, 2005).

While in the past PETA has encouraged people to become vegetarians as a result of the impact meat production has on the environment (PETA2, 2015; PETA Asia-Pacific, 2008), it is now broadening its framing of the problem. On its *Meat's Not Green* website, it no longer focuses purely on meat production and argues that 'raising animals for food' leads to more greenhouse-gas emissions than all forms

Figure 3. The humane society reveals an important fact from the recent UN report
(Deutsch, 2007). Used with permission from The Humane Society of the United States. © 2007 The HSUS. All Rights Reserved.

The Humane Society links
environmental issues and food.

Figure 4. PETA encourages Gore to become a vegetarian
(Deutsch, 2007).
Image courtesy of People for the Ethical Treatment of Animals, www.peta.org.

In case you thought we were just kidding when we wrote to Al Gore urging him to go vegetarian to help stop global warming, maybe this ad will clarify our position for you.

of transport combined. This site now has a *vegan* pledge, where people pledge to try *veganism* for one week (PETA2, 2015). While veganism is not central to PETA's campaigning (Pendergrast, 2010, pp. 14-15), it is significant that PETA is now framing all animal-based foods as a key factor in world environmental problems, with the solution being veganism. Incidentally, though PETA was lobbying Gore to become a vegetarian in previous years, he has recently made the more radical change to a vegan diet, which PETA reported on (Kretzer, 2013).

This changing discourse on the environmental impact of animal agriculture is not limited to PETA. There is a growing move amongst mainstream organisations outside of the animal advocacy movement to frame the problem in a similar way to PETA. This is not totally new – even before the 2006 United Nations report, researchers from the University of Chicago explained in *New Scientist* that their research found that people could do more for the planet by changing to a *vegan* diet than replacing a standard car with an eco-friendly one (New Scientist, 2005). The term 'vegan' is increasingly used in reference to the environmental problems of animal agriculture, as the term gains more mainstream recognition, as was demonstrated in Figures 1 and 2.

Since the 2006 United Nations report, there have been more reports that draw attention to the link between environmentalism and veganism. A report released in 2009 by the environmental organisation, the Worldwatch Institute, argued that the report from the United Nations underestimated the environmental impact of animal agriculture. It found that animal agriculture is actually responsible for at least 51 percent of worldwide greenhouse gas emissions (Goodland & Anhang, 2009). A further United Nations report made a very strong statement advocating 'a substantial worldwide diet change, away from animal products' and 'a global shift towards a vegan diet' as a way of reducing world 'hunger, fuel poverty and the worst impacts of climate change' (Carus, 2010, p. 1). This report was highlighted in the article 'UN Urges Move to Meat and Dairy-Free Diet', which appeared in *The Guardian* (Carus, 2010). *The Guardian* has over a million readers (Media Works, 2009).

Although PETA mentioned the 2009 Worldwatch Institute report among a list of environmental organisations which recognise that animal agriculture is one of the most pressing environmental issues (PETA, 2011a), neither animal advocacy organisations nor mainstream media gave these later reports anywhere near the same attention as was paid to the 2006 report. The reason could lie partly with "climate change fatigue": the notion that over-exposure to the issue has led to the public and media losing interest in the last few years (Nordhaus & Shellenberger, 2009). Climate change was very topical in the media around 2006 when the United Nations report was published, the same year as Al Gore's influential documentary *An Inconvenient Truth* was released. The fact that the earlier report was associated with the United Nations, with its higher profile, may have also contributed to the more limited attention paid to the Worldwatch Institute's findings. While the 2006 report attracted more attention, the findings were not regularly tied to veganism at the time. Vegetarianism was sometimes mentioned, but rarely veganism.

The environmental impact of animal products remains a mainstream environmental issue and coverage of the issue is increasingly linked to veganism. Animal advocacy organisations, including PETA have highlighted findings provided by organisations very separate to animal advocacy such as the Food and Agricultural Organisation of the United Nations, the UK Government (Clover, 2007), and the University of Chicago (New Scientist, 2005). The findings of these reports have been repeated by respected media sources such as *The Guardian* (Carus, 2010), *The New York Times* (Kenter, 2007), and *New Scientist* (2005).

These articles provide an indication of mainstream media's willingness to engage with the environmental problems associated with the consumption of animal products. They also demonstrate the vegan diet being advocated in mainstream media as a response to these problems, and the ways in which the term "vegan" is being recognised and adopted. Cary Williams (2012, pp. 9-10) explains that credibility for claims makers is partly achieved through the 'credibility of frame articulators' and the institutions articulating the frame of animal products being environmentally damaging are widely viewed as credible sources. This coverage is assisting in animal products being viewed more broadly as a legitimate environmental concern, and not just an issue for animal advocates.

It has not just been environmental concerns that have led to a growing interest in veganism. Increasing numbers of high profile people adopting a vegan diet, motivated mainly by awareness of the health benefits, has also assisted in this process. In our celebrity-driven culture, and particularly media, celebrities contribute authority and attention to a cause (Brockington, 2009, p. 26). Celebrities have been very important in bolstering the exposure veganism is given and normalising the concept.

Probably the most publicised example of a person advocating the health benefits of a vegan diet is former US President Bill Clinton. Throughout his presidency, Clinton ate a diet high in animal products. He was overweight and narrowly escaped death due to heart problems, enduring two heart operations. Now that Clinton has dramatically changed his diet, tests show that he is starting to reverse the damage to his heart and blood vessels caused by cardiovascular disease. He has lost over nine kilograms in weight and claims he has more energy than before (Brown, 2011; Martin, 2011). Clinton's story has been covered on *CNN* (Gupta, 2011), *The New York Times* (O'Connor, 2011), *USA Today* (Hellmich, 2011), the *Los Angeles Times* (Brown, 2011), *The Huffington Post* (Shah, 2011) and the American Broadcasting Company (ABC, 2011), to name just a few of the many outlets that have covered this story. There are many other examples of celebrities who have raised the profile of veganism, including talk show host Ellen (DeGeneres, 2011; Hill, 2011) and actress Alicia Silverstone (Hill, 2011; Silverstone, 2011; Vegetarian Star, 2011).

There have also been clear arguments for veganism from an animal rights standpoint published in *The Guardian* (Schonfeld, 2010), *The New York Times* (Steiner, 2009) and *The Sydney Morning Herald* (Fox, 2009). *The Sydney Morning Herald* newspaper has one of the highest circulations in Australia, with a readership of over five million, including an average of over 700,000 people reading the paper each day (Fairfax Media, 2006). The website for the Sydney Morning Herald, smh.com.au, also has over two million unique browsers per month (Ricketson, 2007). While there are a growing number of articles arguing for veganism on animal rights grounds, it has been the environmental and health arguments that have gained the most traction. This is possibly due to the association with respected organisations such as the United Nations in the case of the environmental arguments for veganism and the celebrity association with the health arguments.

GROWING INTEREST IN VEGANISM

Arguments in favour of veganism on environmental, animal rights and health grounds are increasingly being presented in the mainstream media. In an article published in *Bloomberg Businessweek*, Michael Hill (2011) wrote: 'You've come a long way, vegan. Once mocked as a fringe diet for sandal-wearing health food store workers, veganism is moving from marginal to mainstream in the United States'. Similarly, Raman Nijjar (2011) argues, in his article 'From Pro Athletes to CEOs and Donut Cravers, the Rise of the Vegan Diet', published in *CBC News* Canada, that: 'It used to be that vegetarian and vegan diets were looked down upon as almost sect-like fads'. Singer (2012) explains that even the term "vegetarian" had this type of association for the public when he wrote *Animal Liberation* in the 1970s: the best vegetarian restaurant in London gave itself the tongue in cheek title of "Cranks", reflecting the perception of vegetarianism at the time.

As the term 'vegan' and arguments for veganism increasingly enter mainstream media and conscious-ness, it is becoming easier to be a vegan. In Australia, Canada, the United States and many other Western countries, vegan staples like tofu and tempeh, as well as "meat substitutes" such as vegan "chicken" and "bacon" are increasingly available at local grocery stores and restaurants (Hill, 2011; Nadalin, 2012; Nijjar, 2011). These meat substitutes assist some vegans, particularly those making the transition from a non-vegan diet, as they can substitute foods that are more familiar to them with vegan products (Nath & Prideaux, 2011, para. 15-19). Many chain restaurants and university cafeterias in the countries men-tioned above now mark their vegan options and big brands like Kraft Foods now sell products like vegan burgers and crumbles. It was a very different story even ten years ago, with far fewer options available for vegans (Grasgreen, 2011; Hill, 2011; Nijjar, 2011; Vegan Easy, 2012).

As mentioned above, this growing interest in the vegan diet has mainly been a result of environmen-tal and health concerns, rather than a strong sense of ideological commitment to animal rights. As a result, there has been growing awareness of the vegan diet but not necessarily the broader philosophy and lifestyle (Hill, 2011). An example of this occurred when famous musicians Jay-Z and Beyoncé went on a 'Three Week Vegan Diet' as a 'spiritual/physical cleanse' and during this time, Beyoncé was seen wearing fur and suede (ABC, 2013). While veganism is often promoted as an ethical practice beyond diet by animal advocates, outside of the movement there tends to be more trivial representations (Taylor & Twine, 2014, p. 12).

The mainstreaming of the vegan diet for health and environmental reasons is a significant develop-ment, both for the environment as a whole, as well as for non-human animals. However, for non-human animals specifically, these gains are somewhat limited, as they do not incorporate those harmed and killed for other purposes beyond food, such as clothing and entertainment. There has been an attempt by some to differentiate eating a vegan diet for reasons such as health or the environment from the broader philosophy. T. Colin Campbell, author of *The China Study*, which explored the health benefits of plant-based diets, advocates people avoid eating animal products purely to avoid the health risks. He explains why he uses the term "plant-based diet" rather than "vegan":

I don't use the word "vegan" or "vegetarian." I don't like those words. People who chose to eat that way chose to because of ideological reasons. I don't want to denigrate their reasons for doing so, but I want people to talk about plant-based nutrition and to think about these ideas in a very empirical scientific sense, and not with an ideological bent to it (Parker-Pope, 2011; Campbell & Campbell, 2005)

For someone to give up animal products beyond diet, there needs to be this 'ideological bent' involv-ing a commitment to animal rights. No one could argue that wearing a leather jacket is detrimental to people's health and the use of animals in circuses is hardly a huge environmental issue. Those becoming vegans for animal rights reasons will not only boycott animal exploitation beyond food but they are also more likely to stay vegan in the long-term. This is demonstrated by studies which focus on commitment to vegetarianism, though the results are also relevant to veganism.

In their study on exiting vegetarianism, Kenneth Menzies and Judy Sheeshka (2012, p. 163) found that 'ex-vegetarians were more likely to have become vegetarians as a result of concern about the wellbeing of animals and the environment, not animal rights, a value more difficult to compromise'. An example of this is DeGeneres, whose veganism was mentioned above, revealing that she now eats eggs from chickens raised in her neighbour's backyard, 'because they're happy' (Claire, 2012). An animal rights philosophy, which can be summed up by PETA's slogan of 'animals are not ours', totally rules out consuming animal products. More general concern for animals without this animal rights philosophy leaves greater leeway for people to return to consuming those animal products that they view as humane.

Considering the interests of animals (even if not specifically based on animal rights) leads to greater commitment than health concerns. Hoffman, Stallings, Bessinger and Gary Brooks (2013) found that people who became vegetarian for ethical reasons had stronger feelings of conviction than those who did so for health reasons, and for this reason stayed vegetarian longer. Freeman (2013, p. 94) agrees that ethical concerns lead to greater commitment, arguing that those cutting out animal products for health reasons 'may be tempted by the convenience of a meat-based diet and new lower-fat meat items'. She believes that while health and environmental factors are useful in encouraging people to reduce their consumption of animal products, it is concern for animals that is most effective in convincing people to give up animal products all together (Freeman, 2013, p. 95). As discussed above, Bill Clinton has had a dramatic health improvement as a result of significantly reducing the amount of animal products he eats, but he acknowledges that 'once a week or so' he will indulge in a serving of salmon or an omelette (Larson, 2013). Those who are vegan for animal rights reasons are far less likely to have such "indulgences" – as noted above, this value is 'more difficult to compromise' (Menzies & Sheeshka, 2012, p. 163).

It must be acknowledged, however, that environmental and health concerns can be an "entry point" to animal rights, if people who choose to eat a vegan diet for these reasons 'are incorporated into a community where they are exposed to broader concerns of animal well-being' (Waters, 2014). Even though many may only embrace the dietary aspect of veganism, a growing interest in the vegan diet is still very significant, even from an animal standpoint, as an overwhelming majority of non-human animals used and killed by humans is for food (Freeman, 2013, p. 93; Green, 2008; Pearson, 2011; Shapiro, 2004). It also provides a more fertile environment to promote a broader vegan philosophy based around an animal rights position on all forms of animal exploitation, of which diet is a significant part.

CONCLUSION

Most mainstream media focus has been on environmental and health arguments for becoming a vegan rather than a broader animal rights agenda. This increased attention has centred mainly on the dietary aspects of veganism. While this shortcoming needs to be acknowledged by anyone concerned about individual non-human animals and not just the environment as a whole, it is also important to consider the benefits of this mainstreaming of the vegan diet, both for the environment and animals. The increasing prominence afforded to veganism in the media has helped considerably to promote its environmental, health and animal rights benefits. This type of mainstream coverage has also assisted in legitimising claims when they are made by animal advocates.

This chapter has traced the role of environmental concerns in the growing normalisation of veganism in the Western world, focusing particularly on Australia and the United States. While the limited research done to date indicates that there are clearly still a proportionately small number of vegans, there is evidence that veganism is gaining a more prominent place in both the media and community consciousness. Vegetarianism gradually became more widely known and accepted during the mid to latter part of the twentieth century, while veganism remained a "fringe" concept. Now there are signs that this is changing.

There has been a clear increase in the profile that mainstream media has attached to veganism. There is much greater familiarity with the word and the concept amongst the public. With this growing interest in veganism, and demand for vegan options, service providers are beginning to respond to the market potential and create more options, which in turn makes it a less difficult choice to become a vegan. Veganism has made a small contribution towards addressing the seemingly overwhelming environmental issues threatening the planet. There is huge potential to lower the global footprint of the Western world through the greater uptake of veganism.

REFERENCES

American Broadcasting Company (ABC). (2011, August 19). Bill Clinton's vegan diet inspiring others. *ABC News*. Retrieved from http://abcnews.go.com/WNT/video/bill-clintons-vegan-diet-inspiring-14344849

American Broadcasting Company (ABC). (2013, December 8). Jay-Z and Beyonce's three week vegan diet. *ABC News*. Retrieved from http://abcnews.go.com/GMA/video/jay-beyonces-week-vegan-diet-21139910

Animal Liberation Victoria (ALV). (2012). Why vegan? Retrieved from http://veganeasy.org/Why-Vegan

Bianco, A. (2005). The future of The New York Times. *Bloomberg Businessweek*, Retrieved from http://www.bloomberg.com/bw/stories/2005-01-16/the-future-of-the-new-york-times

Bolton, T. (2006). *News on the Net: a critical analysis of the potential of online alternative journalism to challenge the dominance of mainstream news media*. Lilydale, Australia: Swinburne University of Technology.

Boyd-Barrett, O. (2006). Alternative reframing of mainstream media frames. In D. K. Thussu (Ed.), *Media on the move: global flow and contra-flow* (pp. 178–194). London, UK: Routledge.

Brockington, D. (2009). *Celebrity and the environment: fame, wealth and power in conservation*. New York, NY: Zed Books.

Brown, E. (2011, August 18). Bill Clinton talks about being a vegan. *Los Angeles Times*. Retrieved from http://articles.latimes.com/2011/aug/18/news/la-heb-bill-clinton-vegan-20110818 http://articles.latimes.com/2011/aug/18/news/la-heb-bill-clinton-vegan-20110818

Campbell, T. C., & Campbell, T. M. (2005). *The China study*. Dallas, TX: BenBella Books.

Carus, F. (2010, June 3). UN urges global move to meat and dairy-free diet. *The Guardian*. Retrieved from http://www.theguardian.com/environment/2010/jun/02/un-report-meat-free-diet

Chomsky, N. (1989). *Necessary illusions: thought control in democratic societies*. Boston, MA: South End Press.

Claire, J. S. (2012, July 16). Ellen no longer vegan. *Vegans Health*. Retrieved from http://www.vegan-shealth.com/ellen-no-longer-vegan/

Clover, C. (2007, May 30). Go vegan to help climate, says Government. *The Telegraph*. Retrieved from http://www.telegraph.co.uk/news/earth/earthnews/3295795/Go-vegan-to-help-climate-says-Government.html

Cross, L. (1951). Veganism defined. Retrieved from http://www.ivu.org/history/world-forum/1951vegan.html

DeCoux, E. L. (2009). Speaking for the modern Prometheus: The legal and ideological significance of animal sufferring to the abolition movement. *Animal Law, 16*(9), 9–64.

DeGeneres, E. (2011). Change your life by going vegan with Ellen! Retrieved from http://www.ellentv.com/2011/08/02/change-your-life-by-going-vegan-with-ellen/

Deutsch, C. H. (2007). Trying to connect the dinner plate to climate change. *The New York Times*. Retrieved from http://www.nytimes.com/2007/08/29/business/media/29adco.html?_r=0

Easy, V. (2012). Eating out guide. Retrieved from http://www.veganeasy.org/Eating-Out-Guide

Factiva. (2014). Vegan. Retrieved from http://new.dowjones.com/products/factiva/

Ferrante, J. (2011). *Sociology: a global perspective* (8th ed.). Belmont, CA: Wadsworth.

Flew, T. (2008). Not yet the internet election: Online media, political commentary and the 2007 Australian Federal Election. *Media International Australia Incorporating Culture and Policy, 126*, 5–13.

Food and Agriculture Organization of the United Nations (FAO). (2006). Spotlight/2006: Livestock Impacts on the Environment. *FAO Magazine*. Retrieved from http://www.fao.org/ag/magazine/0612sp1.htm

Fox, K. (2009, December 28). Call meat happy, but it is never humane. *The Sydney Morning Herald*. Retrieved from http://www.smh.com.au/it-pro/call-meat-happy-but-it-is-never-humane-20091227-lg61.html

Freeman, C. P. (2013). Stepping up to the veggie plate: framing veganism as living for your values. In E. Plec (Ed.), *Perspectives on human-animal communication: internatural communication* (pp. 93–112). New York, NY: Taylor and Francis.

Glasser, C. L. (2011). *Moderates and radicals under repression: the u.s. animal rights movement, 1990-2010. (Doctor of Philosophy in Sociology)*. Irvine, CA: University of California.

Goggin, G. (2006). The Internet, online and mobile cultures. In S. Cunningham & G. Turner (Eds.), *The media and communications in Australia* (2nd ed., pp. 259–278). Crows Nest, Australia: Allen & Unwin.

Goodland, R., & Anhang, J. (2009, November/December). Livestock and climate change: what if the key actors in climate change are... cows, pigs and chickens? Washington, DC: Worldwatch Institute. *World Watch Magazine, 22*(6), 10-19.

Google. (2013). Google trends: vegan, vegetarian. Retrieved from https://www.google.com.au/trends/explore - q=Vegan%2C%20Vegetarian

Grasgreen, A. (2011, August 18). Southern veganism. *Inside Higher Education*. Retrieved from https://www.insidehighered.com/news/2011/08/18/university_of_north_texas_creates_vegan_dining_hall

Green, C. (2008). How to end 98% of animal abuse in the next 25 years. Retrieved from https://faunalytics.org/how-to-end-98-of-animal-abuse-in-the-next-25-years/

Gupta, S. (2011, August 29). Dr Sanjay Gupta reports: The last heart attack. *CNN*. Retrieved from http://sanjayguptamd.blogs.cnn.com/2011/08/29/sanjay-gupta-reports-the-last-heart-attack/

Harcup, T. (2005). "I'm doing this to change the world": Journalism in alternative and mainstream media. *Journalism Studies*, *6*(3), 361–374. doi:10.1080/14616700500132016

Hellmich, N. (2011). Bill Clinton declares vegan victory. *USA Today*. Retrieved September 9, 2015, from http://usatoday30.usatoday.com/yourlife/food/diet-nutrition/2011-08-24-Bill-Clinton-vegan_n.htm

Hill, M. (2011). Vegan diets becoming more popular, more mainstream. *Bloomberg Businessweek*, Retrieved from http://www.businessweek.com/ap/financialnews/D9KICAT00.htm

Hoffman, S., Stallings, S. F., Bessinger, R. C., & Brooks, G. T. (2013). Differences between health and ethical vegetarians: Strength of conviction, nutrition knowledge, diatry restriction and duration of adherence. *Appetite*, *65*, 139–144. doi:10.1016/j.appet.2013.02.009 PMID:23416470

Kenter, J. (2007, June 6). A vegetarian diet reduces the diner's carbon footprint. *The New York Times*. Retrieved from http://www.nytimes.com/2007/06/06/business/worldbusiness/06iht-greencol07.4.6029437.html?_r=0

Kretzer, M. (2013). A convenient truth: Al Gore goes vegan. Retrieved from http://www.peta.org/blog/al-gore-goes-vegan/

Larson, L. (2013, September 5). Bill Clinton cheats - on his vegan diet. *Daily News*. Retrieved from http://www.nydailynews.com/news/politics/bill-clinton-cheats-vegan-diet-article-1.1446582

Lawson, A. (2008, October 3). Environment turning middle-class vegan. *The Age*. Retrieved from http://www.stockandland.com.au/news/agriculture/agribusiness/general-news/environment-turning-middleclass-vegan/1323831.aspx

Leneman, L. (1999). No animal food: The road to veganism in Britain, 1909-1944. *Society & Animals*, *7*(3), 219–228. doi:10.1163/156853099X00095

Martin, D. S. (2011, August 18). From omnivore to vegan: the dietary education of Bill Clinton. *CNN*. Retrieved from http://edition.cnn.com/2011/HEALTH/08/18/bill.clinton.diet.vegan/

McLaughlin, P., & Khawaja, M. (2000). The organizational dynamics of the U.S. environmental movement: Legitimation, resource mobilization and political opportunity. *Rural Sociology*, *65*(3), 422–439. doi:10.1111/j.1549-0831.2000.tb00037.x

Media, F. (2006). Fairfax annual report 2006. Retrieved from https://www.fairfaxmedia.com.au/ArticleDocuments/191/Fairfax_AR2006_2006.pdf.aspx?Embed=Y

Menzies, K., & Sheeshka, J. (2012). The process of exiting vegetarianism: An exploratory study. *Canadian Journal of Dietetic Practice and Research*, *73*(4), 163–168. doi:10.3148/73.4.2012.163 PMID:23217442

Munro, L. (2012). The animal rights movement in theory and practice: A review of the sociological literature. *Social Compass*, *6*(2), 166–181. doi:10.1111/j.1751-9020.2011.00440.x

Nadalin, T. (2012, February 26). Fake meat tastes so good you can't tell the difference. *Herald Sun*. Retrieved from http://www.heraldsun.com.au/ipad/fake-meat-tastes-so-good-you-cant-tell-the-difference/story-fn6bfm6w-1226281498315

Nath, J., & Prideaux, D. (2011). The civilised burger: Meat alternatives as a conversion aid and social instrument for Australian vegetarias and vegans. *Australian Humanities Review, 51*, 135–151.

Newport, F. (2012). In U.S., 5% consider themselves vegetarians: even smaller 2% say they are vegans. Retrieved from http://www.gallup.com/poll/156215/consider-themselves-vegetarians.aspx

Nijjar, R. (2011, June 4). From pro athletes to CEOs and donut cravers, the rise of the vegan diet. *CBC News Canada*. Retrieved from http://www.cbc.ca/news/canada/from-pro-athletes-to-ceos-and-doughnut-cravers-the-rise-of-the-vegan-diet-1.1049116

Nordhaus, T., & Shellenberger, M. (2009, November 17). Apocalypse fatigue: losing the public on climate change. *The Guardian*. Retrieved from http://www.theguardian.com/environment/2009/nov/17/apocalypse-public-climate-change

O'Connor, A. (2011, August 18). Bill Clinton's vegan journey. *The New York Times*. Retrieved from http://well.blogs.nytimes.com/2011/08/18/bill-clintons-vegan-journey/

Parker-Pope, T. (2011, January 7). Nutrition advice from the China study. *The New York Times*. Retrieved from http://well.blogs.nytimes.com/2011/01/07/nutrition-advice-from-the-china-study/?_r=0

Pearson, M. (2011, January 4) Personal interview with Mark Pearson, Executive Director of Animal liberation in New South Wales (Interviewer N. Pendergrast).

Pendergrast, N. (2010, November 10-11). Veganism, organisational considerations and animal advocacy campaigns. Paper presented at the Voicing the Unseen: Just Write It, Curtin University, Perth, Australia. Retrieved from http://hgsoconference.curtin.edu.au/local/pdf/Pendegrast.pdf

People for the Ethical Treatment of Animals 2 (PETA2). (2015). Meat's not green. Retrieved from http://features.peta2.com/meatsnotgreen/index.asp

People for the Ethical Treatment of Animals Asia-PAcific (PETA Asia-Pacific). (2008). Think you can be a meat-eating environmentalist? Think again! Retrieved from http://www.petaasiapacific.com/resources/literature/leaflets/Think_You_Can_Be_a_Meat-Eating_Environmentalist_Leaflet_GEN_ENG.pdf

People for the Ethical Treatment of Animals (PETA). (2007). PETA to Gore: too chicken to go veg. Retrieved from http://www.peta.org/blog/peta-gore-chicken-go-veg/

People for the Ethical Treatment of Animals (PETA). (2011a). Meat and the Environment. Retrieved from http://www.peta.org/issues/animals-used-for-food/meat-environment/

People for the Ethical Treatment of Animals (PETA). (2011b). *PETA's vegetarian/vegan starter kit*. Sydney, Australia: PETA.

Perez-Peña, R. (2008). Newspaper circulation continues to decline rapidly. *The New York Times*. Retrieved from http://www.nytimes.com/2008/10/28/business/media/28circ.html

Petray, T. L. (2011). Protest 2.0: Online interactions and Aboriginal activists. *Media Culture & Society*, *33*(6), 923–940. doi:10.1177/0163443711411009

Ricketson, M. (2007, February 16). Big jump in online readership. *The Age*. Retrieved September 9, 2015, from http://www.theage.com.au/news/business/big-jump-in-online-readership/2007/02/15/1171405370719.html

Schonfeld, V. (2010, January 18). Five fatal flams of animal activism. *The Guardian*. Retrieved from http://www.theguardian.com/commentisfree/2010/jan/18/five-fatal-flaws-animal-activism

Scientist, N. (2005, December 14). It's better to green your diet than your car. *New Scientist*, *2530*. Retrieved from https://www.google.com.au/search?q=It%E2%80%99s+better+to+green+your+diet+than+your+car&oq=It%E2%80%99s+better+to+green+your+diet+than+your+car&aqs=chrome.69i57j0.461j0j8&sourceid=chrome&es_sm=91&ie=UTF-8

Shah, R. (2011, August 22). Why Bill Clinton went vegan. *The Huffington Post*. Retrieved from http://www.huffingtonpost.com/2011/08/22/bill-clinton-diet_n_932439.html

Shapiro, P. (2004, Summer). On the killing floor. *AV Magazine*, Retrieved from http://aavs.org/our-work/av-magazine/

Silverstone, A. (2011). The kind life with Alica Silverstone: vegan. Retrieved from http://thekindlife.com/?s=Vegan&submit=

Singer, P. (2012). *Personal interview with Peter Singer, Australian ethicist*. Interviewer N. Pendergrast.

Sneijder, P., & te Molder, H. (2009). Normalizing ideological food choice and eating practices: Identity work in online discussions on veganism. *Appetite*, *52*(3), 621–630. doi:10.1016/j.appet.2009.02.012 PMID:19501759

Snow, D. A. (2004). Framing processes, ideology and discursive fields. In D. A. Snow, S. A. Soule, & H. Kriesi (Eds.), *The Blackwell companion to social movements* (pp. 380–412). Malden, MA: Blackwell Publishing. doi:10.1002/9780470999103

Stahler, C. (2011). How often do Americans eat vegetarian meals and how many adults in the U.S. are vegan? *Vegetarian Journal*, *30*(4), 10–11.

Star, V. (2011). Alicia Silverstone starts vegan book club on "The kind life". Retrieved from http://vegetarianstar.com/2011/02/25/alicia-silverstone-starts-vegan-book-club-on-the-kind-life/

Steiner, G. (2009, November 21). Animal, vegetable, miserable. Retrieved from http://www.nytimes.com/2009/11/22/opinion/22steiner.html?pagewanted=all

Steinfeld, H., Gerber, P., Wassenaar, T., Castel, V., Rosales, M., & de Hann, C. (2006). Livestock's long shadow: environmental issues and options. Rome, Italy: Food Agriculture Organisation (FAO). Retrieved from ftp://ftp.fao.org/docrep/fao/010/a0701e/a0701e00.pdf

Taylor, N., & Twine, R. (2014). Introduction: Locating the 'critical' in critical animal studies. In N. Taylor & R. Twine (Eds.), *The rise of critical animal studies: from the margins to the centre* (pp. 1–16). New York, NY: Routledge. doi:10.5744/florida/9780813049205.003.0001

The Vegan Society. (2012). Donald Watson. Retrieved from http://www.foodsforlife.org.uk/people/Donald-Watson-Vegan/Donald-Watson.html

The Vegetarian/Vegan Society of Queensland Incorporated (VVSQ). (2010). A pound of flesh. Brisbane, Australia: VVSQ. Retrieved from https://www.voiceless.org.au/sites/default/files/PoundofFlesh220310.pdf

Vegan Society NSW. (2012). Vegan dining campaign. Retrieved from http://www.vegansocietynsw.com/vs/html/campaigns_vegan_dining.html

Waters, C. (2014). Accommodating the target: veganism, healthfulness, and hegemonic masculinities. Olympia, WA: Faunalytics (Formerly Humane Research Council).

Williams, C. (2012). *The framing of animal cruelty by animal advocacy organizations. (Sociology with Honors)*. ME, USA: The University of Maine.

Works, M. (2009). The Guardian. *Media Works Asia Limited*. Retrieved from http://www.mediaworksasia.com/publication/detail/000000006/page1/index.html.bak

Wrenn, C. L. (2012). Abolitionist animal rights: Critical comparisons and challenges within the animal rights movement. *Interface*, *4*(2), 438–458.

Yates, R. (2015). *On human-nonhuman relations, a sociological exploration of speciesism*. Retrieved from http://onhumannonhumanrelations.tumblr.com

KEY TERMS AND DEFINITIONS

Global Footprint: The global ecological impact of human activities.

PETA: People for the Ethical Treatment of Animals, an American animal rights non-profit organisation.

Vegan Diet: A plant-based diet free of all animal-derived foods.

Veganism: A philosophy and lifestyle that avoids contributing to the exploitation of other animals, whatever the purpose, including food, clothing and entertainment.

Vegetarian: A person who eats mainly a plant-based diet and abstains from meat consumption.

Western World: Includes the countries located in Europe, United States of America, Canada, Australia, New Zealand and parts of Latin America.

ENDNOTE

[1] A 2011 survey conducted by the Vegetarian Resource Group found that 2.5 percent of Americans were vegan. This represented a substantial increase since 2009 where the figure was only 0.8 percent. According to this survey, for the first time, it appears that there may be more vegans than vegetarians in the United States. A total of 2 percent of men and 2 percent of women surveyed avoided flesh but consumed dairy and/or eggs, but 3 percent of men and 2 percent of women were vegan. However, the size of the survey was small (only 2 030) and its results should be treated with caution (Stahler, 2011). A 2012 Gallup poll found 2 percent of Americans were vegan, but again, the sample size was small – it only drew on 1 014 people (Newport, 2012).

Section 2
Diet and Wellbeing

Chapter 7
Meat and Dietary Guidelines

Rosemary Stanton
University of New South Wales, Australia

ABSTRACT

This chapter aims to describe how meat fits into recommended dietary guidelines. In Australia, meat is included in one of the five food groups. However, this food group should not be described as the 'meat group' as it includes alternative choices. These include animal products such as seafood, poultry and eggs but also plant-based alternatives such as legumes, tofu, nuts and seeds. Choosing a range of foods from within this group contributes to a healthy dietary pattern with nuts, seeds and legumes providing extra benefits. Increasing plant-based choices also makes it easier for those who consume meat to keep to the weekly limit recommended to reduce the risk of health problems associated with a high consumption of red meat. Processed meats are not included in any of the five food groups and are now seen as 'discretionary' foods.

INTRODUCTION

In Australia, the National Health and Medical Research Council released its latest Dietary Guidelines in 2013 (NHMRC, 2013). Unlike previous versions, the latest Australian guidelines based their evidence and advice on foods rather than nutrients.

In formulating the guidelines, key sources of evidence included:

1. Previous guidelines and their supporting documentation;
2. An evidence report designed to systematically answer a series of questions about relationships between specific foods and health outcomes. For this report, the expert Working Committee identified areas where new evidence, any changes or uncertainty might be relevant;
3. A Food Modelling System, developed to identify and check a range of combinations of amounts and types of foods that could be consumed to meet nutritional needs as detailed in Australia's Nutrient Reference Values; and
4. Key authoritative reports from governments and august bodies such as the World Cancer Research Fund International.

DOI: 10.4018/978-1-4666-9553-5.ch007

Among the findings of all the deliberations, the benefits of an increased intake of vegetables, fruit, wholegrains, legumes, nuts and seeds stood out. Australians were consuming too little of each of these plant foods to meet what the evidence shows to be a healthful dietary pattern. Since the guidelines were released, results of the 2011-12 national survey, (ABS, 2014) show the situation has worsened with decreases in consumption of vegetables, wholegrains and legumes compared with the previous survey in 1995 (ABS, 1999). Fruit consumption has remained low, at about half the level recommended by the dietary guidelines. Consumption of nuts has increased but from a very low base and remains well below recommendations. This latest survey clearly shows that the major problem in the Australian diet is a high consumption of "discretionary foods and drinks" (also called 'extras' or 'junk foods'). These products now contribute 35% of adults' and 41% of children's kilojoule intake (NHMRC, 2013).

MEAT AND DIETARY GUIDELINES

Where Does Meat Fit?

Meat is a popular component of the typical Australian diet, especially among men, many of whom consume large quantities.

Nutritionally, meat scores well. It is a good source of protein, iron, zinc and several B complex vitamins. Other foods can supply adequate quantities of each of these nutrients, although if meat is not consumed, some other animal product (poultry, milk, yoghurt, cheese, eggs or seafood) will be required to supply vitamin B12, as this vitamin occurs naturally only in animal foods. Those who choose a purely plant-based (vegan) diet need supplementary B12, either from plant foods fortified with sufficient quantities of the vitamin or as a supplement.

Lean red meat was included in one of the five food groups with other choices listed as chicken or other poultry, fish or other seafood, eggs, legumes/beans, tofu, nuts or seeds. The expansion of the named foods to include more plant choices in this food group in the most recent guidelines not only met the nutrient requirements of diets modelled in the Food Modelling System, but also provided extra benefits from the inclusion of nuts, seeds and legumes.

Some objected to the increased emphasis on plant sources of protein in this group and lobbied for a separate set of guidelines for vegetarians. This was not acceptable to the Working Committee, especially as the Evidence Report noted the need to limit red meat and move processed meats out of this food group and into the category of discretionary foods. Those who follow a vegetarian diet were also opposed to separate guidelines and supported the increased variety of foods now included in this group. Perhaps most importantly, a significant number of people are not vegetarians, but do not choose to eat meat every day. On the day prior to the interview for the recent national nutrition survey, for example, 31% of people did not consume meat or poultry (ABS, 2014). This highlights the need for the group of foods that contribute a significant quantity of protein to include alternatives to meat.

Questions about Meat

The NHMRC Working Committee set a series of questions about the role of particular foods in obesity and chronic diseases, including cardiovascular disease and stroke, diabetes, various cancers, hypertension, eye health, bone health, and problems with dental and mental health. Any potential risks were also examined for the general population as well as vulnerable groups. Meat was one of the many foods investigated with an additional assessment of any dose response between consuming fresh red meat and an increased risk of cancer or cardiovascular disease.

Using the NHMRC's strict criteria for evidence statements (NHMRC, 2009), the Evidence Report found Grade B evidence for consumption of red meat and increased risk of colorectal cancer and Grade C evidence for renal cancer and red meat. The difficulty of conducting double blind randomised controlled trials with foods makes it impossible to find Grade A evidence for many links between diet and disease, so the grading of the link between red meat and colorectal cancer is important. A minimum of five high quality studies was required before making any graded evidence statement. Grade B evidence indicates that the body of evidence can be trusted to guide practice in most situations. The World Cancer Research Fund International which has a continual update program now rates the link between red and processed meat and risk of colorectal cancer as 'convincing', its highest level of risk (WCRF & AICR, 2011).

For some health problems, including cardiovascular disease, large studies may fail to adequately distinguish between fresh and processed meat, making it difficult to assign responsibility for relative risk. However a large cohort group followed for long enough to record significant deaths, found modest increases in total mortality, cardiovascular mortality and cancer mortality with red and processed meat (Sinha, Cross, Graubard, Leitzmann & Schatzkin, 2009). A more recent meta-analysis of cohort studies concluded that most of the increased risk for cardiovascular disease can be attributed to processed meat, although this analysis also found that both fresh and processed meats are associated with increased risk of type 2 diabetes (Micha, Michas & Mozaffarian, 2012). The authors of this extensive review also noted that since subsequent results from other large studies have reported an increased risk for cardiovascular disease with red meat, the overall findings suggest that neither unprocessed red nor processed meat consumption is beneficial for cardio-metabolic health, but public health guidance should give priority to reducing processed meat consumption (Micha et al., 2012).

Cohort studies note associations between a food, diet or eating pattern and health outcomes. Further investigations are then required to establish any mechanism that may account for a cause and effect. For the well-established link between red meat and colorectal cancer, researchers continue to examine the effects of breakdown products resulting from the digestion of meat and their possible effect on colorectal bacteria as well as the possibility that heterocyclic amines formed on the surface of meat during some cooking procedures play a role (Wang et al., 2012). The most likely candidate for the damage, however, appears to be the haem iron found in red meats (Fonseca-Nunes, Jakszyn & Agudo, 2014). There is also the likelihood that a person's genetic profile is a risk factor for colorectal cancer (Ananthakrishnan et al., 2015).

Although the Australian Dietary Guidelines include fresh red meat in one of the five food groups, processed meats have been moved to the "discretionary foods" category. This reflects the higher level of risk of these products. To prevent development of overweight and obesity, consumption of discretionary foods is not recommended for the smallest, least active adults in each age and sex group – unless physical activity levels are increased.

How Much Red Meat?

Meat consumption in Australia has previously exceeded recommendations for men. Current mean daily consumption of meat, poultry and game products and dishes is 184g for men and 123g for women. Since many people do not consume meat each day, actual consumption at meals that include meat is well above these levels. Consumption of processed meat, now relegated to the discretionary food category, doubled between 1995 and 2011-12 (ABS, 1999, 2014).

Movement within the meat category shows more chicken and less beef being consumed, compared with previous surveys. This may have advantages from a health viewpoint as the evidence for colorectal cancer clearly identifies the evidence for red meat (which includes pork) as 'convincing' whereas no specific risk for this cancer has been associated with chicken. For those interested in sustainable food production, rearing chickens for meat also uses less water and produces lower greenhouse gas emissions compared with beef (Scarborough et al., 2014; Kristensen, Mogensen, Hermansen, Knudsen & Flysjö, 2006).

Since the 2003 Australian dietary guidelines (NHMRC, 2003), the evidence linking meat consumption with increased risk of disease has strengthened in some areas and remains unclear in others. Overall, however, the evidence for restricting red meat intake as well as the increased evidence favouring consumption of fish, nuts and seeds and legumes/beans led to the guidelines recommending a maximum of 455g of lean meat/week for Australian adults. This could be achieved either by including smaller servings of meat than are usually consumed in Australia – especially by men - or including alternative choices from this food group on more days of the week. These new Australian recommendations are similar to those of the World Cancer Research Fund which suggests a weekly limit of 500g of red meat, defined as beef, pork, lamb and goat (WCRF & AICR, 2011). In 1995, the national nutrition survey reported that mean daily consumption of meat, poultry and game products and dishes for Australian men was 200g, with women consuming 120g. Excluding poultry and dishes where poultry is the main ingredient, mean consumption of red meats was 148g for men and 81g for women. The 2011-12 survey shows a small drop in mean daily intake of meat, poultry and game products with men now consuming 185g and women consuming 125g. Excluding poultry and dishes where poultry is the main ingredient, mean daily consumption of red meats is 112g for men and 72g for women. These figures show some fall in consumption over the 16 year period between national Australian surveys, but consumption remains significantly above recommendations for men and slightly too high for women. It is clear that Australians need to make some alternative plant-based choices among the protein-rich foods included in this food group.

No health risks are likely from reduction of mean meat consumption to the levels proposed, as long as a variety of foods are chosen. Indeed, benefits to health would be expected. It is difficult to quantify whether benefits would ensue from less red meat or more plant-based foods, but current evidence suggests benefits are likely from both changes. The World Cancer Research Fund also notes that plant foods make satisfactory alternative choices to red meat. The range of foods included in the same food group as meat within the Australian Dietary Guidelines provides for safe, healthy and more varied choices. As noted previously, foods other than meat can provide ample quantities of nutrients such as iron and zinc and vegetarians who choose a varied and well-balanced diet have no greater risk of iron deficiency anaemia than those who choose to eat meat (Saunders, Craig, Baines & Posen, 2012).

CONCLUSION

For healthful and sustainable eating patterns, the evidence is mounting that a diet favouring more plant foods and less red meat is the best choice. This is supported by august bodies such as the World Cancer

Research Fund International and the World Health Organisation, who recommend a nutritious diet based on a variety of foods, originating mainly from plants, rather than animals (WHOEU, 2015). Australia's Dietary Guidelines have aimed to promote such choices by encouraging food groups that feature:

- Plenty of vegetables (five serves a day, including different types and colours, and legumes/beans);
- Fruit;
- Grains (cereal) foods, mostly wholegrain and/or high cereal fibre choices;
- Milk, cheese, yoghurt or plant-based alternatives (calcium-enriched soy, rice and oat drinks); and
- A range of plant foods including legumes/beans, tofu, nuts and seeds as alternatives to lean meat, fish, poultry or eggs.

REFERENCES

Ananthakrishnan, A. N., Du, M., Berndt, S. I., Brenner, H., Caan, B. J., Casey, G., & Chan, A. T. et al. (2015). Red meat intake, NAT2 and risk of colorectal cancer: A pooled analysis of 11 studies. *Cancer Epidemiology, Biomarkers & Prevention, 24*(1), 198–205.

Australian Bureau of Statistics (ABS). (1999). National nutrition survey: foods eaten, Australia, 1995. Retrieved February 1, 2015, from http://www.abs.gov.au/AUSSTATS/abs@.nsf/0/9A125034802F94C ECA2568A9001393CE Australian Bureau of Statistics (ABS). (2014). Australian health survey: nutrition first results - food and nutrients, 2011-12. Retrieved February 1, 2015, from http://www.abs.gov. au/ausstats/abs@.nsf/PrimaryMainFeatures/4364.0.55.007?OpenDocument

Fonseca-Nunes, A., Jakszyn, P., & Agudo, A. (2014). Iron and cancer risk--a systematic review and meta-analysis of the epidemiological evidence. *Cancer Epidemiology, Biomarkers & Prevention, 23*(1), 12–31.

Kristensen, T., Mogensen, L., Hermansen, J. E., Knudsen, M. T., & Flysjö, A. (2006). Life cycle assessment of the livestock industry - Greenhouse gas emission. Tjele, Denmark: Aarhus University, Faculty of Science and Technology, Department of Agroecology. Retrieved September 7, 2015, from □http:// pure.au.dk/portal/files/44403731/Life_cycle_analysis_of_the_livestock_industry.doc□

Micha, R., Michas, G., & Mozaffarian, D. (2012). Unprocessed red and processed meats and risk of coronary artery disease and type 2 diabetes--an updated review of the evidence. *Current Atherosclerosis Reports, 14*(6), 515–524. doi:10.1007/s11883-012-0282-8

National Health and Medical Research Council (NHMRC). (2003). Food for Health: Dietary Guidelines for Australian Adults. Canberra, Australia: Commonwealth of Australia, NHMRC. Retrieved September 8, 2015, from https://www.nhmrc.gov.au/_files_nhmrc/publications/attachments/n29.pdf

National Health and Medical Research Council (NHMRC). (2009). NHMRC levels of evidence and grades for recommendations for developers of guidelines. Canberra, Australia: Commonwealth of Australia, NHMRC. Retrieved September 8, 2015, from https://www.nhmrc.gov.au/guidelines-publications/ information-guideline-developers/resources-guideline-developers

National Health and Medical Research Council (NHMRC). (2013). Australian dietary guidelines (2013). Canberra, Australia: Australian Government, NHMRC, Department of Health and Ageing. Retrieved September 8, 2015, from https://www.nhmrc.gov.au/guidelines-publications/n55

Saunders, A. V., Craig, W. J., Baines, S. K., & Posen, J. S. (2012). Iron and vegetarian diets. *The Medical Journal of Australia*, *1*(Suppl 2), 11–16.

Scarborough, P., Appleby, P. N., Mizdrak, A., Briggs, A. D. M., Travis, R. C., Bradbury, K. E., & Key, T. J. (2014). Dietary greenhouse gas emissions of meat-eaters, fish-eaters, vegetarians and vegans in the UK. *Climatic Change*, *125*(2), 179–192.

Sinha, R., Cross, A. J., Graubard, B. I., Leitzmann, M. F., & Schatzkin, A. (2009). Meat intake and mortality: A prospective study of over half a million people. *Archives of Internal Medicine*, *169*(6), 562–571.

Wang, J., Joshi, A. D., Corral, R., Siegmund, K. D., Marchand, L. L., Martinez, M. E., & Stern, M. C. et al. (2012). Carcinogen metabolism genes, red meat and poultry intake, and colorectal cancer risk. *International Journal of Cancer*, *130*(8), 1898–1907.

World Cancer Research Fund (WCRF), & American Institute for Cancer Research (AICR). (2011). Continious update project report. Colorectal cancer 2011: food, nutrition, physical activity and the prevention of colorectal cancer. London, UK and Washington, DC: WCRF and AICR. Retrieved September 8, 2015, from http://www.dietandcancerreport.org/cancer_resource_center/downloads/cu/Colorectal-Cancer-2011-Report.pdf

World Health Organization Regional Office for Europe (WHOEU). (2015). A healthy lifestyle. Retrieved March 1, 2015, from http://www.euro.who.int/en/health-topics/disease-prevention/nutrition/a-healthy-lifestyle

KEY TERMS AND DEFINITIONS

Cohort Study: A clinical study in which a group of people who are free of disease are followed over time to see if they develop a particular disease through their exposure to a particular factor or factors.

Dietary Guidelines: A set of guidelines developed for a population to provide an adequate and healthy dietary pattern. Guidelines may be based on foods, nutrients or a combination of both.

Discretionary Foods: Are foods that provide energy but either have little nutritional value, or have a high content of some dietary risk factor, such as added salt or sugars, or particular types of fat. Also known as 'junk foods and drinks'.

Haem Iron: The type of iron found in animal compounds such as haemoglobin and myoglobin. About 40% of the iron in red meat is in the haem form rather than non-haem iron. Haem iron is absorbed to a greater extent than non-haem iron.

Heterocyclic Amines: A range of compounds, which include some that form when meat, chicken or fish are cooked at high temperature. These may have carcinogenic properties.

National Health and Medical Research Council (NHMRC): Australia's self-governing statutory authority, which provides funding for medical research and develops and maintains health standards, including Dietary Guidelines.

World Cancer Research Fund International (WCRF): The world's major not-for-profit body of experts in cancer prevention. Since it began in 1982, the World Cancer Research Fund network has pioneered the links between food, nutrition, physical activity and prevention of cancer. WCRF also runs continuous update programs that critically examine ongoing research into cancer prevention.

Chapter 8
Red Meat and Health:
Evidence Regarding Red Meat, Health, and Chronic Disease Risk

Kate Marsh
Northside Nutrition and Dietetics, Australia

Angela Saunders
Sanitarium Health and Wellbeing, Australia

Carol Zeuschner
Sydney Adventist Hospital, Australia

ABSTRACT

Despite its nutritional benefits, there is an increasing body of evidence to suggest that regular consumption of red meat may negatively impact health and disease risk, including the risk of most common chronic diseases. This chapter reviews the current evidence linking red and processed meat intakes with chronic disease, obesity and mortality risks and discusses possible mechanisms to explain these associations. Research on the health benefits of diets low in red meat, including vegetarian, vegan, Mediterranean and other plant-based diets, is also reviewed.

INTRODUCTION

While red meat is a good source of protein and other nutrients including iron, zinc and vitamin B12, there is an increasing body of evidence to suggest that higher intakes of red meat may negatively impact health and disease risk.

Epidemiological studies have found a positive relationship between intakes of red meat and processed meats and disease risk including risk of obesity (Wang & Beydoun, 2009; R1osell, Appleby, Spencer & Key, 2006; Vergnaud et al., 2010; Rizzo, Jaceldo-Siegl, Sabate & Fraser, 2013), type 2 diabetes (Micha, Michas & Mozaffarian, 2012; Pan et al., 2011; Pan et al., 2013; Aune, Ursin & Veierod, 2009; Feskens, Sluik & van Woudenbergh, 2013), coronary heart disease (Micha et al., 2012; Clifton, 2011; Micha, Wallace & Mozaffarian, 2010; Bernstein et al., 2010), stroke (Chen, Lv, Pang & Liu, 2013; Kaluza, Wolk

DOI: 10.4018/978-1-4666-9553-5.ch008

& Larsson, 2012; Micha et al., 2010) and some types of cancer (particularly colorectal cancer) (Abid, Cross & Sinha, 2014), independent of other lifestyle factors. All-cause mortality, and cardiovascular and cancer mortality, are also increased with higher intakes of meat, and in particular processed meat (Larsson & Orsini, 2014; Sinha, Cross, Graubard, Leitzmann & Schatzkin, 2009; Pan et al., 2012; Rohrmann et al., 2013; Abete, Romaguera, Vieira, Lopez de Munain & Norat, 2014).On the other hand, plant-based diets which exclude or contain low intakes of red meat, including vegan, vegetarian, Mediterranean, Portfolio and DASH (Dietary Approaches to Stop Hypertension) diets, are associated with improved health outcomes (McEvoy, Temple & Woodside, 2012; Sievenpiper & Dworatzek, 2013).

There are a number of possible explanations for the association between intakes of meat and chronic disease risk. In particular, typical nutrients and other compounds present in red meat including saturated fatty acids, dietary cholesterol, animal protein, haem iron, Advanced Glycation End Products (AGE's), trimethylamine-N-oxide (TMAO), L-carnitine, Neu5Gc, nitrates and nitrosamines in processed and cooked meats are likely to play a role (Uribarri et al., 2010; Feskens et al., 2013).

This chapter reviews the evidence linking high intakes of red meat with chronic disease risk and mortality and discusses the possible mechanisms underlying these associations. It also explores the health benefits of dietary patterns low in red meat. With an increased interest in lower carbohydrate high protein and hunter-gatherer 'paleo' style diets, this is a timely review of the evidence into red meat and health.

BACKGROUND

Chronic diseases including heart disease, stroke, cancer and diabetes, are the leading causes of mortality, contributing to 60% of all deaths worldwide (WHO, 2014b). Cardiovascular disease (CVD) is the leading cause of death with an estimated 17.3 million people dying from CVD in 2008 (30% of all deaths worldwide) a figure that is expected to increase to 23.3 million by 2030 (WHO, 2014b). Of these deaths, the majority were due to coronary heart disease (CHD) (7.3 million) and stroke (6.2 million). Diabetes currently affects an estimated 347 million people worldwide and in 2012 was the direct cause of 1.5 million deaths (WHO, 2014c). The World Health Organisation (WHO) projects that diabetes will be the 7th leading cause of death in 2030. Cancer is also a leading cause of death worldwide, accounting for 8.2 million deaths in 2012, with lung, liver, stomach, colorectal and breast cancers causing the most cancer deaths each year (WHO, 2014a). The WHO estimates that annual cancer cases will rise from 14 million in 2012 to 22 million within the next two decades (WHO, 2014a).

Lifestyle factors, particularly diet, play a significant role in the risk of chronic diseases and diet related chronic disease is one of the most preventable causes of morbidity and mortality worldwide. For example, behavioural risk factors (including an unhealthy diet, physical inactivity, tobacco use and harmful use of alcohol) are responsible for about 80% of coronary heart disease and cerebrovascular disease (WHO, 2011). The World Cancer Research Fund (WCRF) estimates that about a third of the most common cancers are preventable through a nutritious diet, maintaining a healthy weight and regular physical activity (WCRF, 2014). The same lifestyle changes are known to reduce the risk of type 2 diabetes, with studies showing lifestyle intervention programs combining diet, exercise and modest weight loss can reduce the risk of developing diabetes by 58% (Knowler et al., 2002; Tuomilehto et al., 2001). The dietary recommendations for prevention of these common chronic diseases are similar and are consistent with the WHO 2004 *Global Strategy on Diet, Physical Activity and Health* which recommends increasing consumption of fruits, vegetables, legumes, whole grains and nuts, and reducing meat intake (WHO, 2004).

Meat is promoted as an important source of iron, zinc, vitamin B12, protein and omega-3s, and concerns are raised regarding nutritional inadequacies when it is excluded from the diet. Yet each of those nutrients are available from plant foods, with the exception of vitamin B12 (Zeuschner et al., 2013). It is interesting that despite a significant body of evidence showing health benefits of plant-based diets and negative health effects of diets high in red meat, that the focus continues to be on what is 'missing' when meat is removed from the diet. This is particularly the case in the Western world where chronic disease is a much bigger contributor to morbidity and mortality than nutrient deficiencies. Plant-based diets more closely match dietary recommendations for good health and prevention of chronic disease, and there is good evidence to show that plant-based diets containing little or no meat are associated with a reduced risk of chronic disease including cardiovascular disease, diabetes and some cancers. The relationship between red meat and chronic disease risk, the likely mechanisms underlying this relationship, and the reduction in chronic disease risk associated with diets low in red meat are explored in detail in this chapter.

A systematic search was conducted using the Medline database (www.pubmed.com; National Library of Medicine, Bethesda, MD) and Cochrane Reviews (www.cochrane.org/reviews; The Cochrane Collaboration) through to October 2014. Additional studies were obtained from reviewing the reference lists of other original studies and review papers. Due to the large number of studies, in most cases only the results of the most recent systematic reviews and meta-analyses have been included, along with other key papers where relevant and/or where a systematic review wasn't available.

RED MEAT INTAKE, HEALTH, AND CHRONIC DISEASE

Red Meat, Processed Meat, and Chronic Disease Risk

There is increasing evidence of an association between intake of red meat, and particularly processed meats, and chronic disease risk. This section explores the links between red meat intake and common chronic diseases including type 2 diabetes, cardiovascular disease, metabolic syndrome, cancer and obesity.

Type 2 Diabetes

At least 25 studies have been published assessing the relationship between meat intake and type 2 diabetes (T2D) risk, with the majority showing a positive association between red meat and/or processed meat intakes. The results of these studies have been incorporated into four meta-analyses (summarised in Table 1) and two combined analyses (of the Health Professionals Follow-Up Study (HPFS), Nurses' Health Study I (NHS-I) and Nurses' Health Study II (NHS-II)) published between 2009 and 2013.

Aune et al found that intakes of total meat, red meat and processed meat were each associated with an increased risk of T2D (Aune et al., 2009). The relative risks (RR) for the highest versus lowest intakes were 1.17 (95% confidence interval (CI) 0.92–1.48) for total meat, 1.21 (95% CI 1.07–1.38) for red meat and 1.41 (95% CI 1.25–1.60) for processed meat. Looking at dose response, they found that the relative risk per 120g/day red meat was 1.20 (95% CI 1.04–1.38) and per 50g/day processed meat was 1.57 (95% CI 1.28–1.93). A positive association between intakes of specific subgroups of meats

Table 1. Summary of meta-analyses of studies on T2D risk and meat intake

Ref	Inc. Studies	Subjects	Cases	Meat Type	Measure Of Intake	RR	95% CI	I²	Heterogeneity p Value
Aune et al, 2009	n = 12 (all cohort studies)	445 323	6525	Total meat (n=5)	Highest vs lowest	1.17	0.92-1.48	86.9%	p<0.0001
		433 070	12 226	Red meat (n=10)	Highest vs lowest	1.21	1.07-1.38	58.5%	p=0.01
		380 606	9999	Processed meat (n=9)	Highest vs lowest	1.41	1.25-1.60	53.2%	p=0.03
				Total meat (n=4)	Per 120 g/day	1.26	0.84-1.88	90.6%	p<0.0001
				Red meat (n=9)	Per 120 g/day	1.20	1.04-1.38	68.3%	p=0.001
				Processed meat (n=8)	Per 50 g/day	1.57	1.28-1.93	74.0%	p<0.001
				Bacon (n=4)	Highest vs lowest	1.37	1.19-1.57	64.3%	p=0.04
				Hotdogs (n=5)	Highest vs lowest	1.30	1.20-1.42	0.0%	p=0.81
				Other processed meats (n=4)	Highest vs Lowest	1.25	1.09-1.44	63.9%	p=0.04
Micha et al, 2010	n = 7 (all cohort studies)	142 851	5923	Total meat (n=3)	Highest vs lowest	1.12	1.05-1.19	n/a	p=0.29
		298 982	7349	Red meat (n=5)	Highest vs lowest	1.16	0.92-1.46	n/a	p=0.25
		372 279	10 782	Processed meat (n=7)	Highest vs lowest	1.19	1.11-1.27	n/a	p<0.001
				Bacon (n=5)	Per serve (2 slce)/day	2.07	1.40-3.04	n/a	n/a
				Hotdogs (n=4)	Per serve (1 hotdog)/day	1.92	1.33-2.78	n/a	n/a
				Other processed meats (n=4)	Per serve (1 pce)/day	1.66	1.13-2.42	n/a	n/a
Pan et al, 2011	n = 9 (all cohort studies)	442 101	28 228	Total red meat	Highest vs lowest	1.34	1.26-1.43	n/a	n/a
				Unprocessed red meat	Highest vs lowest	1.25	1.17-1.33	n/a	n/a
				Processed meat	Highest vs lowest	1.32	1.24-1.40	n/a	n/a
				Unprocessed red meat (n=10)	Per 100g/day	1.19	1.04-1.37	93.3%	p=0.000
				Processed red meat (n=9)	Per 50g/day	1.51	1.25-1.83	94.3%	p=0.000
Feskens et al, 2013	n=24 (all cohort studies)	n/a	n/a	Total meat (n=14)	Per 100g/day	1.15	1.07-1.24	58%	n/a
				Red meat (n=14)	Per 100g/day	1.13	1.03-1.23	36%	n/a
				Unprocessed red meat (n=11)	Per 100g/day	1.15	0.99-1.33	94%	n/a
				Processed meat (n=21)	Per 50g/day	1.32	1.19-1.48	89%	n/a

n/a - Data not provided.

(hamburger, bacon, hot dogs and other processed meats) and diabetes risk was also demonstrated. Micha et al, on the other hand, found no relationship between red meat intake and diabetes risk (RR 1.16; 95% CI 0.92-1.46), but did demonstrate a 19% higher risk of diabetes with higher intakes of processed meats (RR 1.19; 95% CI 1.11-1.27) (Micha et al., 2010).

In a combined analysis of findings from the HPFS, NHS-I and NHS-II, Pan et al. found a positive association between T2D risk and both unprocessed and processed red meat in each of the three cohorts (Pan et al., 2011). The pooled Hazard Ratios (HR) for a one serving per day increase in unprocessed, processed, and total red meat consumption were 1.12 (95% CI 1.08-1.16), 1.32 (95% CI 1.25-1.40), and 1.14 (95% CI 1.10-1.18), respectively. They estimated that replacing one serving (85g) per day of unprocessed red meat with one serving of nuts, low fat dairy or wholegrains could reduce the risk of T2D by 16-24% and replacing one serving of processed meat (28g bacon, 45g other processed meats) with the same foods could reduce the risk by 29-35%. The authors also conducted a meta-analysis, incorporating their findings into those of previous meta-analyses, which confirmed an increase in risk with both unprocessed (RR per 100g/day 1.19; 95% CI, 1.04-1.37) and processed meat (RR per 50g/day 1.51; 95% CI, 1.25-1.83) (Pan et al., 2011).

More recently, the same researchers explored the association between changes in red meat consumption and development of T2D in the same three study cohorts (Pan et al., 2013). They found that increasing red meat intake during a 4-year interval was associated with an increased risk of developing T2D in the subsequent 4 years. Compared with no change in intake, increasing meat red meat intake by more than 0.5 servings/day was associated with a 48% increased risk (pooled HR 1.48; 95% CI, 1.37-1.59). After adjustment for initial body mass index (BMI) and concurrent weight gain, the association was attenuated but remained significant (HR 1.30; 95% CI, 1.21-1.41) suggesting that the association may be partly mediated by weight. Interestingly, reducing red meat intake by at least 0.5 servings/day had no effect on risk in the first 4 years of follow-up, but reduced the risk by 14% (pooled HR 0.86; 95%CI 0.80-0.93) during a longer follow-up period (12-16 years).

The most recent meta-analysis, published in 2013, similarly found an association between higher intakes of total meat, unprocessed red meat and processed meat and T2D risk (Feskens et al., 2013). The pooled RR (95% CI) were 1.15 (1.07-1.24) per 100g of total meat, 1.13 (1.03–1.23) per 100g unprocessed red meat and 1.32 (1.19–1.48) per 50g processed meats. In addition to research associating red meat intake with risk of T2D, two further studies have shown that a higher intake of red meat prior to falling pregnant may increase the risk of developing gestational diabetes, a known risk factor for T2D which shares the same underlying pathophysiology (Zhang, Schulze, Solomon & Hu, 2006; Bao, Bowers, Tobias, Hu & Zhang, 2013; Bowers et al., 2011).

Cardiovascular Disease

Research has linked higher intakes of red meat and processed meat with an increased risk of cardiovascular disease including coronary heart disease and stroke (Bernstein et al., 2012; Bernstein et al., 2010; de Oliveira Otto et al., 2012; Kaluza, Akesson & Wolk, 2014; Kaluza et al., 2012; Kontogianni, Panagiotakos, Pitsavos, Chrysohoou & Stefanadis, 2008; Micha et al., 2012; Qi, van Dam, Rexrode & Hu, 2007). In 2010, Micha et al published a systematic review and meta-analysis of the effects of meat consumption and CHD risk (Micha et al., 2010). They identified four studies (three cohort and one case-control) evaluating the relationship between unprocessed red meat intake and six cohort studies evaluating processed meat intake and CHD. Their pooled analysis (see Table 2) found no significant

Table 2. Summary of meta-analyses of Studies on CVD and red meat intake

Ref	Inc. Studies	Subjects	Cases	Type of Meat	Measure of Intake	RR	95% CI	I²	Heterogeneity p Value
CHD									
Micha et al, 2010	n = 10 studies (8 cohort and 2 case-control)	635 558	22 562	Total meat (n=5)	Per 100g/day	1.27	1.07-1.89	n/a	P=0.002
		56 311	769	Red Meat (n=4)	Per 100g/day	1.00	0.81-1.23	n/a	P=0.36
		614 062	21 308	Processed meat (n=5	Per 50g/day	1.42	1.07-1.8	n/a	P=0.04
Stroke									
Micha et al, 2010	n = 3 (all cohort studies)	115 500	931	Total meat (n=2)	Per 100g/day	1.24	1.08-1.43	n/a	n/a
		108 898	1700	Red meat (n=2)	Per 100g/day	1.17	0.40-3.43	n/a	n/a
		108 898	1434	Processed meat (n=2)	Per 50g/day	1.14	0.94-1.39	n/a	n/a
Kaluza et al, 2012	n = 6 (all cohort studies)	329 495	10 630	Total meat (n=5)	Per serve (100-120g)/day	1.11	1.06-1.16	0%	P=0.59
				Unprocessed red meat (n=6)	Per serve (100-120g)/day	1.11	1.03-1.20	37.8%	P=0.17
				Processed meat (n=5)	Per serve (50g)/day	1.13	1.03-1.24	0%	P=0.65

n/a - Data not provided.

association between unprocessed red meat intake and CHD risk (RR for each 100g/day serve: 1.00; 95% CI 0.81-1.23) but significantly increased risk with processed meat intake (RR for each 50g/day serve: 1.42, 95% CI 1.07-1.89).

Since this review, Bernstein et al. published findings from the NHS-I evaluating the association between unprocessed red meat and processed meat intake and CHD risk in women. They found a higher risk of both unprocessed red meat (RR per 100g/day: 1.19; 95% CI 1.07-1.32) and processed meat (RR per 50g/day: 1.20; 95% CI 1.03-1.40) intakes, after adjusting for potential confounding factors (Bernstein et al., 2010). Subsequently, Pan et al. combined the findings from the NHS-I, NHS-II and the HPFS to evaluate the relationship between red and processed meat intakes and CVD mortality (Pan et al., 2012). The pooled analysis demonstrated a higher risk of mortality from CVD for both unprocessed red meat (RR 1.18; 95% CI 1.13-1.23) and processed meat (RR 1.21; 95% CI 1.13-1.31) intakes.

Eight studies have explored the relationship between meat intake and stroke risk and the findings of these studies have been summarised in two meta-analyses published in 2010 and 2012 (see Table 2) (Kaluza et al., 2012; Micha et al., 2010). Micha et al. failed to find a relationship between red and processed meat intake and stroke, but identified only three studies, none of which evaluated the same type of meat (i.e. total versus unprocessed versus processed meat) and stroke subtype (Micha et al., 2010). More recently, Kaluza conducted a meta-analysis of six prospective cohort studies published between

2003 and 2012 (Kaluza et al., 2012). Of these, one study assessed mortality from stroke while five looked at stroke incidence, four of which assessed stroke subtypes in addition to total stroke. Combining the results of these studies they found that consumption of red meat, including both unprocessed red meat intake and processed meat was associated with an increased risk of total stroke and ischemic stroke but not haemorrhagic stroke. For ischemic stroke the RR (95% CI) were 1.13 (1.00-1.27) for unprocessed red meat, 1.15 (1.06-1.24) for processed meat and 1.12 (1.05-1.19) for total red meat. Bernstein et al., in their combined analysis of the NHS-I and HPFS, found that replacing one serving/day of red meat with one serving of poultry, fish, nuts or dairy foods (both full fat and low fat) could reduce the risk of stroke by 10-27% (Bernstein et al., 2012).

Several other studies have explored the relationship between red meat intake and cardiovascular risks. In women with T2D, who are at higher risk of CVD, high intakes of both haem iron and red meat were positively associated with a risk of fatal CHD, coronary revascularisation and total CHD (Qi et al., 2007). In the multivariate analysis, the RR (95% CI) of CHD in the highest versus lowest quintiles of red meat intake were 2.05 (1.08-3.90) for fatal CHD, 1.91 (0.96-3.83) for coronary revascularisation and 1.36 (0.97-1.91) for total CHD. A similar increase in risk was seen for haem iron but not total iron intakes. In the Multi-Ethnic Study of Atherosclerosis (MESA), a positive association was found between CVD risk and intakes of haem iron from red meat (HR 1.65; 95%CI 1.10-2.47) and zinc from red meat (HR 1.51; 95% CI 1.02-2.24) (de Oliveira Otto et al., 2012). Finally, a recent study found that consumption of processed but not unprocessed red meat was associated with an increased risk of heart failure (HF) in men. Compared to those who consumed <25g/day, those who consumed more than 75g/day had a 28% higher incidence of HF (RR 1.28; 95% CI 1.10-1.48) and almost a 2.5-fold risk of HF mortality (RR 2.43; 95% CI 1.52-3.88) (Kaluza, Akesson, et al., 2014).

Metabolic Syndrome

While no meta-analyses have been conducted, several studies have demonstrated a link between red meat intake and risk of metabolic syndrome (MetS) (Azadbakht & Esmaillzadeh, 2009; de Oliveira Otto et al., 2012; Babio et al., 2012). Azadbakht et al., in a cross-sectional study of 482 Tehrani female teachers aged 40–60 years, found that compared to those in the lowest quintile of red meat intake, those with the highest intakes had a significantly greater odds of having MetS (odds ratio (OR) 2.15; 95% CI, 1.18-4.01) (Azadbakht & Esmaillzadeh, 2009). In the MESA, de Oliveira et al. followed 6814 adults, aged 45–84 years for up to 8-10 years. They found that intakes of haem iron and zinc from red meat, but not from other sources, were positively associated with risk of MetS (de Oliveira Otto et al., 2012). The HR were 1.25 (95% CI 0.99-1.56) for haem iron intake from red meat and 1.29 (95% CI 1.03-1.61) for zinc intake from red meat. In the PREDIMED (PREvención con DIeta MEDiterránea, or Prevention with Mediterranean Diet) study, subjects in the highest quartile of red meat intake were more likely to meet the criteria for MetS both at baseline (OR 2.3; 95% CI, 1.4-3.9) and after 1 year follow-up (OR 2.2; 95% CI, 1.3-3.7) compared with those with the lowest intakes, after adjusting for potential confounders (Babio et al., 2012). The risk of developing MetS during follow-up was also significantly higher in those with the highest versus lowest red meat intakes (OR 2.7; 95% CI, 1.1-6.8). Consistent with these findings, the CARDIA (Coronary Artery Risk Development in Young Adults) study demonstrated an association between higher intakes of animal but not vegetable protein and incidence of MetS (Pereira et al., 2002).

Table 3. Summary of meta-analyses evaluating the association between red and processed meat intakes and cancer risk

	Ref	Inc Studies	Subjects	Cases	Type of Meat	Measure of Intake	RR	95% CI	I²	Heterogeneity p Value
Bladder Cancer	Wang, 2012	n = 20 (10 cohort, 11 case-control)	1 273740	8488	Total meat (n=11)	Highest vs lowest	1.04	0.80 – 1.27	62.3%	0.003
					Red meat (n=9)	Highest vs lowest	1.17	1.02– 1.34	53.7%	n/a
					Processed meat (n=11)	Highest vs lowest	1.10	1.00 – 1.21	3.5%	n/a
Bladder Cancer	Li, 2014	n = 15 (6 cohort, 9 case-control)	1 558 848	16 646	Red meat (n=14)	Highest vs lowest	1.15	0.97 – 1.36	73.5%	0.0000
					Processed meat (n=11)	Highest vs lowest	1.22	1.04- 1.43	64.9%	0.002
Breast Cancer	Alexander, 2010	n = 11 (includes 16 prospective cohorts (8 pooled) and 2 nested case-control	n/a	n/a	Red meat	Highest vs lowest	1.02 (FE)	0.98- 1.07	n/a	0.001
					Processed meat	Highest vs lowest	1.07 (RE)	0.98- 1.17	n/a	n/a
						Per 100 g	1.04 (FE)	1.00- 1.07	n/a	0.0001
						Per 100g	1.12 (RE)	1.03- 1.23	n/a	n/a
						Highest vs lowest	1.00 (FE)	0.98- 1.01	n/a	0.005
						Highest vs lowest	1·08 (RE)	1.01- 1.16	n/a	n/a
						Per 30g	1·03 (FE)	1.00- 1.06	n/a	<0.0001
						Per 30g	1·06 (RE)	0.99- 1.14	n/a	n/a

Cancer

Research has linked higher intakes of red meat and processed meat with an increased risk of some types of cancer but not others. Numerous meta-analyses have been published exploring the relationship between red and/or processed meat intakes and most types of cancer, the main findings of which are summarised in Table 3.

The most consistent findings are for colorectal cancer with six meta-analyses published between 2009 and 2014 showing an association between red and/or processed meat intake and colorectal cancer (CRC) risk (Aune et al., 2013; Chan et al., 2011; Johnson et al., 2013; Pham et al., 2014; Smolinska & Paluszkiewicz, 2010; Xu et al., 2013). In the largest and most recent of these, Aune et al. combined the results of 26 studies, (19 case-control and 7 prospective cohort studies) as part of the Continuous Update Project of the World Cancer Research Fund (Aune et al., 2013). They found an increased risk of CRC associated with both red (RR per 100g 1.27; 95% CI 1.16-1.40) and processed meat (RR per 50g/day 1.29; 95% CI 1.10-1.53) intakes. These findings are very similar to those of a second meta-analysis published around the same time, incorporating 21 studies (16 case-control and five cohort studies). The

RR of CRC in this analysis was 1.36 (95% CI 1.17-1.58) for every 100g increase in red meat intake and 1.28 (95% CI 1.03-1.60) per 50 g/day increase in intake of processed meat (Xu et al., 2013). An earlier meta-analysis found a higher risk of CRC with an increasing frequency of red meat consumption than with the total amount consumed (Smolinska & Paluszkiewicz, 2010).

An association between red and/or processed meat intakes has also been found for several other gastrointestinal cancers including oesophageal (Hua, Stoohs & Facchini, 2001; Salehi, Moradi-Lakeh, Salehi, Nojomi & Kolahdooz, 2013; Zhu et al., 2014), gastric (Bonequi, Meneses-Gonzalez, Correa, Rabkin & Camargo, 2013; Song et al., 2014; Zhu et al., 2013), and oral and oropharynx (Xu, Yang, Wu, Li & Bai, 2014). The most recent of four meta-analysis for oesophageal cancer combined the findings of 35 studies (7 cohort and 28 case-control), showing an increased risk for the highest versus lowest categories of intake for total meat (RR 1.19; 95% CI 0.98-1.46), red meat (RR 1.55; 95% CI 1.22-1.96) and processed meat (RR 1.33; 95% CI 1.04-1.69) (Zhu et al., 2014). For gastric cancer, the most recent meta-analysis found a significant increase in risk (RR 1.37; 95% CI 1.18-1.59) for the highest versus lowest categories of red meat intake but didn't look separately at processed meats (Song et al., 2014). A previous meta-analysis combining the findings of 12 cohort and 30 case-control studies showed a positive association between gastric cancer risk and intakes of both red and processed meats as well as specific meat subgroups including beef, ham, bacon and sausages (Zhu et al., 2013). Xu et al. found a significant association between intakes of processed meats and oral cavity and oropharynx cancers (RR 1.91, 95% CI 1.19-3.06) but no association with total meat or red meat intakes (Xu et al., 2014).

In addition to gastrointestinal cancers, there is evidence of an association between red meat intakes and endometrial (Bandera, Kushi, Moore, Gifkins & McCullough, 2007), lung (Xue et al., 2014; Yang et al., 2012) and pancreatic (Larsson & Wolk, 2012; Paluszkiewicz, Smolinska, Debinska & Turski, 2012) cancers and between processed meat intakes and bladder, lung (Xue et al., 2014), ovarian, and pancreatic (Larsson & Wolk, 2012) cancers. While an earlier meta-analysis of case-control studies found a relationship between total, red and processed meat intakes and renal cancer, two more recent meta-analyses have failed to show an association (Alexander & Cushing, 2009; Lee et al., 2008). While some individual studies have shown associations with breast and prostate cancer, the most recent meta-analyses don't support a relationship between meat intake and these cancers (Alexander, Morimoto, Mink & Cushing, 2010; Alexander, Mink, Cushing & Sceurman, 2010). However there is some suggestion of a stronger relationship between meat intake and breast cancer in postmenopausal women(Alexander, Morimoto, et al., 2010).

Obesity

Red meat intake has been positively associated with obesity and central obesity (Wang & Beydoun, 2009) and several studies have found an association between a higher intake of red meat and an increased risk of weight gain (Romaguera et al., 2010; Schulz et al., 2002; Vergnaud et al., 2010; Mozaffarian, Hao, Rimm, Willett & Hu, 2011). In the European Prospective Investigation into Cancer and Nutrition–Physical Activity, Nutrition, Alcohol Consumption, Cessation of Smoking, Eating Out of Home, and Obesity (EPIC-PANACEA) study, those with high adherence to a Mediterranean dietary pattern gained less weight and were less likely to develop overweight and obesity, an effect that was largely explained by the low meat content of a Mediterranean style diet (Romaguera et al., 2010). When the researchers assessed the relative contribution of each component of the MD score, the association between the MD score and weight change was no longer significant after excluding meat and meat products. These findings are consistent with research showing that vegetarians tend to have a lower BMI and experience less weight gain than meat-eaters (Rosell et al., 2006; Tonstad, Butler, Yan & Fraser, 2009).

Red Meat Intake and Mortality

Considering the links with chronic disease risk, it is perhaps not surprising that higher intakes of red meat have also been associated with a higher risk of mortality. A large study investigating the association of a wide range of meat intakes with chronic disease mortality found that both red and processed meat intakes were associated with modest increases in total mortality, cancer mortality, and cardiovascular disease mortality (Sinha et al., 2009). The EPIC study showed a moderate positive association between processed meat consumption and mortality, in particular due to cardiovascular diseases, but also to cancer (Rohrmann et al., 2013). A combined analysis of findings from the HPFS and NHS demonstrated an association between red meat consumption and an increased risk of total, CVD, and cancer mortality and that substituting red meat with other protein sources (including fish, poultry, nuts, legumes, low-fat dairy, and whole grains) is associated with a lower mortality risk (Pan et al., 2012).

A recent meta-analysis found that high consumption of red meat, especially processed meat, may increase all-cause mortality (Larsson & Orsini, 2014) while another found that processed meat consumption may increase the risk of mortality from any cause and CVD, while red meat consumption was positively but weakly associated with CVD mortality (Abete et al., 2014). A low meat intake, on the other hand, has been associated with an increased life expectancy, particularly with increasing duration of adherence (Singh, Sabate & Fraser, 2003).

Mechanisms to Explain the Association between Red Meat and Chronic Disease Risk

There is growing evidence to suggest that the adverse effects of meat on health are related not just to saturated fat and cholesterol (as is commonly assumed), but to other substances present in meat. These include nutrients in red meat (such as protein, amino acids, haem iron, L-carnitine and choline), substances produced during the processing and cooking of red meat (including heterocyclic amines, polycyclic aromatic hydrocarbons, nitrates, nitrites, nitrosamines and advanced glycation end products) and other components found almost exclusively in red meat such as NeuGc5 (Feskens et al., 2013; Micha, Michas, Lajous & Mozaffarian, 2013; de Oliveira Otto et al., 2012; Cross et al., 2010). Notably, these different compounds may act synergistically in increasing risk of disease (Kim, Coelho & Blachier, 2013). This section will explore the recent evidence on each of these components to understand how red meat can increase our risk of chronic diseases such as cardiovascular disease, type 2 diabetes mellitus and some cancers.

Saturated Fat

Red meat is a significant source of saturated fat (SF) in the diet and is considered one of the main reasons for the link between red meat intake and disease risk, particularly cardiovascular risk. Saturated fat contributes an average of 11 percent of total energy to the American diet, with meats such as sausages, franks, bacon and ribs contributing 5%, burgers 4.4% and beef/beef mixed dishes 4.1% (NCI, 2013). The American Heart Association recommends aiming for just 5% to 6% of total energy intake from saturated fat (Lichtenstein et al., 2006) and the US Dietary Guidelines recommend consuming less than 10% of calories from saturated fat (USDA & USDHHS, 2010).

The amount and quality of dietary fat and its effects on chronic disease have been studied intensively during the past decades. While previously low-fat diets were recommended without much attention to the type of fat, current guidelines place more emphasis on the quality of fat consumed (Schwab et al., 2014). Despite the recent controversy regarding the role of SF in heart disease, the bulk of evidence continues to recommend replacing SF with PUFA (poly-unsaturated fat) to reduce the risk of heart disease (Kromhout, Geleijnse, Menotti & Jacobs, 2011; Astrup et al., 2011; Willett & Stampfer, 2013; Farvid et al., 2014; Schwab et al., 2014). Considered one of the best systematic reviews examining SF and heart disease, Jakobsen et al. found that substitution of 5% of energy from SF with 5% from PUFA was associated with a significant reduction in CHD events and deaths (Jakobsen et al., 2009). These findings are supported by more recent meta-analyses. Mozaffarian et al. found that a 5% increase in PUFA, in place of SF, reduces CHD events by 10% (Mozaffarian, Micha & Wallace, 2010). Similarly, Farvid et al. showed that substituting 5% of SF with 5% of PUFA lowered risk of CHD events by 9% and risk of death from CHD by 13%(Farvid et al., 2014). Jakobsen et al. also reported that if SF is replaced with low glycemic index (GI) CHO, the risk of CHD is also decreased whereas replacing SF with high GI CHO is associated with higher risk, showing the importance of the overall quality of both fats and CHO in the diet (Jakobsen et al., 2010).

A number of studies have demonstrated an association between high fat diets and risk of T2D (Colditz et al., 1992; Marshall, Hamman & Baxter, 1991; Marshall, Hoag, Shetterly & Hamman, 1994; Tsunehara, Leonetti & Fujimoto, 1990; Lindstrom et al., 2006). In the Finnish Diabetes Study, subjects with the highest fat intakes were more than twice as likely to develop T2D compared to those with the lowest fat intakes, while a higher intake of SF increased the risk of developing diabetes 1.7 times (Lindstrom et al., 2006). The mechanisms linking SF intake and diabetes risk appear to relate to insulin sensitivity, with several studies showing an association between SF intake and reduced insulin sensitivity (Feskens & Kromhout, 1990; Lovejoy & DiGirolamo, 1992; Marshall, Bessesen & Hamman, 1997; Mayer, Newman, Quesenberry & Selby, 1993; Mayer-Davis et al., 1997; Parker et al., 1993). A higher SF intake increases the proportion of SF in muscle cell membranes, which possibly influences insulin action by altering insulin receptor binding or affinity and transport and cell signalling (Marsh & Brand-Miller, 2011; Parillo & Riccardi, 2004; Vessby, 2000).

Dietary Cholesterol

Until recently, dietary cholesterol was thought to play a role in heart disease, but a direct association between dietary cholesterol and risk of CHD has been observed only in a few cohorts (Djousse & Gaziano, 2009). Instead, dietary patterns high in total fat, SF and cholesterol, with a low intake of fibre and PUFA are more consistently correlated with risk of CHD. Whether this is due to the high consumption of SF or cholesterol or both, or perhaps an insufficient supply of one or more protective factors such as fibre or PUFA, has been difficult to determine (Kratz, 2005). A meta-analysis of randomized clinical trials showed only a minimal increase (0.02 units per 100mg) in the total/high-density lipoprotein cholesterol ratio, which suggests the association between dietary cholesterol and CHD risk is, if anything, minor in nature (Kratz, 2005; Weggemans, Zock & Katan, 2001). However, lowering dietary cholesterol content might reduce the risk of CHD considerably in a subgroup of individuals who are highly responsive to changes in cholesterol intake (Kratz, 2005). Furthermore, one study found that the addition of a moderate amount of dietary cholesterol to a reduced fat diet rich in PUFA or SF increased in vitro susceptibility of low density lipoprotein (LDL) cholesterol to oxidation (Schwab et al., 2000).

Trans-Fats

Trans-fats (TF) increase the risk of CHD even more strongly than SF by raising LDL-cholesterol and total: HDL cholesterol and promoting inflammation (Mozaffarian et al., 2004; Baer, Judd, Clevidence & Tracy, 2004; Ascherio, Katan, Zock, Stampfer & Willett, 1999; Mensink, Zock, Katan & Hornstra, 1992; Kromhout et al., 2011). They have also been shown to reduce the particle size of LDL-cholesterol, increase blood levels of Lipoprotetin-a Lp(a) and adversely affect endothelial function (Mozaffarian, Katan, Ascherio, Stampfer & Willett, 2006). Furthermore, TF worsen insulin sensitivity, particularly those with pre-existing insulin resistance, visceral adiposity or lower physical activity (Mozaffarian, Aro & Willett, 2009)

Most TF in the western diet occur as the result of processing, formed during the partial hydrogenation of vegetable oils (converting them to solid or semi-solid fats for cooking or baking). The fat in red meat (and butter) also has small amounts (typically 4-5%) of trans-fatty acids due to bio-hydrogenation in the rumen of cattle and other ruminants (Willett, 2012). As the TF of processed foods have been reduced by manufacturers, TF from ruminant sources has become of greater interest, particularly in relation to the risk of CHD, however four prospective studies did not identify a significant positive association (Willett et al., 1993; Pietinen et al., 1997; Oomen et al., 2001; Jakobsen, Overvad, Dyerberg & Heitmann, 2008). The evidence from observational studies also suggests that higher CHD risk is related to consumption of industrially produced TF rather than ruminant TF. A WHO scientific update (Mozaffarian et al., 2009) on TF stated that the effects of ruminant and industrial TF are similar when consumed in similar quantities, but in the amounts consumed in actual diets, ruminant TF do not appear to be major contributors to CHD risk. However TF from ruminant sources may be of greater importance when consumed on top of TF from industrial sources (Mozaffarian et al., 2009).

Animal Protein

There is some evidence that animal protein itself may increase disease risk, with several studies showing an association between animal protein, but not plant protein, and risk of chronic disease. In healthy men (those without common cardiovascular risk factors) a higher intake of total and animal protein was associated with an increased risk of ischaemic heart disease (IHD), while there was no association with plant protein (Preis, Stampfer, Spiegelman, Willett & Rimm, 2010). Similarly, in an elderly Greek population, higher intakes of total protein and protein from meat and meat products were associated with a higher prevalence of diabetes while there was no association with protein from vegetables or cereals (Pounis et al., 2010) In the EPIC study, subjects with the highest intakes of animal protein had more than twice the risk of developing T2D compared to those with the lowest intakes while there was a 16% non-significant reduction in risk with vegetable protein (Sluijs et al., 2010). The researchers estimated that replacing 5% of energy from fat or carbohydrate with 5% of energy from animal protein would increase diabetes risk by about 20%. Vegetable or plant protein has also been inversely related to blood pressure (Elliott et al., 2006; Altorf-van der Kuil et al., 2012). The reasons for the association between animal protein and disease risk are unclear but emerging data suggests that a higher intake of branched chain amino acids (leucine, isoleucine, valine) may increase insulin resistance, a risk factor underlying many chronic diseases (Beasley & Wylie-Rosett, 2013).

Despite the interest in high protein diets (typically from animal protein) for weight loss, there is some evidence that animal protein (excluding dairy and fish) is positively associated with long-term weight gain (Halkjaer et al., 2010; Clifton, 2011). Furthermore, long-term adherence to high protein diets, without discrimination towards protein source, may have adverse health consequences, as described above. Consistent with these findings, Fung et al (2010), in a combined analysis of the NHS and HPFS, found a higher animal-based low-carbohydrate intake was associated with a 23% higher all-cause mortality, 14% higher cardiovascular mortality and 28% higher cancer mortality comparing extreme deciles of intake, whereas a higher vegetable-based low-carbohydrate intake was associated with a 20% lower all-cause mortality and 23% lower cardiovascular mortality (Fung, van Dam, et al., 2010). A recent systematic review similarly found an increased risk of mortality with low carbohydrate diets (which by default are typically high in animal protein) however they didn't look specifically at the relationship with protein intake (Noto, Goto, Tsujimoto & Noda, 2013).

With respect to CRC risk, animal protein (as well as substances typically found in animal protein including haem iron, *N*-nitroso compounds, heterocyclic amines) can affect the large intestine mucosa with genotoxicity and metabolic disturbances (Kim et al., 2013). Increased bacterial fermentation (putrefaction) of undigested protein and production of bacterial metabolites derived from amino acids may affect colon epithelial homeostasis and renewal. Most colonic cancers are found in the distal colon and rectum where protein fermentation occurs (Kim et al., 2013).

Haem Iron

Red meat is typically recommended as an important source of haem iron, a cofactor for several enzymes and a major component of oxygen transporters in the body. Haem iron (40% of the total iron in meat) is easily absorbed in our body regardless of our body stores, whereas non-haem iron (the type of iron present in plant foods) is absorbed according to need (Saunders, Craig, Baines & Posen, 2012). A strong pro-oxidant and catalyst in many cellular reactions, haem iron causes tissue damage, and increases oxidative stress (Rajpathak, Ma, Manson, Willett & Hu, 2006; Kunutsor, Apekey, Walley & Kain, 2013; Broedbaek et al., 2011). As we have limited means of excreting excess iron, a high intake of haem iron and elevated serum ferritin concentrations (high iron stores) may be a risk factor for chronic disease, including heart disease, diabetes and some types of cancers.

Both high haem iron intake and elevated serum ferritin contribute to the development of atherosclerosis by catalysing the production of free radicals and promoting low density lipoprotein (LDL) oxidation (Hunnicutt, He & Xun, 2014). A recent meta-analysis found that participants with higher haem iron intakes had a 31% increased risk of CHD compared to those with lower intakes. An increase in haem iron intake of 1mg/day was associated with a 27% increase in risk of CHD (Yang et al., 2014). A study of postmenopausal women with T2D found that those with the highest intake of haem iron had a 50% increased risk of total CHD compared to those with the lowest intake (Qi et al., 2007). A high haem iron intake has also been associated with an increased risk of stroke (Kaluza, Wolk & Larsson, 2013) and fatal acute myocardial infarct (AMI) (Kaluza, Larsson, Hakansson & Wolk, 2014).

There is consistent evidence showing a positive association between dietary haem iron intake from red meat and the risk of T2D (Luan de et al., 2008; Jiang et al., 2004; Rajpathak et al., 2006; Lee, Folsom &

Jacobs, 2004; Fernandez-Cao et al., 2013), although one study found haem iron intake to be associated with CVD and metabolic syndrome but not T2D (de Oliveira Otto et al., 2012). Two recent meta-analyses have shown both increased ferritin levels and haem-iron intake are significantly associated with higher risk of T2D (Kunutsor et al., 2013; Zhao et al., 2012) where individuals in the top quintile of ferritin levels and dietary haem iron status have about a 70% and 30% higher risk, respectively, of T2D events compared to the bottom quintile (Kunutsor et al., 2013). Higher intakes of dietary haem iron during pre-pregnancy and early pregnancy have also been associated with an increased risk of gestational diabetes (Bowers et al., 2011; Qiu et al., 2011). The damaging effects of elevated iron stores on pancreatic β-cells and hepatic function impairs glucose and insulin metabolism leading to the development of T2D (Opara, 2004; Rajpathak et al., 2009; Zhao et al., 2012; Kunutsor et al., 2013; Bowers et al., 2011). Catalytic iron damages DNA and the integrity of cell membranes (iron-induced oxidative stress), and also interferes with glucose uptake into skeletal muscles and adipocytes, inducing insulin resistance (Zhao et al., 2012; Andrews, 1999; Lenzen, 2008).

Hereditary hemochromatosis is known to contribute to diabetes due to the excessive accumulation of iron in tissues, but reducing body iron levels using phlebotomy improves insulin sensitivity (Rajpathak et al., 2009). One study found vegetarians, who had lower iron stores (ferritin levels about half that of meat-eaters), were more insulin sensitive than meat-eaters. When six of the male meat-eaters had their iron stores lowered by phlebotomy to levels similar to that seen in vegetarians, they experienced a 40% increase in insulin sensitivity (Hua et al., 2001). Several studies have suggested that elevated levels of ferritin may help identify individuals at high risk of T2D (Kunutsor et al., 2013; Bao, Rong, Rong & Liu, 2012). The National Institutes of Health (NIH) currently defines elevated ferritin levels as ≥300ng/mL for men and ≥150ng/mL for women (NIH, 2014).

A recent meta-analysis found that higher intakes of haem iron have a positive association with cancer risk, particularly colorectal, colon, breast and lung cancers (Fonseca-Nunes, Jakszyn & Agudo, 2014). Dietary haem iron influences cancer progression (Hooda, Shah & Zhang, 2014) mainly due to its pro-oxidant activity (Fonseca-Nunes et al., 2014). Recent studies of lung cancer cells showed they required increased intracellular haem biosynthesis and uptake to meet the increased demands for oxygen-utilizing hemoproteins (Hooda et al., 2014). In the case of CRC, haem iron has a catalytic and toxic effect on both the endogenous formation of carcinogenic N-nitroso compounds (NOC) and the formation of lipid oxidation end products. Most of these NOC can yield alkylating agents during metabolism that cause DNA damage (Bastide, Pierre & Corpet, 2011; Kim et al., 2013). A recent observational study found that haem iron was associated with an increased risk of colorectal tumours and overexpression of P53, suggesting alkylating rather than oxidative DNA-damaging mechanisms are involved in haem-induced colorectal carcinogenesis (Gilsing et al., 2013).

Heterocyclic Amines and Polycyclic Aromatic Hydrocarbons

Heterocyclic amines (HCAs) and polycyclic aromatic hydrocarbons (PAHs) are known mutagens and probable human carcinogens that are formed when cooking meat at high temperatures, such as pan frying or cooking over an open flame (Sinha et al., 2005; Zheng & Lee, 2009). Both HCAs and PAHs occur in grilled meats and poultry, as well as in gravies made from dripping of well-done roasts and fish.

More than 20 HCAs have been identified in red and processed meats and poultry (Kim et al., 2013). They are produced during cooking from the reaction of creatine or creatinine, amino acids and sugars, with greater amounts forming at higher temperatures and longer periods of time. How much is formed also depends on the type of meat and cooking methods. The most mass-abundant HCAs detected in cooked meats are PhIP (2-amino-1-methyl-6-phenylimidazo[4,5-*b*]pyridine) and MeIQx (2-amino-3,8-dimethylimidazo[4,5-*f*]quinoxaline). Animal studies have shown HCAs increase cancer risk in multiple sites, including mammary glands, lung, colon, stomach and prostate. More recently HCA-DNA adducts have been seen in a variety of human tissues, including breast, colorectum and prostate (Zheng & Lee, 2009).

There are 16 PAHs that have potentially genotoxic and carcinogenic effects on humans. They are formed in meats during grilling/roasting/barbequing, when fat and juices from the meat drip onto the fire causing flames. These flames contain PAHs that adhere to the surface of the meat. Other food preparation processes during which PAHs can form are drying, smoking or adding smoke flavour (Cross & Sinha, 2004). The main mechanism involves their effects on enzymes, especially on the cytochrome P450 (CYP) family of enzymes. Benzo(a)pyrene (B(a)P) is the most carcinogenic of the 16 PAHs and is generally used as a marker of the effect of PAHs in foods (Chen et al., 2012). The adducts of PAH-DNA tend to concentrate in the epithelium of the oesophagus, cervix and vulva, the prostate and the placenta (Pratt et al., 2011).

Zheng et al. recently reviewed 21 epidemiological studies published since 1996 specifically evaluating the association of meat carcinogen exposure (by high temperature cooking and meat doneness level) and cancer risk. Most of these studies found that high intakes of well-done meat and high exposure to meat carcinogens, particularly HCAs may increase the risk of human cancers including CRC, breast cancer, prostate cancer, pancreatic cancer, oesophageal and gastric cancers (Zheng & Lee, 2009). A more recent study has highlighted the combined genotoxic potential of B(a)P and PhIP on intestinal cell lines which reflect early steps of colorectal carcinogenesis (Jamin et al., 2013). Cross et al. found those with the highest intake of the heterocyclic amine DiMeIQx were at increased risk for gastric cardia cancer, whereas there was only a suggestive increased risk of oesophageal cancer from other HCAs (MeIQX, PhIP) or haem iron (Cross et al., 2011). Intakes of B(a)P and PhIP in red meat are also associated with an increased risk of renal cell carcinoma (Daniel et al., 2012).

N-Nitroso Compounds

Humans are exposed to N-nitroso compounds (NOCs), also known as nitrosamines, mainly through diet (processed meats), cigarette smoking, occupational exposure and endogenous formation in the human gut, the latter being influenced by the haem iron, nitrate and nitrite content of consumed meats. As previously mentioned, haem iron is a catalyst for the formation of endogenous NOC, causing DNA damage (Bastide et al., 2011; Kim et al., 2013). Although NOC are found in processed meats, they are also formed in the gut endogenously due to the nitrate and nitrite content of processed meat (added for preservation and enhancement of colour and flavour). Hence, red and processed meats are the most significant dietary components linked to the formation of NOC in the gut.

Although nitrate and nitrite are not carcinogenic themselves, under conditions that result in endogenous nitrosation, ingested nitrate and nitrite may possibly be carcinogenic to humans, increasing risk of CRC (Habermeyer et al., 2014; Dellavalle et al., 2014). Studies have shown that high protein diets (either low or moderate in CHO) result in increased intake of N-nitroso compounds with a significant decrease in faecal cancer-protective metabolites and increase in concentrations of hazardous metabolites, with potential for an increased risk of colonic disease in the long-term (Russell et al., 2011). In addition, increased endogenous NOC formation combined with prolonged transit times in the gut may explain the epidemiological associations between high meat/low fibre diets and CRC risk (Hughes, Cross, Pollock & Bingham, 2001). Consumption of a meat-based high protein weight loss diet showed increased faecal NOC, with red meat intake positively correlating with the faecal log NOC concentrations (Holtrop, Johnstone, Fyfe & Gratz, 2012). Bladder cancer risk has also been linked with total dietary nitrite and nitrate plus nitrite from processed meats (Ferrucci et al., 2010) and a Dutch study found that N-nitroso compounds, haem iron intake and nitrite intake increased the risk of oesophageal squamous cell carcinoma in men, but not women (Keszei, Goldbohm, Schouten, Jakszyn & van den Brandt, 2013).

Nitrosamines have been found to be toxic to pancreatic beta cells, increasing the risk of type 1 and type 2 diabetes in animal studies (Helgason, Ewen, Ross & Stowers, 1982; Portha, Giroix, Cros & Picon, 1980) and type 1 diabetes in some epidemiological studies (Dahlquist, Blom, Persson, Sandstrom & Wall, 1990; Sipetic et al., 2005). This is consistent with evidence showing a higher risk of type 2 diabetes associated with processed (red) meat versus unprocessed meats (Feskens et al., 2013; Micha et al., 2012).

L-Carnitine, Choline, and TMAO

Recent research has discovered a link between L-carnitine and choline ingestion, gut microbiota metabolism and CVD risk which suggests gut microbiota play a role in development of atherosclerosis (Koeth et al., 2013). Choline is found in a range of foods but particularly in animal proteins including eggs, red meat, poultry and seafood. The richest source of L-carnitine is red meat. These trimethylamine containing compounds are metabolised by the gut microbiota to trimethylamine (TMA), then converted in the liver to trimethylamine-N-oxide (TMAO). Circulating TMAO levels are strongly associated with atherosclerosis (possibly via suppression of reverse cholesterol transport, cholesterol accumulation and foam cell formation) and may therefore be a potential predictor of CVD risk (Bennett et al., 2013; Wong, 2014; Brugere et al., 2014; Koeth et al., 2013; Wang et al., 2011).

A diet rich in red meat (a significant source of L-carnitine and choline) results in an increased production of TMAO (Koeth et al., 2013). While intakes of choline and L-carnitine are lower in vegetarian diets, a recent study also found that vegans and vegetarian produced less TMAO compared to omnivores following ingestion of L-carnitine due to differences in gut bacteria (Koeth et al., 2013). The same authors found that plasma L-carnitine levels predicted increased risks of both CVD and major cardiac events (myocardial infact, stroke or death) but only among subjects with concurrently high TMAO levels. The lower dietary intake of L-carnitine and total choline by vegans and vegetarians and their reduced capacity to produce TMAO may contribute to their observed protection from CVD (Koeth et al., 2013).

Advanced Glycation End-Products

Endogenous formation of advanced glycation end-products (AGEs) is a normal consequence of metabolism, but may be up-regulated in conditions of metabolic stress (Davis, Prasad, Vijayagopal, Juma & Imrhan, 2014). They are also absorbed by the body from exogenous sources such as cigarette smoke and consumption of highly heated processed foods (Kellow & Savige, 2013). About 10% of AGEs ingested are absorbed, and of these two-thirds are deposited in the body tissues and one third is excreted by the kidneys (van Puyvelde, Mets, Njemini, Beyer & Bautmans, 2014). Advanced glycation end-products are found in high concentrations in high-fat foods and meats cooked at high temperatures, with more formed by dry heat (grilling, frying, roasting, baking, barbecuing). They occur after a reaction between carbonyl groups on sugars and amine groups on proteins, and DNA and lipoproteins to form glycated proteins. This is the first step of the Maillard reaction, which is also responsible for browning reactions between carbohydrates and amines in cooked foods.

AGEs have been shown to contribute to chronic inflammation and oxidative stress, as well as dysfunction and destruction of pancreatic beta cells and insulin resistance, all of which are linked to chronic diseases such as cardiovascular disease and T2D (Jiao et al., 2013; Uribarri et al., 2010; White & Collinson, 2013; Kellow & Savige, 2013). Recent findings from the NIH-AARP Diet and Health Study showed an increased risk of pancreatic cancer in men in the highest quintile of AGE consumption compared to the lowest quintile (HR 1.43, 95% CI 1.06-1.93) (Jiao et al., 2015). Red meat intake was also associated with pancreatic cancer risk (HR 1.35, 95% CI 1.07-1.70) comparing the highest and lowest quintiles and this was attenuated after adjustment for AGE consumption, suggesting that dietary AGEs may partially explain the relationship between red meat intake and pancreatic cancer risk. In animal studies low-AGE diets reverse insulin resistance and chronic inflammation, inhibit the progression of atherosclerosis and prevent experimental diabetic nephropathy and neuropathy (Uribarri et al., 2007), but currently there is insufficient evidence in humans to promote the use of AGE-restricted diets (Kellow & Savige, 2013).

Animal derived foods that are high in fat and protein (particularly red and processed meats) are generally AGE-rich and prone to formation of AGEs during cooking. For example, the AGE content of beef increases from 707 kU/100 g when raw to 6071kU/100g when roasted and 10 058kU/100g when pan-fried. Processed meats, such as bacon and frankfurters have particularly high levels. By contrast, plant foods including fruits, vegetables, grains and legumes, have low levels of AGEs unless prepared with added animal fats such as butter (Uribarri et al., 2010). Guidelines to reduce AGE intake are consistent with established dietary guidelines for disease prevention. Table 4 compares the AGE content of a variety of red and processed meats versus other protein sources (Uribarri et al., 2010).

Neu5Gc

N-glycolylneuraminic acid (Neu5Gc) is an abundant sialic acid found mainly in red meat and to a smaller extent cow's milk but not in plant foods and poultry and only in trace amounts in fish (Tangvoranuntakul et al., 2003; Chen, Pan, Liu, Troy & Wang, 2014). Humans lack the enzyme responsible for its formation, however it can be metabolically incorporated into human tissues from dietary sources (Samraj, Laubli, Varki & Varki, 2014). It has been proposed that once ingested, the interaction between this 'foreign' molecule and anti-Neu5Gc antibodies results in chronic inflammation that promotes carcinogenesis and atherogenesis (Samraj et al., 2014). While further research is needed this could be one explanation for the association between red meat and chronic disease risk that isn't seen with other protein sources.

SOLUTIONS AND RECOMMENDATIONS

Meat Free and Low Meat Diets and Health

Several well-recognised dietary patterns have repeatedly been associated with a reduction in risk of chronic disease including vegetarian and vegan diets, Mediterranean diets, the Dietary Approaches to Stop Hypertension (DASH), and the Portfolio diet. These eating patterns are all rich in plant foods and contain little or no meat.

Table 4. AGE content of red and processed meats versus other protein sources

Food Item—Meat and Meat Substitutes	AGE kU/100g	Serving Size (g)	AGE kU/Serving
Beef, frankfurter, broiled	11270	90	10143
Beef, steak, pan fried with olive oil	10058	90	9052
Beef, steak, strips, stir fried with canola oil	9522	90	8570
Beef, steak, broiled	7479	90	6731
Beef, roast	6071	90	5464
Beef, hamburger (McDonald's Corp)	5418	90	4876
Beef, steak, microwaved	2687	90	2418
Beef, raw	707	90	636
Bacon, fried 5 mins, no oil	91577	13	11905
Bacon, microwaved, 2 slices, 3 mins	9023	13	1173
Sausage, beef and pork, pan fried	5426	90	4883
Pork, chop, pan fried	4752	90	4277
Ham, deli, smoked	2349	90	2114
Lamb, leg, broiled, 30 mins	2431	90	2188
Lamb, leg, microwave, 5 min	1029	90	926
Lamb, leg, raw	826	90	743
Chicken breast, breaded, deep fried	9722	90	8750
Chicken, back/thigh, roasted then BBQ	8802	90	7922
Chicken, breast, pan fried, 13 min, high	4938	90	4444
Chicken breast, skinless, poached	1076	90	968
Chicken, breast, skinless, raw	769	90	692
Salmon, broiled with olive oil	4334	90	3901
Salmon, Atlantic, frozen, poached, 7 mins	1801	90	1621
Tuna, fresh, baked, 25 mins	919	90	827
Salmon fillet, microwaved	912	90	821
Tuna, canned, chunk light, with water	452	90	407
Tofu, sautéed, outside	5877	90	5289
Tofu, broiled	4107	90	3696
Tofu, raw	788	90	709
Soy burger, skillet, with olive oil	437	30	131
Eggs, fried, one large	2749	45	1237
Egg, omelette, pan, low heat, 12 mins	223	30	67
Egg, scrambled, pan, high, olive oil, 1 min	243	30	73
Beans, red kidney, cooked 1 hr	298	100	298
Beans, red kidney, canned	191	100	191
Beans, red kidney, raw	116	100	116

Vegetarian and Vegan Diets

A vegetarian diet consists mostly of plant-based foods including fruit, vegetables, legumes, nuts, seeds and grains. A lacto-ovo vegetarian diet also includes eggs and dairy products, while vegan diets contain no animal products. The complete absence of red meat is a distinguishing factor in both vegan and vegetarian diets.

There is now a significant amount of research to demonstrate the health benefits of vegetarian and vegan diets, particularly with respect to reducing chronic disease risk, including risk of CVD, T2D, MetS, hypertension and obesity. While vegetarians generally have a lower BMI (Tonstad et al., 2009), and tend to be more health conscious than non-vegetarians (Bedford & Barr, 2005), health outcomes remain better even when these factors are taken into account. Furthermore, a number of studies have shown increased longevity among vegetarians (Singh et al., 2003; Fraser & Shavlik, 2001). When compared to lacto-ovo-vegetarian diets, vegan diets appear to offer additional protection for obesity, hypertension, TD2, and cardiovascular mortality (Le & Sabate, 2014).

One of the proposed mechanisms for the protective effect of a vegan diet is the fibre intake and the impact on the intestinal microbial profile. The high fibre intake of a vegan diet may be a pivotal factor on gut microbiota and key measures of inflammation. In addition, the impact of plant *versus* animal protein on microbiota in the gut may provide further insight into variance in microbial profiles, MetS and inflammation variations in vegan and vegetarian diets (Glick-Bauer & Yeh, 2014).

The most consistent evidence for the health benefits of a vegetarian diet is for a reduced risk of CHD and CHD mortality. Several studies have demonstrated a lower incidence of CHD (Appleby, Thorogood, Mann & Key, 1999; Burr & Butland, 1988; Chang-Claude, Hermann, Eilber & Steindorf, 2005; Fraser, 1999; Snowdon, Phillips & Fraser, 1984) and an earlier combined analysis of 5 prospective studies reported a 24% lower risk of mortality from ischaemic heart disease (IHD) in vegetarians compared to meat-eaters, with lacto-ovo vegetarians having a 34% reduced risk and vegans a 26% reduced risk (Key et al., 1998, 1999). The benefit was apparent in those who had followed their diet for at least 5 years and was greater in younger age groups. A more recent meta-analysis of 7 studies involving 124 706 participants found that compared to non-vegetarians, vegetarians had a 9% lower risk of all-cause mortality (RR = 0.91; 95% CI, 66–1.16) and a 29% lower risk of mortality from IHD (RR = 0.71; 95% CI, 0.56–0.87) (Huang et al., 2012).

These findings are perhaps not surprising considering that vegetarians generally have a better cardiovascular risk profile than non-vegetarians with lower total and low density lipoprotein (LDL) cholesterol levels (Appleby, Thorogood, McPherson & Mann, 1995; Burr & Butland, 1988; West & Hayes, 1968; de Biase, Fernandes, Gianini & Duarte, 2007; Padler-Karavani et al., 2008; Nieman et al., 1989) a lower body weight (Rosell et al., 2006; Spencer, Appleby, Davey & Key, 2003; Tonstad et al., 2009), and a lower incidence of diabetes (Snowdon & Phillips, 1985; Tonstad et al., 2009; Vang, Singh, Lee, Haddad & Brinegar, 2008) and hypertension (Appleby, Davey & Key, 2002; Fraser, 2009; Rouse, Armstrong & Beilin, 1983), all of which contribute to CVD risk. A vegetarian diet may also reduce the susceptibility of LDL to oxidation (Lu et al., 2000).

Evidence suggests that a vegetarian diet may help in both the prevention and management of hypertension (Berkow & Barnard, 2005). A meta-analysis of 7 controlled clinical trials and 32 observational studies reviewed the impact of vegetarian diets on blood pressure (BP). While there was only a small number of participants in the controlled trials (311 subjects), consumption of vegetarian diets was associated with a mean systolic BP (-4.8 mm Hg; 95% CI, -6.6 to -3.1) compared with the consumption of omnivorous diets. In the observational studies (a total of 21,604 subjects), vegetarian diets were associated with lower mean systolic BP (-6.9 mm Hg; 95% CI, -9.1 to -4.7) and diastolic BP (-4.7 mm Hg; 95% CI, -6.3 to -3.1) compared with omnivorous diets (Yokoyama et al., 2014). In the EPIC-Oxford study non-meat eaters had a lower prevalence of hypertension and lower systolic and diastolic blood pressures than meat eaters although this was largely due to differences in BMI (Appleby et al., 2002). The age-adjusted prevalence of self-reported hypertension was significantly different between the four diet groups, ranging from 15.0% in male meat eaters to 5.8% in male vegans, and from 12.1% in female meat eaters to 7.7% in female vegans (Appleby et al., 2002). Fish eaters and vegetarians had a similar, intermediate prevalence of hypertension. In the Adventist Health Study 2 (AHS-2), a significant, graded association with blood pressure was also observed (Fraser, 2009). Compared with non-vegetarians, vegans were 75% less likely to be treated for hypertension, lacto-ovo vegetarians 65% less likely, pesco-vegetarians 38% less likely and semi-vegetarians 23% less likely.

Vegetarian and vegan diets also offer significant benefits for the prevention and management of diabetes. Observational studies have demonstrated a significantly lower risk of T2D in individuals following a vegetarian diet compared to non-vegetarians (Snowdon & Phillips, 1985; Tonstad et al., 2009). The original AHS, involving 25 698 Seventh-day Adventist church members who were followed up for 21 years from recruitment in 1960, found that when compared to vegetarians, non-vegetarian women had a 1.4 times higher risk and non-vegetarian men a 1.8 times higher risk of developing T2D (Snowdon & Phillips, 1985). In the more recent AHS-2 a cross-sectional analysis of baseline data from 60 903 Seventh-day Adventist Church members from North America found that those following a vegan diet had a diabetes prevalence approximately one-third that of non-vegetarians (2.9% versus 7.6%), while the lacto-ovo vegetarians, pesco-vegetarians and semi-vegetarians had an intermediate diabetes prevalence of 3.2%, 4.8% and 6.1% respectively (Tonstad et al., 2009). After adjusting for confounding factors, those following a vegan diet had almost a 50% reduction in the risk of developing T2D compared to non-vegetarians. The risk reduction became incrementally smaller as more animal products were consumed - lacto-ovo vegetarians had a 46% reduction in risk, pesco-vegetarians a 30% reduction in risk and semi-vegetarians a 24% reduction in risk (Tonstad et al., 2009) . The researchers also looked at the incidence of diabetes in a subset of 15 200 men and 26 187 women without diabetes at baseline. They were followed for 2 years and after adjustment for confounding factors, vegans were found to have a 62% lower risk of developing type 2 diabetes and lacto-ovo vegetarians a 38% lower risk compared to non-vegetarians (Tonstad et al., 2013).

The AHS-2 also demonstrated a reduced risk of MetS in vegetarians. A cross sectional analysis of 773 adults taking part in the study, just over half of whom were vegetarian (35%) or semi-vegetarian (16%), found the prevalence of MetS to be 23% in the vegetarians, 31% in semi-vegetarians and 37% in non-vegetarians (Rizzo, Sabate, Jaceldo-Siegl & Fraser, 2011). Compared to non-vegetarians, those following a vegetarian diet had 56% lower risk (adjusted OR 0.44, 95% CI 0.30-0.64) of having MetS.

There are only a few intervention studies assessing the effects of a vegetarian diet in people with diabetes, and weight loss has generally been greater on the vegetarian diets, making it difficult to determine the independent effect of the diets (Jenkins, Kendall, Marchie, Jenkins, et al., 2003). However, greater weight loss may be one of the advantages of a vegetarian diet as even small amounts of weight loss can help in both the prevention and management of T2D. One study comparing a low fat vegan diet with a diet based on the American Diabetes Association (ADA) guidelines found that the vegan diet reduced HbA1c levels significantly more than the ADA diet (1.23 versus 0.38 percentage points in those who didn't change medication) (Barnard et al., 2006). Furthermore, 43% of subjects were able to reduce their medication compared to only 26% in the ADA group. Similarly, a study of 74 subjects with T2D randomised to a low vegetarian diet (animal foods restricted to 1 low fat yoghurt/day) or a control diet (both isocaloric and calorie restricted) demonstrated significantly greater weight loss, reductions in waist circumference, visceral and subcutaneous fat, oxidative stress markers and adipokines, and improvements in insulin sensitivity on the vegetarian diet compared to the control diet (Kahleova et al., 2011). Like the previous study, 43% of subjects were able to reduce their use of diabetes medication compared to only 5% of the control group. Earlier studies of low fat vegan diets in patients with type 2 diabetes have also demonstrated significant improvements in blood glucose control and blood fats, as well as considerable reductions in medication use (Barnard, Jung & Inkeles, 1994; Nicholson et al., 1999) although the larger of these studies was not controlled.

Overall, cancer rates in vegetarians appear to be moderately lower than others living in the same communities, and life expectancy appears to be greater (Fraser, 1999; Key et al., 2009; Fraser, 2009). A meta-analysis of 7 prospective cohort studies involving 124 706 participants showed an 18% reduction in incidence of all forms of cancer than non-vegetarians (RR = 0.82; 95% CI, 0.67–0.97) (Huang et al., 2012). Recent results from the AHS-2 found that among 69 120 participants, vegetarians were 8% less likely to develop cancer than non-vegetarians (HR, 0.92; 95% CI, 0.85–0.99) for both genders combined (Tantamango-Bartley, Jaceldo-Siegl, Fan & Fraser, 2013). A statistically significant association was found between vegetarian diets and cancers of the gastrointestinal tract (HR 0.76; 95% CI, 0.63–0.90), while vegan diets were associated with a significant reduction in risk of overall cancer (HR 0.84; 95% CI, 0.72–0.99) in both genders combined and for female-specific cancers (HR 0.66; 95% CI, 0.47–0.92) (Tantamango-Bartley et al., 2013).

Despite the current popularity of high protein, low carbohydrate diets for weight management, research has consistently shown that vegetarians, and particularly vegans, are leaner than their meat-eating counterparts (Berkow & Barnard, 2006; Rosell et al., 2006; Spencer et al., 2003; Tonstad et al., 2009). The EPIC-Oxford study compared weight gain over 5 years in almost 22 000 meat-eating, fish-eating, vegetarian, and vegan men and women and found that weight gain was lowest in the vegan group and those who, during follow-up, had changed to a diet containing fewer animal foods (Rosell et al., 2006). Similarly, the AHS-2 found that mean BMI was lowest in vegans (23.6 kg/m^2) and incrementally higher in lacto-ovo vegetarians (25.7 kg/m^2), pesco-vegetarians (26.3 kg/m^2), semi-vegetarians (27.3 kg/m^2), and non-vegetarians (28.8 kg/m^2) (Tonstad et al., 2009). A recent randomised control trial involving 50 subjects who were assigned to a vegan, vegetarian, pesco-vegetarian, semi-vegetarian or omnivorous diet with unrestricted energy intake for six months to investigate effects of plant-based diets on short-term weight loss found that those on the vegan diet lost significantly more weight than the other dietary patterns. (Turner-McGrievy, Davidson, Wingard & Billings, 2014). Together these studies suggest that plant-based diets may be a useful strategy to combat the obesity epidemic in western countries.

Mediterranean Diets

The most well researched dietary pattern is the Mediterranean diet (MD) which can be described as the diet followed in the olive growing areas of Mediterranean countries. The MD is characterized by a high intake of fruits, vegetables, cereals (mainly wholegrain varieties), nuts and seeds. Fish, seafood, poultry, and eggs are consumed in moderation, while there is low consumption of dairy foods, red meat, processed meats and sweets (Trichopoulou & Lagiou, 1997). A key characteristic of the diet is the abundance of extra virgin olive oil (EVOO), which along with nuts provides the primary sources of fat in the diet.

Foods included in the MD are minimally processed, rich in fibre, antioxidant polyphenols and essential micronutrients (Toledo et al., 2013).

A meta-analysis of 36 prospective cohort studies including a combined study population of over 4 172 412 subjects with a follow-up ranging from 4.9 to 16 years, confirmed the significant and consistent protective effects of the MD on chronic disease risk. A 2-point increase in adherence to the MD was associated with an 8% reduction in overall mortality (RR 0.92, 95% CI: 0.91-0.93), a 10% reduced risk of cardiovascular disease (RR 0.90, 95% CI: 0.87-0.92), and a 4% reduction in cancer incidence (RR 0.96, 95% CI 0.95-0.97) (Sofi, Macchi, Abbate, Gensini & Casini, 2014). A second recent meta-analysis found that a 2-point increase in adherence to a MD was associated with a 13% reduction in the risk of cardiovascular events (pooled RR 0.87, 95% CI 0.85-0.90 (Martinez-Gonzalez & Bes-Rastrollo, 2014). Similarly, a recent systematic review of 37 studies looking at the association between adherence to a MD and prevention of cardiodiabesity (incorporating T2D, obesity, MetS and CVD) found that 33 of these studies provided strong evidence for an association between adherence to a MD and reduced risk of collective cardiodiabesity risk (Garcia-Fernandez, Rico-Cabanas, Rosgaard, Estruch & Bach-Faig, 2014). An earlier meta-analysis of 50 studies (35 clinical trials, 2 prospective and 13 cross-sectional) found that adherence to a MD was associated with a reduced risk of MetS while the clinical studies showed a protective effect of the MD on individual components of the MetS including WC, HDL-cholesterol, TG, systolic and diastolic BP and glucose levels (Kastorini et al., 2011).

The benefits of a MD demonstrated in observational studies have been confirmed in the PREDIMED study, a large randomised controlled trial in individuals at high risk of cardiovascular disease, comparing the effects of a MD enriched with extra virgin olive oil (EVOO) or mixed nuts versus a low fat diet on the primary prevention of CVD. The primary outcome for the study was a reduced risk of CVD and the multivariable –adjusted hazard ratios for cardiovascular disease (a composite of myocardial infarction, stroke and death from cardiovascular causes) were 0.70 (95% CI, 0.54-0.92) and 0.72 (95% CI, 0.54-0.96) for the group assigned to a MD with added EVOO and nuts respectively (Estruch et al., 2013). On the basis of results of an interim analysis, the trial was stopped after a median follow-up of 4.8 years. Combining the results of the PREDIMED study with those of the Lyon Diet Heart Study (an earlier secondary prevention trial), showed that intervention with a MD was associated with a 38% relative reduction in the risk of CVD clinical events (Martinez-Gonzalez & Bes-Rastrollo, 2014).

The PREDIMED study has also shown benefits of a MD for diabetes prevention and management (Salas-Salvadó et al., 2011; Lasa et al., 2014) as well as a greater reversion of the MetS with a MD diet (Babio et al., 2014). In a subgroup of 418 non-diabetic subjects who were randomly assigned to one of the three dietary patterns the incidence rate of diabetes per 1000 person-years was 24.6 (MD plus EVOO), 26.8 (MD plus nuts) and 46.6 for the control group. After adjustment for several confounders, the risk of developing diabetes was reduced by more than 50% in the MD groups compared with the low-fat group (Salas-Salvadó et al., 2011). In a subgroup of 191 study participants with T2D, those following a MD supplemented with either EVOO or nuts, total body weight and glucose metabolism were improved to the same extent as the low-fat diet, despite higher energy intakes than the low fat diet group (Lasa et al., 2014). An analysis of results from 5801 participants in PREDIMED found that after 4.8 years of follow-up, a MD did not reduce the risk of developing MetS but those in the MD groups who had MetS at baseline were more likely to experience reversion compared to the low fat diet group and had significant reductions in central obesity (Babio et al., 2014).

An early analysis of the PREDIMED study showed that after just 3 months, both MD groups had improved metabolic risk factors compared to the low fat diet group, including lower blood pressure, improved lipid profiles, decreased insulin resistance, and reduced concentrations of inflammatory molecules compared with the low-fat diet (Estruch et al., 2006). The mean changes in the MD with olive oil group and the MD with nuts group compared with the low-fat diet, were -0.39mmol/L and -0.30mmol/L, respectively, for plasma glucose levels, -5.9mmHg and -7.1mmHg, respectively for systolic blood pressure, and -0.38 and -0.26, respectively, for the cholesterol–HDL cholesterol ratio. The MD with olive oil also reduced C-reactive protein (CRP) levels by 0.54 mg/L compared with the low-fat diet (Estruch et al., 2006).

DASH Diet

The Dietary Approaches to Stopping Hypertension (DASH) diet is an eating pattern that is low in saturated fat, cholesterol and total fat with an increased intake of fruits, vegetables, and fat-free or low-fat dairy products. The diet is high in dietary fibre and minerals including potassium, magnesium and calcium. It focuses on fruits, vegetables, whole grains, fish, poultry, and nuts and legumes but restricts intake of red meat, sweets, added sugars, and sugar-sweetened drinks.

The original DASH trial found that this dietary intervention can help to prevent hypertension, may be as effective as medication for treating mild hypertension and might prevent or delay the need for medication in people with borderline high blood pressure (Appel et al., 1997). After 8 weeks the DASH diet lowered systolic blood pressure by 5.5/3.0mmHg more than a control diet and to a greater extent than a diet rich in fruit and vegetables. A greater reduction (11.4/5.5mmHg) was observed in those with baseline hypertension. Another study combining the DASH diet with a reduction in sodium intake found that the blood pressure lowering effect of the DASH diet is additive to a reduction in sodium intake (Bray et al., 2004).

A meta-analysis of observational studies has demonstrated that DASH-style dietary interventions can significantly reduce the risk of CVD, CHD, stroke and HF by 20%, 21% 19% and 29% respectively (Salehi-Abargouei, Maghsoudi, Shirani & Azadbakht, 2013). A systematic review of the DASH diet and risk of T2D found that a DASH diet can reduce fasting insulin levels but not fasting glucose levels or measures of insulin resistance (Shirani, Salehi-Abargouei & Azadbakht, 2013). In a combined analysis of the NHS and HPFS, adherence to a DASH-style diet was associated with a lower risk of CRC, with a 20% reduction in risk for those in the top versus bottom quintiles of DASH diet score (Fung, Hu, et al., 2010). Similarly, in the NHS, the DASH diet was associated with a 20% lower risk of estrogen receptor-negative breast cancer in postmenopausal women, comparing the highest and lowest quintiles (Fung, Hu, Hankinson, Willett & Holmes, 2011).

Portfolio Diet

The Portfolio diet was designed by Canadian researchers to combine a number of foods known to have cholesterol-lowering properties – plant sterols (from plant sterol enriched foods), soy protein (from foods such as soy milk and tofu), soluble fibres (from oats, barley, psyllium and certain vegetables such as okra and eggplant), and almonds (a handful each day). The diet is also limited in animal foods (meat and dairy products) and encourages the consumption of 5 to 9 serves of fruit and vegetables each day.

In the original study of the Portfolio diet, 34 hyperlipidaemic subjects all underwent three 1-month treatments in random order as outpatients. Treatment arms included a very low saturated fat diet (control diet), the same control diet with the inclusion of 20mg of lovastatin (statin diet) and the intervention 'portfolio' diet as described above. LDL concentrations decreased by 8.5%, 33.3% and 29.6% in the control, statin and portfolio diets respectively, demonstrating that the diet is as effective as a low dose of statin and can achieve a cholesterol lowering effect more than three times that of a low saturated fat diet over 4 weeks (Jenkins et al., 2002). Several subsequent studies have shown similar benefits (Gigleux et al., 2007; Jenkins, Kendall, Marchie, Faulkner, et al., 2003). The Portfolio diet was also as effective as a first-generation statin in reducing CRP (Jenkins et al., 2005) and has been shown to lower blood pressure, although the latter was associated with the almond content of the diet (Jenkins, Kendall, Nguyen, et al., 2008)

A longer term study of 55 subjects found that after following the Portfolio diet for 3 months, subjects had an average reduction in LDL cholesterol levels of 14.0% and this was maintained at 12.8% at the end of the 12 month study. Almost one-third of participants had reductions in LDL cholesterol of more than 20% after 12 months (Jenkins, Kendall, Faulkner, et al., 2008). Again, these results were similar to that achieved by taking a first generation statin.

Implications and Recommendations for Adopting a Low Meat or Meat Free Diet

While not everyone will choose to adopt a meat-free diet, any reduction in meat intake is likely to benefit health, particularly reducing chronic disease risk. This is evident from the findings of the AHS-2 which, in a population of more than 73 000 men and women, compared health outcomes of non-vegetarians, semi-vegetarians (those who ate meat, poultry or fish less than 1/week), pesco-vegetarians (those who ate fish but no other animal flesh), lacto-ovo vegetarians (those who ate eggs and dairy products but no animal flesh), and vegans (those who ate no animal products). They found that BMI, and risk of diabetes and hypertension (even after controlling for weight) decreased incrementally from the non-vegetarians to vegans, as less animal foods were included in the diet (Orlich et al., 2013).

The World Cancer Research Fund, as part of their cancer prevention recommendations, advise limiting the consumption of red meats (such as beef, pork and lamb) and avoiding processed meats. They recommend the population average consumption of red meat to be no more than 300 g (11 oz) a week, very little if any of which should be processed, and advise individuals who eat red meat to consume less than 500 g (18 oz) a week and to avoid or minimise processed meat (WCRF, 2007). The Australian Dietary Guidelines similarly recommend limiting red meat intake (to 455g per week of cooked lean red meat) on the basis of research showing regular consumption of large quantities of red meat (100–120g/day) may be associated with an increased risk of colorectal cancer (NHMRC, 2013). At the time of writing, the US Dietary Guidelines Advisory Committee had just released their recommendations to the U.S. Department of Agriculture (USDA) and the Department of Health and Human Services (US-DHHS), recommending a reduction in red and processed meat consumption to benefit both health and the environment (USDHHS, 2015).

Three studies looking at the impact of decreasing meat intake to reduce greenhouse gas (GHG) emissions have shown such reductions in intake would also have significant population health benefits. Friel et al. modelled the potential benefits of a 30% reduction in livestock production (to achieve the UK Committee on Climate Change recommended targets) on the burden of IHD (Friel et al., 2009). They estimated that IHD burden would decrease by about 15% in the UK and 16% in Sao Paulo city. Aston et al. conducted a modelling study using dietary intake data from the National Diet and Nutrition Survey of British Adults and found that if there was a doubling of the number of vegetarians (to 4.7% of men and 12.3% of women) and the remainder of the population adopted the current level of meat intake of the lowest quintile of non-vegetarians, there would be a significant reduction in risk of CHD, diabetes and CRC (Aston, Smith & Powles, 2012). Scarborough, while not looking independently at the impact of reduced red meat consumption, estimated that reducing the consumption of meat and dairy foods by 50% and replacing them with fruits, vegetables and cereals, would not only significantly reduce GHG emissions but would result in close to 37 000 deaths delayed or averted per year in the UK (Scarborough, Allender, Clarke, Wickramasinghe & Rayner, 2012). Together these studies show that reducing red meat intake could have significant benefits for both the environment and reducing chronic disease risk.

The main concerns cited in regards to adopting a vegetarian diet are acceptability and nutritional concerns. However several studies have found that low-fat vegetarian or vegan diets were well accepted, and similar to that of standard low fat diets (Barnard, Scialli, Turner-Mcgrievy, & Lanou 2004; Franklin, Kolasa, Griffin, Mayo & Badenhop, 1995; Barnard et al., 2009; Berkow, Bernard, Katcher & Eckart, 2010). For example Barnard et al., in a study of subjects with T2D, found that a vegan diet was equally as acceptable as a more conventional (American Diabetes Association) diet among study participants (Barnard et al., 2009). In fact, after 22 weeks, significantly more of the vegan diet group (67%) met group-specific diet adherence criteria compared to the ADA diet group (44%) (Barnard et al., 2009). In a recent weight loss trial in which overweight subjects were randomised to one of five low fat low GI diets (vegan, vegetarian, pesco-vegetarian, semi-vegetarian and omnivorous) dietary adherence was similar in all groups at two and six months (Turner-McGrievy, Davidson, Wingard, Wilcox & Frongillo, 2015). The introduction of vegetarian meals in schools and workplaces has also shown to be well accepted (de Keyzer et al., 2012; Katcher, Ferdowsian, Hoover, Cohen & Barnard, 2010).

When it comes to nutrition, the American Dietetic Association supports the fact that, in addition to their health benefits, well planned vegetarian diets, including vegan diets, are nutritionally adequate and are appropriate for individuals during all stages of life (Craig & Mangels, 2009). Some nutrients can be more difficult to obtain on a vegetarian diet, but careful planning and in some cases the use of fortified foods or supplements, can ensure that an individual's nutrition needs are met while maximising the health benefits of a vegetarian or vegan diet (Reid, Marsh, Zeuschner, Saunders & Baines, 2013).

A cross-sectional analysis of the US National Health and Nutrition Examination Survey (NHANES) (1999-2004) data looking at diet quality found that lacto-ovo vegetarian diets are nutrient dense and consistent with dietary guidelines. In fact vegetarians had higher mean intakes of fibre, vitamins A, C, and E, thiamin, riboflavin, folate, calcium, magnesium, and iron, and lower intakes of total fat, saturated fat and cholesterol (Farmer, Larson, Fulgoni, Rainville & Liepa, 2011). Overall diet quality, measured by the US Department of Agriculture's Healthy Eating Index 2005 was no different between vegetarians and non-vegetarians. A survey of US vegetarians showed that those who define themselves as vegetarian have dietary patterns that are generally healthier than non-vegetarians, being lower in total fat, saturated fat, and cholesterol and higher in fibre, with a higher intake of grains, legumes, vegetables (green leafy and yellow) and fruit (Haddad & Tanzman, 2003). Vitamin C, vitamin E, thiamine, folate, calcium and magnesium intakes were also higher than in those who ate meat, however iron, zinc and vitamin B12 intakes were lower.

In the study by Barnard et al., both the vegan and ADA dietary groups reduced their intake of energy, protein, fat, cholesterol and sodium. However, the vegan diet group had a greater reduction in fat and cholesterol intakes (Barnard et al., 2009; Turner-McGrievy et al., 2008). The vegan dieters significantly increased their intake of carbohydrate, dietary fibre, vitamins A, C and K, beta-carotene, folate, magnesium and potassium, but reduced their intakes of calcium and vitamin B12 while the ADA diet group reduced their intake of carbohydrate and iron but reported no significant increases in nutrient intake

(Turner-McGrievy et al., 2008). The vegan diet group was also found to have a significant increase in their Alternate Healthy Eating Index (AHEI) score while the ADA diet group did not, and there was a negative correlation between AHEI scores and HbA1c levels (Fung, Mc-Cullough, van Dam & Hu, 2007; Fung et al., 2005; McCullough et al., 2002). The AHEI is a nine-component dietary index used to rate foods and nutrients related to chronic disease risk, and previous research has linked a higher AHEI score with a lower risk of type 2 diabetes and cardiovascular disease (Turner-McGrievy et al., 2008). Similarly, in a weight loss trial comparing diets containing different amounts of animal foods (vegan, vegetarian, pesco-vegetarian, semi-vegetarian and omnivorous) the vegan diet group had the greatest reduction in saturated fat and cholesterol and a greater increase in fibre intake compared to the pesco-vegetarian, semi-vegetarian and omnivorous dieters (Turner-McGrievy et al., 2015).

'Meat-Free Mondays'

One very successful campaign to encourage a population-wide reduction in meat intake is the Meatless Monday movement (http://www.meatlessmonday.com) or Meat-Free Mondays (http://www.meatfree-mondays.com/). The campaign began in the US in 2003 in association with the Johns Hopkins Bloomberg School of Public Health's Center for a Liveable Future and is now a global movement, active in 36 countries including Denmark, France, Germany, Italy, Japan, the United States, Britain, Canada, Israel and Australia, and still growing. The Meat-Free Mondays campaign encourages individuals to give up eating meat one day per week, to benefit both their health and the environment. This encourages participants to learn to enjoy plant-based meals without necessarily becoming vegetarian.

FUTURE RESEARCH DIRECTIONS

Considering the increased interest in low carbohydrate high protein (LCHP) diets for weight management, well-controlled longer term studies comparing such diets with plant-based diets such as vegetarian, vegan and Mediterranean diets are warranted. These studies should investigate the effects of these diets not only on common metabolic and cardiovascular risk factors but also on non-traditional risk factors reflective of chronic disease risk. Animal studies have shown deleterious effects of low carbohydrate high protein diets on metabolic and cardiovascular function. In one study mice fed a LCHP diet developed adverse vascular effects including increased aortic atherosclerosis and an impaired ability to generate new vessels in response to tissue ischemia, independent of traditional risk factors including cholesterol levels, inflammatory markers and markers of oxidative stress (Foo et al., 2009). A second study found that adult (but not younger) mice fed a LCHP diet developed marked glucose intolerance and arterial dysfunction resulting in cardiovascular changes similar to those seen in old age (National Institute of Health, 2014). The same authors similarly found adverse effects on glucose tolerance and major arterial dysfunction in older mice fed a LCHP diet (Baron et al., 2014). Together with the wealth of information showing health benefits, reduced chronic disease risk and lower mortality with plant-based diets, these studies highlight the need to consider the longer-term health effects of diets high in animal protein and fat.

CONCLUSION

There is now a significant body of evidence demonstrating an association between red and processed meat intakes and chronic disease risk, and a number of plausible mechanisms to explain this association. On the other hand, low meat or meat-free diets are associated with improve health outcomes and reduced disease risk. In Western countries in particular, plant-based diets low in red meat may present a significant advantage over meat-based diets, not only for the environment but also for reducing chronic disease risk and increasing longevity.

REFERENCES

Abete, I., Romaguera, D., Vieira, A. R., Lopez de Munain, A., & Norat, T. (2014). Association between total, processed, red and white meat consumption and all-cause, CVD and IHD mortality: A meta-analysis of cohort studies. *The British Journal of Nutrition, 112*(5), 762–775.

Abid, Z., Cross, A. J., & Sinha, R. (2014). Meat, dairy, and cancer. *The American Journal of Clinical Nutrition, 100*(Supplement 1), 386S–393S.

Alexander, D. D., & Cushing, C. A. (2009). Quantitative assessment of red meat or processed meat consumption and kidney cancer. *Cancer Detection and Prevention, 32*(5-6), 340–351.

Alexander, D. D., Mink, P. J., Cushing, C. A., & Sceurman, B. (2010). A review and meta-analysis of prospective studies of red and processed meat intake and prostate cancer. *Nutrition Journal, 9*(50), 1–17.

Alexander, D. D., Morimoto, L. M., Mink, P. J., & Cushing, C. A. (2010). A review and meta-analysis of red and processed meat consumption and breast cancer. *Nutrition Research Reviews, 23*(2), 349–365.

Altorf-van der Kuil, W., Engberink, M. F., Vedder, M. M., Boer, J. M., Verschuren, W. M., & Geleijnse, J. M. (2012). Sources of dietary protein in relation to blood pressure in a general dutch population. *Public Library of Science (Plos). One, 7*(2), 1–8.

Andrews, N. C. (1999). Disorders of iron metabolism. *The New England Journal of Medicine, 341*(26), 1986–1995.

Appel, L. J., Moore, T. J., Obarzanek, E., Vollmer, W. M., Svetkey, L. P., Sacks, F. M., & Karanja, N. et al. (1997). A clinical trial of the effects of dietary patterns on blood pressure. DASH Collaborative Research Group. *The New England Journal of Medicine, 336*(16), 1117–1124.

Appleby, P. N., Davey, G. K., & Key, T. J. (2002). Hypertension and blood pressure among meat eaters, fish eaters, vegetarians and vegans in EPIC-Oxford. *Public Health Nutrition, 5*(5), 645–654.

Appleby, P. N., Thorogood, M., Mann, J. I., & Key, T. J. (1999). The Oxford Vegetarian Study: An overview. *The American Journal of Clinical Nutrition, 70*(3Suppl), 525S–531S.

Appleby, P. N., Thorogood, M., McPherson, K., & Mann, J. I. (1995). Associations between plasma lipid concentrations and dietary, lifestyle and physical factors in the Oxford Vegetarian Study. *Journal of Human Nutrition and Dietetics, 8*(5), 305–314.

Ascherio, A., Katan, M. B., Zock, P. L., Stampfer, M. J., & Willett, W. C. (1999). Trans fatty acids and coronary heart disease. *The New England Journal of Medicine, 340*(25), 1994–1998.

Aston, L. M., Smith, J. N., & Powles, J. W. (2012). Impact of a reduced red and processed meat dietary pattern on disease risks and greenhouse gas emissions in the UK: A modelling study. *British Medical Journal, 2*(5), 1–9.

Astrup, A., Dyerberg, J., Elwood, P., Hermansen, K., Hu, F. B., Jakobsen, M. U., & Willett, W. C. et al. (2011). The role of reducing intakes of saturated fat in the prevention of cardiovascular disease: Where does the evidence stand in 2010? *The American Journal of Clinical Nutrition, 93*(4), 684–688.

Aune, D., Chan, D. S., Vieira, A. R., Navarro Rosenblatt, D. A., Vieira, R., Greenwood, D. C., & Norat, T. et al. (2013). Red and processed meat intake and risk of colorectal adenomas: A systematic review and meta-analysis of epidemiological studies. *Cancer Causes & Control, 24*(4), 611–627.

Aune, D., Ursin, G., & Veierod, M. B. (2009). Meat consumption and the risk of type 2 diabetes: A systematic review and meta-analysis of cohort studies. *Diabetologia, 52*(11), 2277–2287.

Azadbakht, L., & Esmaillzadeh, A. (2009). Red meat intake is associated with metabolic syndrome and the plasma C-reactive protein concentration in women. *The Journal of Nutrition, 139*(2), 335–339.

Babio, N., Sorli, M., Bullo, M., Basora, J., Ibarrola-Jurado, N., Fernandez-Ballart, J., & Salas-Salvado, J. et al. (2012). Association between red meat consumption and metabolic syndrome in a Mediterranean population at high cardiovascular risk: Cross-sectional and 1-year follow-up assessment. *Nutrition, Metabolism, and Cardiovascular Diseases, 22*(3), 200–207.

Babio, N., Toledo, E., Estruch, R., Ros, E., Martinez-Gonzalez, M. A., Castaner, O., & Salas-Salvado, J. et al. (2014). Mediterranean diets and metabolic syndrome status in the PREDIMED randomized trial. *Canadian Medical Association Journal, 186*(17), E649–E657.

Baer, D. J., Judd, J. T., Clevidence, B. A., & Tracy, R. P. (2004). Dietary fatty acids affect plasma markers of inflammation in healthy men fed controlled diets: A randomized crossover study. *The American Journal of Clinical Nutrition, 79*(6), 969–973.

Bandera, E. V., Kushi, L. H., Moore, D. F., Gifkins, D. M., & McCullough, M. L. (2007). Consumption of animal foods and endometrial cancer risk: A systematic literature review and meta-analysis. *Cancer Causes & Control, 18*(9), 967–988.

Bao, W., Bowers, K., Tobias, D. K., Hu, F. B., & Zhang, C. (2013). Prepregnancy dietary protein intake, major dietary protein sources, and the risk of gestational diabetes mellitus: A prospective cohort study. *Diabetes Care, 36*(7), 2001–2008.

Bao, W., Rong, Y., Rong, S., & Liu, L. (2012). Dietary iron intake, body iron stores, and the risk of type 2 diabetes: A systematic review and meta-analysis. *BioMed Central Medicine, 10,* 1–13.

Barnard, N. D., Cohen, J., Jenkins, D. J., Turner-McGrievy, G., Gloede, L., Jaster, B., & Talpers, S. et al. (2006). A low-fat vegan diet improves glycemic control and cardiovascular risk factors in a randomized clinical trial in individuals with type 2 diabetes. *Diabetes Care, 29*(8), 1777–1783.

Barnard, N. D., Gloede, L., Cohen, J., Jenkins, D. J., Turner-McGrievy, G., Green, A. A., & Ferdowsian, H. (2009). A low-fat vegan diet elicits greater macronutrient changes, but is comparable in adherence and acceptability, compared with a more conventional diabetes diet among individuals with type 2 diabetes. *Journal of the American Dietetic Association, 109*(2), 263–272.

Barnard, N. D., Scialli, A. R., Turner-Mcgrievy, G., & Lanou, A. J. (2004). Acceptability of a low-fat vegan diet compares favourably to a step II diet in a randomized, controlled trial. *Journal of Cardiopulmonary Rehabilitation, 24*(4), 229–235.

Baron, S., Bedarida, T., Cottart, C. H., Vilbert, F., Vessieres, E., Ayer, A.,… Nivet-Antoine, V. (2014). Dual effects of resveratrol on arterial damage induced by insulin resistance in aged mice. *The Journal of Gerontology: Series A Biological Sciences and Medical Sciences, 69*(3), 260-269. doi: 2.10.1093/gerona/glt1081

Barnard, R. J., Jung, T., & Inkeles, S. B. (1994). Diet and exercise in the treatment of NIDDM. The need for early emphasis. *Diabetes Care, 17*(12), 1469–1472.

Bastide, N. M., Pierre, F. H., & Corpet, D. E. (2011). Heme iron from meat and risk of colorectal cancer: A meta-analysis and a review of the mechanisms involved. *Cancer Prevention Research (Philadelphia, Pa.), 4*(2), 177–184.

Beasley, J. M., & Wylie-Rosett, J. (2013). The role of dietary proteins among persons with diabetes. *Current Atherosclerosis Reports, 15*(9), 1–8.

Bedford, J. L., & Barr, S. I. (2005). Diets and selected lifestyle practices of self-defined adult vegetarians from a population-based sample suggest they are more 'health conscious'. *The International Journal of Behavioral Nutrition and Physical Activity, 2*(1), 1–11.

Bennett, B. J., de Aguiar Vallim, T. Q., Wang, Z., Shih, D. M., Meng, Y., Gregory, J., & Lusis, A. J. et al. (2013). Trimethylamine-N-oxide, a metabolite associated with atherosclerosis, exhibits complex genetic and dietary regulation. *Journal of Cell Metabolism, 17*(1), 49–60.

Berkow, S. E., & Barnard, N. (2006). Vegetarian diets and weight status. *Nutrition Reviews, 64*(4), 175–188.

Berkow, S. E., & Barnard, N. D. (2005). Blood pressure regulation and vegetarian diets. *Nutrition Reviews, 63*(1), 1–8.

Berkow, S. E., Bernard, N., Katcher, H., & Eckart, J. (2010). Four Theraputic Diets: Adherance and Acceptanility. *Canadian Journal of Dietetic Practice and Researc, 71*(4), 199–204.

Bernstein, A. M., Pan, A., Rexrode, K. M., Stampfer, M., Hu, F. B., Mozaffarian, D., & Willett, W. C. (2012). Dietary protein sources and the risk of stroke in men and women. *Stroke, 43*(3), 637–644.

Bernstein, A. M., Sun, Q., Hu, F. B., Stampfer, M. J., Manson, J. E., & Willett, W. C. (2010). Major dietary protein sources and risk of coronary heart disease in women. *Circulation, 122*(9), 876–883.

Bonequi, P., Meneses-Gonzalez, F., Correa, P., Rabkin, C. S., & Camargo, M. C. (2013). Risk factors for gastric cancer in Latin America: A meta-analysis. *Cancer Causes & Control, 24*(2), 217–231.

Bowers, K., Yeung, E., Williams, M. A., Qi, L., Tobias, D. K., Hu, F. B., & Zhang, C. (2011). A prospective study of prepregnancy dietary iron intake and risk for gestational diabetes mellitus. *Diabetes Care, 34*(7), 1557–1563.

Bray, G. A., Vollmer, W. M., Sacks, F. M., Obarzanek, E., Svetkey, L. P., & Appel, L. J. (2004). A further subgroup analysis of the effects of the DASH diet and three dietary sodium levels on blood pressure: Results of the DASH-Sodium Trial. *The American Journal of Cardiology, 94*(2), 222–227.

Broedbaek, K., Siersma, V., Andersen, J. T., Petersen, M., Afzal, S., & Hjelvang, B., cPoulsen, H.E. (2011). The association between low-grade inflammation, iron status and nucleic acid oxidation in the elderly. *Free Radical Research, 45*(4), 409–416.

Brugère, J. F., Borrel, G., Gaci, N., Tottey, W., O'Toole, P. W., & Malpuech-Brugere, C. (2014). Archaebiotics: Proposed therapeutic use of archaea to prevent trimethylaminuria and cardiovascular disease. *Gut Microbes, 5*(1), 5–10.

Burr, M. L., & Butland, B. K. (1988). Heart disease in British vegetarians. *The American Journal of Clinical Nutrition, 48*(3Suppl), 830–832.

Chan, D. S., Lau, R., Aune, D., Vieira, R., Greenwood, D. C., Kampman, E., & Norat, T. (2011). Red and processed meat and colorectal cancer incidence: Meta-analysis of prospective studies. *Public Library of Science (PLoS). One, 6*(6), 1–11.

Chang-Claude, J., Hermann, S., Eilber, U., & Steindorf, K. (2005). Lifestyle determinants and mortality in German vegetarians and health-conscious persons: Results of a 21-year follow-up. *Cancer Epidemiology, Biomarkers & Prevention, 14*(4), 963–968.

Chen, G. C., Lv, D. B., Pang, Z., & Liu, Q. F. (2013). Red and processed meat consumption and risk of stroke: A meta-analysis of prospective cohort studies. *European Journal of Clinical Nutrition, 67*(1), 91–95.

Chen, Y., Pan, L., Liu, N., Troy, F. A., & Wang, B. (2014). LC-MS/MS quantification of N-acetylneuraminic acid, N-glycolylneuraminic acid and ketodeoxynonulosonic acid levels in the urine and potential relationship with dietary sialic acid intake and disease in 3- to 5-year-old children. *The British Journal of Nutrition, 111*(2), 332–341.

Chen, Y. H., Xia, E. Q., Xu, X. R., Li, S., Ling, W. H., Wu, S., & Li, H. B. et al. (2012). Evaluation of benzo[a]pyrene in food from China by high-performance liquid chromatography-fluorescence detection. *International Journal of Environmental Research and Public Health, 9*(11), 4159–4169.

Clifton, P. M. (2011). Protein and coronary heart disease: The role of different protein sources. *Current Atherosclerosis Reports, 13*(6), 493–498.

Colditz, G. A., Manson, J. E., Stampfer, M. J., Rosner, B., Willett, W. C., & Speizer, F. E. (1992). Diet and risk of clinical diabetes in women. *The American Journal of Clinical Nutrition, 55*(5), 1018–1023.

Craig, W. J., & Mangels, A. R. (2009). Position of the American Dietetic Association: Vegetarian diets. *Journal of the American Dietetic Association, 109*(7), 1266–1282.

Cross, A. J., Ferrucci, L. M., Risch, A., Graubard, B. I., Ward, M. H., Park, Y., & Sinha, R. et al. (2010). A large prospective study of meat consumption and colorectal cancer risk: An investigation of potential mechanisms underlying this association. *Cancer Research and Clinical Oncology, 70*(6), 2406–2414.

Cross, A. J., Freedman, N. D., Ren, J., Ward, M. H., Hollenbeck, A. R., Schatzkin, A., & Abnet, C. C. et al. (2011). Meat consumption and risk of esophageal and gastric cancer in a large prospective study. *The American Journal of Gastroenterology, 106*(3), 432–442.

Cross, A. J., & Sinha, R. (2004). Meat-related mutagens/carcinogens in the etiology of colorectal cancer. *Environmental and Molecular Mutagenesis, 44*(1), 44–55.

Dahlquist, G. G., Blom, L. G., Persson, L. A., Sandstrom, A. I., & Wall, S. G. (1990). Dietary factors and the risk of developing insulin dependent diabetes in childhood. *British Medical Journal, 300*(6735), 1302–1306.

Daniel, C. R., Cross, A. J., Graubard, B. I., Park, Y., Ward, M. H., Rothman, N., & Sinha, R. et al. (2012). Large prospective investigation of meat intake, related mutagens, and risk of renal cell carcinoma. *The American Journal of Clinical Nutrition, 95*(1), 155–162.

Davis, K. E., Prasad, C., Vijayagopal, P., Juma, S., & Imrhan, V. (2014). Advanced glycation end products, inflammation, and chronic metabolic diseases: Links in a chain? [Epub ahead of print]. *Critical Reviews in Food Science and Nutrition*, (Sep): 26.

de Biase, S. G., Fernandes, S. F., Gianini, R. J., & Duarte, J. L. (2007). Vegetarian diet and cholesterol and triglycerides levels. *Arquivos Brasileiros de Cardiologia, 88*(1), 35–39.

de Keyzer, W., van Caneghem, S., Heath, A. L., Vanaelst, B., Verschraegen, M., de Henauw, S., & Huybrechts, I. (2012). Nutritional quality and acceptability of a weekly vegetarian lunch in primary-school canteens in Ghent, Belgium: 'Thursday Veggie Day'. *Public Health Nutrition, 15*(12), 2326–2330.

de Oliveira Otto, M. C., Alonso, A., Lee, D. H., Delclos, G. L., Bertoni, A. G., Jiang, R., & Nettleton, J. A. et al. (2012). Dietary intakes of zinc and heme iron from red meat, but not from other sources, are associated with greater risk of metabolic syndrome and cardiovascular disease. *The Journal of Nutrition, 142*(3), 526–533.

Dellavalle, C. T., Xiao, Q., Yang, G., Shu, X. O., Aschebrook-Kilfoy, B., Zheng, W., & Ward, M. H. et al. (2014). Dietary nitrate and nitrite intake and risk of colorectal cancer in the Shanghai Women's Health Study. *International Journal of Cancer, 134*(12), 2917–2926.

Djoussé, L., & Gaziano, J. M. (2009). Dietary cholesterol and coronary artery disease: A systematic review. *Current Atherosclerosis Reports, 11*(6), 418–422.

Elliott, P., Stamler, J., Dyer, A. R., Appel, L., Dennis, B., Kesteloot, H., & Zhou, B. et al. (2006). Association between protein intake and blood pressure: The INTERMAP Study. *Journal of the American Medical Association Internal Medicine, 166*(1), 79–87.

Estruch, R., Martinez-Gonzalez, M. A., Corella, D., Salas-Salvado, J., Ruiz-Gutierrez, V., Covas, M. I., & Ros, E. et al. (2006). Effects of a Mediterranean-style diet on cardiovascular risk factors: A randomized trial. *Annals of Internal Medicine, 145*(1), 1–11.

Estruch, R., Ros, E., Salas-Salvado, J., Covas, M. I., Corella, D., Aros, F., & Martinez-Gonzalez, M. A. et al. (2013). Primary prevention of cardiovascular disease with a Mediterranean diet. *The New England Journal of Medicine, 368*(14), 1279–1290.

Farmer, B., Larson, B. T., Fulgoni, V. L. III, Rainville, A. J., & Liepa, G. U. (2011). A vegetarian dietary pattern as a nutrient-dense approach to weight management: An analysis of the national health and nutrition examination survey 1999-2004. *Journal of the American Dietetic Association, 111*(6), 819–827.

Farvid, M. S., Ding, M., Pan, A., Sun, Q., Chiuve, S. E., Steffen, L. M., & Hu, F. B. et al. (2014). Dietary linoleic acid and risk of coronary heart disease: A systematic review and meta-analysis of prospective cohort studies. *Circulation, 130*(18), 1568–1578.

Fernandez-Cao, J. C., Arija, V., Aranda, N., Bullo, M., Basora, J., Martinez-Gonzalez, M. A., & Salas-Salvado, J. et al. (2013). Heme iron intake and risk of new-onset diabetes in a Mediterranean population at high risk of cardiovascular disease: An observational cohort analysis. *BioMed Central Public Health, 13*, 1042–1048.

Ferrucci, L. M., Sinha, R., Ward, M. H., Graubard, B. I., Hollenbeck, A. R., Kilfoy, B. A., & Cross, A. J. et al. (2010). Meat and components of meat and the risk of bladder cancer in the NIH-AARP Diet and Health Study. *Cancer, 116*(18), 4345–4353.

Feskens, E. J., & Kromhout, D. (1990). Habitual dietary intake and glucose tolerance in euglycaemic men: The Zutphen Study. *International Journal of Epidemiology, 19*(4), 953–959.

Feskens, E. J., Sluik, D., & van Woudenbergh, G. J. (2013). Meat consumption, diabetes, and its complications. *Current Diabetes Reports, 13*(2), 298–306.

Fonseca-Nunes, A., Jakszyn, P., & Agudo, A. (2014). Iron and cancer risk--a systematic review and meta-analysis of the epidemiological evidence. *Cancer Epidemiology, Biomarkers & Prevention, 23*(1), 12–31.

Foo, S. Y., Heller, E. R., Wykrzykowska, J., Sullivan, C. J., Manning-Tobin, J. J., & Moore, K. J., … Rosenzweig, A. (2009). Vascular effects of a low-carbohydrate high-protein diet. *Proceedings of the National Academy of Sciences of the United States (PNAS), 106*(36), 15418-15423.

Franklin, T. L., Kolasa, K. M., Griffin, K., Mayo, C., & Badenhop, D. T. (1995). Adherance to very-low-fat diet by a group of cardiac rehabilitation patients in the rural southeastern United States. *Archives of Family Medicine, 4*(6), 551–554.

Fraser, G. E. (1999). Associations between diet and cancer, ischemic heart disease, and all-cause mortality in non-Hispanic white California Seventh-day Adventists. *The American Journal of Clinical Nutrition, 70*(3), 532S–538S.

Fraser, G. E. (2009). Vegetarian diets: What do we know of their effects on common chronic diseases? *The American Journal of Clinical Nutrition, 89*(5), 1607S–1612S.

Fraser, G. E., & Shavlik, D. J. (2001). Ten years of life: Is it a matter of choice? *Journal of the American Medical Association Internal Medicine, 161*(13), 1645–1652.

Friel, S., Dangour, A. D., Garnett, T., Lock, K., Chalabi, Z., Roberts, I., & Haines, A. et al. (2009). Public health benefits of strategies to reduce greenhouse-gas emissions: Food and agriculture. *Lancet*, *374*(9706), 2016–2025.

Fung, T. T., Hu, F. B., Hankinson, S. E., Willett, W. C., & Holmes, M. D. (2011). Low-carbohydrate diets, dietary approaches to stop hypertension-style diets, and the risk of postmenopausal breast cancer. *American Journal of Epidemiology*, *174*(6), 652–660.

Fung, T. T., Hu, F. B., Wu, K., Chiuve, S. E., Fuchs, C. S., & Giovannucci, E. (2010). The Mediterranean and Dietary Approaches to Stop Hypertension (DASH) diets and colorectal cancer. *The American Journal of Clinical Nutrition*, *92*(6), 1429–1435.

Fung, T. T., McCullough, M., van Dam, R. M., & Hu, F. B. (2007). A prospective study of overall diet quality and risk of type 2 diabetes in women. *Diabetes Care*, *30*(7), 1753–1757.

Fung, T. T., McCullough, M. L., Newby, P. K., Manson, J. E., Meigs, J. B., Rifai, N., & Hu, F. B. et al. (2005). Diet-quality scores and plasma concentrations of markers of inflammation and endothelial dysfunction. *The American Journal of Clinical Nutrition*, *82*(1), 163–173.

Fung, T. T., van Dam, R. M., Hankinson, S. E., Stampfer, M., Willett, W. C., & Hu, F. B. (2010). Low-carbohydrate diets and all-cause and cause-specific mortality: Two cohort studies. *Annals of Internal Medicine*, *153*(5), 289–298.

García-Fernández, E., Rico-Cabanas, L., Rosgaard, N., Estruch, R., & Bach-Faig, A. (2014). Mediterranean diet and cardiodiabesity: A review. *Nutrients*, *6*(9), 3474–3500.

Gigleux, I., Jenkins, D. J., Kendall, C. W., Marchie, A., Faulkner, D. A., Wong, J. M., & Lamarche, B. et al. (2007). Comparison of a dietary portfolio diet of cholesterol-lowering foods and a statin on LDL particle size phenotype in hypercholesterolaemic participants. *The British Journal of Nutrition*, *98*(6), 1229–1236.

Gilsing, A. M., Fransen, F., de Kok, T. M., Goldbohm, A. R., Schouten, L. J., de Bruine, A. P., & Weijenberg, M. P. et al. (2013). Dietary heme iron and the risk of colorectal cancer with specific mutations in KRAS and APC. *Carcinogenesis*, *34*(12), 2757–2766.

Glick-Bauer, M., & Yeh, M. C. (2014). The Health Advantage of a Vegan Diet: Exploring the Gut Microbiota Connection. *Nutrients*, *6*(11), 4822–4838.

Habermeyer, M., Roth, A., Guth, S., Diel, P., Engel, K. H., Epe, B., & Eisenbrand, G. et al. (2014). Nitrate and nitrite in the diet: How to assess their benefit and risk for human health. *Molecular Nutrition & Food Research*, *59*(1), 106–128.

Haddad, E. H., & Tanzman, J. S. (2003). What do vegetarians in the United States eat? *The American Journal of Clinical Nutrition*, *78*(3), 626S–632S.

Halkjaer, J., Olsen, A., Overvad, K., Jakobsen, M. U., Boeing, H., Buijsse, B., & Tjonneland, A. et al. (2011). Intake of total, animal and plant protein and subsequent changes in weight or waist circumference in European men and women: The Diogenes project. *International Journal of Obesity*, *35*(8), 1104–1113.

Helgason, T., Ewen, S. W., Ross, I. S., & Stowers, J. M. (1982). Diabetes produced in mice by smoked/cured mutton. *Lancet, 320*(8306), 1017–1022.

Holtrop, G., Johnstone, A. M., Fyfe, C., & Gratz, S. W. (2012). Diet composition is associated with endogenous formation of N-nitroso compounds in obese men. *The Journal of Nutrition, 142*(9), 1652–1658.

Hooda, J., Shah, A., & Zhang, L. (2014). Heme, an essential nutrient from dietary proteins, critically impacts diverse physiological and pathological processes. *Nutrients, 6*(3), 1080–1102.

Hua, N. W., Stoohs, R. A., & Facchini, F. S. (2001). Low iron status and enhanced insulin sensitivity in lacto-ovo vegetarians. *The British Journal of Nutrition, 86*(4), 515–519.

Huang, T., Yang, B., Zheng, J., Li, G., Wahlqvist, M. L., & Li, D. (2012). Cardiovascular disease mortality and cancer incidence in vegetarians: A meta-analysis and systematic review. *Annals of Nutrition & Metabolism, 60*(4), 233–240.

Hughes, R., Cross, A. J., Pollock, J. R., & Bingham, S. (2001). Dose-dependent effect of dietary meat on endogenous colonic N-nitrosation. *Carcinogenesis, 22*(1), 199–202.

Hunnicutt, J., He, K., & Xun, P. (2014). Dietary iron intake and body iron stores are associated with risk of coronary heart disease in a meta-analysis of prospective cohort studies. *The Journal of Nutrition, 144*(3), 359–366.

Jakobsen, M. U., Dethlefsen, C., Joensen, A. M., Stegger, J., Tjønneland, A., Schmidt, E. B., & Overvad, K. (2010). Intake of carbohydrates compared with intake of saturated fatty acids and risk of myocardial infarction: Importance of the glycemic index. *The American Journal of Clinical Nutrition, 91*(6), 1764–1768.

Jakobsen, M. U., O'Reilly, E. J., Heitmann, B. L., Pereira, M. A., Bälter, K., Fraser, G. E., & Ascherio, A. et al. (2009). Major types of dietary fat and risk of coronary heart disease: A pooled analysis of 11 cohort studies. *The American Journal of Clinical Nutrition, 89*(5), 1425–1432.

Jakobsen, M. U., Overvad, K., Dyerberg, J., & Heitmann, B. L. (2008). Intake of ruminant trans fatty acids and risk of coronary heart disease. *International Journal of Epidemiology, 37*(1), 173–182.

Jamin, E. L., Riu, A., Douki, T., Debrauwer, L., Cravedi, J. P., Zalko, D., & Audebert, M. (2013). Combined genotoxic effects of a polycyclic aromatic hydrocarbon (B(a)P) and an heterocyclic amine (PhIP) in relation to colorectal carcinogenesis. *Public Library of Science (PLoS). One, 8*(3), 1–11.

Jenkins, D. J., Kendall, C. W., Faulkner, D., Vidgen, E., Trautwein, E. A., Parker, T. L., & Connelly, P. W. et al. (2002). A dietary portfolio approach to cholesterol reduction: Combined effects of plant sterols, vegetable proteins, and viscous fibers in hypercholesterolemia. *Metabolism: Clinical and Experimental, 51*(12), 1596–1604.

Jenkins, D. J., Kendall, C. W., Faulkner, D. A., Kemp, T., Marchie, A., Nguyen, T. H., & Singer, W. et al. (2008). Long-term effects of a plant-based dietary portfolio of cholesterol-lowering foods on blood pressure. *European Journal of Clinical Nutrition, 62*(6), 781–788.

Jenkins, D. J., Kendall, C. W., Marchie, A., Faulkner, D. A., Josse, A. R., Wong, J. M., & Connelly, P. W. et al. (2005). Direct comparison of dietary portfolio vs statin on C-reactive protein. *European Journal of Clinical Nutrition*, *59*(7), 851–860.

Jenkins, D. J., Kendall, C. W., Marchie, A., Faulkner, D. A., Wong, J. M., de Souza, R., & Connelly, P. W. et al. (2003). Effects of a dietary portfolio of cholesterol-lowering foods vs lovastatin on serum lipids and C-reactive protein. *Journal of the American Medical Association*, *290*(4), 502–510.

Jenkins, D. J., Kendall, C. W., Marchie, A., Jenkins, A. L., Augustin, L. S., Ludwig, D. S., & Anderson, J. W. et al. (2003). Type 2 diabetes and the vegetarian diet. *The American Journal of Clinical Nutrition*, *78*(3), 610S–616S.

Jenkins, D. J., Kendall, C. W., Nguyen, T. H., Marchie, A., Faulkner, D. A., Ireland, C., & Singer, W. et al. (2008). Effect of plant sterols in combination with other cholesterol-lowering foods. *Metabolism: Clinical and Experimental*, *57*(1), 130–139.

Jiang, R., Manson, J. E., Meigs, J. B., Ma, J., Rifai, N., & Hu, F. B. (2004). Body iron stores in relation to risk of type 2 diabetes in apparently healthy women. *Journal of the American Medical Association*, *291*(6), 711–717.

Jiao, L., Kramer, J. R., Chen, L., Rugge, M., Parente, P., Verstovsek, G., & El-Serag, H. B. et al. (2013). Dietary consumption of meat, fat, animal products and advanced glycation end-products and the risk of Barrett's oesophagus. *Alimentary Pharmacology & Therapeutics*, *38*(7), 817–824.

Jiao, L., Stolzenberg-Solomon, R., Zimmerman, T. P., Duan, Z., Chen, L., Kahle, L., & Sinha, R. et al. (2015). Dietary consumption of advanced glycation end products and pancreatic cancer in the prospective NIH-AARP Diet and Health Study. *The American Journal of Clinical Nutrition*, *101*(1), 126–134.

Johnson, C. M., Wei, C., Ensor, J. E., Smolenski, D. J., Amos, C. I., Levin, B., & Berry, D. A. (2013). Meta-analyses of colorectal cancer risk factors. *Cancer Causes & Control*, *24*(6), 1207–1222.

Kahleova, H., Matoulek, M., Malinska, H., Oliyarnik, O., Kazdova, L., Neskudla, T., & Pelikanova, T. et al. (2011). Vegetarian diet improves insulin resistance and oxidative stress markers more than conventional diet in subjects with Type 2 diabetes. *Diabetic Medicine*, *28*(5), 549–559.

Kaluza, J., Akesson, A., & Wolk, A. (2014). Processed and unprocessed red meat consumption and risk of heart failure: Prospective study of men. *Circulation: Heart Failure*, *7*(4), 552–557.

Kaluza, J., Larsson, S. C., Hakansson, N., & Wolk, A. (2014). Heme iron intake and acute myocardial infarction: A prospective study of men. *International Journal of Cardiology*, *172*(1), 155–160.

Kaluza, J., Wolk, A., & Larsson, S. C. (2012). Red meat consumption and risk of stroke: A meta-analysis of prospective studies. *Stroke*, *43*(10), 2556–2560.

Kaluza, J., Wolk, A., & Larsson, S. C. (2013). Heme iron intake and risk of stroke: A prospective study of men. *Stroke*, *44*(2), 334–339.

Kastorini, C. M., Milionis, H. J., Esposito, K., Giugliano, D., Goudevenos, J. A., & Panagiotakos, D. B. (2011). The effect of Mediterranean diet on metabolic syndrome and its components: A meta-analysis of 50 studies and 534,906 individuals. *Journal of the American College of Cardiology*, *57*(11), 1299–1313.

Katcher, H. I., Ferdowsian, H. R., Hoover, V. J., Cohen, J. L., & Barnard, N. D. (2010). A worksite vegan nutrition program is well-accepted and improves health-related quality of life and work productivity. *Annals of Nutrition & Metabolism, 56*(4), 245–252.

Kellow, N. J., & Savige, G. S. (2013). Dietary advanced glycation end-product restriction for the attenuation of insulin resistance, oxidative stress and endothelial dysfunction: A systematic review. *European Journal of Clinical Nutrition, 67*(3), 239–248.

Keszei, A. P., Goldbohm, R. A., Schouten, L. J., Jakszyn, P., & van den Brandt, P. A. (2013). Dietary N-nitroso compounds, endogenous nitrosation, and the risk of esophageal and gastric cancer subtypes in the Netherlands Cohort Study. *The American Journal of Clinical Nutrition, 97*(1), 135–146.

Key, T. J., Appleby, P. N., Spencer, E. A., Travis, R. C., Roddam, A. W., & Allen, N. E. (2009). Mortality in British vegetarians: Results from the European Prospective Investigation into Cancer and Nutrition (EPIC-Oxford). *The American Journal of Clinical Nutrition, 89*(5), 1613S–1619S.

Key, T. J., Fraser, G. E., Thorogood, M., Appleby, P. N., Beral, V., Reeves, G., & McPherson, K. et al. (1998). Mortality in vegetarians and non-vegetarians: A collaborative analysis of 8300 deaths among 76,000 men and women in five prospective studies. *Public Health Nutrition, 1*(1), 33–41.

Key, T. J., Fraser, G. E., Thorogood, M., Appleby, P. N., Beral, V., Reeves, G., & McPherson, K. et al. (1999). Mortality in vegetarians and nonvegetarians: Detailed findings from a collaborative analysis of 5 prospective studies. *The American Journal of Clinical Nutrition, 70*(3Suppl), 516S–524S.

Kim, E., Coelho, D., & Blachier, F. (2013). Review of the association between meat consumption and risk of colorectal cancer. *Nutrition Research (New York, N.Y.), 33*(12), 983–994.

Knowler, W. C., Barrett-Connor, E., Fowler, S. E., Hamman, R. F., Lachin, J. M., Walker, E. A., & Nathan, D. M. (2002). Reduction in the incidence of type 2 diabetes with lifestyle intervention or metformin. *The New England Journal of Medicine, 346*(6), 393–403.

Koeth, R. A., Wang, Z., Levison, B. S., Buffa, J. A., Org, E., Sheehy, B. T., & Hazen, S. L. et al. (2013). Intestinal microbiota metabolism of L-carnitine, a nutrient in red meat, promotes atherosclerosis. *Nature Medicine, 19*(5), 576–585.

Kontogianni, M. D., Panagiotakos, D. B., Pitsavos, C., Chrysohoou, C., & Stefanadis, C. (2008). Relationship between meat intake and the development of acute coronary syndromes: The CARDIO2000 case-control study. *European Journal of Clinical Nutrition, 62*(2), 171–177.

Kratz, M. (2005). Dietary cholesterol, atherosclerosis and coronary heart disease. *Handbook of Experimental Pharmacology, 170*, 195–213.

Kromhout, D., Geleijnse, J. M., Menotti, A., & Jacobs, D. R. Jr. (2011). The confusion about dietary fatty acids recommendations for CHD prevention. *The British Journal of Nutrition, 106*(5), 627–632.

Kunutsor, S. K., Apekey, T. A., Walley, J., & Kain, K. (2013). Ferritin levels and risk of type 2 diabetes mellitus: An updated systematic review and meta-analysis of prospective evidence. *Diabetes/Metabolism Research and Reviews, 29*(4), 308–318.

Larsson, S. C., & Orsini, N. (2014). Red meat and processed meat consumption and all-cause mortality: A meta-analysis. *American Journal of Epidemiology, 179*(3), 282–289.

Larsson, S. C., & Wolk, A. (2012). Red and processed meat consumption and risk of pancreatic cancer: Meta-analysis of prospective studies. *British Journal of Cancer, 106*(3), 603–607.

Lasa, A., Miranda, J., Bullo, M., Casas, R., Salas-Salvado, J., Larretxi, I., & Portillo, M. P. et al. (2014). Comparative effect of two Mediterranean diets versus a low-fat diet on glycaemic control in individuals with type 2 diabetes. *European Journal of Clinical Nutrition, 68*(7), 767–772.

Le, L. T., & Sabate, J. (2014). Beyond meatless, the health effects of vegan diets: Findings from the Adventist cohorts. *Nutrients, 6*(6), 2131–2147.

Lee, D. H., Folsom, A. R., & Jacobs, D. R. (2004). Dietary iron intake and Type 2 diabetes incidence in postmenopausal women: The Iowa Women s Health Study. *Diabetologia, 47*(2), 185–194.

Lee, J. E., Spiegelman, D., Hunter, D. J., Albanes, D., Bernstein, L., van den Brandt, P. A., & Smith-Warner, S. A. et al. (2008). Fat, protein, and meat consumption and renal cell cancer risk: A pooled analysis of 13 prospective studies. *Journal of the National Cancer Institute, 100*(23), 1695–1706.

Lenzen, S. (2008). Oxidative stress: The vulnerable beta-cell. *Biochemical Society Transactions, 36*(3), 343–347.

Lichtenstein, A. H., Appel, L. J., Brands, M., Carnethon, M., Daniels, S., Franch, H. A., & Wylie-Rosett, J. et al. (2006). Summary of American Heart Association Diet and Lifestyle Recommendations revision 2006. *Arteriosclerosis, Thrombosis, and Vascular Biology, 26*(10), 2186–2191.

Lindström, J., Peltonen, M., Eriksson, J. G., Louheranta, A., Fogelholm, M., Uusitupa, M., & Tuomilehto, J. (2006). High-fibre, low-fat diet predicts long-term weight loss and decreased type 2 diabetes risk: The Finnish Diabetes Prevention Study. *Diabetologia, 49*(5), 912–920.

Lovejoy, J., & DiGirolamo, M. (1992). Habitual dietary intake and insulin sensitivity in lean and obese adults. *The American Journal of Clinical Nutrition, 55*(6), 1174–1179.

Lu, S. C., Wu, W. H., Lee, C. A., Chou, H. F., Lee, H. R., & Huang, P. C. (2000). LDL of Taiwanese vegetarians are less oxidizable than those of omnivores. *The Journal of Nutrition, 130*(6), 1591–1596.

Luan de, C., Li, H., Li, S. J., Zhao, Z., Li, X., & Liu, Z. M. (2008). Body iron stores and dietary iron intake in relation to diabetes in adults in North China. *Diabetes Care, 31*(2), 285–286.

Marsh, K., & Brand-Miller, J. (2011). Vegetarian Diets and Diabetes. *American Journal of Lifestyle Medicine, 5*(2), 135–143.

Marshall, J. A., Bessesen, D. H., & Hamman, R. F. (1997). High saturated fat and low starch and fibre are associated with hyperinsulinaemia in a non-diabetic population: The San Luis Valley Diabetes Study. *Diabetologia, 40*(4), 430–438.

Marshall, J. A., Hamman, R. F., & Baxter, J. (1991). High-fat, low-carbohydrate diet and the etiology of non-insulin-dependent diabetes mellitus: The San Luis Valley Diabetes Study. *American Journal of Epidemiology, 134*(6), 590–603.

National Cancer Institute (NCI). (2013). Sources of saturated fat in the diet of the U.S. population ages 2 years and older, NHANES 2005-2006. Retrieved from http://appliedresearch.cancer.gov/diet/food-sources/sat_fat.html

Marshall, J. A., Hoag, S., Shetterly, S., & Hamman, R. F. (1994). Dietary fat predicts conversion from impaired glucose tolerance to NIDDM. The San Luis Valley Diabetes Study. *Diabetes Care, 17*(1), 50–56.

Martinez-Gonzalez, M. A., & Bes-Rastrollo, M. (2014). Dietary patterns, Mediterranean diet, and cardiovascular disease. *Current Opinion in Lipidology, 25*(1), 20–26.

Mayer, E. J., Newman, B., Quesenberry, C. P., & Selby, J. V. (1993). Usual dietary fat intake and insulin concentrations in healthy women twins. *Diabetes Care, 16*(11), 1459–1469.

Mayer-Davis, E. J., Monaco, J. H., Hoen, H. M., Carmichael, S., Vitolins, M. Z., Rewers, M. J., & Karter, A. J. et al. (1997). Dietary fat and insulin sensitivity in a triethnic population: The role of obesity. The Insulin Resistance Atherosclerosis Study (IRAS). *The American Journal of Clinical Nutrition, 65*(1), 79–87.

McCullough, M. L., Feskanich, D., Stampfer, M. J., Giovannucci, E. L., Rimm, E. B., Hu, F. B., & Willett, W. C. et al. (2002). Diet quality and major chronic disease risk in men and women: Moving toward improved dietary guidance. *The American Journal of Clinical Nutrition, 76*(6), 1261–1271.

McEvoy, C. T., Temple, N., & Woodside, J. V. (2012). Vegetarian diets, low-meat diets and health: A review. *Public Health Nutrition, 15*(12), 2287–2294.

Mensink, R. P., Zock, P. L., Katan, M. B., & Hornstra, G. (1992). Effect of dietary cis and trans fatty acids on serum lipoprotein[a] levels in humans. *Journal of Lipid Research, 33*(10), 1493–1501.

Micha, R., Michas, G., Lajous, M., & Mozaffarian, D. (2013). Processing of meats and cardiovascular risk: Time to focus on preservatives. *BioMed Central Medicine, 11*, 136. doi:10.1186/1741-7015-11-136

Micha, R., Michas, G., & Mozaffarian, D. (2012). Unprocessed red and processed meats and risk of coronary artery disease and type 2 diabetes--an updated review of the evidence. *Current Atherosclerosis Reports, 14*(6), 515–524.

Micha, R., Wallace, S. K., & Mozaffarian, D. (2010). Red and processed meat consumption and risk of incident coronary heart disease, stroke, and diabetes mellitus: A systematic review and meta-analysis. *Circulation, 121*(21), 2271–2283.

Mozaffarian, D., Aro, A., & Willett, W. C. (2009). Health effects of trans-fatty acids: Experimental and observational evidence. *European Journal of Clinical Nutrition, 63*(2), S5–S21.

Mozaffarian, D., Hao, T., Rimm, E. B., Willett, W. C., & Hu, F. B. (2011). Changes in diet and lifestyle and long-term weight gain in women and men. *The New England Journal of Medicine, 364*(25), 2392–2404.

Mozaffarian, D., Katan, M. B., Ascherio, A., Stampfer, M. J., & Willett, W. C. (2006). Trans fatty acids and cardiovascular disease. *The New England Journal of Medicine, 354*(15), 1601–1613.

Mozaffarian, D., Micha, R., & Wallace, S. (2010). Effects on coronary heart disease of increasing poly-unsaturated fat in place of saturated fat: A systematic review and meta-analysis of randomized controlled trials. *Public Library of Science (PLoS). Medicine*, *7*(3), 1–10.

Mozaffarian, D., Pischon, T., Hankinson, S. E., Rifai, N., Joshipura, K., Willett, W. C., & Rimm, E. B. (2004). Dietary intake of trans fatty acids and systemic inflammation in women. *The American Journal of Clinical Nutrition*, *79*(4), 606–612.

National Cancer Institute (NCI). (2013). Sources of saturated fat in the diets of the U.S. population ages 2 years and older, NHANES 2005-2006. Retrieved November 17, 2014, from http://appliedresearch.cancer.gov/diet/foodsources/sat_fat/sf.html

National Health and Medical Research Council (NHMRC). (2013). Australian Dietary Guidelines. Canberra: Australian Government, NHMRC, Department of Health and Ageing. Retrieved September 8, 2015, from https://www.nhmrc.gov.au/guidelines-publications/n55

National Institute of Health (NIH). (2014). Ferritin blood tests. Retrieved 17 November, 2014, from http://www.nlm.nih.gov/medlineplus/ency/article/003490.htm

Nicholson, A. S., Sklar, M., Barnard, N. D., Gore, S., Sullivan, R., & Browning, S. (1999). Toward improved management of NIDDM: A randomized, controlled, pilot intervention using a lowfat, vegetarian diet. *Preventive Medicine*, *29*(2), 87–91.

Nieman, D. C., Sherman, K. M., Arabatzis, K., Underwood, B. C., Barbosa, J. C., Johnson, M., & Lee, J. et al. (1989). Hematological, anthropometric, and metabolic comparisons between vegetarian and nonvegetarian elderly women. *International Journal of Sports Medicine*, *10*(4), 243–251.

Noto, H., Goto, A., Tsujimoto, T., & Noda, M. (2013). Low-carbohydrate diets and all-cause mortality: A systematic review and meta-analysis of observational studies. *Public Library of Science (PLoS). One*, *8*(1), 1–10.

Oomen, C. M., Ocke, M. C., Feskens, E. J., van Erp-Baart, M. A., Kok, F. J., & Kromhout, D. (2001). Association between trans fatty acid intake and 10-year risk of coronary heart disease in the Zutphen Elderly Study: A prospective population-based study. *Lancet*, *357*(9258), 746–751.

Opara, E. C. (2004). Role of oxidative stress in the etiology of type 2 diabetes and the effect of antioxidant supplementation on glycemic control. *Journal of Investigative Medicine*, *52*(1), 19–23.

Orlich, M. J., Singh, P. N., Sabate, J., Jaceldo-Siegl, K., Fan, J., Knutsen, S., & Fraser, G. E. et al. (2013). Vegetarian dietary patterns and mortality in Adventist Health Study 2. *Journal of the American Medical Association Internal Medicine*, *173*(13), 1230–1238.

Padler-Karavani, V., Yu, H., Cao, H., Chokhawala, H., Karp, F., Varki, N., & Varki, A. et al. (2008). Diversity in specificity, abundance, and composition of anti-Neu5Gc antibodies in normal humans: Potential implications for disease. *Glycobiology*, *18*(10), 818–830.

Paluszkiewicz, P., Smolinska, K., Debinska, I., & Turski, W. A. (2012). Main dietary compounds and pancreatic cancer risk. The quantitative analysis of case-control and cohort studies. *Cancer Epidemiology*, *36*(1), 60–67.

Pan, A., Sun, Q., Bernstein, A. M., Manson, J. E., Willett, W. C., & Hu, F. B. (2013). Changes in red meat consumption and subsequent risk of type 2 diabetes mellitus: Three cohorts of US men and women. *Journal of the American Medical Association Internal Medicine, 173*(14), 1328–1335.

Pan, A., Sun, Q., Bernstein, A. M., Schulze, M. B., Manson, J. E., Stampfer, M. J., & Hu, F. B. et al. (2012). Red meat consumption and mortality: Results from 2 prospective cohort studies. [formerly Archives of Internal Medicine]. *Journal of the American Medical Association Internal Medicine, 172*(7), 555–563.

Pan, A., Sun, Q., Bernstein, A. M., Schulze, M. B., Manson, J. E., Willett, W. C., & Hu, F. B. (2011). Red meat consumption and risk of type 2 diabetes: 3 cohorts of US adults and an updated meta-analysis. *The American Journal of Clinical Nutrition, 94*(4), 1088–1096.

Parillo, M., & Riccardi, G. (2004). Diet composition and the risk of type 2 diabetes: Epidemiological and clinical evidence. *The British Journal of Nutrition, 92*(1), 7–19.

Parker, D. R., Weiss, S. T., Troisi, R., Cassano, P. A., Vokonas, P. S., & Landsberg, L. (1993). Relationship of dietary saturated fatty acids and body habitus to serum insulin concentrations: The Normative Aging Study. *The American Journal of Clinical Nutrition, 58*(2), 129–136.

Pereira, M. A., Jacobs, D. R. Jr, Van Horn, L., Slattery, M. L., Kartashov, A. I., & Ludwig, D. S. (2002). Dairy consumption, obesity, and the insulin resistance syndrome in young adults: The CARDIA Study. *Journal of the American Medical Association, 287*(16), 2081–2089.

Pham, N. M., Mizoue, T., Tanaka, K., Tsuji, I., Tamakoshi, A., Matsuo, K., & Sasazuki, S. et al. (2014). Meat consumption and colorectal cancer risk: An evaluation based on a systematic review of epidemiologic evidence among the Japanese population. *Japanese Journal of Clinical Oncology, 44*(7), 641–650.

Pietinen, P., Ascherio, A., Korhonen, P., Hartman, A. M., Willett, W. C., Albanes, D., & Virtamo, J. (1997). Intake of fatty acids and risk of coronary heart disease in a cohort of Finnish men. The Alpha-Tocopherol, Beta-Carotene Cancer Prevention Study. *American Journal of Epidemiology, 145*(10), 876–887.

Portha, B., Giroix, M. H., Cros, J. C., & Picon, L. (1980). Diabetogenic effect of N-nitrosomethylurea and N-nitrosomethylurethane in the adult rat. *Annales de la Nutrition et de l'Alimentation, 34*(5-6), 1143–1151.

Pounis, G. D., Tyrovolas, S., Antonopoulou, M., Zeimbekis, A., Anastasiou, F., Bountztiouka, V., & Panagiotakos, D. B. et al. (2010). Long-term animal-protein consumption is associated with an increased prevalence of diabetes among the elderly: The Mediterranean islands (MEDIS) study. *Diabetes & Metabolism, 36*(6), 484–490.

Pratt, M. M., John, K., MacLean, A. B., Afework, S., Phillips, D. H., & Poirier, M. C. (2011). Polycyclic aromatic hydrocarbon (PAH) exposure and DNA adduct semi-quantitation in archived human tissues. *International Journal of Environmental Research and Public Health, 8*(7), 2675–2691.

Preis, S. R., Stampfer, M. J., Spiegelman, D., Willett, W. C., & Rimm, E. B. (2010). Dietary protein and risk of ischemic heart disease in middle-aged men. *The American Journal of Clinical Nutrition, 92*(5), 1265–1272.

Qi, L., van Dam, R. M., Rexrode, K. M., & Hu, F. B. (2007). Heme iron from diet as a risk factor for coronary heart disease in women with type 2 diabetes. *Diabetes Care, 30*(1), 101–106.

Qiu, C., Zhang, C., Gelaye, B., Enquobahrie, D. A., Frederick, I. O., & Williams, M. A. (2011). Gestational diabetes mellitus in relation to maternal dietary heme iron and nonheme iron intake. *Diabetes Care, 34*(7), 1564–1569.

Rajpathak, S., Ma, J., Manson, J., Willett, W. C., & Hu, F. B. (2006). Iron intake and the risk of type 2 diabetes in women: A prospective cohort study. *Diabetes Care, 29*(6), 1370–1376.

Rajpathak, S. N., Crandall, J. P., Wylie-Rosett, J., Kabat, G. C., Rohan, T. E., & Hu, F. B. (2009). The role of iron in type 2 diabetes in humans. *Biochimica et Biophysica Acta, 1790*(7), 671–681.

Reid, M. A., Marsh, K. A., Zeuschner, C. L., Saunders, A. V., & Baines, S. K. (2013). Meeting the nutrient reference values on a vegetarian diet. *The Medical Journal of Australia, 199*(4), S33–S40.

Rizzo, N. S., Jaceldo-Siegl, K., Sabate, J., & Fraser, G. E. (2013). Nutrient profiles of vegetarian and nonvegetarian dietary patterns. *Journal of the Academy of Nutrition and Dietetics, 113*(12), 1610–1619.

Rizzo, N. S., Sabate, J., Jaceldo-Siegl, K., & Fraser, G. E. (2011). Vegetarian dietary patterns are associated with a lower risk of metabolic syndrome: The adventist health study 2. *Diabetes Care, 34*(5), 1225–1227.

Rohrmann, S., Overvad, K., Bueno-de-Mesquita, H. B., Jakobsen, M. U., Egeberg, R., Tjonneland, A., & Linseisen, J. et al. (2013). Meat consumption and mortality--results from the European Prospective Investigation into Cancer and Nutrition. *BioMed Central Medicine, 11*(63), 1–12.

Romaguera, D., Norat, T., Vergnaud, A. C., Mouw, T., May, A. M., Agudo, A., & Peeters, P. H. et al. (2010). Mediterranean dietary patterns and prospective weight change in participants of the EPIC-PANACEA project. *The American Journal of Clinical Nutrition, 92*(4), 912–921.

Rosell, M., Appleby, P., Spencer, E., & Key, T. (2006). Weight gain over 5 years in 21,966 meat-eating, fish-eating, vegetarian, and vegan men and women in EPIC-Oxford. *International Journal of Obesity, 30*(9), 1389–1396.

Rouse, I. L., Armstrong, B. K., & Beilin, L. J. (1983). The relationship of blood pressure to diet and lifestyle in two religious populations. *Journal of Hypertension, 1*(1), 65–71.

Russell, W. R., Gratz, S. W., Duncan, S. H., Holtrop, G., Ince, J., Scobbie, L., & Flint, H. J. et al. (2011). High-protein, reduced-carbohydrate weight-loss diets promote metabolite profiles likely to be detrimental to colonic health. *The American Journal of Clinical Nutrition, 93*(5), 1062–1072.

Salas-Salvadó, J., Bulló, M., Babio, N., Martínez-González, M. A., Ibarrola-Jurado, N., Basora, J., & Ros, E. et al. (2011). Reduction in the Incidence of Type 2 Diabetes with the Mediterranean Diet. *Diabetes Care, 34*(1), 14–19.

Salehi, M., Moradi-Lakeh, M., Salehi, M. H., Nojomi, M., & Kolahdooz, F. (2013). Meat, fish, and esophageal cancer risk: A systematic review and dose-response meta-analysis. *Nutrition Reviews, 71*(5), 257–267.

Salehi-Abargouei, A., Maghsoudi, Z., Shirani, F., & Azadbakht, L. (2013). Effects of Dietary Approaches to Stop Hypertension (DASH)-style diet on fatal or nonfatal cardiovascular diseases--incidence: A systematic review and meta-analysis on observational prospective studies. *Nutrition (Burbank, Los Angeles County, Calif.)*, *29*(4), 611–618.

Samraj, A. N., Laubli, H., Varki, N., & Varki, A. (2014). Involvement of a non-human sialic acid in human cancer. *Frontiers in Oncology*, *4*, 1–13.

Saunders, A. V., Craig, W. J., Baines, S. K., & Posen, J. S. (2012). Iron and vegetarian diets. *The Medical Journal of Australia*, *1*(2), 11–16.

Scarborough, P., Allender, S., Clarke, D., Wickramasinghe, K., & Rayner, M. (2012). Modelling the health impact of environmentally sustainable dietary scenarios in the UK. *European Journal of Clinical Nutrition*, *66*(6), 710–715.

Schulz, M., Kroke, A., Liese, A. D., Hoffmann, K., Bergmann, M. M., & Boeing, H. (2002). Food groups as predictors for short-term weight changes in men and women of the EPIC-Potsdam cohort. *The Journal of Nutrition*, *132*(6), 1335–1340.

Schwab, U., Lauritzen, L., Tholstrup, T., Haldorssoni, T., Riserus, U., Uusitupa, M., & Becker, W. (2014). Effect of the amount and type of dietary fat on cardiometabolic risk factors and risk of developing type 2 diabetes, cardiovascular diseases, and cancer: A systematic review. *Food & Nutrition Research*, *58*, 1–26.

Schwab, U. S., Ausman, L. M., Vogel, S., Li, Z., Lammi-Keefe, C. J., Goldin, B. R., & Lichtenstein, A. H. et al. (2000). Dietary cholesterol increases the susceptibility of low density lipoprotein to oxidative modification. *Atherosclerosis*, *149*(1), 83–90.

Shirani, F., Salehi-Abargouei, A., & Azadbakht, L. (2013). Effects of Dietary Approaches to Stop Hypertension (DASH) diet on some risk for developing type 2 diabetes: A systematic review and meta-analysis on controlled clinical trials. *Nutrition (Burbank, Los Angeles County, Calif.)*, *29*(7-8), 939–947.

Sievenpiper, J. L., & Dworatzek, P. D. (2013). Food and dietary pattern-based recommendations: An emerging approach to clinical practice guidelines for nutrition therapy in diabetes. *Canadian Journal of Diabetes*, *37*(1), 51–57.

Singh, P. N., Sabate, J., & Fraser, G. E. (2003). Does low meat consumption increase life expectancy in humans? *The American Journal of Clinical Nutrition*, *78*(3), 526S–532S.

Sinha, R., Cross, A. J., Graubard, B. I., Leitzmann, M. F., & Schatzkin, A. (2009). Meat intake and mortality: A prospective study of over half a million people. *Journal of the American Medical Association Internal Medicine*, *169*(6), 562–571.

Sinha, R., Peters, U., Cross, A. J., Kulldorff, M., Weissfeld, J. L., Pinsky, P. F., & Hayes, R. B. et al. (2005). Meat, meat cooking methods and preservation, and risk for colorectal adenoma. *Cancer Research*, *65*(17), 8034–8041.

Sipetic, S. B., Vlajinac, H. D., Kocev, N. I., Marinkovic, J. M., Radmanovic, S. Z., & Bjekic, M. D. (2005). The Belgrade childhood diabetes study: A multivariate analysis of risk determinants for diabetes. *European Journal of Public Health*, *15*(2), 117–122.

Sluijs, I., & Beulens, J. W., van der A, D. L., Spijkerman, A. M., Grobbee, D. E., & van der Schouw, Y. T. (2010). Dietary intake of total, animal, and vegetable protein and risk of type 2 diabetes in the European Prospective Investigation into Cancer and Nutrition (EPIC)-NL study. *Diabetes Care*, *33*(1), 43–48.

Smolinska, K., & Paluszkiewicz, P. (2010). Risk of colorectal cancer in relation to frequency and total amount of red meat consumption. Systematic review and meta-analysis. *Archives of Medical Science*, *6*(4), 605–610.

Snowdon, D. A., & Phillips, R. L. (1985). Does a vegetarian diet reduce the occurrence of diabetes? *American Journal of Public Health*, *75*(5), 507–512.

Snowdon, D. A., Phillips, R. L., & Fraser, G. E. (1984). Meat consumption and fatal ischemic heart disease. *Preventive Medicine*, *13*(5), 490–500.

Sofi, F., Macchi, C., Abbate, R., Gensini, G. F., & Casini, A. (2014). Mediterranean diet and health status: An updated meta-analysis and a proposal for a literature-based adherence score. *Public Health Nutrition*, *17*(12), 2769–2782.

Song, P., Lu, M., Yin, Q., Wu, L., Zhang, D., Fu, B., & Zhao, Q. et al. (2014). Red meat consumption and stomach cancer risk: A meta-analysis. *Journal of Cancer Research and Clinical Oncology*, *140*(6), 979–992.

Spencer, E. A., Appleby, P. N., Davey, G. K., & Key, T. J. (2003). Diet and body mass index in 38000 EPIC-Oxford meat-eaters, fish-eaters, vegetarians and vegans. *International Journal of Obesity and Related Metabolic Disorders*, *27*(6), 728–734.

Tangvoranuntakul, P., Gagneux, P., Diaz, S., Bardor, M., Varki, N., Varki, A., & Muchmore, E. (2003). Human uptake and incorporation of an immunogenic nonhuman dietary sialic acid. [PNAS]. *Proceedings of the National Acadamy of Sciences of the United States*, *100*(21), 12045–12050.

Tantamango-Bartley, Y., Jaceldo-Siegl, K., Fan, J., & Fraser, G. (2013). Vegetarian diets and the incidence of cancer in a low-risk population. *Cancer Epidemiology, Biomarkers & Prevention*, *22*(2), 286–294.

Toledo, E., Hu, F. B., Estruch, R., Buil-Cosiales, P., Corella, D., Salas-Salvado, J., & Martinez-Gonzalez, M. A. et al. (2013). Effect of the Mediterranean diet on blood pressure in the PREDIMED trial: Results from a randomized controlled trial. *BioMed Central Medicine*, *11*, 207.

Tonstad, S., Butler, T., Yan, R., & Fraser, G. E. (2009). Type of vegetarian diet, body weight, and prevalence of type 2 diabetes. *Diabetes Care*, *32*(5), 791–796.

Tonstad, S., Stewart, K., Oda, K., Batech, M., Herring, R. P., & Fraser, G. E. (2013). Vegetarian diets and incidence of diabetes in the Adventist Health Study-2. *Nutrition, Metabolism, and Cardiovascular Diseases*, *23*(4), 292–299.

Trichopoulou, A., & Lagiou, P. (1997). Healthy traditional Mediterranean diet: An expression of culture, history, and lifestyle. *Nutrition Reviews*, *55*(11), 383–389.

Tsunehara, C. H., Leonetti, D. L., & Fujimoto, W. Y. (1990). Diet of second-generation Japanese-American men with and without non-insulin-dependent diabetes. *The American Journal of Clinical Nutrition*, *52*(4), 731–738.

Tuomilehto, J., Lindstrom, J., Eriksson, J. G., Valle, T. T., Hamalainen, H., Ilanne-Parikka, P., & Uusitupa, M. et al. (2001). Prevention of type 2 diabetes mellitus by changes in lifestyle among subjects with impaired glucose tolerance. *The New England Journal of Medicine, 344*(18), 1343–1350.

Turner-McGrievy, G. M., Barnard, N. D., Cohen, J., Jenkins, D. J., Gloede, L., & Green, A. A. (2008). Changes in nutrient intake and dietary quality among participants with type 2 diabetes following a low-fat vegan diet or a conventional diabetes diet for 22 weeks. *Journal of the American Dietetic Association, 108*(10), 1636–1645.

Turner-McGrievy, G. M., Davidson, C. R., Wingard, E. E., & Billings, D. L. (2014). Low glycemic index vegan or low-calorie weight loss diets for women with polycystic ovary syndrome: A randomized controlled feasibility study. *Nutrition Research (New York, N.Y.), 34*(6), 552–558.

Turner-McGrievy, G. M., Davidson, C. R., Wingard, E. E., Wilcox, S., & Frongillo, E. A. (2015). Comparative effectiveness of plant-based diets for weight loss: A randomized controlled trial of five different diets. *Nutrition (Burbank, Los Angeles County, Calif.), 31*(2), 350–358.

United States Department of Agriculture (USDA) and United States Department of Health and Human Services. (USDHHS). (2010). Dietary Guidelines for Americans, 2010. Washington DC: USDA and USDHHS. Retrieved September 9, 2015, from http://health.gov/dietaryguidelines/dga2010/dietary-guidelines2010.pdf

United States Department of Agriculture (USDA) and United States Department of Health and Human Services. (USDHHS). (2015). Scientific Report of the 2015 Dietary Guidelines Advisory Committee. United States Department of Health & Human Services. Retrieved September 19, 2015, from http://health.gov/dietaryguidelines/2015-scientific-report/pdfs/scientific-report-of-the-2015-dietary-guidelines-advisory-committee.pdf

Uribarri, J., Cai, W., Peppa, M., Goodman, S., Ferrucci, L., Striker, G., & Vlassara, H. (2007). Circulating glycotoxins and dietary advanced glycation endproducts: Two links to inflammatory response, oxidative stress, and aging. *The Journals of Gerontology. Series A, Biological Sciences and Medical Sciences, 62*(4), 427–433.

Uribarri, J., Woodruff, S., Goodman, S., Cai, W., Chen, X., Pyzik, R., & Vlassara, H. et al. (2010). Advanced glycation end products in foods and a practical guide to their reduction in the diet. *Journal of the American Dietetic Association, 110*(6), 911–916.

van Puyvelde, K., Mets, T., Njemini, R., Beyer, I., & Bautmans, I. (2014). Effect of advanced glycation end product intake on inflammation and aging: A systematic review. *Nutrition Reviews, 72*(10), 638–650.

Vang, A., Singh, P. N., Lee, J. W., Haddad, E. H., & Brinegar, C. H. (2008). Meats, processed meats, obesity, weight gain and occurrence of diabetes among adults: Findings from Adventist Health Studies. *Annals of Nutrition & Metabolism, 52*(2), 96–104.

Vergnaud, A. C., Norat, T., Romaguera, D., Mouw, T., May, A. M., Travier, N., & Peeters, P. H. et al. (2010). Meat consumption and prospective weight change in participants of the EPIC-PANACEA study. *The American Journal of Clinical Nutrition, 92*(2), 398–407.

Vessby, B. (2000). Dietary fat and insulin action in humans. *The British Journal of Nutrition*, *83*(1), S91–S96.

Wang, Y., & Beydoun, M. A. (2009). Meat consumption is associated with obesity and central obesity among US adults. *International Journal of Obesity*, *33*(6), 621–628.

Wang, Z., Klipfell, E., Bennett, B. J., Koeth, R., Levison, B. S., Dugar, B., & Hazen, S. L. et al. (2011). Gut flora metabolism of phosphatidylcholine promotes cardiovascular disease. *Nature*, *472*(7341), 57–63.

Weggemans, R. M., Zock, P. L., & Katan, M. B. (2001). Dietary cholesterol from eggs increases the ratio of total cholesterol to high-density lipoprotein cholesterol in humans: A meta-analysis. *The American Journal of Clinical Nutrition*, *73*(5), 885–891.

West, R. O., & Hayes, O. B. (1968). Diet and serum cholersterol levels. A comparision between vegetarians and nonvegetarians in a Seventh-day Adventist group. *The American Journal of Clinical Nutrition*, *21*(8), 853–862.

White, D. L., & Collinson, A. (2013). Red meat, dietary heme iron, and risk of type 2 diabetes: The involvement of advanced lipoxidation endproducts. *Advances in Nutrition*, *4*(4), 403–411.

Willett, W. C. (2012). Dietary fats and coronary heart disease. *Journal of Internal Medicine*, *272*(1), 13–24.

Willett, W. C., & Stampfer, M. J. (2013). Current evidence on healthy eating. *Annual Review of Public Health*, *34*, 77–95.

Willett, W. C., Stampfer, M. J., Manson, J. E., Colditz, G. A., Speizer, F. E., Rosner, B. A., & Hennekens, C. H. et al. (1993). Intake of trans fatty acids and risk of coronary heart disease among women. *Lancet*, *341*(8845), 581–585.

Wong, J. M. W. (2014). Gut microbiota and cardiometabolic outcomes: Influence of dietary patterns and their associated components. *The American Journal of Clinical Nutrition*, *100*(1), 369S–377S.

World Cancer Research Fund (WCRF). (2007). Our Cancer Recommendations - Animal Foods. Retrieved February, 2015, from http://www.wcrf.org/int/research-we-fund/cancer-prevention-recommendations/animal-foods

World Cancer Research Fund. (2014). Our Cancer Prevention Recommendations. Retrieved November 17, 2014, from http://www.wcrf.org/int/research-we-fund/our-cancer-prevention-recommendations

World Health Organization (WHO). (2004). Global Strategy of Diet, Physical Activity and Health. Geneva, Switzerland: WHO. Retrieved September 10, 2015, from http://www.who.int/dietphysicalactivity/strategy/eb11344/strategy_english_web.pdf

World Health Organization (WHO). (2011). Global status report on noncommunicable diseases 2010. Geneva, Switzerland: WHO. Retrieved Septemver 10, 2015, from http://www.who.int/nmh/publications/ncd_report2010/en/

World Health Organization (WHO). (2014a). Cancer (Fact Sheet). Retrieved November 17, 2014, from http://www.who.int/mediacentre/factsheets/fs297/en/

World Health Organization (WHO). (2014b). Cardiovascular diseases (Fact sheet). Retrieved November 17, 2014, from http://www.who.int/mediacentre/factsheets/fs317/en

World Health Organization (WHO). (2014c). Diabetes (Fact sheet). Retrieved November 17, 2014, from http://www.who.int/mediacentre/factsheets/fs312/en/

Xu, J., Yang, X. X., Wu, Y. G., Li, X. Y., & Bai, B. (2014). Meat consumption and risk of oral cavity and oropharynx cancer: A meta-analysis of observational studies. *Public Library of Science (PLoS). One, 9*(4), 1–9.

Xu, X., Yu, E., Gao, X., Song, N., Liu, L., Wei, X., & Fu, C. et al. (2013). Red and processed meat intake and risk of colorectal adenomas: A meta-analysis of observational studies. *International Journal of Cancer, 132*(2), 437–448.

Xue, X. J., Gao, Q., Qiao, J. H., Zhang, J., Xu, C. P., & Liu, J. (2014). Red and processed meat consumption and the risk of lung cancer: A dose-response meta-analysis of 33 published studies. *International Journal of Clinical and Experimental Medicine, 7*(6), 1542–1553.

Yang, W., Li, B., Dong, X., Zhang, X. Q., Zeng, Y., Zhou, J. L., & Xu, J. J. et al. (2014). Is heme iron intake associated with risk of coronary heart disease? A meta-analysis of prospective studies. *European Journal of Nutrition, 53*(2), 395–400.

Yang, W. S., Wong, M. Y., Vogtmann, E., Tang, R. Q., Xie, L., Yang, Y. S., & Xiang, Y. B. et al. (2012). Meat consumption and risk of lung cancer: Evidence from observational studies. *Annals of Oncology, 23*(12), 3163–3170.

Yokoyama, Y., Nishimura, K., Barnard, N. D., Takegami, M., Watanabe, M., Sekikawa, A., & Miyamoto, Y. et al. (2014). Vegetarian diets and blood pressure: A meta-analysis. *Journal of the American Medical Association Internal Medicine, 174*(4), 577–587.

Zeuschner, C. L., Hokin, B. D., Marsh, K. A., Saunders, A. V., Reid, M. A., & Ramsay, M. R. (2013). Vitamin B12 and vegetarian diets. *The Medical Journal of Australia, 199*(4), S27–S32.

Zhang, C., Schulze, M. B., Solomon, C. G., & Hu, F. B. (2006). A prospective study of dietary patterns, meat intake and the risk of gestational diabetes mellitus. *Diabetologia, 49*(11), 2604–2613.

Zhao, Z., Li, S., Liu, G., Yan, F., Ma, X., Huang, Z., & Tian, H. (2012). Body iron stores and heme-iron intake in relation to risk of type 2 diabetes: A systematic review and meta-analysis. *Public Library of Science (PLoS). One, 7*(7), 1–17.

Zheng, W., & Lee, S. A. (2009). Well-done meat intake, heterocyclic amine exposure, and cancer risk. *Nutrition and Cancer, 61*(4), 437–446.

Zhu, H., Yang, X., Zhang, C., Zhu, C., Tao, G., Zhao, L., & Sun, X. et al. (2013). Red and processed meat intake is associated with higher gastric cancer risk: A meta-analysis of epidemiological observational studies. *Public Library of Science (PLoS). One, 8*(8), 1–10.

Zhu, H. C., Yang, X., Xu, L. P., Zhao, L. J., Tao, G. Z., Zhang, C., & Sun, X. C. et al. (2014). Meat consumption is associated with esophageal cancer risk in a meat- and cancer-histological-type dependent manner. *Digestive Diseases and Sciences, 59*(3), 664–673.

KEY TERMS AND DEFINITIONS

Advanced Glycation End-Products (AGEs): AGEs are formed when foods high in fat and protein (particularly meat) are cooked at high temperatures resulting in the browning or charring of food.

Haem Iron: The form of iron found in animal products including red and processed meats, which is readily absorbed by the body; a catalyst or pro-oxidant in many cellular reactions, including the endogenous formation of n-nitroso compounds.

Heterocyclic Amines (HCA): Chemicals formed in meat as a result of cooking meat at high temperatures. HCAs are known carcinogens.

N-Glycolylneuraminic Acid (Neu5Gc): A sialic acid (mainly in red meat) that promotes chronic inflammation and development of chronic diseases.

N-Nitroso Compounds: Carcinogenic substances formed in the gut following ingestion of haem iron (in red meat) and preservatives nitrites and nitrates (added to processed meats).

Polycyclic Aromatic Hydrocarbons (PAH): Chemicals formed when meat is charred during the cooking process. PAHs are also found in exhaust fumes and cigarette smoke and are known carcinogens.

Saturated Fat: A non-essential fatty acid found mostly in animal products, including red and processed meats and dairy, as well as coconut oil/milk, palm kernel oil and palm oil.

Trans-Fats: A non-essential fatty acid formed during the partial hydrogenation of vegetable oils to improve stability and shelf life; small amounts are naturally occurring in ruminant animals.

Trimethylamine-N-Oxide (TMAO): A toxic compound associated with increased cardiovascular disease risk, produced by the action of gut bacteria on L-carnitine and choline.

Chapter 9
The Future of Antibiotics and Meat

Talia Raphaely
Curtin University, Australia

Dora Marinova
Curtin University, Australia

Mira Marinova
Notre Dame University, Australia

ABSTRACT

This chapter discusses antibiotic use in the livestock industry and potential ramifications for human health. Antibiotics are routinely administered to food animals, primarily at sub-therapeutic levels. The extensive use of antibiotics in global animal husbandry in quantities greater than used for humans is creating antibiotic resistance. There is evidence that antibiotic resistant organisms emerging in food animals transfer to humans through the food chain, environmental contamination, direct association with animals or through mobile resistant genetic elements resulting in co-resistance to other antibiotics. No new classes of antibiotics have been developed since the 1980s. Intensifying use of existing antibiotics for meat production poses new challenges for treating humans, needs to be taken seriously and dealt with urgently. This chapter argues that reduced meat consumption is an under-considered but essential part in any suite of solutions aimed at preserving the use of antibiotics for human treatment.

INTRODUCTION

Demand for animal protein for human consumption is escalating worldwide at a record pace. To meet this challenge, intensification of production has occurred. Contemporary intensive animal production processes are associated with systematic, therapeutic and non-therapeutic use of antibiotics to control infectious diseases or to promote efficient production. Antibiotic use in meat production is likely to pose new challenges for treating human illness and infections. There is an increasingly likely possibility of a post-antibiotic world where bacterial infections, now largely treatable, become the life-threatening

DOI: 10.4018/978-1-4666-9553-5.ch009

scourges they once were. Antibacterial resistance is becoming a " ticking time bomb in our midst, which needs to be taken seriously and urgently dealt with (HAIAP & TWN, 2013, p. 2).

This chapter describes the potential impending international crisis of antimicrobial resistance in relation to increasing meat consumption. It argues for the urgent need to take action through reducing both, meat consumption and antibiotic use.

ANTIBIOTIC USE

Before we engage with the threats posed by the wide application of antibiotics, it is useful to define some of the terms used in this chapter. This is followed by a brief history of the antibiotic era.

Terminology

For the layperson, the terminology around the use of antibiotics may be unclear. Antibiotics are defined as substances, which can kill or inhibit the growth of bacteria and can be produced by microorganisms or synthetically (but still chemically related to natural versions). Antimicrobials include antibiotics as well as other substances (such as fungicides and disinfectants) that kill or inhibit the growth of micro-organisms. In the medical literature the two terms are used interchangeably.

Antibiotic or antimicrobial resistance is when the targeted bacteria are no longer controlled or killed by the presence of antibiotics but are able to survive and even multiply. In the literature antibacterial or antibiotic resistance (ABR) and antimicrobial resistance (AMR) are also used synonymously. In this chapter we refer only to antibiotics and thus use the acronym ABR.

A Brief History of the Antibiotic Era

Antibiotics made a significant contribution to the control of infectious and bacterial diseases that for much of humanity's existence were the leading causes of human mortality and morbidity. Although it is common belief that exposure to these substances has been confined to the modern antibiotic era, research reveals antibiotics have been in use since antiquity. Traces of tetracycline found in Sudanese human skeletal remains (from 350-550 CE) point to tetracycline-containing materials in the diets of these ancient peoples (Bassett, Keith, Armelagos, Martin & Villanueva, 1980; Nelson, Dinardo, Hochberg & Armelagos, 2010). Although it is not clear whether these were used specifically for disease control, histological samples taken from late Roman period skeletal remains from the Dakhleh Oasis in Egypt also evidence the presence of tetracycline in the diet at that time (Cook, Molto & Anderson, 1989). Interestingly, rates of bacterial infection documented in these two population groups were low (Armelagos, 1969; Cook et al., 1989). Anecdotes about red soils in Jordan historically used for their antibiotic-like properties in treating skin infections have led to discovering antibiotic-producing bacteria and concomitant antibiotic production in these soils. Traditional Chinese medicine used remedies for millennia based on plants containing powerful antimicrobial substances (Cui & Su, 2009) also offering evidence of human exposure in the pre-antibiotic era.

The modern "antibiotic era" is associated first with Paul Ehrlich and then with Alexander Fleming. Ehrlich in 1904 envisaged a magic bullet that selectively targeted only disease-causing microbes and not the host (Aminov, 2010). This became the cornerstone of the pharmaceutical industry's drug research strategy resulting in a variety of antimicrobial drugs identified and translated into clinical practice. Penicillin, somewhat serendipitously discovered by Fleming in 1928 and mass produced and distributed from 1945 heralded a golden era (1950 -1970) of discovery of novel antibiotic groups. No new classes have been discovered since the 1980s (Alliance to Save Our Antibiotics, 2014). The existing mainstream approach for the development of new drugs to combat emerging and re-emerging resistance of pathogens to antibiotics has been the modification of existing antibiotics (Chopra, Hesse & O'Neill, 2002). For more than 60 years antibiotics have been the cornerstone of modern medicine, suppressing and combating myriad potentially deadly infections (Todar, 2012).

Projections about the future of antibiotic development highlight there are few potential avenues that offer benefits over existing antibiotics. There are no magic bullets in the antibiotic pipeline that will eradicate bacteria resistance to existing antibiotics, including superbugs (Butler & Cooper, 2012).

ANTIBIOTIC RESISTANCE

Antibiotics are essential for combating human illness and various infections. Recently, however, we are witnessing the emergence and spread of antibiotic resistance which poses a serious challenge and threat to humanity.

The Potential Impact of Antibiotic Resistance (ABR)

Antibiotics are commonly used for the treatment of a wide range of bacterial infections from pneumonia and strep throat to meningitis and food poisoning making them treatable conditions rather than life-threatening scourges. Now much of the progress achieved in recent decades towards improving human health is at risk (WHO, 2014a). Rather than declining, bacterial infections and diseases have become an increasingly serious threat to global public health with once life-saving drugs now worthless and bacterial infections failing to respond to antibiotic treatment. This increasing ineffectiveness of antibiotics in treating common infections is on the rise (Laxminarayan et al., 2013) and we are at the dawn of a post-antibiotic era (CDC, 2013c; WHO, 2014a). Some experts warn that "antibiotic resistance is a catastrophic threat and a nightmare on par with terrorism and climate change" (Davies and Frieden in Sample, Harvey & Campbell, 2013, n.p.).

The World Health Organization (WHO) estimates that, since their development, antibiotics have added around 20 years of life expectancy — a benefit that we are now in danger of losing (Grant & Taylor, 2013). "If we fail to act, we are looking at an almost unthinkable scenario where antibiotics no longer work and we are cast back into the dark ages of medicine where treatable infections and injuries will kill once again" (Cameron in Walsh, 2014, n.p.).

As with the threat of climate change, ABR has no borders and humans are largely responsible for its causes. The immense potential scale of disease and resistance evolution does not respect national boundaries, cultures or geographies. International trade and travel are contributing to the problem. Although the contribution of this problem is largely unknown and empirical research is lacking, national responses by individual countries taken to protect their populations stand defunct in the face of these challenges. Only an understanding of the causes underlying ABR can assist in successfully dealing with the threats.

How Does It Happen?

Simply explained, antibiotics are used to kill (bacteriocidal) or inhibit (bacteriostatic) bacterial growth. Bacteria that don't succumb to the applied antibiotic, survive to propagate and become resistant. Therefore the use of antibiotics drives the selection process for the development of bacterial resistance.

Drug resistance is a particular problem in bacteria because it is readily transmitted, even to different bacteria species. One normally thinks of inherited characteristics passing only to successive generations of that particular organism. This is not the case for bacteria. It makes sense to think of the entire bacterial world "as one huge multicellular organism in which the cells interchange their genes" (Levy cited in Miller, 1998, p. 67).

The speed with which bacteria propagate makes the problem of drug resistance particularly acute. The generation time of bacteria can be measured in minutes (Goforth & Goforth, 2000) and a single bacterium produces over a million new bacteria in under a day. One Escherichia coli (E.coli) bacterium can produce well over a million progeny (20 generations) in a mere seven hours (Harrison & Svec, 1998; Witte, 1998). Just one drug-resistant bacterium can create millions of resistant bacteria first in the host and then in other hosts or in the environment.

Resistance which develops in animal populations may emerge in humans through organisms transferring to people either through the food chain, environment contamination or through direct association with these animals. Once the resistant bacteria are in the human population they can spread quickly and easily amongst increasing numbers of people. The interconnected ecosystems within the microbial environment "allow exchange of DNA, promoting the spread of resistance from one genus to another. The combination of increased bacterial virulence and increased drug resistance creates a potential for increased risk of morbidity and mortality for animals and humans that some have extrapolated to a catastrophic potential" (NRC, 1999, p. 70).

Further, the continued use of a single antibiotic can lead to resistance to multiple structurally unrelated antimicrobials. Pathogenic bacteria can be either zoonotic (which transfer between animals and humans) or sapronotic (which are in the wider environment) or both (Woolhouse, Ward, van Bunnik & Farrar, 2015). This interplay between different ecologies is particularly important in the context of ABR (Woolhouse et al., 2015). Mobile resistant genetic elements within bacteria can be transmitted between bacteria of the same or different species thereby resulting in co-resistance to other antibiotics (Rushton, Ferreira & Stark, 2014; Woolhouse et al., 2015).

There are multiple links between animal, environmental and human ecologies and because of the movement of genetic elements, livestock animals act as a vast "reservoir of resistant genes" (Woolhouse et al., 2015). Scientific evidence shows "[t]his reservoir of resistance can be transmitted directly or indirectly to humans through food consumption and direct or indirect contact… and… antibiotic-resistant strains can be disseminated to the environment via animal waste" (Economou & Gousia, 2015). The key causes of antimicrobial resistance are further explored as part of natural evolution, misuse in human medicine, overuse for livestock production and under the influence of vested interests.

Natural Evolution

Bacteria are one of the most successful colonisers of the planet and are found in almost all environments including the human body. We inhabit a world of bacteria – there are more bacteria in our bodies than we have human cells (Laxminarayan, 2015). Many of the bacteria that live on and in the human body are beneficial to the host; some are even essential. Antibiotics originated in soil where microbes have been producing antibiotics for millennia (Smith & Coast, 2013). For billions of years, before antibiotics were considered or used for medicinal purposes, bacteria have been developing resistance to these natural compounds (D'Costa et al., 2011; Whitman, Coleman & Wiebe, 1998).

Pathogenic bacteria are those capable of causing disease (Goforth & Goforth, 2000). It is these pathogens we are concerned about and attack with antibiotics. Factory mass-produced antibiotics are now presenting bacteria with a chemical attack they have been evolving for billions of years to overcome. Since the 1940s and the commencement of the industrial manufacture of antibiotics, production of antibiotics has been on a rapid increase. Today 20 tonnes of antibiotics are produced every hour by a global industry worth more than US$30 billion per annum (Woolhouse & Farrar, 2014).

The deployment of antibiotics is fighting against well-established evolution. Resistance will inevitably arise after any new antibiotic is introduced. This has been illustrated time and time again and it is known now beyond doubt that there is a continuing race between the discovery and use of new antibiotics and bacterial response through emerging resistance mechanisms.

Any use of antibiotics places selection pressure for resistant bacteria to endure – Darwinian selection at play. Antibiotics kill off the sensitive bacteria, letting the resistant ones survive. These mechanisms naturally select stronger and more resistant bacteria whether in humans or animals. However, "the selection pressure caused by the use of millions of tonnes of antibiotics over the past 75 years since antibiotics were introduced has made almost all disease-causing bacteria resistant to antibiotics commonly used to treat them" (Laxminarayan et al., 2013, p. 1).

As a result of the evolutionary programming of bacteria, use of antibiotics is the single most important factor leading to antibiotic resistance around the world (CDC, 2013b). Indiscriminate use of antibiotics exacerbates and accelerates the problem of resistance. Wherever antibiotics are administered in large volumes, whether in humans or animals, there is a risk of having infections that are caused by organisms that do not respond to antibiotics.

Human Medicine

There is a strong correlation between levels of human antibiotic consumption and antibiotic-resistant infections (Goossens, Ferech, vander Stichele, Elseviers & ESAC Project Group, 2005). This relationship has been relatively well researched and publicised. Antibiotics are essential therapeutic tools but there are cases when they are used sub-optimally. Reasons include frequent intake due to regular prescription or access without prescription as well as incomplete or "blind" treatment.

Over-Prescription

Antibiotics are often prescribed when not needed:

- In countries such as Japan, USA and Australia patient expectations often drive antibiotic prescriptions (Laxminarayan et al., 2013) requesting antibiotics when they are not needed (e.g. for simple viral infections);

- Hospitals, in countries including China (Daemmrich, 2013) rely on pharmaceutical sales for income generation, creating an incentive for medical staff to prescribe and overprescribe;
- Antibiotic providers such as pharmacies, may offer commission to doctors for directing patients to their businesses. Pharmacies in some countries, such as India, routinely compensate doctors for sending patients to them as customers (Sweidan et al., 2005); and
- Patients with health cover are less likely to be price sensitive, thus antibiotic sales and prescriptions increase with insurance status (Dong, Bogg, Rehnberg & Diwan, 1999).

Non-Prescription Use

Non-prescription use, including self-medication, varies from country to country and can account for between 19% and 100% of all antibiotics consumed (Morgan, Okeke, Laxminarayan, Perencevich & Weisenberg, 2011). Whilst generally consumers seem to have positive attitudes about antibiotics, they also paradoxically have a limited knowledge about these drugs and diseases (Hawkings, Butler & Wood, 2008). Studies in Europe show a strong positive correlation between a higher prevalence of antibiotic resistance and lack of knowledge about antibiotics (Grigoryan et al., 2007).

The availability of antibiotics without prescription is driven primarily by the following factors (Grigoryan et al., 2008; Planta, 2007; Buchman et al., 2008; Gartin, Brewis & Schwartz, 2010):

1. Absence of prescription-only regulations;
2. Lack of access to appropriate medical facilities;
3. Ineffective law enforcement; and
4. Poverty-driven practice, cultural and social norms.

Such self-medication lacks key attributes of successful therapy such as proper diagnosis, suitable antibiotic choice, correct usage, compliance and treatment efficiency monitoring, thus contributing to the mounting resistance problem (Aminov, 2010).

Sub-Optimal Compliance

Sub-optimal compliance includes (Zarb & Goossens, 2012; Aminov, 2010):

- Taking left-over antibiotics from previous treatment courses;
- Not properly complying with drug use regimens;
- Sharing unused drugs with other people;
- Starting "blind" antibiotic treatment before diagnosis can be confirmed; and
- Selecting antibiotics "empirically" because determining the bacteria, which are causing the illness, and the antibiotics to which they are sensitive, can take a few days and can be expensive.

Clearly use of antibiotics in human medicine is making us vulnerable. However, there could be a larger problem in use of antibiotics in animal agriculture (Laxminarayan, 2015) to which far less attention is paid.

Livestock Production

There are four uses of antibiotics in animal agriculture:

1. **Therapeutic:** For treatment of manifested illness in individual animals or groups;
2. **Metaphylactic:** Application involves medicating at risk groups, with no clinical disease, at therapeutic concentrations to prevent expected outbreaks;
3. **Prophylactic:** Prolonged, mass administration of antibiotics at sub-therapeutic levels for intensively raised animals; and
4. **Growth Promotion:** Prolonged, mass administration of antibiotics at sub-therapeutic levels to promote growth of livestock animals and enhance feed efficiency.

Both prophylactic use and administration for growth promotion have the sole purpose of increasing meat production output. Yet it is increasingly difficult to discount the scientific evidence indicating sub-therapeutic dosing of livestock is a contributing factor in the spread of antibiotic resistant organisms.

Commercial Interests

In intensive factory farming, commercial interests often take precedence over human and animal welfare and wellbeing. Below are some examples of the power of commercial influence:

- **High Antibiotic Sales for Use for Livestock:** Evidence gathered by Foer (2009) in the US shows that for every dose of antibiotics taken by a sick human, eight doses are given to a "healthy" animal;
- **Rejecting the Link between Antibiotic Use in Livestock and Antibiotic Resistance:** The World Organization for Animal Health states that concerns for bacterial resistance in humans is over-blown (de la Hamaide, 2012); the Responsible Use of Medicine in Agriculture Alliance (RUMA), a trade alliance representing the interests of the pharmaceutical and intensive-farming industries, dismisses claims that the overuse of antibiotics in intensive farming adds to the serious public-health threat from antibiotic resistance calling this a "myth" (RUMA, 2012, 2013);
- **Safeguarding Economic Viability of Animal Agriculture:** The use of subtherapeutic doses of antibiotics makes factory farm practices feasible" (O'Brien, 1996, p. 413); "one reason large confinement systems have worked so well is because of antibiotics. Without the antibiotics it would be hard to have these larger systems" (Goforth & Goforth, 2000, p. 56); antibiotics enable cost-effective economies of scale (Scully, 2002);
- **Successfully Lobbying for Tax Exemptions and Subsidies:** Farming, including livestock production, "is protected by deference and vested interest, excluded from the regulations, planning conditions and taxes other businesses must observe" (Mombiot, 2015, n.p.); livestock industry is feasting from the federal trough (Shapiro, 2015) and;
- **Blocking Regulations Limiting Antibiotic Use in Livestock:** The monitoring and regulation of antibiotic use in industrialized countries have been difficult because the meat industry is enormously powerful and able to successfully thwart attempts to collect data let alone create effective regulation" (Sharma, 2015, n.p.).

Magnitude of the Threat

Given that antibiotic resistance is scientifically undisputed, it is not unreasonable to expect policy awareness and urgent responses. Despite this, the facts point to the contrary as discussed below.

Although governments haven't kept records of quantities of antibiotics issued for farm animals, it is conservatively estimated that well over 50% of total global antibiotic use is attributable to the international livestock industry. Limited data available vary from country to country, animal species and stage of production. In the non-medical areas of animal agriculture some countries use four times the amount of antibiotics administered in human medicine; estimates are that 80% - 90% of all antibiotic sales in the USA is used in livestock production (Philpott, 2011a, 2011b; FDA, 2010, 2013a, 2013b; Shea, Florini & Barlam, 2001; Mellon, Benbrook & Benbrook, 2001).

The practice of antibiotic use in sub-therapeutic concentrations began as a developed world phenomenon and it would appear that this is where the problem lies. For example the USA uses more antibiotics per kilogram of meat produced than any other country in the world (Aarestrup & Wegener, 2009).

With increases in affluence, the middle-income countries are experiencing unprecedented and rising demand for meat causing significant increase in antibiotic use (Myers & Kent, 2003; Tilman, Balzar, Hill & Befort, 2011). To meet growth in demand for meat, Brazil, Russia, India, China and South Africa (the BRICS economies) have shifted towards more intensive livestock production systems, which rely on antibiotics instead of good husbandry practices. Calculations show that China's swine and poultry production industries alone used over 38.5 million kg of antibiotics in 2012 (Wu, 2012; Krishnasamy, Otte & Silbergeld, 2015).

A study by van Boeckel et al. (2014) estimates that the global consumption of antibiotics in food animal production is projected to rise by 67% between 2010 and 2030. Two thirds of this growth is due to the burgeoning numbers of animals raised for food production and the remaining third is attributable to a shift to increased intensive farming practices. By 2030 Asian consumption will represent 82% of the 2010 global antibiotic consumption in food animals.

In 2010, the five countries with the largest shares of global antimicrobial consumption in food animal production were China (23%), the United States (13%), Brazil (9%), India (3%) and Germany (3%). Whilst China and Brazil are currently amongst the largest consumers of antimicrobials for livestock production they have already initiated a shift towards more intensified livestock production and are using antibiotics to maintain animal health and increase productivity. The five countries with the greatest projected percentage increases in antibiotics consumption by 2030 are likely to by Myanmar (205%), Indonesia (202%), Nigeria (163%), Peru (160%) and Vietnam (157%) (van Boeckel et al., 2014, p. 5650; WPAPM, 2015). For BRICS the increase in antimicrobial consumption will be 99% (van Boeckel et al., 2014), up to seven times the projected population growth (13%) in this group of countries (World Bank, 2015).

As far back as the 1960s scientists warned against extensive and non-therapeutic use of antibiotics in farmed animal feed (e.g. The Swann Committee Report, Parliament of the United Kingdom, 1969). The World Health Organisation (WHO), United States Food and Drug Administration (FDA), the US Department of Agriculture, the Centres for Disease Control and Prevention (CDC) and the Institute of Medicine (a division of the National Academy of Sciences) have testified to the link between routine non-therapeutic use of antibiotics in food animals and antibiotic resistance in humans and called for a ban (Rogers, 2012). According to a former FDA commissioner, "[w]e have more than enough scientific evidence to justify curbing the rampant use of antibiotics for livestock, yet the food and drug industries are not only fighting proposed legislation to reduce these practices, they also oppose collecting the data"

(Kessler, 2013, n.p.). In fact, global data on antibiotic use in livestock are scarce, stemming from the absence of surveillance systems combined with lack of support from the livestock sector to report antibiotic consumption and reluctance of pharmaceutical companies to provide accurate information on sales.

It is clear that those with vested interests argue that it can be difficult to accurately trace or prove the link between antibiotic use in animals and antibiotic resistance in humans. However, we are already confronted with examples of resistance in humans directly related to antibiotic use for livestock. Nonetheless there remains a lack of essential research on the evidence how these specific links occur.

Examples of Resistance Related to Livestock

Studies have confirmed and traced the spread of resistance determinants in bacteria found in animals to those found in humans. For example nourseothricin was used for growth promotion in pigs but not given directly to humans. Nonetheless within a few years of its introduction in sub-therapeutic doses into pig feed, resistant E.coli were observed not only in pigs but in employees at the pig farms, their family members and members of the surrounding community (Hummel, Tschäpe & Witte, 1986). In this study, the resistant bacteria in humans was not attributable to the direct use or misuse of antibiotics in the human population as the antibiotic was only fed to pigs (Hummel et al., 1986). Other cases have been recorded of human infections with bacteria originally acquired from farm animals, such as Staphylococcus, without any contact with livestock or meat products (Chang, Wang, Regev-Yochay, Lipstich & Hanage, 2015).

Most classes of antibiotics that are prescribed for humans are also used sub-therapeutically in food animals. Tetracycline, penicillin, streptomycin and bacitracin are common additives in livestock feed. Similarly, critically important classes of drugs such as third and fourth generation cephalosporins and fluoroquinolone are also given to livestock (Aarestrup, Wegener & Collignon, 2008). Whilst some restrictions apply in specific countries –for example, the use of fluoroquinolone in food animals is banned in Australia but permitted in numerous other countries, including the United States (Shaban, Cruickshank, Christiansen & Antimicrobial Resistance Standing Committee, 2013), all important classes of antibiotics are being used in livestock production somewhere around the world. This renders every class of antibiotic susceptible to development of resistance (Silbergeld, Graham & Price, 2008).

Examples of resistant bacteria linked to antibiotics commonly used in livestock (Knox, 2012) include:

- "Common bacteria (for example, Escherichia coli [E.coli], Klebsiella pneumonia and Staphylococcus aureus) that cause common health-care associated and community-acquired infections (urinary tract infections, wound infections, bloodstream infections and pneumonia)" (WHO, 2014a, p. 12) . In particular, E.coli is spread by food with for example, meat becoming contaminated with bacteria during the slaughtering process;
- The use of fluoroquinolones (egenrofloxacin) in food animals resulted in the development of ciprofloxacin-resistant Salmonella, Campylobacter and E.coli, which have caused human infections, and spread worldwide through travel and food trade (Wegener, 2003);
- Enterobacteriaceae resistant to carbapenems have been linked to antibiotic use in livestock feed (Carlet et al., 2012). Carbapenems are a class of antibiotics considered a drug of last resort and carbapenem-resistant enterobacteriaceae (CRE) are seen as a new "superbug" (Eisler, 2013), a nightmare bacteria (Breslow, 2014) which currently kills half of those who contract bloodstream infections (CDC, 2013c);

- A CDC study recently found over half of chicken meat in the US were contaminated with the Enterococcus bacteria resistant to the antibiotic virginiamycin (CSPI, 2014, n.p.);
- Multi-drug resistant strains of Listeria monocytogenes were found in hamburger patties in supermarkets and other retail shops in Malaysia (Wong et al., 2012);
- A study by the FDA (2000) found that 20% of supermarket samples of ground beef, pork, chicken and turkey were contaminated with Salmonella; 84% of these were resistant to at least one antibiotic, and more than half were resistant to at least three antibiotics;
- Campylobacter infections that are resistant to fluoroquinolone antibiotics are a serious concern as they cause more food poisoning than any other bacteria (Collingon, 2005). In the United States, 19% of Campylobacter isolated from humans were ciprofloxacin resistant in 2001 (Gupta et al., 2004) but in 2010 this percentage increased to 22% (NARMS, 2010). Resistance rates higher than 90% were seen in Spain (Luangtongkum et al., 2009). By contrast, in Australia, where fluoroquinolones were never approved for use in food animals, domestically acquired infections with fluoroquinolone-resistant Campylobacter spp. are rarely found in humans (Helms, Simonsen, Olsen & Mølbak, 2005); and
- Vancomycin-resistant enterococci can spread from animals to humans (Wegener, 2003; Baragona, 2010). This antibiotic is on the World Health Organisation's List of Essential Medicines – the most important medications needed in a basic health system (WHO, 2013).

These examples illustrate that driven by increasing meat consumption, excessive antibiotic usage in intensive animal husbandry practices ABR will become a more global problem in the coming years. Ironically, whilst there is growing awareness of the threats presented by growing antibiotic resistance, while medical use is decreasing farm use of critically important antibiotics (WHO, 2012) is on the rise with more animals being raised intensively (Antibiotics Alliance, 2014).

Whilst these threats may still in cases be located within countries, no-one country is immune. Antibiotic resistance "is happening right now in every region of the world and has the potential to affect anyone, of any age, in any country" (WHO, 2014b, n.p.) . Antibiotic resistance can cross international boundaries and spread between continents with remarkable speed and ease. World health leaders have described antibiotic resistant microorganisms as "nightmare bacteria" that "pose a catastrophic threat" to people in every country in the world (CDC, 2013b, p. 11).

SUSTAINABILITY IMPLICATIONS OF ANTIBIOTIC RESISTANCE

As well as the obvious public and animal health impacts antibiotic resistance also carries significant economic, social and environmental impacts.

Economic Implications

A 2014 report commissioned by the UK's Review on Antimicrobial Resistance (Taylor et al., 2014) paints a bleak picture about what might happen to the world's economy if the problem of antibiotic resistance is left unchecked (Zuraw, 2014). The overall message of the Review is that the cost of taking action now to slow the spread of resistance is much smaller than what it could cost over time.

It is unavoidably clear in light of the existing body of research that there is a significant externalised economic impact arising from antibiotic resistance. Yet there is only limited information qualifying and quantifying this impact. However, studies have shown that costs are directly proportional to the prevalence of resistance. Experience with bacteria such as Streptococcus pneumonia have shown that resistance rises exponentially over time (Turnridge, Nimmo, Francis & Australian Group on Antimicrobial Resistance, 1996) and the financial costs of management will also increase exponentially.

In Australia the financial costs of antibiotic resistance (such as use of more expensive antibiotics, multiple courses of antibiotics, increased length of hospital stay and increased mortality) are largely borne by the government and the patient (JETACAR, 1999; Holmberg, Solomon & Blake, 1987). In the UK drug-resistant strains of bacteria are responsible for 5000 deaths a year while in Europe, drug-resistant bacteria are responsible for 25000 deaths a year, with related healthcare costs and productivity losses of €1.5 billion (US $2 billion) (Aminov, 2010; Norrby et al., 2009; Walsh, 2014; Grant & Taylor, 2013). In the US, a recent assessment of antibiotic resistance by the CDC found more than two million people each year get infections that are resistant to antibiotics and at least 63000 people die as a result of hospital-acquired bacterial infections (Norrby et al., 2009; NIH, 2014) again incurring related healthcare costs and productivity losses. In the EU, the costs of treating hospital-acquired infections from just six species of antibiotic-resistant bacteria were estimated to be at least US $1.87 billion in 2006 dollars – more than the annual spending on influenza (Laxminarayan, Malani, Howard & Smith, 2011). Another economic consequence of resistance surrounds the inability to do other interventions such as surgery, biopsies, transplants and chemotherapy (Laxminarayan, Malani, Howard & Smith, 2007).

No reliable estimates are available for low and middle-income countries but the higher burden of infectious disease and limited access to new antibiotics suggest a bigger problem than in high-income countries (Kayange, Kamugisha, Mwizamholya, Jeremiah & Mshana, 2010). This is exacerbated by the limited ability in low and middle income countries to pay for or access second-line drugs (Laxminarayan, van Boeckel & Teillant, 2015).

As a result of market failure no new class of antibiotics has been developed since the 1980s. On average, pharmaceutical companies spend $5 billion to develop any new drug, including new antibiotics (Krans, 2015). "In the current world of complex treatments and interventions, pharmaceutical companies pursue more profitable causes than the development of new types of antibiotics... it is also difficult to justify the expenditure required for research and development in a commercial environment when it has been demonstrated that resistance to a new antimicrobial is likely to emerge within a foreseeable timeframe, rendering the new product less marketable" (Australian Commission on Safety and Quality in Health Care, 2013, p. 5). Evidence shows that the time lag between a new antibiotic being introduced and the first instance of resistance to it has drastically shortened in the 2000s (Australian Commission on Safety and Quality in Health Care, 2013). Pharmaceutical companies are more interested in developing other drugs that can be used regularly without losing their effectiveness, such as antidepressants or anti-inflammatories (Krans, 2015). With the pipeline drying up, stewardship of antibiotics in health and farming is increasingly economically important.

Environmental Implications

Antimicrobial use in food animals is a primary source of aquatic and terrestrial antibiotic contamination mainly through excreted biological matter (usually faeces and urine). Antibiotics reach the environment through application of antibiotic laden manure or slurry on agricultural lands or direct deposition of manure by grazing animals. Surface run-off, driftage and leaching into deeper layers of the earth often follow (Kemper, 2008). A proportion of the antibiotics that reach the environment this way remains biologically active. Such sub-therapeutic build-up of antibiotics accumulates over time effecting ecosystems. Antibiotic concentrations may also exert pressure on environmental bacteria or foster the transference of resistant genes creating an environmental mixing pot of genetic ABR traits.

A growing number of studies provides evidence of high concentrations in soils of antibiotics used in livestock (Rooklidge, 2004; Kay, Blackwell & Boxall, 2004; Aust et al., 2008), such terracumulation (the concentration of antibiotics in soil) runs-off or leaches into surface and/or ground water. A study of 47 groundwater sites tested in the US found that 23% of them were contaminated with sulfamethoxazole – an antibiotic used in both humans and livestock and one of the most frequently detected chemical compounds in national surveys of waste water contaminants (Michigan State University, 2011). The bacteria in these water and terrestrial environments move to new areas and new hosts by many routes, through contact with other animals and insects, as well as with food produce (Levy, 2002).

It has also been shown that airborne particulate matter from feedlots disperses and spreads several antibiotics used in livestock. The microbial communities containing antibiotic resistant genes are also carried in airborne particulate matter. A recent study shows that airborne particulates downwind of feed yards had significantly greater amounts of genes resistant to tetracycline antibiotics (McEachran et al., 2015).

Antibiotics are designed to kill or inhibit bacterial pathogens in animals and people but they can also be hazardous to many types of non-targeted environmental microorganisms (Martinez, 2008). Antibiotic residues have been reported to markedly hurt plant growth and development, causing inhibition of germination, root and shoot growth (Brian, 1957). Residues have also shown toxicity to aquatic organisms (Wollenberger, Halling-Sorensen & Kusk, 2000) affecting places of ecological significance.

Social Implications

There is a number of significant social impacts associated with intensifying livestock production, increasing meat consumption and public health inequities and injustices. People in developed countries generally can afford better medical care, including new and more expensive antibiotics. By comparison, the average person in a developing country can barely afford even older, cheaper but often less effective, antibiotics. Increasing access to antibiotics is morally and critically important particularly as previously antibiotics were unavailable to treat infections to populations of developing countries. There is no quantification of the increased human morbidity and mortality occurring in developing countries due to treatment failure with older antibiotics, such as tetracyclines and penicillins that may be the only antibiotics available to people living in poverty. The tragedy of resistance is that when people in these countries become wealthy enough to afford antibiotics, these drugs won't work (Laxminarayan, 2015) and that increasing global meat consumption is associated with building antibiotic resistance.

Similarly, there is also the question about intergenerational justice. Antibiotic resistance is condemning those ahead of us to an era of infection that for all intents and purposes could and should be behind us.

PROTECTING THE ANTIBIOTICS

"Unless we take significant actions to… change how we produce, prescribe and use antibiotics, the world will lose more and more of these global public health goods and the implications will be devastating" (Fukuda cited in WHO, 2014b, n.p.). Clearly economic vested interests are stronger and more protected than the public good. Surveillance of drug use in livestock is still relatively poor and looks set to remain this way due to vested interests, the huge growth in demand for meat and the concomitant growth of intensive factory farming operations around the world.

No doubt surveillance and reporting are essential but they do not offer solutions. The WHO accepted in 2014 that considerations around ABR might fall within the remit of the international health regulations (implemented in 2007 to deal with events such as influenza pandemics). Many calls to action have been made over the last few decades and scientists have been sounding alarm bells about the inevitability of the evolution of disease-causing microbes to become resistant to drugs for decades but there has been too little progress (Woolhouse & Farrar, 2014; Woolhouse et al., 2015). Now committed and coordinated action and response are urgently required on the root causes of resistance, primarily the misuse of antibiotics and the paucity of development of new drugs. Guidelines must be implemented to improve the use of existing drugs and a better regulatory environment, coupled with education is needed to help those working in both the human and livestock sectors to help them change their ways. Woolhouse and Farrar (2014, p. 556) call for the creation of an organisation similar to the Intergovernmental Panel on Climate Change (IPCC) to marshal evidence and catalyse policy across governments and stakeholders. Such an independent organisation, namely the Intergovernmental Panel on Antimicrobial Resistance (IPAMR) (Woolhouse & Farrar, 2014), would be able to foster further research and evaluate barriers and obstacles to mitigating ABR. It could set targets and limits for antibiotic use in livestock. Others argue that we should return to more humane farming methods with less need for sub-therapeutic use of antibiotics (CIWF, 2011). This could be included in the options explored by an independent IPAMR.

Although all these calls are important to mitigate the growing threat, there is a much more logical and immediately accessible solution. It is simply to reduce the amount of meat in human diets through appropriate food policies both in developed and transitioning countries. This will reduce meat related production and consumption consequences, including uncontrolled use of antibiotics. As western diets exceed healthy levels of meat consumption (Marsh, Saunders & Zeuschne, 2015), there is ample space for reduction in the demand for meat and hence, the use of antibiotics.

Developing countries are projected to have the greatest increase in meat consumption and antibiotics used based on them emulating the western path of development. This is where the largest potential for protecting antibiotics lies. Growing affluence in developing countries should be decoupled from growing meat consumption and concomitant growth in antibiotic use. This would allow the preservation of traditional diets which are typically low in meat. It will also facilitate a number of other significant benefits concerned with public goods.

Increasingly affluent consumers should benefit from their wealth. This is a given, especially in light of the meagre lifestyles that many of them may have previously experienced and the far greater consumption of long-affluent countries. However, it is in the self-interests of all countries, as well as in the interests of any individual that we learn from our mistakes, restrict further damage and protect the antibiotics for present and future generations.

CONCLUSION

Clearly the problem of antibiotic resistance is complex. This complexity does not however diminish the dimensions of the problems posed by antibiotic resistance nor in any way detract from the urgent need to find ways to address the threats. Whilst some might consider that sub-therapeutic dosing of animals is not *the* major cause of the problem, it is clearly *a* cause. Increased and globally collaborative data collection and surveillance informing management and mitigation measures are unquestionably essential, perhaps through an IPAMR. Even more critical are development and widespread adoption of appropriate national and global food policies moderating the currently unmitigated demand for meat that underlies the most widespread application and consumption of antibiotics.

Whether such appropriate national and international food policy will solve the problem is not the key question. It is whether the existing evidence suggests that such food policy and dietary intervention are an equitable, affordable and judicious part of the solution. There seems no cause for doubt that antibiotics must be used judiciously in humans and livestock because both uses unquestioningly contribute to the emergence, persistence and spread of antibiotic-resistant bacteria (AVMA, 2015; CDC, 2013a, 2013b, 2013c). "It's time to get serious about preserving antibiotics for people, instead of using them on healthy livestock" (Tomson, 2013).

REFERENCES

Aarestrup, F. M., & Wegener, H. (2009). Hearing on H.R. 1549, the Preservation of Antibiotics for Medical Treatment Act of 2009. Washington, DC: Government of the United States of America, US House of Representatives Committee on Rules. Retrieved from http://goo.gl/4YI8mv

Aarestrup, F. M., Wegener, H. C., & Collignon, P. (2008). Resistance in bacteria of the food chain: Epidemiology and control strategies. *Expert Review of Anti-Infective Therapy*, *6*(5), 733–750. doi:10.1586/14787210.6.5.733 PMID:18847409

Alliance to Save Our Antibiotics. (2014). Antimicrobial resistance - why the irresponsible use of antibiotics in agriculture must stop. The Soil Association. Retrieved from http://www.soilassociation.org/LinkClick.aspx?fileticket=G9q4uEb5deI%3D&tabid=1841

American Veterinary Medical Association. (2015). Judicious therapeutic use of antimicrobials: position statement. Retrieved from https://www.avma.org/KB/Policies/Pages/Judicious-Therapeutic-Use-of-Antimicrobials.aspx

Aminov, R. I. (2010). A brief history of the antibiotic era: lessons learned and challenges for the future (Article 134). Frontiers in Microbiology, 1, 1-7. doi:10.3389/fmicb.2010.00134

Antibiotics Alliance. (2014). *Antimicrobial resistance - why the irresponsible use of antibiotics in agriculture must stop: A briefing from the Alliance to Save Our Antibiotics*. London, UK: Antibiotics Alliance.

Armelagos, G. J. (1969). Disease in ancient Nubia. *Science*, *163*(3864), 225–258. doi:10.1126/science.163.3864.255 PMID:5762604

Aust, M. O., Godlinski, F., Travis, G. R., Hao, X., McAllister, T. A., Leinweber, P., & Thiele-Bruhn, S. (2008). Distribution of sulfamethazine, chlorotetracycline and tylosin in manure and soil of Canadian feedlots after subtherapeutic use in cattle. *Environmental Pollution*, *156*(3), 1243–1251. doi:10.1016/j.envpol.2008.03.011 PMID:18440678

Australian Commission on Safety and Quality in Health Care. (2013). Australian one health antimicrobial resistance colloquium. Background paper. Retrieved from http://www.safetyandquality.gov.au/wp-content/uploads/2013/07/Briefing-paper-for-One-Health-AMR-Colloquium-participants-Final-Jul-2013.pdf

Baragona, S. (2010). Use of antibiotics in livestock debated. Retrieved from http://www.voanews.com/content/use-of-antbiotics-in-livestock-debated-93683324/169529.html

Bassett, E. J., Keith, M. S., Armelagos, G. J., Martin, D. L., & Villanueva, A. R. (1980). Tetracycline labelled human bone from ancient Sudanese Nubia (A.D. 350). *Science*, *209*(4464), 1532–1534. doi:10.1126/science.7001623 PMID:7001623

Breslow, J. M. (2014, January 8). Illinois "nightmare bacteria" outbreak raises alarms. *Frontline*. Retrieved September 6, 2015, from http://www.pbs.org/wgbh/pages/frontline/health-science-technology/hunting-the-nightmare-bacteria/illinois-nightmare-bacteria-outbreak-raises-alarms/

Brian, P. W. (1957). Effects of antibiotics on plants. *Annual Review of Plant Physiology*, *81*(1), 413–426. doi:10.1146/annurev.pp.08.060157.002213

Buchman, T. G., Dushoff, J., Effron, M. B., Ehrlich, P. R., Fitzpatrick, S., Laxminarayan, R., & Solomkin, J. S. et al. (2008). Antibiotic overuse: The influence of social norms. *Journal of the American College of Surgeons*, *207*(2), 265–275. doi:10.1016/j.jamcollsurg.2008.02.035 PMID:18656057

Butler, M., & Cooper, M. (2012, December 7). New antibiotics: what's in the pipeline? The Conversation, Retrieved from https://theconversation.com/new-antibiotics-whats-in-the-pipeline-10724

Carlet, J., Jarlier, V., Harbarth, S., Voss, A., Goossens, H., & Pittet, D. (2012). Ready for a world without antibiotics? The Pensières Antibotic Risistance Call to Action. *Antimicrobial Resistance and Infection Control*, *1*(11), 1–13. doi:10.1186/2047-2994-1-11 PMID:22958725

Centre for Science in the Public Interest (CSPI). (2014). Antibiotic Resistance Project. Retrieved February 21, 2015, from http://www.cspinet.org/ar/

Centres for Disease Control and Prevention (CDC). (2013a, March 5). Action needed now to halt spread of deadly bacteria: data show more inpatients suffering infections from bacteria resistant to all or nearly all antibiotics. CDC Newsroom. Retrieved from http://www.cdc.gov/media/releases/2013/p0305_deadly_bacteria.html

Centres for Disease Control and Prevention (CDC). (2013b). Antbiotic resistance threats in the United States 2013. Atlanta, GA: CDC, Division of Healthcare Quality Promotion, National Center for Emerging and Zoonotic Infectious Diseases. Retrieved from http://www.cdc.gov/drugresistance/pdf/ar-threats-2013-508.pdf

Centres for Disease Control and Prevention (CDC). (2013c). Vital signs: Carbapenem-resistant enterobacteriaceae. *Morbidity and Mortality Weekly Report*, *62*(9), 165–170. PMID:23466435

Chang, Q., Wang, W., Regev-Yochay, G., Lipstich, M., & Hanage, W. P. (2015). Antibiotics in agriculture and the risk to human health: How worried should we be? *Evolutionary Applications*, *8*(3), 240–247. doi:10.1111/eva.12185 PMID:25861382

Chopra, I., Hesse, L., & O'Neill, A. (2002). Discovery and development of new anti-bacterial drugs. In H. van der Goot (Ed.), Pharmacochemistry library (Vol. 32) (pp. 213-225). Amsterdam, the Netherlands: Elsevier. doi:10.1016/S0165-7208(02)80022-8

Collingon, P. (2005). Fluroquinolone use in food animals. *Emerging Infectious Diseases*, *11*(11), 1789–1792. doi:10.3201/eid1111.040630

Compassion in World Farming (CIWF). (2011). Antibiotics in animal farming: public health and animal welfare. Surrey, UK: CIWF. Retrieved from https://www.ciwf.org.uk/media/3758863/Antibiotics-in-Animal-Farming-Public-Health-and-Animal-Welfare.pdf

Cook, M., Molto, E., & Anderson, C. (1989). Flurochrome labelling in Roman period skeletons from Dakhleh Oasis, Egypt. *American Journal of Physical Anthropology*, *80*(2), 137–143. doi:10.1002/ajpa.1330800202 PMID:2679120

Cui, L., & Su, X. Z. (2009). Discovery, mechanisms of action and combination theraphy of artemisinin. *Expert Review of Anti-Infective Therapy*, *7*(8), 999–1013. doi:10.1586/eri.09.68 PMID:19803708

D'Costa, V. M., King, C. E., Kalan, L., Morar, M., Sung, W. W., Schwarz, C., & Wright, G. D. et al. (2011). Antibiotic resistance is ancient. *Nature*, *477*(7365), 457–461. doi:10.1038/nature10388 PMID:21881561

Daemmrich, A. (2013). The political economy of healthcare reform in China: Negotiating public and private. *SpringerPlus*, *2*(448), 1–13. PMID:24052932

de la Hamaide, S. (2012, January 11). Antibiotics for livestock vital to feed world: OIE. Reuters. Retrieved from http://www.reuters.com/article/2012/01/11/us-antibiotics-livestock-idUSTRE80A1JF20120111

Dong, H., Bogg, L., Rehnberg, C., & Diwan, V. (1999). Association between health insurance and antibiotics prescribing in four counties in rural China. *Health Policy (Amsterdam)*, *48*(1), 29–45. doi:10.1016/S0168-8510(99)00026-3 PMID:10539584

Economou, V., & Gousia, P. (2015). Agriculture and food animals as a source of antimicrobial-resistant bacteria. *Infection and Drug Resistance*, *8*, 49–61. doi:10.2147/IDR.S55778 PMID:25878509

Eisler, P. (2013, March 6). Deadly superbugs invade U.S. health care facilities. USA Today. Retrieved from http://www.usatoday.com/story/news/nation/2012/11/29/bacteria-deadly-hospital-infection/1727667/

Foer, J. S. (2009, October 28). Eating animals is making us sick. CNN. Retrieved from http://edition.cnn.com/2009/OPINION/10/28/opinion.jonathan.foer/index.html?iref=allsearch

Food and Drug Administration (FDA). (2000). The human health impact of fluroquinolone resistant Campylobacter attributed to the consumption of chicken. Rockville, MD: FDA, Center for Veterinary Medicine. Retrieved from http://goo.gl/iZhFU7

Food and Drug Administration (FDA). (2010). CVM Updates - DVM Reports on Antimicrobials Sold or Distributed for Food-Producing Animals. Silver Spring, MD: FDA. Retrieved from http://www.fda.gov/AnimalVeterinary/NewsEvents/CVMUpdates/ucm236143.htm

Food and Drug Administration (FDA). (2013a). Guidance for industry, new animal drugs and new animal drug combination products administered in or on medicated feed or drinking water of food-producing animals: recommendations for drug sponsors for voluntarily aligning product use conditions with GFI#209. Rockville, MD: Center for Veterinary Medicine, FDA. Retrieved from http://goo.gl/iZhF

Food and Drug Administration (FDA). (2013b). Summary report on antimicrobials sold or distributed for use in food-producing animals. Silver Spring, MD: Government of the United States of America, Department of Health and Human Services, FDA. Retrieved February 21, 2015, from

Gartin, M., Brewis, A. A., & Schwartz, N. A. (2010). Nonprescription antibiotic therapy: Cultural models on both sides of the counter and both sides of the border. *Medical Anthropology Quarterly*, *24*(1), 85–107. doi:10.1111/j.1548-1387.2010.01086.x PMID:20420303

Goforth, R., & Goforth, C. (2000). Appropriate Regulation of Antibiotics in Livestock Feed. *Boston College Environmental Affairs Review*, *28*(1), 39–77.

Goossens, H., Ferech, M., Vander Stichele, R., & Elseviers, M.European Survelliance of Antimicrobial Consumption (ESAC) Project Group. (2005). Outpatient antibiotic use in Europe and association with resistance: A cross-national database study. *Lancet*, *365*(9459), 579–587. doi:10.1016/S0140-6736(05)70799-6 PMID:15708101

Grant, J., & Taylor, J. (2013, September 18). Who killed Mrs X? The global threat of antimicrobial resistance. The Rand Blog. Retrieved from http://www.rand.org/blog/2013/09/who-killed-mrs-x-the-global-threat-of-antimicrobial.html

Grigoryan, L., Burgerhof, J. G., Degener, J. E., Deschepper, R., Lundborg, C., Monnet, D., & Haaijer-Ruskamp, F. et al. (2008). Determinants of self-medication with antibiotics in Europe: The impacts of beliefs, country wealth and the healthcare system. *The Journal of Antimicrobial Chemotherapy*, *61*(5), 1172–1179. doi:10.1093/jac/dkn054 PMID:18296694

Grigoryan, L., Burgerhof, J. G., Degener, J. E., Deschepper, R., Lundborg, C. S., & Monnett, D. L., ... SAR Consortium. (2007). Attitudes, beliefs and knowledge concerning antibiotic use and self-medication: A comparative European Study. *Pharmacoepidemiology and Drug Safety*, *16*(11), 1234–1243. PMID:17879325

Gupta, A., Nelson, J. M., Barrett, T. J., Tauxe, R. V., Rossiter, S. P., Friedman, C. R., & Angulo, F. J. et al. (2004). Antimicrobial resistance among Campylobacter strains, United States, 1997-2001. *Emerging Infectious Diseases*, *10*(6), 1102–1109. doi:10.3201/eid1006.030635 PMID:15207064

Harrison, J. W., & Svec, T. A. (1998). The beginning of the end of the antibiotic era? Part II Proposed solutions to antibiotic abuse. *Quintessence International*, *29*(4), 223–229. PMID:9643260

Hawkings, N. J., Butler, C. C., & Wood, F. (2008). Antibiotics in the community: A typology of user behaviours. *Patient Education and Counseling*, *73*(1), 146–152. doi:10.1016/j.pec.2008.05.025 PMID:18640805

Health Action International Asia Pacific (HAIAP), & Third World Network (TWN) Penang in Association with Consumers Association of Penang. (2013). Antibiotic Use and Antibiotic Resistance in Food Animals in Malaysia: A Threat to Human and Animal Health. Penang, Malaysia: HAIAP and TWN.

Helms, M., Simonsen, J., Olsen, K. E., & Mølbak, K. (2005). Adverse halth events associated with antimicrobial drug resistance in Campylobacter species: A registry-based cohort study. *The Journal of Infectious Diseases*, *191*(7), 1050–1055. doi:10.1086/428453 PMID:15747238

Holmberg, S. D., Solomon, S. L., & Blake, P. A. (1987). Health and economic impacts of antimicrobial resistance. *Reviews of Infectious Diseases*, *9*(6), 1065–1078. doi:10.1093/clinids/9.6.1065 PMID:3321356

Hummel, R., Tschäpe, H., & Witte, W. (1986). Spread of plasmid-mediated nourseothricin Resistance due to antibiotic use in animal husbandry. *Journal of Basic Microbiology*, *26*(8), 461–466. doi:10.1002/jobm.3620260806 PMID:3033194

Joint Expert Technical Advisory Committee on Antibiotic Resistance (JETACAR). (1999). *The use of antibiotics in food-producing animals: antibiotic-resistant bacteria in animals and humans.* Canberra, Australia: Government of Australia, Commonwealth Department of Health and Aged Care and Commonwealth Department of Agriculture, Fisheries and Forestry.

Kay, P., Blackwell, P. A., & Boxall, A. B. A. (2004). Fate of veterinary antibiotics in a macroporous tile drained clay soil. *Environmental Toxicology and Chemistry*, *23*(5), 1136–1144. doi:10.1897/03-374 PMID:15180364

Kayange, N., Kamugisha, E., Mwizamholya, D. L. M., Jeremiah, S., & Mshana, S. E. (2010). Predictors of positive blood culture and deaths among neonates with suspected neonatal sepsis in a tertiary hospital, Mwanza-Tanzania. *BioMed Central Pediatrics*, *10*(39), 1–9. PMID:20525358

Kemper, N. (2008). Veterinary antibiotics in the aquatic and terrestrial environment. *Ecological Indicators*, *8*(1), 1–13. doi:10.1016/j.ecolind.2007.06.002

Kessler, D. A. (2013, March 27). Antibiotics and the meat we eat. The New York Times. Retrieved from http://www.nytimes.com/2013/03/28/opinion/antibiotics-and-the-meat-we-eat.html

Knox, R. (2012). How using antibiotics in animal feed creates superbugs. Retrieved from http://www.npr.org/sections/thesalt/2012/02/21/147190101/how-using-antibiotics-in-animal-feed-creates-superbugs

Krans, B. (2015). Few new drugs: why the antibiotic pipeline is running dry. Retrieved from http://www.healthline.com/health/antibiotics/why-pipeline-running-dry

Krishnasamy, V., Otte, J., & Silbergeld, E. (2015). Antimicrobial use in Chinese swine and broiler poultry production. *Antimicrobial Resistance and Infection Control*, *4*(17), 1–9. doi:10.1186/s13756-015-0050-y PMID:25922664

Laxminarayan, R. (2015, April 3). Growth in livestock antibiotics raises risks for humans. Living on Earth. Retrieved from http://loe.org/shows/segments.html?programID=15-P13-00014&segmentID=2

Laxminarayan, R., Duse, A., Wattal, C., Zaidi, A. K. M., Wertheim, H. F. L., Sumpradit, N., & Cars, O. et al. (2013). Antibiotic resistance - the need for global solutions. *The Lancet Infectious Diseases Commission*, *13*(12), 1057–1098. doi:10.1016/S1473-3099(13)70318-9 PMID:24252483

Laxminarayan, R., Malani, A., Howard, D., & Smith, D. L. (2007). *Extending the cure: policy responses to the growing threat of antibiotic resistance.* Washington, DC: Resources for the Future.

Laxminarayan, R., Malani, A., Howard, D., & Smith, D. L. (2011). Extending the cure: policy response to the growing threat of antibiotic resistance. Washington, DC/New Delhi, India: Centre for Disease Dynamics, Economics and Policy (CDDEP).

Laxminarayan, R., van Boeckel, T., & Teillant, A. (2015). The economic costs of withdrawing antimicrobial growth promoters from the livestock sector. OECD Food, Agriculture and Fisheries Paper No. 78. Paris, France: Organisation for Economic Cooperation and Development (OECD). doi:10.1787/18156797

Levy, S. B. (2002). *The antibiotic paradox: how the misuse of antibiotics destroys their curative powers* (2nd ed.). Cambridge, MA: Perseus Publishing.

Luangtongkum, T., Jeon, B., Han, J., Plummer, P., Logue, C. M., & Zhang, Q. (2009). Antibiotic resistance in Campylobacter: Emergence, transmission and persistence. *Future Microbiology, 4*(2), 189–200. doi:10.2217/17460913.4.2.189 PMID:19257846

Marsh, K., Saunders, A., & Zeuschne, C. (2015). Red meat and health: evidence regarding red meat, health and chronic disease risk. In T. Raphaely & D. Marinova (Eds.), *Impacts of meat consumption on health and environmental sustainability*. Hershey, PA: IGI Global.

Martinez, J. L. (2008). Antibiotics and antibiotic resistance genes in natural environments. *Science, 321*(5887), 365–367. doi:10.1126/science.1159483 PMID:18635792

McEachran, A. D., Blackwell, B. R., Delton Hanson, J., Wooten, K. J., Mayer, G. D., Cox, S. B., & Smith, P. N. (2015). Antibiotics, bacteria and antibiotic resistance genes: Arial transport from cattle feed yards via particulate matter. *Environmental Health Perspectives, 123*(4), 337–343. PMID:25633846

Mellon, M., Benbrook, C., & Benbrook, K. L. (2001). *Hogging it: estimates of antimicrobial abuse in livestock*. Cambridge, MA: Union of Concerned Scientists.

Michigan State University. (2011). The environmental impact of imprudent antimicrobial use in animals: Open source teaching modules. Retrieved from http://amrls.cvm.msu.edu/veterinary-public-health-module/iii.-the-environmental-impact-of-imprudent-antimicrobial-use-in-animals

Miller, R. V. (1998). Bacterial gene swapping in nature. *Scientific American, 278*(1), 66–71. doi:10.1038/scientificamerican0198-66 PMID:9418300

Mombiot, G. (2015, May 19). Feces, bacteria, toxins: welcome to the chicken farm. The Guardian. Retrieved from http://www.theguardian.com/commentisfree/2015/may/19/chicken-welfare-human-health-meat

Morgan, D. J., Okeke, I. N., Laxminarayan, R., Perencevich, E. N., & Weisenberg, S. (2011). Non-perscription antimicrobial use worldwide: A systematic review. *The Lancet Infectious Diseases Commission, 11*(9), 692–701. doi:10.1016/S1473-3099(11)70054-8 PMID:21659004

Myers, N., & Kent, J. (2003). New Consumers: The influence of affluence on the environment. *Proceedings of the National Academy of Sciences of the United States of America of the United States of America, 100*(8), 4963–4968. doi:10.1073/pnas.0438061100 PMID:12672963

National Antimicrobial Resistance Monitoring System (NARMS). (2010). 2010 Executive report. Retrieved from http://www.fda.gov/downloads/AnimalVeterinary/SafetyHealth/AntimicrobialResistance/NationalAntimicrobialResistanceMonitoringSystem/UCM312360.pdf

National Institute of Health (NIH). (2014, July/August). Global leaders raise alarm on antibiotic resistance. Global Health Matters, 13. Retrieved from http://www.fic.nih.gov/News/GlobalHealthMatters/july-august-2014/Pages/antimicrobial-resistance.aspx

National Research Council (NRC). (1999). *The use of drugs in food animals: Benefits and risks*. Washington, DC: National Academic Press.

Nelson, M. L., Dinardo, A., Hochberg, J., & Armelagos, G. J. (2010). Brief communication: Mass spetroscopic characterization of tetracycline in the skeletal remains of an ancient population from Sudanese Nubia 350-550 CE. *American Journal of Physical Anthropology, 143*(1), 151–154. doi:10.1002/ajpa.21340 PMID:20564518

Norrby, R., Powell, M., Aronsson, B., Monnet, D. L., Lustar, I., & Bocsan, I. S. ... Gyssens, I. C. (2009). The bacterial challenge time to react: A call to narrow the gap between multidrug-resistant bacteria in the EU and the development of new antibacterial agents. Technical Report. Stockholm, Sweden: European Centre for Disease Prevention Control (ECDC) and European Medicines Agency (EMEA).

O'Brien, B. (1996). Animal welfare reform and the magic bullet: The use and abuse of subtherapeutic doses of antibiotics in livestock. *University of Colorado Law Review, 67*, 407–442.

Parliament of the United Kingdom. (1969). Joint committee on the use of antibiotics in animal husbandry and veterinary medicine: Swann Committee Report. London, UK: Government of the United Kingdom, House of Commons, Her Majesty's Stationary Office (HMSO).

Philpott, T. (2011a). Meat industry still denying antibiotic resistance. Retrieved from http://www.motherjones.com/tom-philpott/2011/09/meat-industry-antibiotic-resistance

Philpott, T. (2011b). What the USDA doesn't want you to know about antibiotics and factory farms. Retrieved from http://www.motherjones.com/tom-philpott/2011/07/what-usda-doesnt-want-you-know-about-antibiotics-and-factory-farms

Planta, M. B. (2007). The role of poverty in antimicrobial resistance. *Journal of the American Board of Family Medicine, 20*(6), 533–539. doi:10.3122/jabfm.2007.06.070019 PMID:17954860

Responsible Use of Medicines in Agriculture Alliance (RUMA). (2012). RUMA comment on the Early Day Motion in the House of Commons on the use of antibiotics in intensive farming. Retrieved from http://www.ruma.org.uk/news/20121105.htm

Responsible Use of Medicines in Agriculture Alliance (RUMA). (2013). *Written evidence submitted to Science and Technology Committee by the Responsible Use of Medicines in Agriculture Alliance (RUMA) (AMR0045)*. London, UK: RUMA.

Rogers, L. (2012). Pew comments on proposed antibiotics legislation. Retrieved from http://www.pewtrusts.org/en/about/news-room/press-releases/0001/01/01/pew-comments-on-proposed-antibiotics-legislation

Rooklidge, S. J. (2004). Environmental antimicrobial contamination from terracumulation and diffuse pollution pathways. *The Science of the Total Environment, 325*(1), 1–13. doi:10.1016/j.scitotenv.2003.11.007 PMID:15144773

Rushton, J., Ferreira, J. P., & Stark, K. D. (2014). Antimocrobial resistance: the use of antimicrobial in the livestock sector (No. 68). Paris, France: Organisation for Economic Co-operation and Development (OECD).

Sample, I., Harvey, F., & Campbell, D. (2013, June 11). UK raises alarm on deadly rise of superbugs. The Guardian. Retrieved from http://www.theguardian.com/society/2013/jun/11/uk-urge-global-clampdown-antibiotics-g8

Scully, M. (2002). *Dominion: the power of man, the suffering of animals and the call to mercy*. New York, NY: St Martin's Press.

Shaban, R. Z., Cruickshank, M., & Christiansen, K.Antimicrobial Resistance Standing Committee. (2013). *National surveillance and reporting of antimicrobial resistance and antibiotic usage for human health in Australia*. Canberra, Australia: Government of Australia, Antimicrobial Resistance Standing Committee, Australian Health Protection Principal Committee.

Shapiro, P. (2015). Feasting from the Federal Trough: how the meat, egg and dairy industries gorge on taxpayer dollars while fighting modest rules. In T. Raphaely & D. Marinova (Eds.), *Impacts of meat consumption on health and environmental sustainability*. Hershey, PA: IGI Global.

Sharma, S. (2015). Superbugs a grave global threat by 2030 due to industrial meat production. Retrieved from http://www.iatp.org/blog/201504/superbugs-a-grave-global-threat-by-2030-due-to-industrial-meat-productions

Shea, K., Florini, K., & Barlam, T. (2001). When wonder drugs don't work: how antibiotic resistance threatens children, seniors and the medically vaulnerable. Washington, DC: Environmental Defense Fund (EDF).

Silbergeld, E. K., Graham, J., & Price, L. B. (2008). Industrial food animal production, antimicrobial resistance and human health. *Annual Review of Public Health*, 29(1), 151–169. doi:10.1146/annurev.publhealth.29.020907.090904 PMID:18348709

Smith, R., & Coast, J. (2013). The true cost of antimicrobial resistance. *British Medical Journal*, 346(1493), 1–5. PMID:23479660

Sweidan, M., Zhang, Y., Harvey, K., Yang, Y.-H., Shen, X., & Yao, K. (2005, November 28–30). Report on Proceedings of the 2nd National Workshop on Rational Use of Antibiotics in China. Paper presented at the 2nd National Workshop on Rational Use of Antibiotics in China, Beijing Children's Hospital, Beijing, China.

Taylor, J., Hafner, M., Yerushalmi, E., Smith, R., Bellasio, J. M., & Vardavas, R. ... Rubin, J. (2014). Estimating the economic costs of antimicrobial resistance: model and results. Cambridge, UK: RAND Europe.

Tilman, D., & Balzar, C., Hill, & Befort, B. L. (2011). Global food demand and the sustainable intensification of agriculture. Proceedings of the National Academy of Sciences of the United States of America of the Unites States of America, 108(50), 20260-20264. doi:10.1073/pnas.1116437108

Todar, K. (2012). *The good, the bad, and the deadly.* Online Textbook of Bacteriology. Retrieved September 6, 2015, from http://textbookofbacteriology.net/resantimicrobial.html

Tomson, B. (2013). Study: livestock drugs, hardier bugs. Retrieved from http://www.politico.com/story/2013/09/livestock-antibiotic-study-proof-97078.html

Turnridge, J., Nimmo, G., & Francis, G., & Australian Group on Antimicrobial Resistance. (1996). Evolution of resistance in Staphylococcus aureus in Australian teaching hospitals. *The Medical Journal of Australia, 164*(2), 68–71. PMID:8569574

van Boeckel, T. P., Brower, C., Gilbert, M., Bryan, T., Grenfell, B. T., & Levina, S. A., ... Laxminarayan, R. (2014). Global trends in antimicrobial use in food animals. Proceedings of the National Academy of Sciences (PNAS), 112(18), 5649-5654.

Walsh, F. (2014, July 2). Antibiotic resistance: Cameron warns of medical 'dark ages'. BBC News. Retrieved from http://www.bbc.com/news/health-28098838

Wegener, H. C. (2003). Antibiotics in animal feed and their role in resistance development. *Current Opinion in Microbiology, 6*(5), 439–445. doi:10.1016/j.mib.2003.09.009 PMID:14572534

Whitman, W. B., Coleman, D. C., & Wiebe, W. J. (1998). Prokaryotes: The unseen majority. [PNAS]. *Proceedings of the National Academy of Sciences of the United States of America, 95*(12), 6578–6583. doi:10.1073/pnas.95.12.6578 PMID:9618454

Witte, W. (1998). Medical consequences of antibiotic use in agriculture. *Science, 279*(5353), 996–997. doi:10.1126/science.279.5353.996 PMID:9490487

Wollenberger, L., Halling-Sorensen, B., & Kusk, K. O. (2000). Acute and chronic toxicity of veterinary antibiotics to Daphnia magna. *Chemosphere, 40*(7), 723–730. doi:10.1016/S0045-6535(99)00443-9 PMID:10705550

Wong, W. C., Pui, C. F., Tunung, R., Ubong, A., Noor Hidayah, M. S., Farinazleen, M. G., & Son, R. et al. (2012). Antibiogram patterm among cultures of listeria monocytogenes isolated from frozen burger patties in Malaysia. *Pertanika. Journal of Tropical Agricultural Science, 35*(4), 793.

Woolhouse, M., & Farrar, J. (2014). Policy: An intergovernmental panel on antimicrobial resistance. *Nature, 509*(7502), 555–557. doi:10.1038/509555a PMID:24877180

Woolhouse, M., Ward, M., van Bunnik, B., & Farrar, J. (2015). Antimicrobial resistance in humans, livestock and the wider environment. Biological Sciences, 370(1670), 1-7.

Working Party on Agricultural Policies and Markets (WPAPM). (2015). Global antimicrobial use in the livestock sector: final report. Paris, France: Organisation for Economic Co-operation and Development (OECD).

World Bank. (2015) Population Estimates and Projections. Retrieved from http://data.worldbank.org/data-catalog/population-projection-tables

World Health organisation (WHO). (2012). *Critically important antimicrobials for human medicine. 3rd Revision.* Geneva, Switzerland: WHO.

World Health Organisation (WHO). (2013). *WHO model list of essential medicines. 18th List*. Geneva, Switzerland: WHO.

World Health Organisation (WHO). (2014a). *Antimicrobial resistance: global report on surveillance 2014*. Geneva, Switzerland: WHO.

World Health Organisation (WHO). (2014b). WHO's first global report on antibiotic resistance reveals serious, worldwide threat to public health. Retrieved from http://www.who.int/mediacentre/news/releases/2014/amr-report/en/

Wu, A. (2012, January 12). Livestock in China given too many antibiotics. The Epoch Times. Retrieved from https://kristinasaid.wordpress.com/2012/01/24/livestock-in-china-given-too-many-antibiotics/

Zarb, P., & Goossens, H. (2012). Human use of antimicrobial agents. *Revue Scientifique et Technique (International Office of Epizootics)*, *31*(1), 121–133. PMID:22849272

Zuraw, L. (2014). UK review makes staggering antibiotic resistance predictions. Retrieved from http://www.foodsafetynews.com/2014/12/uk-review-releases-staggering-antibiotic-resistance-predictions/VcAwspOqqko

KEY TERMS AND DEFINITIONS

Antibiotic: A type of medicine made from bacteria or mould. Kills or slows growth of other bacteria. Examples include penicillin and streptomycin.

Antibiotic Growth Promotion: Prolonged mass administration of antibiotics at sub-therapeutic levels to promote growth of livestock animals and enhance feed efficiency in order to produce more meat.

Antibiotic Resistance: The result of bacteria changing in ways that reduce or eliminate the effectiveness of antibiotics. Antibiotic resistance is one type of antimicrobial resistance.

Antimicrobial Resistance: The result of microorganisms changing in ways that reduce or eliminate the effectiveness of drugs, chemicals, or other agents used to cure or prevent infections. In this chapter, the focus is on antibiotic resistance, which is one type of antimicrobial resistance.

Bacteria: Single-celled organisms living in and around us. Some bacteria are beneficial, but others can be pathogenic causing diseases and infections.

Food Animals: Animals reared for meat consumption.

Metaphylactic Use: Application of medicine at therapeutic levels to risk groups with no clinical symptoms to prevent disease outbreak.

Prophylactic Use: Application of medicine at sub-therapeutic levels to prevent disease from occurring.

Sapronotic: Comes from a source of infection in an abiotic substrate (soil, water, decaying plants, or animal corpses, excreta, and other substrata) or non-living environment.

Therapeutic Use: Application of medicine for treatment of manifested illness.

Zoonotic: Infectious diseases in animals which can be transmitted to humans.

Chapter 10
Impact of Meat Consumption on the Health of People, the Animals They Eat, and the Earth's Resources

Joyce D'Silva
Ambassador for Compassion in World Farming, UK

ABSTRACT

Food consumption has both direct, immediate impacts and longer-lasting effects on human health and wellbeing. This chapter considers the evidence behind the impact of animal foods on human health. It also reviews the impact of intensive animal production on the health and wellbeing of the animals themselves, including their breeding and conditions in which they are reared. The potential for factory farms to contribute to the global threat of antibiotic resistance and to be the breeding ground for a viral pandemic is also considered. Intensive animal farming is further adversely affecting the earth's resources of cereals, soy and water. The conclusion is that there is overwhelming evidence for a reduction in consumption of animal products on the grounds of health, the use of the earth's precious resources and the wellbeing of the animals consumed for food.

INTRODUCTION

We know that what we eat and drink has both direct, immediate impacts and longer-lasting effects on our health and wellbeing. In this chapter I investigate the evidence behind the animal foods we eat and how our consumption can affect not only us but also people and places many thousands of miles away. The story involves the animals themselves, how they live, what we feed them and how the medications we give them can affect us all. This also raises an alarm bell regarding the potential for factory farms to be the breeding ground for a viral pandemic.

DOI: 10.4018/978-1-4666-9553-5.ch010

The evidence for a reduction in consumption of animal products on the grounds of health, our use of the earth's precious resources and the wellbeing of the animals we consume is overwhelming. Below is an overview of some of the many problems triggered by excessive meat consumption.

MEAT CONSUMPTION AND HEALTH

I focus firstly on the latest evidence regarding diets and health and then investigate other issues arising from consumption of animal products, such as the threat to public health from antibiotic use in farm animals and the danger from zoonoses – diseases passed to humans from animals.

Consumption of Animal Products and Impact on Individual Health

A remarkable scientific paper on health and diet was published in the Nutrition Journal in July 2014 (McDougall et al., 2014). Over three years, more than 1,000 people took part in a one-week programme where they ate only a low fat, totally vegan diet. In just that one week, all their biomarkers for future cardiovascular and metabolic disease improved, including blood pressure, weight, blood sugar and cholesterol levels (McDougall et al., 2014).

So should we all be vegans? A simplistic response might be "yes", but no health practitioner would give such advice based on one short-term study. However a long-term larger study of the impacts of such a diet, with follow-up over several years, might possibly lead to that conclusion. We shall have to wait and see.

Having said that, a different 2014 paper looking at a wide range of possibly health-promoting diets, also pronounced that well-constructed vegan diets had a range of health benefits (Katz & Meller, 2014). The paper pointed out that intervention trials of short to moderate duration suggested benefits related to overall diet quality, inflammation, cardiac risk measures, cancer risk, anthropometry, and insulin sensitivity (Katz & Meller, 2014). Katz and Meller (2014) looked at a wide range of dietary patterns, including the Mediterranean diet. They pointed out that both the scientific literature and the links between native diet and adaptation for all species led them to conclude that a diet of foods mostly direct from nature and predominantly plants is supportive of health across one's life span (Katz & Meller, 2014). Another 2014 publication compared the health and sustainability of six common diets (van Dooren, Marinussen, Blonk, Aiking & Vellinga, 2014). On health factors, the Mediterranean and vegan diets scored the highest. The paper also concluded that diets with lower proportions of animal products score better on sustainability (van Dooren et al., 2014).

The Mediterranean diet is commonly understood to include plenty of cereals, mostly whole-grain, vegetables, fruit, nuts, olive oil and fish, small amounts of red meat, some poultry and low-fat dairy products, plus small amounts of red wine. Many view the Mediterranean diet as a good example of a healthy diet, as it is mainly plant-based yet not extreme in any way, reduces overall meat consumption and substitutes olive oil for butter and margarine, thus further reducing saturated fat intake.

A paper published in the British Medical Journal in December 2014 (Crous-Bou et al., 2014) studied telomere length in over 4,000 middle-aged women, who were already participating in the famous "Nurses Study". Telomere length is a biological marker of aging. As we get older, telomere length shortens. Crous-Bou et al. (2014) concluded that the women who adhered the most to a Mediterranean style diet had longer telomere length. The suggestion from this study is that a healthy diet can help postpone the aging process.

Over the last few years, more and more evidence is being published in the academic and specialist journals showing that reduced consumption of red meat has health benefits. A 2012 paper based on a large group study stated: "We estimated that substitutions of 1 serving per day of other foods (including fish, poultry, nuts, legumes, low-fat dairy, and whole grains) for 1 serving per day of red meat were associated with a 7% to 19% lower mortality risk. We also estimated that 9.3% of deaths in men and 7.6% in women in these cohorts could be prevented at the end of follow-up if all the individuals consumed fewer than 0.5 servings per day (approximately 42 g/d) of red meat" (Pan et al., 2012, p. 555). The study also showed that red meat consumption is associated with an increased risk of cardiovascular disease (CVD), cancer and total mortality and that substitution of other healthy protein sources for red meat is associated with a lower mortality risk (Pan et al., 2012).

Red meat, such as beef, lamb, pork etc., is often distinguished from "white meat", i.e. meat from chickens, turkeys, geese and ducks. In 1987, the National Pork Board in the US ran a $7million campaign promoting "Pork. The other white meat," Pork sales soared (National Pork Board, n.d.)! This was a clever campaign, but was not based on science. All medical and dietary professionals recognise that pork is a red, not a white meat.

The World Cancer Research Fund (WCRF) and the American Institute for Cancer Research (AICR) carry out a massive review of scientific evidence on diet and cancer every 10 years. Their 2007 Report (WCRF & AICR, 2007) concluded that although fat intake in itself doesn't seem to increase colon cancer risk, there is convincing evidence that high consumption of red meat (beef, pork and lamb) and processed meat (hot dogs, bacon and deli meats) does increase colon cancer risk. The Report's dietary recommendations include: "Eat mostly foods of plant origin" and "Limit intake of red meat and avoid processed meat" (WCRF & AICR, 2007, p. 11 and p. 12).

The WCRF/AICR also states that the public health goal should be an average consumption of red meat of no more than 16 kilograms per year, of which little to none should be processed meats (WCRF & AICR, 2007). This means a daily consumption of not more than 1.4 ozs (less than 40g) red meat, a very, very small portion! The UK Department of Health advises people who eat more than 90 grams (just over 3 ozs) (cooked weight) of red and processed meat a day to cut down to 70 grams (2.4 ozs) (NHS Choices, 2013). A portion of 90 grams is the equivalent of about three thin-cut slices of roast beef, lamb or pork, where each slice is about the size of half a piece of sliced bread (NHS Choices, 2013). Although the specific advice does vary, the message is clear – eat only small amounts of red and processed meat. For many people this will mean eating considerably less meat than they may currently consume.

Although the link between processed meats and some cancers is clear (WCRF & AICR, 2007), the reason is not established, but may lie in the use of the processing salts themselves. Processed meats such as ham, bacon, salami, pastrami and sausages are usually preserved using salts, nitrates and nitrites. The health advice from the WCRF/AICR is to avoid processed meats (WCRF & AICR, 2007).

Another paper looked at the relationship of consumption of animal products and animal fats with the development of different cancers. It identified that when people derive a larger proportion of energy from consumption of animal products it is an important risk factor for several of the more common cancers such as colon, kidney, breast, ovary, pancreas, and prostate (Grant, 2013). A UK study of post-menopausal women showed a lower risk of breast cancer among vegetarians (Taylor, Burley, Greenwood & Cade, 2007). It showed that each increase of 50 g/day in red meat intake increased breast cancer risk by 11% (between 1.04 and 1.18 with a 95% confidence interval), even after adjustment for lifestyle confounders, including menopausal status (Taylor et al., 2007).

In 2010, Friends of the Earth UK published a paper by Dr Mike Rayner, Director of the British Heart Foundation Health Promotion Research Group based within the Nuffield Department of Population Health of the University of Oxford, which says that more than 45,000 lives a year in the UK could be saved if everyone ate meat no more than two or three times a week (Rayner, Clarke, Thomas & Scarborough, 2010). Widespread switching to low-meat diets would stop 31,000 people dying early from heart disease, 9,000 from cancer and 5,000 from strokes (Rayner et al., 2010).

In another 2014 study, published in *Nature*, comparing a typical omnivorous diet with three alternative diets, Mediterranean, pescetarian and vegetarian diets, the three alternatives showed significant health benefits, with incidence rates of type II diabetes reduced by 16%–41% and of cancer by 7%–13%, while relative mortality rates from coronary heart disease were 20%–26% lower and overall mortality rates for all causes combined were 0%–18% lower (Tilman & Clark, 2014). A pooled analysis of five prospective cohort studies, involving approximately 76,000 subjects from the USA, UK and Germany, over a mean follow-up period of 10.6 years, reported that vegetarians had a 24% (between 6% and 38% with a 95% confidence interval) reduction in mortality from coronary heart disease (CHD) compared with regular meat-eaters (Key et al., 1999).

Vegetarian diets are also associated with lower body mass index (BMI) (Tonstad, Butler, Yan & Fraser, 2009). This is linked to lower blood pressure levels (Drøyvold, Midthjell, Nilsen & Holmen, 2005). A large study of the health of Seventh Day Adventists, many of whom follow a vegetarian diet, showed lower risk of colon and prostate cancer among the vegetarians (Fraser, 1999). Epidemiological studies further show that vegetarians have a lower risk of developing type-2 diabetes, which is a serious condition increasing in frequency across much of the world (Jenkins et al., 2003).

Of course it's not just cutting meat consumption which is important, but eating more healthy foods as replacement. One paper cites evidence for the protective effect of high consumption of vegetables and fruits as well as regular consumption of nuts and soya, all of which are associated with vegetarian and vegan diets (McEvoy, Temple & Woodside, 2012). Another 2014 paper shows that if people swap just 5% of the calories they consume from saturated fat from meat, butter and other dairy products with foods containing linoleic acid from nuts, seeds and vegetables, they lower their risk of CHD events by 9% and their risk of death from CHD by 13% (Farvid et al., 2014).

Whilst the World Health Organisation (WHO) has been cautious in advocating cuts to meat consumption, it is clear on the fat issue: "An unhealthy diet is one of the major risk factors for a range of chronic diseases, including cardiovascular diseases, cancer, diabetes and other conditions linked to obesity. Specific recommendations for a healthy diet include: eating more fruit, vegetables, legumes, nuts and grains; cutting down on salt, sugar and fats. It is also advisable to choose unsaturated fats, instead of saturated fats and towards the elimination of trans-fatty acids" (WHO, 2015a, n.p.).

However their health advice linked to carbon footprinting does say: "Industrialized countries need to reduce their meat consumption from the current 224 g/person/day. Global convergence to 90 g/person/day would have a significant effect on carbon levels and health" (WHO, 2008, p. 2). The American Heart Association (AHA) says: "Saturated fats occur naturally in many foods. The majority come mainly from animal sources, including meat and dairy products. Examples are fatty beef, lamb, pork, poultry with skin, beef fat (tallow), lard and cream, butter, cheese and other dairy products made from whole or reduced-fat (2 percent) milk (AHA, 2015, n.p.). The advice points out that many baked goods and fried foods can contain high levels of saturated fats and that some plant foods, such as palm oil, palm kernel oil and coconut oil, also contain primarily saturated fats, but do not contain cholesterol, unlike the animal foods mentioned above. Even foods cited as healthier, like chicken and nuts, do contribute some saturated fat to the diet, though their content is much lower in saturated fat than beef, cheese and ice cream.

Since all mammals can produce saturated fat in their own bodies, we do not need to consume it from external sources. However we do not produce polyunsaturated fats, so these we must consume. Omega-3 fatty acids are an important type of polyunsaturated fat. These are essential fats and can help prevent heart disease and stroke, may help control lupus, eczema, and rheumatoid arthritis, and may play protective roles in cancer and other conditions (Kris-Etherton, Harris, Appel, AHA Nutrition Committee & AHA, 2003). Omega-6 fatty acids play a crucial role in brain function, as well as normal growth and development (University of Maryland Medical Centre, 2013). They help stimulate skin and hair growth, maintain bone health, regulate metabolism, and maintain the reproductive system (University of Maryland Medical Centre, 2013). There should be a balance between omega-6 and omega-3 fatty acids. The ratio of omega-6 to omega-3 should be in the range of 2:1 – 4:1 (Simopoulous, 2008). The average diet provides plenty of omega-6 fatty acids, in fact the ratio in modern western diets is often skewed towards omega-6, sometimes by as much as 15:1 (Simopoulous, 2008).

All meat has a high omega-6 to omega-3 fatty acid ratio, although it is more favourable in beef from grass-fed animals – between 35% and 85% lower in pasture-reared beef compared with intensively reared beef (Alfaia et al., 2009; Garcia et al., 2008). The majority of studies report a lower ratio of omega-6 to omega-3 fatty acids for free-range pig meat, ranging from a 7% to 42% reduction compared with intensively reared pig meat (Soto et al., 2009; Ventanas, Ventanas, Tovar, García & Estévez, 2007; Bee, Guex & Herzog, 2004). However, the omega-6 to omega-3 ratio is relatively high for pig meat from all systems, ranging from 11.8 to 18.1 for extensive systems and from 12.4 to 31.2 for intensive systems (Pickett, 2012, p. 20).

Chicken meat contains both saturated and unsaturated fats. A whole uncooked chicken will have about 15 grams fat per 100 grams chicken, of which 3 grams will be saturated (National Chicken Council, 2014). Upon cooking, those figures come down slightly, as some fat drains out during the cooking process (National Chicken Council, 2014). Research in the UK has shown that a typical supermarket chicken today contains 2.7 times as much fat as in 1970 and 30 per cent less protein (Wang, Lehane, Ghebremeskel & Crawford, 2009). The proportion of omega-3 fatty acids tends to be higher in chicken meat from free-range systems compared with indoor systems but the differences are not consistent across all studies. This could be because meat chickens (broilers) have been bred for fast growth and lead comparatively sedentary lives as the huge weight of their bodies precludes them running around. Only layer type birds, who grow much more slowly and can live active lives, display much higher proportions of omega-3s in their fat (Sirri et al., 2011). It would seem wise, on health grounds alone, when consuming meat, dairy products and eggs, to choose the highest possible welfare option, as free-ranging animals, eating a mainly pasture-based diet, have a healthier fat composition.

Cholesterol is a waxy substance, which is made in the body by the liver but is also found in some foods. It plays a vital role in how every cell works and is also needed to make Vitamin D, some hormones and bile for digestion. However, too much cholesterol in the blood can increase the risk of heart and circulatory disease (HSPH, 2015). There are two main forms of cholesterol, low-density lipoprotein (LDL) and high-density lipoprotein (HDL). The LDL carries cholesterol from the liver to the cells that need it. If there is too much cholesterol for the cells to use, it can build up in the artery walls, leading to disease of the arteries. This LDL cholesterol is often referred to as "bad cholesterol" because too much of it is unhealthy. The HDL carries LDL cholesterol away from the cells and back to the liver, where it is either broken down or passed out of the body as a waste product. This HDL is often referred to as "good cholesterol" because it is protective. Eating too much saturated fat raises the levels of "bad" LDL cholesterol. However, research has shown that cutting down on saturated fat and replacing it with everyday foods that contain more unsaturated fat can improve our cholesterol levels (HSPH, 2015).

There is no doubt that milk contains useful amounts of calcium, an important mineral to support healthy bones. However milk also contains significant amounts of saturated fat, and also hormonal factors such as IGF-1, which can adversely affect human health. These hormonal factors are not surprising as a lot of milk is produced by pregnant cows, who are lactating at the same time. In fact it is the presence of IGF-1, which has led to associations between dairy consumption and cancers of the prostate, breast and ovaries (Qin, Xu, Wang, Tong & Hoshi, 2007; Song et al., 2013; Kroenke, Kwan, Sweeney, Castillo & Caan, 2013; Larsson, Bergkvist & Wolk, 2004). It is well known that exercise is one of the best ways to maintain bone density and there are many plant-based calcium-rich foods such as kale, broccoli and other green leafy vegetables, beans and nuts if people wish to cut down on dairy products. One of the world's most respected nutritionists, Dr Walter Willett, and his colleague declare that milk consumption is a comparatively recent addition to the human diet, and that consuming too much dairy is not a good thing (Ludwig & Willett, 2013). Perhaps the best overall advice as far as animal products and health are concerned, comes from Michael Pollan in his book *In Defense of Food: An Eater's Manifesto:* "Eat food. Not too much. Mostly plants" (Pollan, 2007, p. 1).

Antibiotic Risks

There are other important health issues associated with consumption of animal products, perhaps the most obvious being the overuse of antibiotics in the farming of animals. This is not primarily an issue of residues of antibiotics in animal products, although that can be an issue in countries where regulation and veterinary knowledge are absent. The vital issue is that bacteria become resistant to the effects of antibiotics if they are overused, either in humans or animals. Then when a person becomes seriously ill and may need a life-saving antibiotic, it may not work. Doctors will have to seek, often with desperation, for a substitute antibiotic that is effective for that particular infection.

In 2013 the UK's Chief Medical Officer, Professor Dame Sally Davies, told the Members of Parliament at Westminster: "Antimicrobial resistance poses a catastrophic threat. If we don't act now, any one of us could go into hospital in 20 years for minor surgery and die because of an ordinary infection that can't be treated by antibiotics" (Government of the United Kingdom, 2013, May 12, n.p.).

Antibiotic resistance to foodborne infections such as salmonella and campylobacter are nearly always of farm animal origin. In its 2011 report on foodborne antibiotic resistance, the WHO said: "Resistance in the foodborne zoonotic bacteria Salmonella and Campylobacter is clearly linked to antibiotic use in food animals, and foodborne diseases caused by such resistant bacteria are well documented in people" (WHO, 2011, p. xiii). In 2008, the European Food Safety Authority declared: "'Resistant [bacteria] involved in human disease are mostly spread through foods. With regards to salmonella, contaminated poultry meat, eggs, pork and beef are prominent in this regard. For campylobacter, contaminated poultry meat is prominent. Cattle are a major reservoir for E coli [verotoxigenic Escherichia coli] and resistant strains may colonize humans via contaminated meat of bovine origin more commonly than from other foods. Animal-derived products remain a potential source of methicillin-resistant staphylococcus aureus (MRSA). Food-associated MRSA, therefore, may be an emerging problem" (Andreoletti et al., 2008, p. 2). The American Centers for Disease Control (CDC) published the report "*Antibiotic resistance threat in the United States*" (CDC, 2013), quoting data from the Food and Drug Administration, that in the US more antibiotics are sold for farm animal use than for treating humans and that this contributes to the rise of antibiotic resistance. Compassion in World Farming has also produced, with its partners in the Antibiotics Alliance, a useful, detailed report on this subject: *Antimicrobial resistance - why the irresponsible use of antibiotics in agriculture must stop* (Antibiotics Alliance, 2014).

Sadly, industrial-type factory farming has become hugely dependent on antibiotic use. If one visualises a typical broiler chicken shed, with 20,000 birds crowded together inside for weeks at a time, or a pig factory farm where hundreds of pigs are housed within one shed, then it is obvious that disease is likely to develop rapidly in these hot-house conditions and that it is probable that it will spread from one creature to the other very quickly. In addition, for many years farmers used certain antibiotics as growth-promoters. Although this practice has now been outlawed or advised against, antibiotics are still being used to prevent disease and disease spread, not just to treat it. We would rarely take such action in human medicine. If you were prescribed an antibiotic for, say, a bacterial throat infection, you would not expect your doctor to give antibiotics for your whole family/household at the same time. This is exactly what is happening in factory farming and this poses serious long-term risks to the use of antibiotics for the treatment of humans.

Zoonoses

The US Centers for Disease Control say that approximately 75% of recently emerging infectious diseases affecting humans are diseases of animal origin and approximately 60% of all human pathogens are zoonotic, that is, they can spread from animals to humans (CDC, 2014). Most worrying of these may be the viral infections, in particular avian and swine flu. An intensive poultry farm provides the optimum conditions for viral mutation and transmission - thousands of birds are crowded together in a closed, warm and dusty environment, which is highly conducive to the transmission of a contagious disease. Each chicken-to-chicken or pig-to-pig transmission is an opportunity for the virus to mutate, possibly into a form, which is highly pathogenic to humans.

Selecting generation after generation of birds for their faster growth rates and higher meat yields has left the birds' immune systems less able to cope with infections. There is also a high degree of genetic uniformity in the population, making spread of the virus all the more easy. It is possible for an influenza virus to become highly infectious to humans within an animal host - it has already happened in the case of the 1918 Spanish flu epidemic (Billings, 1997). The H5N1 avian flu virus spread across Asia, the Middle East, Europe and Africa. The WHO states that from 2003 to 2015, 784 cases of H5N1 have been reported and 429 people have died from this disease (WHO, 2015b), which is passed primarily from poultry to humans (WHO, 2015c). This represents a mortality rate of around 60%.

The greater longer-term risk to people is that as the virus infects more people it could mutate to become much more easily transmitted between humans, leading to an influenza pandemic (a global epidemic). Public health experts writing in *The Lancet* have estimated that such a flu pandemic could kill as many as 62 million people, mostly in developing countries (Murray, Lopez, Chin, Feehan & Hill, 2006). New variants of these influenza strains seem to be popping up across the world at an alarming rate, especially in Asia, where intensive poultry farming is growing rapidly. However, it's not just in Asia. There is a massive global increase in consumption of poultry, resulting in a 300% increase in production across the world in the last two decades (FAOSTAT, 2015). It is surely true to say that the greater the amount of chicken we consume, the more likely they are to be factory-farmed and the more possibilities there will be for a pathogenic strain of avian flu to develop.

CONSUMPTION OF ANIMAL PRODUCTS AND IMPACT ON EARTH'S RESOURCES

Having looked at the research on the impact of meat consumption and related issues on our own health, I now discuss the impact of meat consumption on the earth's resources of land, edible crops and water.

Sustainability of Natural Resources

It is estimated that around 70% of farm animals globally are now reared indoors in intensive, industrial-type farms (Steinfeld et al., 2006). As they can no longer graze on pasture or forage in local vegetation, their food is delivered to them and contains a mix of legumes including soya meal, cereals such as wheat, barley and corn (maize), as well as ingredients such as fish meal or fermented grass silage. Globally 33% of the world's cereal harvest and 97% of the world's soymeal is used as animal feed (FOE, 2008). This is occurring when around 800 million people are still suffering from hunger and many more are malnourished (World Hunger Education Service, 2015).

Feeding cereals and soy to animals is inefficient as much of their food value is lost during conversion from plant to animal matter. Research shows that several kilograms of cereals are needed to produce 1 kg of edible meat (Trostle, 2008). The nutritional value consumed by animals in eating a given quantity of cereals is much greater than that delivered for humans by the resultant meat (Lundqvist, de Fraiture & Molden, 2008). Lundqvist et al. (2008) point out that for every 100 calories that we feed to animals in the form of crops, we receive on average just 30 calories in the form of meat and milk.

A report by the UN Environment Programme shows that the world's total edible crop harvest could supply 4,600 kcal per person per day (Nellemann et al., 2009). However, of this 600 kcal disappear in post-harvest losses and 800 kcal are lost or wasted during distribution or in households, that is 1,400 kcal (30.5%) in all are lost or wasted. A further 1,700 kcal are used as animal feed. However, the resultant meat and dairy products only provide 500 kcal for human consumption. In effect, therefore 1,200 kcal (26%) are lost in the poor return achieved by feeding human-edible crops to farm animals (Nellemann et al., 2009). Hence, more than half of the harvest (i.e. 56% or 3,600 kcal) is not used directly for human consumption. The UNEP report calculates that the cereals that, on a business-as-usual basis, are expected to be fed to livestock by 2050 could, if they were instead used to feed people directly, provide the necessary food energy for more than 3.5 billion additional people (Nellemann et al., 2009). If a target were adopted of just halving the amount of cereals that, on a business-as-usual basis, would be used for animal feed by 2050, an extra 1.75 billion people could be fed.

To put it another way, researchers at Minnesota University calculated that 36% of the world's crop calories are used to feed animals and of those calories, only 12% actually nourish humans (Cassidy, West, Gerber & Foley, 2013). They estimate that if all those feed crops were used instead to feed people, it would result in an extra 70% calories, which would sustain an extra 4 billion people. Even halving the consumption of grain-fed animal products would release enough calories to feed an extra 2 billion people (Cassidy et al., 2013).

The UK government's Foresight Report concluded: "Demand for the most resource-intensive types of food must be contained", adding that "major increases in the consumption of meat, particularly grain-fed meat, would have serious implications for competition for land, water and other inputs" (Government of the United Kingdom, 2011, p. 53). Further global industrialisation of livestock production is unsustainable, as it will entail a substantial increase in demand for feed crops. This in turn will lead to expansion and intensification of feed crop production which will involve land degradation and deforestation, increased use of artificial fertilisers, pollution and overuse of water, increased greenhouse gas emissions and loss of biodiversity (Raphaely & Marinova, 2014). None of this is sustainable. Referring to the need to feed over 9 billion people by 2050, UNEP stresses that "simply cranking up the fertilizer and pesticide-led production methods of the 20th Century is unlikely to address the challenge" as it will increasingly undermine the critical natural inputs on which agriculture depends (Nellemann et al., 2009, p. 5).

Water is probably, along with soil itself, the earth's most precious resource. There are estimates that by 2025, 64% of humanity will be living in areas of water scarcity (Steinfeld et al., 2006). In recent years, water experts at the University of Twente in the Netherlands have developed a methodology for measuring the amount of water used to produce various agricultural products. This is known as the water footprint. It includes the use of rainwater which nourishes grass and crops, and which is labelled green water, blue water which describes the volume of surface and groundwater consumed (that is evaporated after withdrawal) and refers to irrigation, and grey water which refers to polluted water resulting from the production of each product (Water Footprint Network, 2014b). By combining the figures of each type of water, green, grey and blue, a water footprint for a product is reached.

Mekonnen and Hoekstra (2012) estimate the water footprint of meat from beef cattle (15,400 litre/kg as a global average), which is considerably larger than the footprints of meat from sheep (10,400 litre/kg), pig (6,000 litre/kg), goat (5,500 litre/kg) or chicken (4,300 litre/kg).[1] All these water footprints are considerable and per kilogram of product, animal products generally have a larger water footprint than crop products. The same is true when we look at the water footprint per calorie of protein. Estimates show that the average water footprint per calorie for beef for example is twenty times larger than for cereals and starchy roots (Mekonnen & Hoekstra, 2012). The average water footprint per gram of protein again in the case of beef is six times larger than for pulses (Mekonnen & Hoekstra, 2012). These two leading experts reckon that if all meat were replaced by an equivalent amount of crop products such as pulses and nuts this will result in a 30% reduction of the food-related water footprint of the average (American) citizen (Mekonnen & Hoekstra, 2012).

Perhaps the most shocking statistic about water use, is the following. The basic water requirement for a human per year has been calculated at 18,250 litres (Liu & Savenije, 2008). It includes water for hygiene, cooking, sanitation, basic household needs and is just a little more than the water footprint for 1 kilogram of beef - 15,400 litres (Water Footprint Network, 2014a). Yes, just one kilogram of beef today could provide for a person's total water needs for the best part of a year!

Impact on Climate

Apart from the impact of meat consumption on our own health and on the earth's resources, evidence continues to build of its impact on greenhouse gas emissions (GHG). A 2014 study examined the impact on GHG emissions of six different diets and shows that a high meat diet (>100g/day) is responsible for much higher GHG emissions than a low meat diet (<50g/day) and that the production and consumption of animal-based foods is associated with higher GHG emissions than plant-based foods (Scarborough et al., 2014).

CONSUMPTION OF ANIMAL PRODUCTS AND IMPACT ON ANIMAL WELLBEING

Having discussed the impact of meat consumption on human health and on the earth's precious resources, it is important, in fact vital from an ethical perspective, to investigate the impact of meat consumption on the lives of the actual animals who are farmed for food.

Factory Farms

Globally, over 60 billion animals are slaughtered for our food each year (Steinfeld et al., 2006). Each one of these animals is, in scientific terms, a sentient being, capable of experiencing pain and suffering, yet the majority of these animals are bred for productivity and reared in "factory farms". Their quality of life is poor and many may live in a state of physical and/or psychological misery. This kind of factory farming is the biggest cause of animal suffering on the planet.

I invite you to look into the factory farm. See the 20,000 chickens being reared for meat, crammed into a barren shed, bred for unnatural, super-fast growth, over a quarter of them suffering from lameness as their skeletons can no longer bear the weight of their fast-growing bodies. They are transformed from fluffy chick to chicken dinner in just 6 weeks (Poultry Site, 2015b).

There is fairly widespread understanding that factory farming cages and confines farm animals, and that this is an ethically questionable practice. It is not so well understood that the animals themselves have been changed, and often damaged physiologically, through the selective breeding process, which has become increasingly sophisticated.

Breed

Ninety per cent of the breeding stock for the 58 billion broiler chickens slaughtered globally per annum (CIWF, 2013) come from just three companies – Hubbard, Aviagen and Cobb (Poultry Site, 2015a). Many breed lines are developed to reach market weight one day earlier each year (Cruickshank, 2003, p. 22). A range of studies shows that fast growth rates in broilers contribute to leg disorders and to metabolic disorders such as ascites and sudden death syndrome (SDS) in broilers (Knowles et al., 2008). In ascites, the abdominal cavity fills with fluid, leading to heart failure. In SDS, the birds just fall over and die. The fast growth rates promoted by the industry rise the risk of these two conditions through an increased demand for oxygen putting pressure on the heart and lungs (Bøtner et al., 2010). Such fast growing chickens find walking difficult so they spend much more time sitting on the soiled wood shavings on the floor, thereby allowing the ammonia from the floor to burn their hocks.

Other poultry, pigs, egg-laying hens, beef cattle and dairy cows also suffer from being bred for high productivity. The milk yield of dairy cows has doubled in many countries over the last 40 years, partly as a result of genetic selection (Algers et al., 2009). Cows have been changed in shape and increased in size, with longer, thinner bodies and larger udders. A modern high-yielding dairy cow can now produce ten times more milk than her calf would have suckled from her (DairyCo, 2014). Moreover, the calf is now removed once it has suckled for long enough to have taken in health-protecting colostrum from its mother, usually at a day old. The cow is then milked to capacity for human consumption. Breeding for high milk yield is the major factor causing poor welfare to cows, especially in relation to health problems. It is associated with increased risk of lameness, mastitis (infection of the teat/udder), infertility and metabolic disorders such as fatty liver syndrome and acidosis, which can lead to laminitis and immunosuppression (Algers et al., 2009).

In a bid to speed up so-called breed "improvements" many experiments have taken place to genetically engineer and/or clone farm animals. The process has resulted in many animals born with deformities and poor immune systems, leading to infection and early death (Barlow et al., 2008). The scandalous truth is that we have bred millions of animals whose own bodies have become their worst enemy. The problems for these sentient beings however do not end here.

Separation of Mother/Offspring

Another aspect of factory farming - and indeed of some free-range farming too – is the removal of the young from the mother. Chickens are hatched in a hatchery, so they never see their mothers at all. The male chicks of laying hens cannot be reared for egg-laying so they are killed at day one, being suffocated with gas or macerated.

Piglets would naturally wean themselves from their mother at about 11 weeks of age, but in modern farming are usually taken away from her at 3-4 weeks old for fattening up for slaughter. The sow can then be inseminated, often artificially, and set about producing her next litter of piglets.

In modern dairy farms, calves are removed from their mother within a few hours to a day of birth. Cows usually call out for their calves for quite some time and calves can also be heard calling for their mothers.

The male calves born to dairy cows are unsuitable for milk production and are often of a breed not ideal for meat production. The long, thinner bodies they inherit do not make for good beef. They too are disposed of at a couple of weeks old for low quality "bobby" calf meat or are reared in industrial conditions for veal. Since 2007 it has been illegal in the European Union (EU) to keep these calves in the notorious narrow "veal crates" where they were isolated with no regular physical contact with any other being and were kept unable to turn round, sometimes chained by the neck, and where they were fed only on a low-iron liquid diet throughout their lives (EC, 1997). Now they must be able to turn round and be in groups from 8 weeks old, and must have roughage in their diet, but they do not get to graze or have straw bedding in most farms (although in the UK, this is required). These notorious veal crates are also slowly being abandoned in some US states, but are still allowed elsewhere.

Isolation, Confinement and Crowding

Factory farms confine animals individually, away from their fellow-creatures, or crowd them together in massive sheds. Although all farm animals are descended from wild ancestors who lived on plain, hillside or in the forest or jungle, factory-farmed animals are not allowed outside at all. In fact their only glimpse of daylight and sunshine may be through the ventilation spaces on the trucks taking them to slaughter.

Laying hens in factory farms are kept in cages. In the barren battery cage they would spend their productive lives (from about 18 weeks of age) until slaughter, about a year later, standing on a sloping wire mesh floor, unable to stretch or flap their wings, unable to scratch at the soil and dust-bathe, unable to lay their eggs in a nest – which science shows to be very important for the hen (Dawkins, 1998). Since 2012, the barren cage has been banned throughout the EU, but most hens are still caged in so-called "modified" cages (EC, 1999). They now have a token nest space to share and a length of wooden rod to stand on if they want to be off the wire mesh floor. Space allowances have been marginally increased too and the larger "colony" cages can contain 60-80 hens in a larger area (CIWF, 2014b).

Until 2013 most breeding pigs in the EU were kept in sow stalls – narrow metal-barred crates with concrete-and-slats floors - in which they were unable to turn round throughout their sixteen-and-a-half week pregnancies. A partial ban on the system was agreed in 2001 and a 12-year phase out was given as a concession to the pig industry and to give farmers time to change their sow housing systems. Since 2013 it has become illegal to keep sows in these after the first 28 days of pregnancy, so more sows are now being housed in groups (EC, 2001).

Some commercial companies and other countries are now following Europe's lead – and the sooner the better! In a 1997 report, *The welfare of intensively kept pigs*, the Scientific Veterinary Committee (SVC) (Jensen et al., 1997) of the European Union commented that inactivity and unresponsiveness are abnormal and it is likely that crated sows become clinically depressed. The SVC criticized sow stalls/gestation crates, and concluded: "No individual pen should be used which does not allow the sow to turn around easily" (Jensen et al., 1997, p. 100).

A couple of days before they are due to give birth, the sows are moved to another narrow stall, the farrowing crate. Here they farrow (give birth) and suckle their piglets for the first few weeks. However as they can't turn around they have limited contact with their young. The young piglets are then reared in groups indoors, often on concrete, until they are ready to be slaughtered for meat at a few months old. In natural conditions pigs would spend up to 50% of their time using their snouts to root in the soil, seeking tubers and grubs to eat and another 23% in foraging behaviour (Stolba & Wood-Gush, 1989). Such behaviour is totally impossible in the factory farm.

The trend in dairy farming is towards "zero-grazing", which is just what it describes – the cattle are permanently housed or allowed only into a yard. They are not permitted to go out into the fields to exercise and eat grass, their natural food. Factory farmed animals are frequently mutilated to make them fit into their unnatural environments. For example, pigs, being highly inquisitive rooting animals, become frustrated and often proceed to bite each other's tails, presumably out of boredom. To prevent this, they are routinely tail-docked in infancy, usually without anaesthesia or analgesia. Many hens still have their beak tips cut back to prevent them pecking each other.

Farm animals are often transported on long journeys just to achieve a better price. These journeys can be national or international – even global. Around 3.4 million animals a year are exported from EU countries to Turkey, North Africa and the Middle East (Eurostat, 2015). Millions of farm animals are exported live from Australia every year to markets in the Middle East and, increasingly, in Southeast Asia. Conditions are often poor, with animals suffering from heat stress, trampling, hunger and thirst, and being held up at border posts. Apart from the stress of the long journeys, frequently accompanied by excessive heat and unfamiliar feedstuffs, the animals are often slaughtered in horrendous conditions upon arrival (CIWF, 2014a).

The law in the EU and many other countries requires animals to be slaughtered humanely. Research and undercover filming show that even in countries where respect for the law is high, failures are routine and many animals suffer from rough handling and inadequate stunning procedures (Animals Australia, 2014).

CONCLUSION

In view of the long list of adverse impacts of meat and dairy production, it does seem like an anomaly that every day people in many societies – who can afford to – chomp their way through large quantities of meat, imbibe large amounts of dairy products and often consume significant numbers of eggs too – without a thought for their origin or their effect. Maybe this is just another human emotional survival mechanism – let's not think about it because it's too awful to contemplate – just as we block out images of poverty and hunger, war and torture, our own aging and inevitable death. Let's not go there…

I think it's time we grew up. It's time for all of us to recognise our impact on our own health, on our planetary environment and on our fellow beings, be they human or animal. What we consume is a huge part of this and consumption of certain foods comes at, or near, the top of the list for impact.

It's also time governments, international institutions and food and farming companies took action. If they don't act now, then the health consequences of unlimited meat and dairy consumption may force them to act. If not health, then the issues of environmental degradation, global warming and scarce crop and water resources may push them to take action. Why wait?

So, if you are concerned about your own long-term health and that of your loved ones, please take action to reduce your meat consumption. If self-interest is not your driving force (and sadly it is for most of us) then perhaps altruism can be your guide. If we continue to increase our global meat consumption, then water will become more scarce, global warming will increase more speedily, more forests and other priceless ecosystems will be turned over to cattle ranching or growing soya for animal feed, more factory farms will spring up, impacting the livelihoods of local small-scale farmers who cannot compete, more soils and waterways will be polluted, more animals will suffer and it is possible that while the rich will continue to eat, the numbers of hungry may increase, as there simply will not be enough food for all.

REFERENCES

Alfaia, C. P. M., Alves, S. P., Martins, S. I. V., Costa, A. S. H., Fontes, C. M. G. A., Lemos, J. P. C., & Prates, J. A. M. et al. (2009). Effect of the feeding system on intramuscular fatty acids and conjugated linoleic acid isomers of beef cattle, with emphasis on their nutritional value and discriminatory ability. *Food Chemistry*, *114*(3), 939–946. doi:10.1016/j.foodchem.2008.10.041

Algers, B., Blokhuis, H. J., Botner, A., Broom, D. M., Costa, P., Domingo, M., & Wierup, M. et al. (2009). Scientific opinion on the overall effects of farming systems on dairy cow welfare and disease. Scientific Opinion of the Panel on Animal Health and Animal Welfare (Question No EFSA-Q- 2006-113). *European Food Safety Authority*, *1143*, 1–38.

American Heart Association (AHA). (2015). Saturated fats. Retrieved April 2, 2015, from http://www.heart.org/HEARTORG/GettingHealthy/NutritionCenter/HealthyEating/Saturated-Fats_UCM_301110_Article.jsp

Andreoletti, O., Budka, H., Buncic, S., Colin, P., Collins, J. D., & de Koijer, A. ... Vanopdenbosch, E. (2008). Foodborne antimicrobial resistance as a biological hazard. Scientific Opinion of the Panel on Biological Hazards (Question No EFSA-Q-2007-089). Agreement by the BIOHAZ Panel for public consultation 5-6 March 2008. Public consultation 17 April - 27 May 2008. *European Food Safety Authority, 765*, 1-87. Retrieved September 8, 2015, from http://www.efsa.europa.eu/sites/default/files/scientific_output/files/main_documents/biohaz_op_ej765_antimicrobial_resistance_en,3.pdf?ssbinary=true

Animals Australia. (2014). Exposing live export cruelty. Retrieved April 8, 2015, from http://www.animalsaustralia.org/investigations/live-export/

Antibiotics Alliance. (2014). Antimicrobial resistance - why the irresponsible use of antibiotics in agriculture must stop. A briefing from the Alliance to Save Our Antibiotics. London, UK: Antibiotics Alliance. Retrieved September 8, 2015, from http://www.soilassociation.org/LinkClick.aspx?fileticket=G9q4uEb5deI%3D&tabid=1841

Barlow, S., Chesson, A., Collins, J. D., Flynn, A., Hardy, A., & Jany, K.-D. ... Vannier, P. (2008). Scientific opinion of the Scientific Committee. Food safety, animal health and welfare and environmental impact of animals derived from cloning by somatic cell nucleus transfer (scnt) and their offspring and products obtained from those animals (Question No EFSA-Q-2007-092). *European Food Safety Authority, 767*, 1-49. Retrieved September 8, 2015, from http://www.efsa.europa.eu/sites/default/files/scientific_output/files/main_documents/sc_op_ej767_animal_cloning_en.pdf

Bee, G., Guex, G., & Herzog, W. (2004). Free range rearing of pigs during the winter: Adaptations in muscle fiber characteristics and the effects on adipose tissue composition and meat quality traits. *Journal of Animal Science, 82*(4), 1206–1218. PMID:15080344

Billings, M. (1997). The influenza pandemic of 1918. Retrieved April 8, 2015, from https://virus.stanford.edu/uda/

Bøtner, A., Broom, D. M., Doherr, M. G., Domingo, M., Hartung, J., & Keeling, L. ... Wierup, M. (2010). Scientific opinion on the influence of genetic parameters on the welfare and the resistance to stress of commercial broilers. EFSA Panel on Animal Health and Welfare. *European Food Safety Authority, 8*(7), 1-82. Retrieved September 8, 2015, from http://www.efsa.europa.eu/sites/default/files/scientific_output/files/main_documents/1666.pdf

Cassidy, E. S., West, P. C., Gerber, J. S., & Foley, J. A. (2013). Redefining agricultural yields: From tonnes to people nourished per hectare. *Environmental Research Letters, 8*(3), 1–8. doi:10.1088/1748-9326/8/3/034015

Centers for Disease Control and Prevention (CDC). (2014). CDC and zoonotic disease. Retrieved December 17, 2014, from http://www.cdc.gov/about/facts/cdcfastfacts/zoonotic.html

Centrers for Disease Control and Prevention (CDC). (2013). Antbiotic resistance threat in the United States, 2013. Atlanta, GA: CDC, Division of Healthcare Quality Promotion, National Center for Emerging and Zoonotic Infectious Diseases. Retrieved September 8, 2015, from http://www.cdc.gov/drugresistance/pdf/ar-threats-2013-508.pdf

Choices, N. H. S. (2013). Meat in your diet. Retrieved December 17, 2014, from http://www.nhs.uk/Livewell/Goodfood/Pages/meat.aspx

Compassion in World Farming (CIWF). (2013). Statistics: Broiler chickens. Retrieved May 10, 2015, from https://www.ciwf.org.uk/media/5235303/Statistics-Broiler-chickens.pdf

Compassion in World Farming (CIWF). (2014a). Victory: notorious slaughterhouse is shutdown. Retrieved November 20, 2014, from http://www.philiplymbery.com/2014/11/victory-notorious-slaughterhouse-is-shutdown/

Compassion in World Farming (CIWF). (2014b). Welfare issues for egg laying hens. Retrieved April 8, 2015, from http://www.ciwf.org.uk/farm-animals/chickens/egg-laying-hens/welfare-issues/

Crous-Bou, M., Fung, T. T., Prescott, J., Julin, B., Du, M., Sun, Q., & de Vivo, I. et al. (2014). Mediterranean diet and telomere length in Nurse's Health Study: Population based cohort study. *British Medical Journal*, *349*(7986), 1–11. PMID:25467028

Curickshank, G. (2003). Cobb focuses on bottom line performance. *Poultry World*, *157*(7), 22–25.

DairyCo. (2014). Average UK milk yield. Retrieved April 8, 2015, from http://www.dairyco.org.uk/market-information/farming-data/milk-yield/average-milk-yield/ -. VU8cvGBi6Ct

Dawkins, M. S. (1998). *Through our eyes only? The search for animal consciousness*. Oxford, UK: Oxford University Press. doi:10.1093/acprof:oso/9780198503200.001.0001

Drøyvold, W. B., Midthjell, K., Nilsen, T. I. L., & Holmen, J. (2005). Change in body mass index and its impact on blood pressure: A prospective population study. *Obesity (Silver Spring, Md.)*, *29*, 650–655. PMID:15809666

European Commission (EC). (1997). *Council Directive 97/2/EC of 20 January 1997 amending Directive 91/629/EEC laying down minimum standards for the protection of calves*. Brussels, Belgium: EC. Retrieved September 8, 2015, from http://ec.europa.eu/food/fs/aw/aw_legislation/calves/97-2-ec_en.pdf

European Commission (EC). (1999). *Council Directive 1999/74/EC laying down minimum standards for the protection of laying hens*. Brussels, Belgium: EC. Retrieved from http://eur-lex.europa.eu/LexUriServ/LexUriServ.do?uri=OJ:L:1999:203:0053:0057:EN:PDF

European Commission (EC). (2001). *Commission directive 2001/93/EC of 9 November 2001 amending Directive 91/630/EEC laying down minimum standards for the protection of pigs*. Brussels, Belgium: EC. Retrieved September 8, 2015, from http://faolex.fao.org/cgi-bin/faolex.exe?rec_id=029442&database=faolex&search_type=link&table=result&lang=eng&format_name=@ERALL

Eurostat. (2015). *Meat production and foreign trade - head - monthly data*. Luxembourg: Eurostat. Retrieved September 8, 2015, from http://appsso.eurostat.ec.europa.eu/nui/show.do?dataset=apro_mt_pheadm&lang=en

Farvid, M. S., Ding, M., Pan, A., Sun, Q., Chiuve, S. E., Steffen, L. M., & Hu, F. B. et al. (2014). Dietary linoleic acid and risk of coronary heart disease: A systematic review and meta-analysis of prospective cohort studies. *Circulation*, *130*(18), 1568–1578. doi:10.1161/CIRCULATIONAHA.114.010236 PMID:25161045

Food and Agriculture Organization of the United Nations Statistics Division (FAOSTAT). (2015). Livestock Primary. Production: Meat Indigenous Poultry. Retrieved May 10, 2015, from http://faostat3. fao.org/browse/Q/QL/E

Fraser, G. E. (1999). Associations between diet and cancer, ischemic heart disease, and all-cause mortality in non-Hispanic white California Seventh-day Adventists. *The American Journal of Clinical Nutrition*, *70*(Suppl 3), 532S–538S. PMID:10479227

Friends of the Earth (FOE). (2008). What's feeding our food? The environmental and social impacts of the livestock sector. London, UK: FOE. Retrieved September 8, 2015, from http://www.foe.co.uk/sites/default/files/downloads/livestock_impacts.pdf

Garcia, P. T., Pensel, N. A., Sancho, A. M., Latimori, N. J., Kloster, A. M., Amigone, M. A., & Casal, J. J. (2008). Beef lipids in relation to animal breed and nutrition in Argentina. *Meat Science*, *79*(3), 500–508. doi:10.1016/j.meatsci.2007.10.019 PMID:22062910

Government of the United Kingdom. (2011). The future of food and farming: challenges and choices for global sustainability. London, UK: Government Office for Science. Retrieved September 8, 2015, from https://www.gov.uk/government/uploads/system/uploads/attachment_data/file/288329/11-546-future-of-food-and-farming-report.pdf

Government of the United Kingdom. (2013, May 12). *Antimicrobial resistance poses 'catastrophic threat,' says Chief Medical Officer*. Press Release. Retrieved September 8, 2015, from https://www.gov.uk/government/news/antimicrobial-resistance-poses-catastrophic-threat-says-chief-medical-officer--2

Grant, W. B. (2013). A multicountry ecological study of cancer incidence rates in 2008 with respect to various risk-modyfying factors. *Nutrients*, *6*(1), 163–189. doi:10.3390/nu6010163 PMID:24379012

Harvard School of Public Health (HSPH). (2015). Fats and cholesterol: out with the bad, in with the good. Retrieved September 8, 2015, from http://www.hsph.harvard.edu/nutritionsource/what-should-you-eat/fats-and-cholesterol/

Jenkins, D. J. A., Kendall, C. W. C., Marchie, A., Jenkins, A. L., Augustin, L. S. A., Ludwig, D. S., & Anderson, J. W. et al. (2003). Type 2 diabetes and the vegetarian diet. *The American Journal of Clinical Nutrition*, *78*(Suppl 3), 610S–616S. PMID:12936955

Jensen, P., Blokhuis, H., Broom, D. M., Bubna-Littitz, H., Canali, E., & Dantzer, R. … Zygoyiannis, D. (1997). The Welfare of intensively kept pigs. Brussels, Belgium: European Commission, Scientific Veterinary Committee, Animal Welfare Section.

Katz, D. L., & Meller, S. (2014). Can we say what diet is best for health? *Annual Review of Public Health*, *35*(1), 83–103. doi:10.1146/annurev-publhealth-032013-182351 PMID:24641555

Key, T. J., Fraser, G. E., Thorogood, M., Appleby, P. N., Beral, V., Reeves, G., & McPherson, K. et al. (1999). Mortality in vegetarians and nonvegetarians: Detailed findings from a collaborative analysis of 5 prospective studies. *The American Journal of Clinical Nutrition*, *70*(Suppl 3), 516S–524S. PMID:10479225

Knowles, T. G., Kreslin, S. C., Haslam, S. M., Brown, S. N., Green, L. E., Butterworth, A., & Nicol, C. J. et al. (2008). Leg disorders in broiler chickens: Prevalence, risk factors and prevention. *PLoS ONE*, *3*(2), 1–7. doi:10.1371/journal.pone.0001545 PMID:18253493

Kris-Etherton, P. M., Harris, W. S., & Appel, L. J.AHA Nutrition Committee. (2003). Omega-3 fatty acids and cardiovascular disease: New recommendations from the American Heart Association. *Arteriosclerosis, Thrombosis, and Vascular Biology*, *23*(2), 151–152. doi:10.1161/01.ATV.0000057393.97337. AE PMID:12588750

Kroenke, C. H., Kwan, M. L., Sweeney, C., Castillo, A., & Caan, B. J. (2013). High- and low-fat dairy intake, recurrence, and mortality after breast cancer diagnosis. *Journal of the National Cancer Institute*, *105*(9), 616–623. doi:10.1093/jnci/djt027 PMID:23492346

Larsson, S. C., Bergkvist, L., & Wolk, A. (2004). Milk and lactose intakes and ovarian cancer risk in the Swedish Mammography Cohort. *The American Journal of Clinical Nutrition*, *80*(5), 1353–1357. PMID:15531686

Liu, J., & Savenije, H. H. G. (2008). Food consumption patterns and their effect on water requirement in China. *Hydrology and Earth System Sciences*, *12*(3), 887–898. doi:10.5194/hess-12-887-2008

Ludwig, D. S., & Willett, W. C. (2013). Three daily servings of reduced-fat milk an evidence-based recommendation? *The Journal of the American Medical Association (JAMA). Pediatrics*, *167*(9), 788–789.

Lundqvist, J., de Fraiture, C., & Molden, D. (2008). Saving water: from field to fork, curbing losses and wastage in the food chain. Huddinge, Sweden: Stockholm International Water Institute (SIWI). Retrieved September 8, 2015, from http://www.siwi.org/documents/Resources/Policy_Briefs/PB_From_Filed_to_Fork_2008.pdf

McDougall, J., Thomas, L. E., McDougall, C., Moloney, G., Saul, B., Finnell, J. S., & Petersen, K. M. et al. (2014). Effects of 7 days on ad libitum low-fat vegan diet: The McDougall Program cohort. *Nutrition Journal*, *13*(1), 1–7. doi:10.1186/1475-2891-13-99 PMID:25311617

McEvoy, C. T., Temple, N., & Woodside, J. V. (2012). Vegetarian diets, low-meat diets and health: A review. *Public Health Nutrition*, *15*(12), 2287–2294. doi:10.1017/S1368980012000936 PMID:22717188

Mekonnen, M. M., & Hoekstra, A. Y. (2012). A global assessment of the water footprint of farm animal products. *Ecosystems (New York, N.Y.)*, *15*(3), 401–415. doi:10.1007/s10021-011-9517-8

Murray, C. J. L., Lopez, A. D., Chin, B., Feehan, D., & Hill, K. H. (2006). Estimation of potential global pandemic influenza mortality on the basis of vital registry data from the 1918-20 pandemic: A quantitative analysis. *Lancet*, *368*(9554), 2211–2218. doi:10.1016/S0140-6736(06)69895-4 PMID:17189032

National Chicken Council. (2014). The nutritional value of chicken. Retrieved December 17, 2014, from http://www.nationalchickencouncil.org/chicken-the-preferred-protein-for-your-health-and-budget/the-nutritional-value-of-chicken/

National Pork Board. (n.d.). Pork, the other white meat brand. Retrieved April 27, 2015, from http://www.porkbeinspired.com/about-the-national-pork-board/the-other-white-meat-brand/

Nellemann, C., MacDevette, M., Manders, T., Eickhout, B., Svihus, B., Prins, A. G., & Kaltenborn, B. P. (2009). The environmental food crisis: the environment's role in averting future food crises. A UNEP Rapid Response Assessment. Arendal, Norway: United Nations Environment Programme. Retrieved September 8, 2015, from http://www.grida.no/files/publications/FoodCrisis_lores.pdf

Pan, A. (2012). Red meat consumption and mortality: Result from two prospective cohort studies. *The Journal of the American Medical Association (JAMA). Archives of Internal Medicine, 172*(7), 555–563. doi:10.1001/archinternmed.2011.2287 PMID:22412075

Pickett, H. (2012). Nutritional benefits of higher welfare animal products. Surrey, UK: Compassion in World Farming (CIWF).

Pollan, M. (2007). *In defense of food, an eater's manifesto: eat food, not too much mostly plants*. London, UK: Penguin Books.

Poultry Site. (2015a). Breeding and genetics products and services. Retrieved April 8, 2015, from http://www.thepoultrysite.com/products/2/breeding-and-genetics/

Poultry Site. (2015b). Influence of increasing slaughter age of chickens on meat quality, welfare and technical and economic results. Retrieved April 8, 2015, from http://www.thepoultrysite.com/articles/2540/influence-of-increasing-slaughter-age-of-chickens-on-meat-quality-welfare-and-technical-and-economic-results/

Qin, L. Q., Xu, J. Y., Wang, P. Y., Tong, J., & Hoshi, K. (2007). Milk consumption is a risk factor for prostate cancer in Western countries: Evidence from cohort studies. *Asia Pacific Journal of Clinical Nutrition, 16*(3), 467–476. PMID:17704029

Raphaely, T., & Marinova, D. (2014b). Flexitarianism: A more moral dietary option. *International Journal of Sustainable Society, 6*(1-2), 189–211. doi:10.1504/IJSSOC.2014.057846

Rayner, M., Clarke, D., Thomas, P., & Scarborough, P. (2010). Healthy planet eating: how lower meat diets can save lives and the planet. Oxford, UK: Friends of the Earth & Oxford University, British Heart Foundation Health Promotion Research Group. Retrieved September 8, 2015, from http://www.foe.co.uk/sites/default/files/downloads/healthy_planet_eating.pdf

Scarborough, P., Appleby, P. N., Mizdrak, A., Briggs, A. D. M., Travis, R. C., Bradbury, K. E., & Key, T. J. (2014). Dietary greenhouse gas emissions of meat-eaters, fish-eaters, vegetarians and vegans in the UK. *Climatic Change, 125*(2), 179–192. doi:10.1007/s10584-014-1169-1 PMID:25834298

Simopoulous, A. P. (2008). The importance of the omega-6/omega-3 fatty acid ratio in cardiovascular disease and other chronic diseases. *Experimental Biology and Medicine, 233*(6), 674–688. doi:10.3181/0711-MR-311 PMID:18408140

Sirri, F., Castellini, C., Bianchi, M., Petracci, M., Meluzzi, A., & Franchini, A. (2011). Effect of fast-, medium- and slow-growing strains on meat quality of chickens reared under the organic farming method. *Animal, 5*(2), 312–319. doi:10.1017/S175173111000176X PMID:22440776

Song, Y., Chavarro, J. E., Cao, Y., Qiu, W., Mucci, L., Sesso, H. D., & Ma, J. et al. (2013). Whole milk intake is associated with prostate cancer-specific mortality among U.S. male physicians. *The Journal of Nutrition*, *143*(2), 189–196. doi:10.3945/jn.112.168484 PMID:23256145

Soto, E., de La Hoz, L., Ordonez, J., Herranz, B., Hierro, E., Lopez-Bote, C. J., & Cambero, M. (2009). The feeding and rearing systems of Iberian pigs affect the lipid composition and texture profile of dry-cured loin. *Journal of Animal and Feed Sciences*, *18*(1), 78–89.

Steinfeld, H., Gerber, P., Wassenaar, T., Castel, V., Rosales, M., & de Hann, C. (2006). Livestock's long shadow: environmental issues and options. Rome, Italy: Food and Agriculture Organisation (FAO).

Stolba, A., & Wood-Gush, D. G. M. (1989). The behaviour of pigs in a semi-natural environment. *Animal Production*, *48*(2), 419–425. doi:10.1017/S0003356100040411

Taylor, E. F., Burley, V. J., Greenwood, D. C., & Cade, J. E. (2007). Meat Consumption and risk of breast cancer in the UK Women's Cohort Study. *British Journal of Cancer*, *96*(7), 1139–1146. doi:10.1038/sj.bjc.6603689 PMID:17406351

Tilman, D., & Clark, M. (2014). Global diets link environmental sustainability and human health. *Nature*, *515*(7528), 518–522. doi:10.1038/nature13959 PMID:25383533

Tonstad, S., Butler, T., Yan, R., & Fraser, G. E. (2009). Type of vegetarian diet, body weight, and prevalence of type 2 diabetes. *Diabetes Care*, *32*(5), 791–796. doi:10.2337/dc08-1886 PMID:19351712

Trostle, R. (2008). Global agricultural supply and demand: factors contributing to the recent increase in food commodity prices. Washington, DC: United States Department of Agriculture (USDA), Economic Research Service. Retrieved September 8, 2015, from http://www.ers.usda.gov/media/218027/wrs0801_1_.pdf

University of Maryland Medical Centre. (2013). Omega-6 fatty acids. Retrieved April 2, 2015, from http://umm.edu/health/medical/altmed/supplement/omega6-fatty-acids

van Dooren, C., Marinussen, M., Blonk, H., Aiking, H., & Vellinga, P. (2014). Exploring dietary guidelines based on ecological and nutritional values: A comparison of six dietary patterns. *Food Policy*, *44*, 36–46. doi:10.1016/j.foodpol.2013.11.002

Ventanas, S., Ventanas, J., Tovar, J., García, C., & Estévez, M. (2007). Extensive feeding versus oleic acid and tocopherol enriched mixed diets for the production of Iberian dry-cured hams: Effect on chemical composition, oxidative status and sensory traits. *Meat Science*, *77*(2), 246–256. doi:10.1016/j.meatsci.2007.03.010 PMID:22061597

Wang, Y., Lehane, C., Ghebremeskel, K., & Crawford, M. A. (2009). Modern organic and broiler chickens sold for human consumption provide more energy from fat than protein. *Public Health Nutrition*, *13*(3), 400–408. doi:10.1017/S1368980009991157 PMID:19728900

Water Footprint Network. (2014a). Product gallery: Beef. Retrieved December 17, 2014, from http://waterfootprint.org/en/resources/interactive-tools/product-gallery/

Water Footprint Network. (2014b). What is water footprint? Retrieved December 17, 2014, from http://waterfootprint.org/en/water-footprint/what-is-water-footprint/

World Cancer Research Fund (WCRF), & American Institute for Cancer Research (AICR). (2007). Food, nutrition, physical activity and the prevention of cancer: a global perspective. Washington, DC: AICR.

World Health organization (WHO). (2008). *Protecting health from climate change, World Health Day 2008: Annex 1 Reducing your carbon footprint can be good for your health, a list of mitigating actions.* Geneva, Switzerland: WHO.

World Health Organization (WHO). (2011). *Tackling antibiotic resistance from a food safety perspective in Europe.* Copenhagen, Denmark: WHO.

World Health Organization (WHO). (2015a). Health topics: Diet. Retrieved April 2, 2015, from http://www.who.int/topics/diet/en/

World Health Organization (WHO). (2015b). Statistics: A(H5N1) reported human cases. Retrieved April 8, 2015, from http://www.who.int/influenza/human_animal_interface/EN_GIP_20150303cumulativeNumberH5N1cases.pdf?ua=1

World Health Organization (WHO). (2015c). H5N1 Influenza. Retrieved April 8, 2015, from http://www.who.int/influenza/human_animal_interface/avian_influenza/h5n1_research/faqs/en/

World Hunger Education Service. (2015). 2015 world hunger and poverty facts and statistics. Retrieved May 10, 2015, from http://www.worldhunger.org/articles/Learn/world hunger facts 2002.htm

KEY TERMS AND DEFINITIONS

Antibiotic Resistance: When infectious bacteria cease to be affected by antibiotics which previously killed them or limited their growth. This makes it difficult to treat infections in humans and animals.

Broiler (Chicken): A chicken raised primarily for its meat, rather than for eggs. The most commonly factory farmed animal on earth.

Factory Farms/Factory Farming: Farming for maximum profit, breeding animals for high productivity, often keeping them indoors, confined and crowded or isolated. Based on cereal-and-soy feeding.

Pescetarian: Someone who does not eat meat, but does eat fish. Also used as a description of such a diet.

Sentient (Being): One who is sensitive both to physiological pain and to mental/emotional suffering. Used of humans and a wide range of animals, including aquatic animals.

Vegan: Someone who eats no animal products (meat, fish, dairy, eggs) and whose diet is entirely plant-based. Also used as a description of such a diet.

Vegetarian: Someone who eats no meat or fish, but consumes dairy products and eggs. Also used as a description of such a diet.

Zoonosis/Zoonotic: A disease, which spreads from animals to humans, such as rabies and salmonellosis.

ENDNOTE

[1]   The calculations can be seen at: http://www.waterfootprint.org/?page=files/productgallery

Chapter 11
China's Growing Meat Demands:
Implications for Sustainability

Xiumei Guo
Curtin University, Australia

Talia Raphaely
Curtin University, Australia

Dora Marinova
Curtin University, Australia

ABSTRACT

The chapter examines China's growing meat demand and its implications. Australia and China are currently set to expand trade in meat and livestock facilitated by a government negotiated Free Trade Agreement. China is already the world's largest meat consumer and with the increasing consumerism and wealth of its rapidly growing middle and upper class, the demand for animal products is likely to grow. This country's unprecedented appetite for animal proteins has stimulated the Australian livestock and related sectors, potentially enabling vast growth and profitability within these industries. Chinese customers have strong purchasing power and are eager to buy imported frozen and locally slaughtered Australian meat. While Australian farmers are capitalising on these economic opportunities, only the animal welfare sector voices any concern. This chapter highlights the ignored health and environmental costs.

INTRODUCTION

China as a nation consumes more meat than any other country (see Figure 1) and is also already responsible for more than half of the world's total pork intake (Winglee, 2014). Rising individual meat consumption is being supported by rapidly increasing urbanisation, rising incomes and a growing middle and upper class demographic, coupled with increasing emulation of western dietary habits and food styles.

DOI: 10.4018/978-1-4666-9553-5.ch011

Figure 1. Income per person for Chinese rural and urban residents, 2009-2050
Compiled from Hamshere et al., 2014.

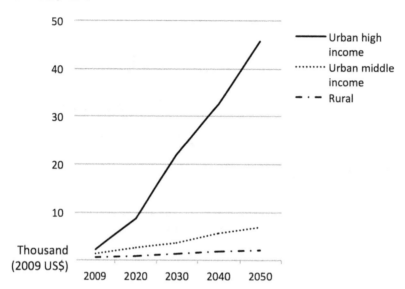

In China (as in other transitioning countries) this increasing affluence and resulting consumerism include an unprecedented demand for meat (Myers & Kent, 2004; Tilman, Balzar, Hill & Befort, 2011; OECD, 2014). The proportion of the population living in urban areas in China has grown from 25% in 1990 to 50% (half of the entire population) in 2011 and is projected to increase even further with 75% of the total Chinese population residing in urban areas by 2050 (Hamshere, Sheng, Moir, Syed & Gunning-Trant (2014). Furthermore, the percentage of urbanites in the middle class population has climbed from 4% in 2000 to 68% in 2012 (Barton, 2013). By 2022, the population of China's urban middle class is expected to comprise 650 million people or 76% of all urban Chinese residents (Barton, 2013).

As a result from China's fast economic expansion, GDP has grown on average over 10% annually for the last 35 years with the per capita income increasing in tandem. Personal income among Chinese residents has drastically increased since 2009 with urban dwellers rapidly becoming richer (see Figure 1). This rise in per capita affluence has greatly enhanced purchasing power and demand for animal proteins including for Australian meat believed to be of a higher quality than local produce. Rising incomes and increasing urbanization have played and continue to play a significant role in China's growing meat demands (Liu, Parton, Zhou & Cox, 2012), particularly amongst wealthy local middle class consumers (Cao et al., 2013).

CHINA'S GROWING MEAT DEMAND

The growing Chinese urban middle class is pursuing western dietary styles including increasingly excessive meat consumption. As a consequence, per capita meat consumption (see Figure 2) has grown from 13.7 kg per person in 1980, to 27 kg in 2000 and 49 kg per person in 2013 (National Bureau of Statistics of China, 2014). Rapid urbanization brings rural population to the cities where they adopt the urban and more affluent consumption behaviours, including dietary habits such as eating less food grains and vegetables and more livestock products.

Figure 2. Annual meat consumption per capita, China, 1980–2013
Compiled from National Bureau of Statistics of China. (1980–2013). China statistical yearbook, Beijing, China: China Sta-tistical Press.

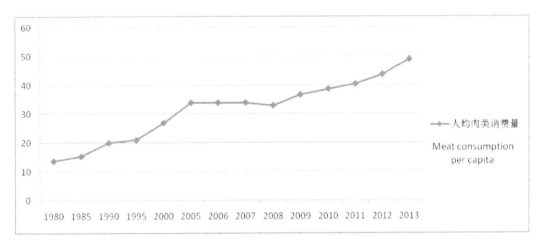

China responds to this growing demand for animal proteins by increasing local production and ex-panding imports (China Meat, 2014). Food scandals around domestic products are making the task more difficult. For example, China's largest meat processor Shuanghui was found to use the illegal steroid additive clenbuterol in pork meat which resulted in loss of confidence in local livestock producers and products. This has become a key driver behind more affluent Chinese consumers' preference to purchase food, including meat, from international markets (ACC, 2014). This growing demand for foreign livestock products has created substantial opportunities for Australia's livestock producers and exporters. Existing mineral resources relationships between China and Australia paved the way for industry and government to seek new opportunities, encouraged by the sharply growing Chinese demand for animal protein.

It is not only demand for safe meat that is driving exports between China and Australia. China is unable to supply sufficient quantities of meat products to meet its growing demand. Until recently, the country relied primarily on extensive grazing systems and farming models which required vast tracts of land and were labour intensive. Industrialization and urbanization changed this. China's increasingly intensified meat industry is focused on maximising supply but the country's emerging obsession with high meat consumption is creating a demand that cannot be met locally.

Australian livestock companies, in response, are poised to supply the massive Chinese market, which is increasingly opening the doors for meat, live and frozen for exports. In 2014 China was Australia's fourth largest beef export market (Locke, 2014a, 2014b; MLA, 2015). The economic and demographic changes in China have had a significant effect on global markets and particularly on the Australian meat industry where the demand provides numerous profitable opportunities (MLA, 2014). Australia's rich agricultural resources and already well-established trade relationship with China have firmly positioned Australian meat as Chinese consumers' first choice (Australian Government, 2012, 2013).

Australian meat processors currently export boxed beef to China and expect to double the amounts of boxed, frozen and chilled meat exported into China in 2015 (Locke, 214a, 2014b). Australia's good reputation as a significant contributor to food industry innovation on food security and food sovereignty (ACC, 2014; Paddenburg, 2014, July 12) means Australian meat is increasingly available and popular in Chinese supermarkets (see Picture 1). In addition to the already established frozen meat export trade,

Figure 3. Beef from Australia for sale in one of Tianjin's supermarkets, China

Australian farmers and exporters are pursuing live export deals with their Chinese counterparts and transforming the Australian cattle industry to meet this burgeoning demand.

Australia is striving to accommodate, perpetuate and supply the seemingly insatiable growth in demand for meat (ACC, 2014). With a recent trade liberalisation agreement consolidated between the two countries, China is expected to become the biggest importer of Australian meat and livestock products (Australian Government, 2012, 2013). The meat trade opportunities have full support from the Australian government and livestock sector. This chapter explains the favourable conditions created for opening of the meat trade and then analyses the sustainability implications suggesting that decoupling improvement in living standards from meat consumption is a better strategy for China, Australia and the world.

Promoting the Meat Trade

The Australian government is actively supporting Australia-China trade. In the 2012 Australian Government White Paper, Australian businesses are encouraged to adjust their way of engagement with Asia, and to recognise that meat and live cattle are a significant growth industry (Australian Government, 2012). China is also a major part of Australia's Asian Century strategy (Australian Government, 2013). The food industry, particularly meat products, is recognized as an important export engagement focus in the Asian region. Although Chinese people consume more pork as a traditional choice, beef consumption has been increasing rapidly in recent years. A Free Trade Agreement was announced in 2014 and the Australian livestock sector is expected to profit by $11 billion by 2030 due to the now-removed trade tariffs (MLA, 2014). Australia's Prime Minister publically announced its government's intention to export live 1 million cattle to China for slaughter doubling current live-export cattle trade (Cullen, 2014; Neales, 2014). In order to supply China's rising demand for meat, Queensland growers are now transitioning to livestock industry planning or building huge abattoirs, feedlots and export facilities (Brown, 2014). Australian pork exporters are also preparing to supply the booming pork demand in China (Locke, 2014b).

Australian meat companies believe China's growing demand will boost prices significantly because it will make the market more competitive. In response Australian cattle farmers have to extend all their available export capacity and services without considering concomitant environmental damage. Australia is also witnessing increasing acquisition of national farmland by Chinese investors eager to capitalise on the meat trade opportunities (Cranston, 2015).

According to Hamshere et al. (2014), China's demand for agrifood products will double by 2050. Projections show beef consumption will increase 236% and sheep and goat meat consumption will both increase over 70% (Hamshere et al., 2014). Such an expansion of the Australian cattle industry will increase domestic environmental pressures and impacts such as land degradation and water shortages.

The government has described the live cattle deal as a "breakthrough" for economic outcomes (Cullen, 2014) and the potential increasing meat trade with China is fast becoming a core agricultural and financial reality. Considering the profits and vested interest involved it is hard to argue against such a powerful economic and political lobby or to even look for ways to counter the vast health and environmental cost such growth in meat production and consumption will have. Yet the long-term real costs of this production and trade need to be made clear.

IMPLICATIONS FOR SUSTAINABILITY

Meat production and consumption have serious environmental and social impacts and irrespective as to where the animals are raised or slaughtered the global ecology still bears the brunt. Excessive meat consumption also has significant public health implications (Raphaely, Marinova, Crisp & Panayotov, 2013). Given the localised and global health and environmental impacts of high meat production and consumption both Australia and China are poised to contribute towards a potential global disaster should the predicted growth in production and consumption eventuate.

Environmental Impacts

China's shifting diet pattern and its meat demands have a great effect on the Australian and global environment (Winglee, 2014). Meat products have serious ramifications for climate change and the ecology (Friel, 2010). Already livestock companies and producers are questioning whether they will be able to provide the cattle needed by the new Chinese market particularly with the regular condition of drought and water shortage in Australia, a trend that is only expected to worsen with climate change (Quiggin, 2010).

Australia's carbon emission figures are already among the highest in the world on a per capita basis while China is the highest national greenhouse gases (GHG) emitter. Given that livestock is estimated to be responsible for up to 51% of all anthropogenic GHG emissions (Goodland & Anhang, 2009), a further increase in meat production will push climate change even further. Anthropogenic carbon dioxide and methane emissions into the atmosphere such as the levels associated with cattle production are a negative externality, which causes a market failure (Rezai, Foley & Taylor, 2009). It will be interesting to see which country will be held accountable for the increase in emissions caused by Australian livestock. China's ecological footprint already exceeds the country's biocapacity (WWF, 2014). Adding more meat to the Chinese diet will only make things worse for the country and the world.

With Australia being an increasingly dry continent, it is important to recognise that livestock farming uses large volumes of water (NCGRT, 2013). By raising and exporting cattle Australia is shipping its precious water to China. Furthermore, supplying China's growing demands for red meat means increasingly bigger herds with associated land, water, feeding requirements and waste problems. Land used for livestock produced to supply the Chinese demand will create further agriculturally driven deforestation and habitat destruction.

Social Externalities

The vast increase in meat production inherent in China's growing demand not only carries significant ecological costs such as contribution to climate change, water use, land depletion and pollution. There are also numerous threats to human health and wellbeing. Public health alarms are being raised around the impacts of global warming, increased severe weather events such as droughts, floods and fires (Hewitson, 2012). All of these directly and indirectly impact human health and wellbeing.

However, excessive meat intake per se is already causing a rise in non-communicable diseases in China, including cancers and cardiovascular disease (Campbell & Campbell, 2006) replacing ailments of poverty with lifestyle illnesses. In 2003, Chiu et al. (2003) studied incidents of colon cancer in Shanghai, China and found that diets high in red meat were linked with increased risk. This is in line with the colorectal cancer study by Friel (2010) in Australia. Furthermore, research by Song et al. (2014) suggests that a diet of just 100 grams of red meat per day raises the risk of gastric cancer by 17%.

Chinese consumers, particularly the upper and middle class group, have increasing opportunities to experience western lifestyles, including meat-based diets. Like the West, China is starting to experience the negative health externalities of excessive consumption of animal proteins, including higher body mass index (BMI), childhood and adult obesity of the Chinese adult population (Wang et al., 2013) due to changing dietary styles away from the traditional plant-based nutritional model (The Economist, 2014; Xue, Liu, Duan, Zhou & Cheng, 2014).

According to FAO (2014), the per capita meat supply in China is half the level of that in the US; "[i] f each Chinese were ever to consume as much beef as today's average American, and if the cattle were to be raised largely on grain, that would require more grain than the entire U.S. grain harvest" (Myers & Kent, 2004, p. 70). Both global hunger and overconsumption continue to rise whilst western countries feed grains to animals that might instead be used better for people. According to calculations by the United Nations Environment Programme (UNEP), " the calories that are lost by feeding cereals to animals, instead of using them directly as human food, could theoretically feed an extra 3.5 billion people" (Global Agriculture, 2015, n.p.). The most populous nation on earth, China, is increasingly contributing to this injustice at the expense of its own health.

A reduction in meat production and consumption both in Australia and China would create environmental and social benefits. It should be encouraged in both countries.

Animal Welfare Ethical Issues

The significant scale of predicted exports will double Australia's current live cattle exports to China and will also increase overall meat production. There are strong voices that the Free Trade Agreement "would be a dreadful deal that will result in more cruelty to Australian animals" (Cullen, 2014, n.p.). Both in Australia and China (as in most parts of the world) food animals are given little or no protection by law. Many are justifiably anxious about the live cattle export to China (LSNSWYL, 2014) and assurance systems for animal welfare standards which to date have failed. More attention should be paid both in Australia and China to moral considerations relating to animal ethics, and moral considerations relating to animal welfare and animal rights.

Rather than considering how to address welfare standards for food animals, dietary interventions reducing meat intake should be prioritised. Given the environmental and health threats associated with meat, the important consideration should be whether exporting Australian meat and livestock to China is a smart or sustainable strategy.

CONCLUSION

China's population of nearly 1.4 billion, and 20% of the global population, can play a significant role in global environmental and social wellbeing or destruction and this largely depends on present and future dietary choices in this country. The rapid urbanisation, growing affluence and resulting rise in consumerism in China have driven massive opportunities for increased meat production and export from Australia. The country's growing demand for animal proteins has generated unprecedented trade opportunities for Australian livestock producers.

It may be in Australia's economic interests, and particularly in the interest of the livestock and related sectors, to continue to promote increasing consumption and demand for Australian meat products in China. However, this is where the benefit stops. The direct consequences on Chinese public health and on the Australian and global environment should not be ignored or underestimated. Decoupling improvement in living standards from meat consumption is an important imperative that should accompany production and consumption opportunities for healthy, plant-based, high protein, alternatives. This will result in enhanced environmental and social outcomes in China, Australia and globally and mitigate negative externalities.

REFERENCES

Australia China Connections (ACC). (2014). Food safety in China opens doors for Australia's agri sector. Retrieved from http://www.chinaconnections.com.au/en/magazine/current-issue/1940-food-safety-in-china-opens-doors-for-australia%E2%80%99s-agri-sector

Australian Government. (2012). Australia in the Asian Century. White paper. Canberra, Australia: Department of the Prime Minister and Cabinet. Retrieved from http://www.murdoch.edu.au/ALTC-Fellowship/_document/Resources/australia-in-the-asian-century-white-paper.pdf

Australian Government. (2013). China country strategy. Australia in the Asian Century: towards 2025. Canberra, Australia: Department of Foreign Affairs and Trade. Retrieved from http://apo.org.au/files/Resource/dfat_chinacountrystrategy_aug_2013.pdf

Barton, D. (2013). The rise of the middle class in China and its impact on the Chinese and world economies. In China-United States Exchange Foundation, *US-China economic relations in the next ten years: towards deeper engagement and mutual benefit* (Ch. 7, pp. 1-12). Retrieved from http://www.chinausfocus.com/us-china-2022/

Brown, C. (2014, November 11). Chinese demand for Australian beef drives $75 million Townsville abattoir plan. *ABC News*. Retrieved from http://www.abc.net.au/news/2014-11-10/second-abattoir-proposed-for-north-queensland-port-town/5877062

Campbell, T. C., & Campbell, T. M. (2006). *The China study*. Dallas, TX: BenBella Books.

Cao, L., Tian, W., Wang, J., Malcom, B., Liu, H., & Zhou, Z. (2013). Recent food consumption trends in China and trade implications to 2020. *Australasian Agribusiness Review*, *21*, 15–44.

China Meat. (2014). Statistical data on the pork, beef and lamb imports in the first half year. Retrieved from http://www.chinameat.cn/html/n1/1_1_0_0/2014-08-05/2014080509363507.shtml

Chiu, B. C.-H., Ji, B.-T., Dai, Q., Gridley, G., McLaughlin, J. K., Gao, Y.-T., & Chow, W.-H. et al. (2003). Dietary factors and risks of colon cancer in Shanghai, China. *Cancer Epidemiology, Biomarkers & Prevention*, *12*(3), 201–208. PMID:12646508

Cranston, M. (2015, May 4). Chinese agri giant wants $100m of Australian farmland. *Financial review*. Retrieved from http://www.afr.com/real-estate/chinese-agri-giant-wants-100m-of-australian-farmland-20150430-1mwvtw

Cullen, S. (2014, November 7). Beef deal: Australia, China close to finalising $1 billion agreement that would see 1 million cattle exported. *Abc News*. Retrieved from http://www.abc.net.au/news/2014-11-07/billion-dollar-beef-deal-china/5873496

Food and Agriculture Organization of the United Nations (FAO). (2014). FAOSTAT. Retrieved from http://faostat.fao.org/site/610/DesktopDefault.aspx?PageID=610#ancor

Friel, S. (2010). Climate change, food insecurity and chronic disease: Sustainable and healthy policy opportunities for Australia. *New South Wales Public Health Bulletin*, *21*(6), 129–133. doi:10.1071/NB10019 PMID:20637169

Global Agriculture. (2015). Meat and animal feed. *Agriculture at a crossroads: findings and recommendations for future farming*. Retrieved from http://www.globalagriculture.org/report-topics/meat-and-animal-feed.html

Goodland, R., & Anhang, J. (2009, November/December). Livestock and climate change: What if the key actors in climate change are... cows, pigs and chickens? *World Watch Magazine*, *22*(6), 10–19.

Hamshere, P., Sheng, Y., Moir, B., Syed, F., & Gunning-Trant, C. (2014). *Analysis of China's food demand to 2050*. Paper presented at the 42nd Australian Bureau of Agriculture and Resource Economics and Science (ABARES) Outlook 2015 Conference, Canberra, Australia. Retrieved from http://data.daff.gov.au/data/warehouse/9aat/2014/WhatChinaWants/AnalysisChinaFoodDemandTo2050_v.1.0.0.pdf

Hewitson, G. (2012, February 14). Will consumer horror undo the meat industry? *Food Magazine*, Retrieved from http://www.foodmag.com.au/News/Will-consumer-horror-undo-the-meat-industry-

Liu, H., Parton, K. A., Zhou, Z., & Cox, R. (2012). Away-from-home meat consumption in China. *Asian Journal of Agriculture and Development*, 8(2), 1–15.

Locke, S. (2014a, November 17). China already swallows $2.3 billion worth of Australian meat before a Free Trade deal. *ABC News*. Retrieved from http://www.abc.net.au/news/2014-11-17/nrn-china-meat-consumption-prefta/5896162

Locke, S. (2014b, November 18). Beef and lamb exporters winners in the free trade agreement with China. *ABC News*. Retrieved from http://www.abc.net.au/news/2014-11-17/nrn-hold-beef-fta-china/5896494

Meat and Livestock Australia (MLA). (2014). China-Australia Free Trade Agreement announced. Retrieved from http://www.mla.com.au/News-and-resources/Industry-news/China-Australia-Free-Trade-Agreement-announced

Meat and Livestock Australia (MLA). (2015). Beef. Retrieved from http://www.mla.com.au/Prices-and-markets/Overseas-markets/South-Asia/Beef

Myers, N., & Kent, J. (2004). *New consumers: the influence of affluence on the environment*. Washington, DC: Island Press.

National Bureau of Statistics of China. (2014). *2014 China statistical yearbook*. Beijing, China: China Statistical Press.

National Centre for Groundwater Research and Training (NCGRT). (2013). Future mining and dining booms depend on water. Retrieved from http://www.groundwater.com.au/news_items/media-release-future-mining-dining-booms-depend-on-water

Neales, S. (2014, November 7). China deal on cattle exports. *The Australian*. Retrieved from http://www.theaustralian.com.au/national-affairs/foreign-affairs/china-deal-on-cattle-exports/story-fn-59nm2j-1227115289349

Organisation for Economic Co-operation and Development (OECD). (2014). Agricultural output. Retrieved from https://data.oecd.org/agroutput/meat-consumption.htm

Paddenburg, T. (2014, July 12). Good times for WA agriculture as farmers ride a food boom fueled by two billion middle-class Asians. *Sunday Times: Perth Now*. Retrieved from http://www.perthnow.com.au/news/western-australia/good-times-for-wa-agriculture-as-farmers-ride-a-food-boom-fuelled-by-two-billion-middleclass-asians/story-fnhocxo3-1226986783946

Quiggin, J. (2010). Draught, climate change and food prices in Australia: Working Paper. Brisbane, Australia: University of Queensland, School of Economics and School of Political Science and International Studies.

Raphaely, T., Marinova, D., Crisp, G., & Panayotov, J. (2013). Flexitarianism (flexible or part-time vegetarianism): A user-based dietary choice for improving personal, population and planetary wellbeing. *International Journal of User-Driven Healthcare, 3*(3), 34–58. doi:10.4018/ijudh.2013070104

Rezai, A., Foley, D. K., & Taylor, L. (2009). Global warming and economic externalities. Working Paper 2009-3. New York, NY: The New School for Social Research, Department of Economics, Schwartz Center for Economic Policy Analysis.

Song, P., Lu, M., Yin, Q., Wu, L., Zhang, D., Fu, B., & Zhao, Q. et al. (2014). Red meat consumption and stomach cancer risk: A meta-analysis. *Cancer Research and Clinical Oncology, 140*(6), 979–992. doi:10.1007/s00432-014-1637-z PMID:24682372

The Economist. (2014, June 14). Obesity. Chubby little emperors: why China is under- and over-nourished at the same time. *From the print edition: China.* Retrieved from http://www.economist.com/news/china/21604221-why-china-under-and-over-nourished-same-time-chubby-little-emperors

The Law Society of New South Wales Young Lawyers (LSNSWYL). (2014). Submission on the Agriculture Competitiveness Green Paper. Sydney, Australia: LSNSWYL, Animal Law Committee. Retrieved from https://www.lawsociety.com.au/cs/groups/public/documents/internetyounglawyers/920563.pdf

Tilman, D., Balzar, C., Hill, J., & Befort, B. L. (2011). Global food demand and the sustainable intensification of agriculture. [PNAS]. *Proceedings of the National Academy of Sciences of the United States of America of the Unites States of America, 108*(50), 20260–20264. doi:10.1073/pnas.1116437108 PMID:22106295

Wang, Z., Zhang, B., Wang, H., Zhang, J., Du, W., & Su, C. … Zhai, F. (2013). Multi-level and longitudinal association between red meat consumption and changes in BMI, body weight and risk of incident overweight among Chinese adults (In Chinese). Paper presented at the 11th National Nutritional Science Conference of Chinese Nutrition Society, Hangzhou, China. *Zhonghua Liu Xing Bing Xue Za Zhi, 34*(7):661-667. Retrieved from http://www.ncbi.nlm.nih.gov/pubmed/24257164

Winglee, M. (2014). China's meat, processed food consumption is rising: what does that mean for China? *Food Tank.* Retrieved from http://foodtank.com/news/2014/06/chinas-rising-appetite

World Wide Fund for Nature (WWF). (2014). Ecological footprint and sustainable consumption in China. Vaud/Beijing, Switzerland/China: WWF and China-ASEAN Environmental Cooperation Centre. Retrieved from http://www.footprintnetwork.org/images/article_uploads/China_EF_Sustainable_Consumption_2014_English.pdf

Xue, H., Liu, Y., Duan, R., Zhou, X., & Cheng, G. (2014). Trends in children's overweight and obesity in China and their influencing factors. *China School of Health, 35*(8), 1258–1262.

KEY TERMS AND DEFINITIONS

Affluent Population: Population with increased disposable income.

Externality: A cost that is transferred to an entity other than the one that generates it.

Food Animals: Animals raised to produce food for humans.

Free Trade Agreement: Reduction in trade barriers between two or more countries to help protect local markets and industries and facilitate increased exports.

Live Export: Food animals transported alive across national borders for commercial purposes.

Urbanization: Population movement away from rural and agricultural lifestyles through processes of industrialization and migration to cities, towns and conurbations.

White Paper: A government report about a complex issue presenting a particular position.

Section 3
Society and Politics

Chapter 12
Exploring Flexitarianism:
Meat Reduction in a Meat-Centred Food Culture

Hans Dagevos
Wageningen University, the Netherlands

ABSTRACT

Broad scholarly consensus exists nowadays that high meat consumption is particularly critical from an ecological perspective. Traditionally, technological progress and efficiency innovations in food supply processes are identified as key to solving food sustainability problems. However, it is increasingly recognised that technological innovation and efficiency gains alone are not enough to reduce the environmental impacts of growing meat production and consumption. Therefore, this chapter's point of view is consumption-oriented. Are consumers part of the solution by making transitions towards more sustainable consumption patterns in general and less meat-centric diets specifically? This chapter explores flexitarianism as a present-day food style that consists of different forms or levels, ranging from minor adjustments to regular meat consumption patterns to fundamental departure from habitual meat eating practices.

Plenty of people are attempting to change their daily lives in ways that do reduce their consumption. (Klein, 2014, p. 90)

[A] diet higher in plant-based foods (...) and lower in (...) animal-based foods promotes personal health and is associated with less environmental impact. (USDA & USDHHS, 2015, p. 7)

DOI: 10.4018/978-1-4666-9553-5.ch012

INTRODUCTION

As a result of several decades of scientific work, broad scholarly consensus exists nowadays that meat production and consumption account for a significant share of our ecological footprint (Smil, 2013; Steinfeld et al., 2006; Westhoek et al., 2011). One of the major causes of the "excessively high environmental cost of meat," to use Vaclav Smil's (2014, p. 67) words, is that the production of meat for human consumption requires a lot of plant material – and in turn, vast amounts of land, water and resources. In addition to arguments for the reduction of meat intakes centred around animal welfare or human health, the impact of meat production and consumption on environmental sustainability and climate change is addressed as a topic of attention and urgency. A wide scholarly consensus currently exists that reduction of meat in our diets is a significantly more sustainable option – both in terms of the environment and human health – than meat-rich diets (see van Dooren, Marinussen, Blonk, Aiking & Vellinga, 2014; Scarborough et al., 2014; Westhoek et al., 2014 for other recent studies). This state of the art is gradually making its way into the establishment: it has been put forward by the Health Council of the Netherlands (HCN) (2011), and, more recently, has reached the columns of *Nature* (Tilman & Clark, 2014) and is even subscribed to by the Scientific Advisory Committee of the 2015 U.S. Dietary Guidelines (USDA & USDHHS, 2015). Time will tell whether the latter report will become a comparable landmark study with respect to meat consumption in the way FAO's *Livestock's Long Shadow* (Steinfeld et al., 2006) previously was in relation to meat production.

Having said this, two conflicting tendencies demand attention. The first one is that low-meat or meatless diets are contrary to the worldwide dietary transition towards a higher intake of animal proteins (this includes, besides meat, dairy, eggs, and fish). This nutrition transition is a global phenomenon manifested since the 1950s. Between then and now meat consumption has doubled in many affluent and developing countries. Per capita meat consumption has increased significantly particularly over recent decades. The meat-rich diet is advancing not so much in high-income countries (which show frequently high but relatively constant levels of meat consumption since the 1990s) but especially in a number of Asian countries, such as China and Thailand. It is widely expected that other Asian countries such as Indonesia, Vietnam and the Philippines will follow this meat consumption pattern in the coming years, due to a growing number of people in the middle classes who can afford to purchase and eat meat for functional (nutritional), sensory (taste) and symbolic (status) reasons. Overall, the global trend of rising meat consumption is fuelled by the combination of increasing global population and growing prosperity in many developing countries (Henchion, McCarthy, Resconi & Troy, 2014; Smil, 2014). Put differently, the forecasted future of meat consumption is based on increasing meat production and growing meat demand. Flexitarianism, the topic of this chapter, opposes this future because it is about eating less (frequently) meat.

The second trend concerns the problem-solving capacity of technological progress and efficiency innovations in the food supply processes. Despite deeply-ingrained expectations of technology to solve food sustainability problems, the premise here is that technological improvements in productivity processes cannot keep pace and are not enough to reduce the environmental impacts of growing meat production and consumption. The market for meat is, as every market, an interplay of both supply and demand. If only because of this it is important to pay attention to consumption. We live in a consumer society in which consumption is vital to people's identities as well as to nations' economies. It is unjust and unwise however to present consumption simply and solely as the cause of sustainability problems without considering its potential role in solving these challenges. Fortunately, it is increasingly recognised that transitions towards more sustainable consumption patterns in general and less meat-centred food habits specifically are essential components of a more sustainable food system (Bailey, Froggatt & Wellesley, 2014; Dagevos & Voordouw, 2013; DEFRA, 2013; EEA, 2014; Raphaely & Marinova, 2014; Reisch, Eberle & Lorek, 2013). This chapter has as its starting point that developing a more sustainable food system will not be possible without a change towards more sustainable consumption patterns. Therefore, its focus is on how and to what extent contemporary food consumers reduce their meat intake.

Moderation of meat consumption is about flexitarianism insofar that it is interpreted as ways of eating which include eating meat occasionally. In contrast to the original (and Anglo-Saxon) interpretation of flexitarianism that takes vegetarianism as its starting point, the (European) definition of flexitarianism used here looks from a 'meat eating' perspective. It sees flexitarianism as abstaining from eating meat for one or more days of the week rather than as vegetarianism combined with eating meat occasionally (for definition questions, see Forestell, Spaeth & Kane, 2012; Raphaely & Marinova, 2014; Verain, Dagevos & Antonides, 2015). This chapter explores flexitarianism as a present-day food style, or rather food styles, because flexitarianism not only exists amid a meat-dominated food culture, but also in different forms or levels, ranging from minor adjustments to regular meat consumption patterns to fundamental departure from habitual meat eating practices and conventional meat-centric diets.

EXPLORING FLEXITARIANISM

Flexitarianism amid Carnivorism

While environmental and health benefits of moderate (as apposed to excessive) meat consumption become increasingly obvious and convincing, the actual moderation of meat consumption remains difficult. The latter is hardly surprising considering that meat is a key component of the prevalent food culture in most affluent societies. Meat dominates most restaurant menus with little difference between fast food restaurants and high-end restaurants. Most main meals at home also tend to be centred around meat. In retail, special offers on meat products often serve to attract customers: this includes bottom-price butchery products sold by supermarkets as well as places to eat with special offers on meat dishes, such as 'all you can eat', for items such as spare ribs, schnitzels or steaks. Such examples are part of a broader development in which meat is promoted and marketed in terms of special offers or price as well as in such intangibles as what is considered normal, acceptable and even desirable. Neither the physical environment nor the confirmatory bias of the socio-cultural environment are much in favour of changing our meat consumption habits and levels. In other words, consumers' choices are strongly influenced by a food environment that is as yet hardly enabling of the low-meat or meatless choices.

Taking a consumer-oriented perspective on meat eating practices should therefore always take into account the limited availability of an enabling food environment for restrained meat consumption. With consumer choices and habits being largely context-dependent or situationally driven, the importance of the physical and socio-cultural environment to food behaviour cannot be easily overstated. Food consumers' intentions and decisions whether or not to shift towards diets with a lower meat content must always be seen against the backdrop of the cultural dominance of the meat-based diet. This so-called ecological approach helps to identify the many obstacles which must be overcome to establish a food system in which low to moderate consumption of meat evolves into the personal or social norm (Vinnari & Vinnari, 2014).

This 'carnivorous' consumer cultural context corresponds with the afore-sketched global situation of more people eating more meat. Meat has now become a product for the masses. Although up to approximately the mid-20th century meat was still a luxury item ('meat as a treat') for many people of the Western world, a break with tradition gradually took place and the consumption of animal proteins overtook plant proteins. Meat intakes have risen continually since then to the current high levels. In many affluent countries the levels of meat consumption stabilized more or less at the amounts reached during the 1990s. Recently studies find evidence that meat consumption in some high-income countries is reaching its saturation point or even has passed 'peak meat' (Cole & McCoskey, 2013; Kayser, Nitzko & Spiller, 2013; Vranken, Avermaete, Petalios & Mathijs, 2014).

The market developments seem to match with flexitarianism. However, caution is very much called for. Claiming empirical proof for flexitarianism in dropping meat demand implies a zero-sum relationship between flexitarianism and the meat-centred food culture. The current situation yet is that flexitarianism peacefully co-exists with 'carnivorism' or meat-centred omnivorism (for those who take carnivorism literally as a diet consisting only of meat) rather than on fighting battles with the 'carnivorous' or meat-based world of food.

MEAT REDUCTION IN A MEAT-CENTRED FOOD CULTURE

Forms of Flexitarianism

Reducing meat consumption is a form of sustainable food consumption behaviour. Within the framework of sustainable consumptive behaviour, different variations in change are addressed ranging from gradual or incremental transition to fundamental or systemic transformation. Distinctions have been made between an (eco-)efficiency perspective and a sufficiency or demand restrained approach. Other dichotomies used in this respect are green growth and degrowth, weak sustainable consumption (wSC) and strong sustainable consumption (sSC) (Dagevos & Voordouw, 2013; Garnett, 2013; Lorek & Fuchs, 2013; Verain et al., 2015). In each of these conceptual pairs a distinction is made between change that leaves the status quo intact and change that is more fundamentally different from the vested situation.

These distinctions are helpful in exploring the new field of research on flexitarianism. Meat reduction can be approached in various ways. For example, 'weak' strategies are:

1. Reducing meat portion size;
2. Replacing part of the meat in meat-based products with plant-based alternatives (so-called hybrid meats);
3. Consumption of meat substitutes (e.g. vegetarian burgers);
4. Replacing meat by other (animal or plant based) protein rich foods;
5. Leaving out meat from the dish without a replacement; and
6. Possible strategy when the focus is on environmental and animal welfare concerns when alone - then meat could be replaced by more environmentally friendly (white meat instead of red meat) or animal-friendly meat products (e.g. organic or free-range meat)

'Weak' strategies signify that food consumers in essence continue to eat as they are used to but achieve meat reduction by eating meat-free or low-meat products (or meals, for that matter). Meat reduction is primarily product-driven, so to speak. 'Weak' strategies and accompanying food products are most helpful to many consumers in choosing a flexitarian food style.

More far-reaching strategies of meat reduction are predominantly behaviour-driven. 'Strong' strategies imply that consumers do not just follow the mainstream but actively address and shape alternative food styles which involve change and modification of the existing eating habits. In many ways these changes break with the reigning social norm and consumption practices related to meat. Often, there is a correlation with more holistic perspectives on food, such as Slow Food (a more generic movement that wishes to bring about fundamental changes in the current agricultural and food systems) or the adoption of a vegetarian or vegan diet. Although supporters of these alternative food styles may be modest in numbers and their direct impact on the conventional food system may be limited, they do encourage critical reflection and stimulate discussions about the food sector and food consumption, environmental pollution and energy use. Therefore, their transformative contribution may lie in the fact that they prepare the minds of more mainstream food consumers for a (drastic) decrease in their meat intake. In this case 'strong' sustainable lifestyles generally and 'strong' food styles particularly could affect forms of flexitarianism as well as help their acceptance, popularization or familiarization. Hence, both the weak and strong poles of the spectrum can have a positive effect on flexitarianism.

As variations on the weak-strong theme, several pathways towards moderating meat consumption have been identified. de Bakker and Dagevos (2012) termed 'sustainability by stealth' and 'moderate involvement' as weak routes while 'cultural change' is positioned as 'strong'. 'Sustainability by stealth' entails the acceptance of rather unnoticeable sustainable food innovations. An example is the incorporation of plant component in meat-based products to form more sustainable hybrid products. This rather technological route is particularly suitable for uninvolved or passive consumers. Moderately involved consumers take the second route in which small practical steps towards meat reduction are made through moderating meat portion size and/or incorporating meatless days. The third route, 'cultural change', includes lifestyle alternatives that are structurally different from the current consumption practices. It requires high involvement in food choices and consumer concerns.

These dietary transitions to less meat in the diet were applied or adapted in other research focusing on eating meat and flexitarianism, such as the pathways discerned by Schösler, de Boer and Boersema (2012). The path of cultural change is similar to the 'strong' strategies and deviates the most from the cultural norm and common eating habits. A second pathway entails an incremental change towards more health-conscious vegetarian meals by replacing meat with other regular products such as fish, eggs and cheese. As these meat alternatives are still animal-based, this route is not very encurging in terms of sustainability, but can be a first step in a shift towards more plant-based dietary patterns. Other potential routes presented by Schösler et al. (2012) include a lowered meat proportion in convenience products and smaller meat portions in dishes. The convenience-oriented path and the meat portion size strategy refer to meat consumption practices that belong respectively to 'sustainability by stealth,' and 'moderate involvement' (Verain et al., 2015).

Empirical research by Vanhonacker, van Loo, Gellynck and Verbeke (2013) is also structured along similar lines. They successively address hybrid meat products, meat types with lower environmental impact such as chicken, plant-based meat substitutes (ranging from vegetarian burgers to tofu), sustainable farmed fish, organic meat, proteins from insects and moderated meat consumption. A recent paper by de Boer, Schösler and Aiking (2014) likewise presents a number of similar strategies for stimulating consumption of other or fewer animal proteins.

In addition to the recent attention for changes in meat consumption behaviour and flexitarianism in the Netherlands and Belgium (de Backer & Hudders, 2014, 2015), this topic has also attracted attention in Northern European research (Latvala et al., 2012; Nordgren, 2012; Pohjolainen, Vinnari & Jokinen, 2015; Zur & Klöckner, 2014) and a South European study (Graca, Oliveira & Calheiros, 2015) as well as – at the other side of the globe – in newly-published Australian studies (Hayley, Zinkiewicz & Hardiman, 2015; Raphaely & Marinova, 2014) and research from New Zealand (Tucker, 2014). However modest, today's scholarly attention differs from previous decades when meat reducers were almost ignored in academic circles.

Empirical research from recent years yields an overall view that in various European countries consumers are somehow changing their meat consumption and restricting their intake. Substantial consumer groups of 'heavy' and 'medium' flexitarians who abstain from eating meat for multiple days per week now exist. It is becoming increasingly common to shift to low-meat or meatless days or to opt for vegetarian dishes – both at home and when eating out. In the Netherlands a group of chefs has taken the initiative of 'Dutch Cuisine' with one of its main principles being that dishes consist of vegetables or fruits for 50 to 80 percent and include meat and fish for a maximum of 50 to 20 percent. Flexitarian eating seems to be gradually settling down. This evidence gives reason to paraphrase Naomi Klein's words (which have been used as a motto of this chapter to the motto of this chapter) and subscribe to the opinion that indeed plenty of people are attempting to change their daily food choices in ways that do reduce their meat consumption.

On the other hand current empirical research also confirms the central place of meat on our plates and in our food patterns. Studies find large segments of consumers who eat meat in the majority of days during the week and have generally no intention to change their established meat consumption pattern. With a little goodwill these food consumers can be transformed into 'light' flexitarians.

Is Flexitarianism Enough?

The advantage of flexitarianism is that it allows for significant environmental and health benefits at a minimum dietary modification. A relatively small reduction in the consumption of animal proteins is likely to be more readily accepted by a greater number of consumers than would a call for a fully plant-based diet. Voluntary vegetariansm is very unlikely to become a food style popular with the broad public in the foreseeable future. Abstaining from eating meat for a few days a week however yields major environmental benefits and is feasible for large groups of consumers. Flexitarianism means that many will be able to take (small) steps toward increased sustainability of the food system without having to ban meat completely from their diets. Therefore, flexitarianism may be termed "rational meat eating," (Smil, 2013). As a 'reasonable' dietary option flexitarianism offers food consumers viable opportunities to participate in and contribute to the process of sustainability in the world of food through reduced meat intakes.

Such a dietary transition in which many consumers transform into moderate carnivores leads us to the ever-poignant question of whether the diet-environment-health trilemma is best served by large groups taking small steps or by small groups taking giant leaps. Is consumer contribution as light and medium flexitarians sufficient? Based on the urgency and scope of the food-related environmental sustainability and health problems, 'weak' flexitarianism may be defined as addressing only the symptoms, and as too little, too late. Such qualifications are based on the argument that real improvement in the world of food requires treating the very root of the problem, demanding a radical instead of a gradual approach. This philosophy would consider flexitarianism a necessary, but insufficient step.

The reality however is harsh. From a sustainability perspective, 'weak' flexitarian food styles may represent small steps, but within the scope of the dominant neat-based food culture, those steps can be considered substantial. The very presence of flexitarianism within a food consumer culture that is mainly devoted to encouraging, continuing and facilitating the eating of meat, should not be taken for granted and is extremely noteworthy. It is important to keep in mind that many structural conditions are not in favour of reduced meat consumption and given their impact on food preferences and consumer choices, it would be naive to determine the strength and contribution of flexitarianism solely or predominantly on the intrinsic motivation of consumers. Frustration about progress is easily born when all responsibility for change is put on individual consumers. It is furthermore problematic that politicians and policy makers as well as NGOs rarely explicitly encourage reduced meat consumption or take up this issue seriously (Bailey et al., 2014; Dagevos & Voordouw, 2013; Laestadius, Neff, Barry & Frattaroli, 2014) see, with respect to this topic, also (Morris, Kirwan & Lally, 2014).

Thus, in answering the question whether flexitarianism is enough it is reasonable to refer to the opportunities offered or hindrances raised by structural, institutional and cultural conditions. As long as meat takes a central role in our food culture where eating meat is the norm, is promoted in numerous ways, and, last but not least, is widely available at very low prices, it is realistic to maintain modest expectations for a general breakthrough of 'strong' flexitarian food styles. Therefore, flexitarians should not be blamed for the decision not to ban meat completely from their plate. Their contributions may be qualified as small in the light of the worldwide environmental and human health problems, but it is counterproductive to belittle these reformist dietary changes or to ignore the practical and fundamental efforts involved. On the other hand, however, it is also counterproductive to present medium flexitarianism as the highest achievable result because our meat-based preferences and existing dietary practices leave us no room for meat consumption choices beyond moderate intake.

The answer to the question whether flexitarianism is enough depends on one's point of view. From human health and environmental sustainability perspectives, flexitarianism may be regarded as inadequate and at best a transitional step in the right direction. From a consumer cultural perspective however flexitarianism should be considered neither simple nor self-evident, but a meaningful improvement on the current reality.

CONCLUSION

Consumer studies in recent years yield an overall view that in various countries contemporary food consumers are somehow changing their meat consumption and restricting their intake. Substantial consumer groups of 'heavy' and 'medium' flexitarians who abstain from eating meat for multiple days per week now exist. Such empirical results indicate that flexitarianism exists. It is becoming increasingly common for many modern food consumers to have low-meat or meatless days for several days a week as well as to opt for plant-centric dishes and diets. Such findings are promising regarding consumers and consumption as being not only causing sustainability problems but also being part of solutions to make the food system of the near future more sustainable from an ecological and health-related point of view.

REFERENCES

Bailey, R., Froggatt, A., & Wellesley, L. (2014). *Livestock - climate change's forgotten sector: global public opinion on meat and dairy consumption.* London, UK: Chatham House, The Royal Institute of International Affairs. Retrieved September 7, 2015, from http://www.chathamhouse.org/sites/files/chathamhouse/field/field_document/20141203LivestockClimateChangeBaileyFroggattWellesley.pdf

Cole, J. R., & McCoskey, S. (2013). Does global meat consumption follow an environmental kuznets curve? *Sustainability: Science, Practice, &. Policy*, *9*(2), 26–36.

Dagevos, H., & Voordouw, J. (2013). Sustainability and meat consumption: is reduction realistic. *Sustainability: Science, Practice, &. Policy*, *9*(2), 60–69.

de Backer, C. J. S., & Hudders, L. (2014). From Meatless Mondays to Meatless Sundays: Motivations for meat reduction among vegetarians and semi-vegetarians who mildly or significantly reduce their meat intake. *Ecology of Food and Nutrition*, *53*(6), 639–657.

de Backer, C. J. S., & Hudders, L. (2015). Meat morals: Relationship between meat consumption consumer attitudes towards human and animal welfare and moral behaviour. *Meat Science*, *99*, 68–74.

de Bakker, E., & Dagevos, H. (2012). Reducing meat consumption in today's consumer society: Questioning the citizen-consumer gap. *Journal of Agricultural & Environmental Ethics*, *25*(6), 877–894.

de Boer, J., Schösler, H., & Aiking, H. (2014). 'Meatless days' or 'less but better?': Exploring strategies to adapt Western meat consumption to health and sustainability challenges. *Appetite*, *76*, 120–128.

Department for Environment. Food & Rural Affairs (DEFRA). (2013). Sustainable consumption report: follow up to the green food project. London, UK: DEFRA. Retrieved September 7, 2015, from https://www.gov.uk/government/publications/sustainable-consumption-report-follow-up-to-the-green-food-project

European Environment Agency (EEA). (2014). *Environmental indicator report 2014: environmental impacts of production-consumption systems in Europe.* Luxembourg: Publications Office of the European Union. Retrieved September 7, 2015, from http://www.eea.europa.eu/publications/environmental-indicator-report-2014

Forestell, C. A., Spaeth, A. M., & Kane, S. A. (2012). To eat or not to eat red meat: A closer look at the relationship between restrained eating and vegetarianism in college females. *Appetite*, *58*(1), 319–325.

Garnett, T. (2013). Food sustainability: Problems, perspectives and solutions. *The Proceedings of the Nutrition Society*, *72*(1), 29–39.

Graca, J., Oliveira, A., & Calheiros, M. M. (2015). Meat beyond the plate: Data-driven hypotheses for understanding consumer willingness to adopt a more plant-based diet. *Appetite*, *90*, 80–90.

Hayley, A., Zinkiewicz, L., & Hardiman, K. (2015). Values, attitudes and frequency of meat consumption: Predicting meat-reduced diets in Australia. *Appetite, 84*, 98–106.

Health Council of the Netherlands (HCN). (2011). Guidelines for a healthy diet: the ecological perspective. The Hague, the Netherlands: HCN. Retrieved September 7, 2015, from http://www.gezondheidsraad. nl/sites/default/files/201108E.pdf

Henchion, M., McCarthy, M., Resconi, V. C., & Troy, D. (2014). Meat consumption: Trends and quality matters. *Meat Science, 98*, 561–568.

Kayser, M., Nitzko, S., & Spiller, A. (2013). Analysis of differences in meat consumption patterns. *International Food and Agribusiness Management Review, 16*, 43–56.

Klein, N. (2014). *This changes everything: capatilism vs the climate*. New York, NY: Simon and Schuster.

Laestadius, L. I., Neff, R. A., Barry, C. L., & Frattaroli, S. (2014). "We don't tell people what to do": An examination of the factors influencing NGO decisions to campaign for reduced meat consumption in light of climate change. *Global Environmental Change, 29*, 32–40.

Latvala, T., Niva, M., Mäkelä, J., Pouta, E., Heikkilä, J., Kotro, J., & Forsman-Hugg, S. (2012). Diversifying meat consumption patterns: Consumers' self-reported past behaviour and intentions for change. *Meat Science, 92*(1), 71–77.

Lorek, S., & Fuchs, D. (2013). Strong sustainable consumption governance - precondition for a degrowth path? *Journal of Cleaner Production, 38*, 36–43.

Morris, C., Kirwan, J., & Lally, R. (2014). Less meat initiatives: An initial exploration of a diet-focused social inovation in transitions to a more sustainable regime of meat provision. *International Journal of Sociology of Agriculture and Food, 21*(2), 189–208.

Nordgren, A. (2012). Ethical issues in mitigation of climate change: The option of reduced meat production and consumption. *Journal of Agricultural & Environmental Ethics, 25*, 563–584.

Pohjolainen, P., Vinnari, M., & Jokinen, P. (2015). Consumers' perceived barriers to following a plant-based diet. *British Food Journal, 117*(3), 1150–1167.

Raphaely, T., & Marinova, D. (2014). Flexaririanism: Decarbonising through flexible vegetarianism. *Renewable Energy, 67*, 90–96.

Reisch, L., Eberle, U., & Lorek, S. (2013). Sustainable food consumption: an overview of contemporary issues and policies. *Sustainability: Science, Practice, &. Policy, 9*(2), 7–25.

Scarborough, P., Appleby, P. N., Mizdrak, A., Briggs, A. D. M., Travis, R. C., Bradbury, K. E., & Key, T. J. (2014). Dietary greenhouse gas emissions of meat-eaters, fish-eaters, vegetarians and vegans in the UK. *Climatic Change, 125*(2), 179–192.

Schösler, H., de Boer, J., & Boersema, J. J. (2012). Can we cut out the meat of the dish? Constructing consumer-orientated pathways towards meat substitution. *Appetite, 58*, 39–47.

Smil, V. (2013). *Should we eat meat? Evolution and consequence of modern carnivory*. Chichester, UK: Wiley-Blackwell.

Smil, V. (2014). Eating meat: Constants and changes. *Global Food Security, 3*(2), 67–71.

Steinfeld, H., Gerber, P., Wassenaar, T., Castel, V., Rosales, M., & de Hann, C. (2006). Livestock's long shadow: environmental issues and options. Rome, Italy: Food and Agriculture Organisation (FAO).

Tilman, D., & Clark, M. (2014). Global diets link environmental sustainability and human health. *Nature, 515*, 518–522.

Tucker, C. A. (2014). The significance of sensory appeal for reduced meat consumption. *Appetite, 81*, 168–179.

United States Department of Agriculture (USDA) and United States Department of Health and Human Services. (USDHHS). (2015). Scientific report of the 2015 Dietary Guidelines Advisory Committee. Retrieved September 7, 2015, from http://health.gov/dietaryguidelines/2015-scientific-report/pdfs/scientific-report-of-the-2015-dietary-guidelines-advisory-committee.pdf

van Dooren, C., Marinussen, M., Blonk, H., Aiking, H., & Vellinga, P. (2014). Exploring dietary guidelines based on ecological and nutritional values: A comparison of six dietary patterns. *Food Policy, 44*, 36–46.

Vanhonacker, F., van Loo, E. J., Gellynck, X., & Verbeke, W. (2013). Flemish consumer attitudes towards more sustainable food choices. *Appetite, 62*, 7–16.

Verain, M., Dagevos, H., & Antonides, G. (2015). Flexitarianism: a range of sustainable food styles. In L. A. Reisch & J. Thøgersen (Eds.), *Handbook of research on sustainable consumption* (pp. 209–223). Cheltenham, UK: Edward Elgar Publishing.

Vinnari, M., & Vinnari, E. (2014). A framework for sustainable transition: The case of plant-based diets. *Journal of Agricultural & Environmental Ethics, 27*, 369–396.

Vranken, L., Avermaete, T., Petalios, D., & Mathijs, E. (2014). Curbing global meat consumption: Emerging evidence of a second nutrition transition. *Environmental Science & Policy, 39*, 95–106.

Westhoek, H., Lesschen, J. P., Rood, T., Wagner, S., De Marco, A., Murphy-Bokern, D., & Oenema, O. et al. (2014). Food choices, health and environment: Effects of cutting Europe's meat and dairy intake. *Global Environmental Change, 26*, 196–205.

Westhoek, H., Rood, T., van den Berg, M., Janse, J., Nijdam, D., Reudink, M., & Stehfest, E. (2011). The protein puzzle: the consumption and production of meat, dairy and fish in the European Union. The Hague, the Netherlands: Netherlands Environmental Assessment Agency (NEAA). Retrieved September 7, 2015, from http://www.pbl.nl/en/publications/2011/meat-dairy-and-fish-options-for-changes-in-production-and-consumption

Zur, I., & Klöckner, C. A. (2014). Individual motivations for limiting meat consumption. *British Food Journal, 116*(4), 629–642.

KEY TERMS AND DEFINITIONS

Dietary Change: A transition in food consumption pattern that evolves slowly or quickly and lasts.

Enabling Environment: Conditions in the physical ('hard ware') or socio-cultural ('soft ware') context that implicitly or explicitly promote certain consumption habits and consumer choices and, as a consequence, hinder others.

Environmental Consequences: Impacts of economic activities or consumer choices on water use, natural resources or land erosion.

Flexitarianism: A term coined in the beginning of the 20th century that originally referred to vegetarians who occasionally eat meat, but is more recently also interpreted from a 'meat eating' perspective, i.e., food consumers who abstain from eating meat for one or more days of the week.

Food Style: A combination of life style and food with a focus on salient characteristics of food consumption habits of individuals or groups.

Strong Sustainable Consumption (sSC): Emphasizes sufficiency principles and issues of affluence – with de-growth as an ideological background – addressing more explicitly changes in conventional consumption patterns and reductions in volumes particularly as well as limits to economic growth generally.

Weak Sustainable Consumption (wSC): Concentrates on efficiency improvements and the supply side – with technological optimism as an ideological background – as well as the quality of consumer products ('improving consumption efficiency') as major drivers of sustainable development without addressing economic growth and/or overconsumption in absolute terms.

Chapter 13

Feasting From the Federal Trough:
How the Meat, Egg and Dairy Industries Gorge on Taxpayer Dollars While Fighting Modest Rules

Paul Shapiro
The Humane Society of the United States, USA

ABSTRACT

The animal agribusiness industries often proclaim a libertarian mantra when asked to accept rules for their conduct in regard to animal welfare, the environment, and food safety. However, in this chapter, the author explores how when these industries suffer from lack of demand, their clamor toward socialism is stark. They consistently come to the US Congress and the United States Department of Agriculture with outstretched arms and cupped palms, seeking to defy the normal laws of economics that other businesses must navigate. In fact, the meat, egg, and dairy industries are enormous beneficiaries of generous federal subsidies, research and development, and even surplus buy-ups of unwanted product. Such a reliance on federal handouts by animal agribusiness calls into question their proclamation of libertarianism and free market principles.

INTRODUCTION

Picture this: It's 2008. The financial crisis is at its peak. The plight of U.S. automakers can hardly be grimmer. Facing double-digit sales declines as fuel costs soar and Americans turn from (formerly popular) US-made SUVs and trucks to smaller, more efficient cars from Japan and Europe, the federal government has to make a choice: Throw the Big Three automakers—Ford, General Motors, and Chrysler—to the harsh fate of market forces and allow them to go bankrupt at a time when the economy is in free fall, or step in and provide aid from the hand of Uncle Sam to keep Detroit rolling.

DOI: 10.4018/978-1-4666-9553-5.ch013

We all know the outcome. When the auto industry got its federal bailout, serious strings were attached, such as improving fuel efficiency, eliminating corporate jet budgets, and more. Automakers of course had to repay the US Treasury. The move, in the end, made American taxpayers a handsome profit and helped keep the nation's auto industry on the road.

Contrast this approach with the one the US federal government takes when the industries allegedly in need aren't producing cars, but rather cows—or pigs, chickens, or turkeys. In these cases, the government pork seemingly flows freely and faster than a Corvette, leaving taxpayers to foot the bill for corporate welfare and aiding industries with a notoriously sordid track record on important social issues.

FEASTING FROM THE FEDERAL TROUGH

The Subsidy-Stuffed-Pizza

Now, envision a pizza brimming with sausage and pepperoni—a bed of extra cheese, melted over a hot doughy base beneath. What lies beneath that crust, helping deliver the pie to your doorstep?

Take the pepperoni, for example. Imagine that you're producing pigs, parts of whom are destined to be atop Domino's double-stuffed pizzas nationwide. What if you produce more than the market demands? Let's say more consumers are switching from meat to mushrooms as toppings, leaving you with more pork than the people want.

Most businesses operate under established economic laws. They produce enough to supply what demand there may be, and if they want to increase demand, they drum up business through the normal routes, such as advertising. If they produce more than what the market demands, they shrink to better accommodate the circumstance. This is not the case though, if you happen to be in the business of producing pigs. In that case, the basic rules of our economic system need not apply.

In fact, the federal government spends millions to bail out its friends in pork production – an industry notoriously resistant to even basic standards of ethical business conduct – when they produce more pigs than consumers want to eat. For example, when the industry took a hit from all the swine flu media attention which depressed demand for pork, Time Magazine reported on the $30 million federal buy-up of surplus pork, noting: "If you're in prison, now might be a good time to develop a taste for pork" (Kluger, 2009, para. 1).

The swine flu surplus buy-up was unfortunately not an isolated incident. In fact, when the country started experiencing drought, the US Department of Agriculture (USDA) stepped in to buy an additional $100 million of unwanted pork in order to boost demand when supply was too high (IPPA, 2012). Were public school kids, service-members, prisoners, and other recipients of federally purchased food demanding more pig on their pizzas? No. The US Agriculture Secretary Tom Vilsack was clear about the motivation: "President Obama and I will continue to take swift action to get help to America's farmers and ranchers," (IPPA, 2012, para. 4). No doubt many other industries wouldn't mind the government paying them for products they're having a tough time selling.

It's not just the pepperoni on the pizza getting a helping hand from Uncle Sam. Imagine you want to forgo the federal pork by switching from Domino's pepperoni pie to its Buffalo Chicken option. Sadly, you won't be much better off.

For example, in 2013, despite lagging demand for chicken products, the industry continued to overproduce. Rather than adjust to the marketplace, it simply asked its federal friends for a handout. After the National Chicken Council, an industry lobbying group, lamented to the USDA with a letter stating "over-supply of dark meat is a burden to producers/processors," (MPS, 2013, para. 3) and pleading for a buy-up, the federal government agreed to buy 139 million pounds of otherwise-unwanted chicken products (Shapiro, 2013). In their announcement, USDA was very clear why they'd do this: "for surplus removal" (AMS, 2013, para. 1). In 2014 exactly the same explanation was given for another purchase of chicken products (AMS, 2014). In other words, since the poultry industry produced more chickens than it could sell—just as its pork industry pals did with pigs—the government stepped in and provided a cushy taxpayer bailout.

So now you think you may just want to skip the toppings and go for a simple, welfare-free cheese pizza. Unsurprisingly, you're not in luck.

The dairy producers supplying that bed of cheese benefit tremendously from federally supervised programs that spend millions of dollars to "get people to eat more pizza" (Plumber, 2014). That's right. As the First Lady encourages us to eat healthier and get active (Grier, 2010), the administration is at the same time signing off on a dairy check-off program (Plumber, 2014) that levies an amount on milk sold raising millions to be spent in urging Americans to pick up another slice of Three Cheese Stuffed Crust Pizza Extreme. Do national pizza chains really need federal assistance to advertise? After all, as author David Simon points out in his book *Meatonomics*, the dairy industry spends more on advertising in one week than the blueberry, mango, watermelon, and mushroom industries spend together in a year (Simon, 2013).

Also, it's not just pizza. As US Senator Tom Coburn points out in an entertaining section of his annual report on wasteful government spending, "We All Scream for Federally Funded Ice Cream," (Coburn, 2014, p. 72). The USDA gave out more than a million dollars to private dairy companies to help them market new flavors of herbal ice creams, Greek yogurts, and rabbinically-supervised milks (Coburn, 2014).

To be clear, unlike with the auto-bailouts, these agri-bailouts aren't loans; they're gifts. There is no need to repay the government—ever. Also unlike the Detroit deal, there are no strings attached, no required improvements in these industries' sordid animal welfare records, no reduction in environmental harm, no new food safety rules – just a completely free lunch. Pizza, most likely.

Subsidizing Soy (Not Tofu) and Corn

When most Americans think soy, they may conjure images of chic Japanese restaurants offering steamy bowls of edamame, or perhaps they think about the braised tofu "Sofritas" burrito they had at Chipotle, or maybe a soy cappuccino at Starbucks. However most soy that Americans consume doesn't come from such health-halo-inducing items. Rather, the overwhelming majority of soy grown in the US and around the world is fed to farm animals.

In fact, feeding animals is the number one cost of raising them for food (Haris, 2013). If animal producers could reduce feed costs, they could substantially reduce operating costs. Indeed, certain technologies have reduced the cost of producing farm animal feed, but it's not just laboratory science agribusinesses are relying on; it's politics too. Instead of simply being content with innovation that leads to cheaper food inputs, the American meat, egg, and dairy industries lobby lawmakers with a voracious appetite for feed crop subsidies. As a consequence, this political acumen has led to a perverse federal subsidy system that doles out billions to major agribusiness nationwide.

When agricultural subsidies began in the 1930s, the then-Secretary of Agriculture Henry Wallace called them "a temporary solution to deal with an emergency" (WS, 2013). That emergency was the

Great Depression and the Dust Bowl, which really did threaten the American food production system. Yet that temporary solution has now lasted nearly a century, in both good and bad times for farmers, and even today when the average farm household incomes greatly exceed the national average (WS, 2013).

The other big ingredient fed to farm animals – corn, is perhaps the king of the subsidy game, yielding billions in federal handouts. To be clear, only a tiny percentage of corn grown in the US ends up on grills at backyard barbecues. Just like with soy, a large portion of corn we grow ends up in the stomachs of the chickens, turkeys, pigs, and cattle who wind up on those barbecue grills instead (Foley, 2013).

In other words, the meat industry receives billions of dollars in indirect agricultural subsidies that artificially reduce the cost of the most expensive part of their business: the corn and soy grown to feed billions of animals. Tufts University estimates that the chicken industry alone saved $1.25 billion in feed costs from 1997 to 2005 just from taxpayer-funded subsidies (GDAE, 2013). In short, while meat companies themselves don't receive the subsidy directly, they benefit enormously from federally subsidized feed in what amount to massive indirect subsidy.

Devouring Federal Funds, Opposing Common Sense Reform

Ironically, despite their reliance on big government handouts, animal producers desperately fight any proposed rules governing their conduct, especially when they relate to animal welfare. Take the pork industry, for example. Sadly, it's an example that reminds us that not everyone views the Dark Ages in the past tense.

Most American breeding pigs are confined day and night during their four-month pregnancies in gestation crates on factory farms. These cages are roughly the same size as the animals' bodies and designed to prevent them from even turning around. The pigs are subsequently transferred into another crate to give birth, and are then re-impregnated and put back into a gestation crate. This happens pregnancy after pregnancy for their entire lives, adding up to years of immobilization. Many become so worn out that they're slaughtered after only a couple years.

Nine states have passed laws to ban gestation crates, and nearly all of the major food retailers in the country recently have announced their plans to eliminate it from their supply chains. Animal science experts like Temple Grandin, condemn the practice, arguing that "confining an animal for most of its life in a box in which it is not able to turn around does not provide a decent life" (Grandin, 2010, p. 15). Grandin further states: "We've got to treat animals right, and the gestation stalls have got to go" (FMI, n.d., p. 1). It's difficult to imagine a more miserable existence for an animal. Yet when asked about this standard but abusive practice, the National Pork Producers Council (NPPC) defended the severe, virtual lifetime immobilization of breeding pigs by scoffing: "So our animals can't turn around for the 2.5 years that they are in the stalls producing piglets. I don't know who asked the sow if she wanted to turn around" (Friedrich, 2012, para. 1).

So even though its own experts encourage the pork industry to abandon gestation crates, the NPPC still defends what most Americans know is indefensible. That defense goes so far as to involve aggressive lobbying against all state and federal rules that would reduce the suffering of pigs. Every state in which legislation is introduced to curb the use of gestation crates, for example, the NPPC is in the statehouse pressing its case that government should stay out of farmers' business, unless it involves giving them billions in handouts, of course (Colvin, 2014, para. 11).

In fact, the NPPC takes its disregard for animal welfare rules even further than protesting pig protection efforts. It's not only working furiously to prevent improvements for sows on factory farms, it actively works to crush efforts to protect any farm animals. For example, NPPC waged a successful campaign to squelch a federal bill to improve the treatment of egg-laying hens despite the fact that both egg industry leaders and animal protection groups backed the bill (Wheat, 2015). The NPPC argued that if there were federal rules protecting chickens, pigs would be next. In other words, the NPPC wants no rules protecting pigs from abuse, and it's against even modest anti-cruelty regulations in other agribusiness sectors in which it holds no stake, even when those farmers who would be affected by the new rules support them.

The Only Regulation the Meat Industry Loves: Ag-Gag

The meat industry may not be big on rules requiring basic decency in its treatment of farm animals, but there are some regulations it does actually favor. In recent years, whistleblowing exposés by groups like The Humane Society of the United States, Mercy For Animals, and Compassion Over Killing have repeatedly documented inhumane treatment of animals, unsafe working conditions and food safety problems inside of US factory farms and slaughter plants. These investigations are helping to shine a bright light on notorious agribusiness practices such as confining animals in tiny cages where they can barely move an inch their entire lives. The videotaped evidence often has led to meat recalls (including the largest in US history), slaughter plant shutdowns, criminal convictions, Congressional hearings and even new federal policies.

Yet the meat industry's response to these exposés hasn't been to try to prevent these abuses from occurring. Rather, it's simply been to try to prevent the public from finding out about them. As a result of the industry's lobbying efforts, it's now a crime in Utah to photographically document someone abusing an animal in a slaughter plant. In Iowa, if an agribusiness employer asks applicants if they're a member of an animal welfare charity and they say no, but actually are—that's not just grounds for firing; it's a jailable offense (Gerke, 2012). Another iteration of these whistleblower suppression "ag-gag" bills is to require that anyone documenting inhumane treatment of farm animals "out" themselves nearly immediately and turn over all their evidence before any pattern of abuse can possibly be established—or face criminal action.

Countless newspapers and even some industry commentators have panned this effort. For example, the *Washington Post* condemned ag-gag bills, noting, "As you next cut into a steak or crack an egg, ask yourself why an industry that claims it has nothing to hide demands protections afforded to no other" (Editorial Board, 2013, para. 6)

You know that an industry has a lot to hide when it wants to make it a crime to document what it's doing. One thing the meat industry seems very happy to reveal however is that its disregard for animal welfare is just about as strong as its disregard for freedom of speech and whistleblower protection.

Not Just Ag-Gag: Also the Name Game

As is fairly evident, industrialized animal agribusiness is facing a crisis, especially since meat consumption in the US is on the decline (Bittman, 2012). Dependent on federal handouts but opposed to any government say about how it conducts its business, the industry relies on lobbying efforts to kill any and all rules for farm animal treatment. However that leaves producers with a vulnerability: with no rules of the road, standard practices have become so extreme and inhumane that they're just out of step with mainstream values about how animals ought to be treated. The industry's leaders, having failed to accept that factory farming practices are actually at the root of the problem, are frantically looking in every direction for quick fixes. Stunned by repeated whistleblowing exposés revealing inhumane treatment of animals throughout the meat, egg and dairy industries, as noted, ag lobbyists are working to enact ag-gag bills. They may not be literally shooting the messenger, but they do want to imprison her.

Similarly, a revealing column by pork industry veteran Linden Olson unashamedly advises fellow producers to simply change the way they talk about their most abusive practices, rather than changing the practices themselves (Olson, 2013). For example, rather than ceasing to immobilize mother pigs in tiny cages, Olson's recommendation is simply to stop calling the cages "gestation crates"—which they've been called for decades—and start referring to them as "individual maternity pens". Here again, we see agribusiness fail to understand the root of its crisis, for it's not what the crates are called that's the problem, but rather the extreme degree to which they confine animals.

According to Olson (2013), it's not just extreme confinement practices that need an extreme makeover: The industry "harvests" animals rather than slaughters them; animals are "neutered" rather than castrated without pain relief; and confinement barns in which animals never set foot on a blade of grass are really just "environmentally controlled housing,". Olson will likely find a sympathetic ear at the US Farmers and Ranchers Alliance. In a survey it commissioned, pollsters found that the ag industry's go-to messages just aren't resonating with consumers (Muirhead, 2011). Rather than discussing what factory farming actually does to animals, the pollster simply "advised not going into a lot of detail about current practices when talking to consumers as it may generate more concern than necessary," (Muirhead, 2011, para. 15).

How much concern is "necessary" though, when billions of chickens, pigs, turkeys and other animals are raised in systems so inhumane, merely discussing it makes people uncomfortable enough that Orwellian euphemisms like "individual maternity pens" and "harvest" are being pushed as so-called solutions? How much concern is "necessary" when standard meat industry practices are so abusive that your industry wants to make it a crime simply to photograph them?

Meat producers are trying to do everything they can to keep Americans in the dark about how they abuse animals. Whether through ag-gag laws to prevent videos of animal abuse from surfacing or through playing the name game, this is an industry that knows it has a lot to hide. After all, "one of the best things modern animal agriculture has going for it is that most people...haven't a clue how animals are raised and processed," wrote an editor of the Journal of Animal Science in an animal agriculture textbook (Cheeke, 1999 in Pond, Bazer & Rollin, 2011, p. 112). Cheeke aptly concluded: "For modern animal agriculture, the less the consumer knows about what's happening before the meat hits the plate, the better" (Cheeke, 1999 in Pond, Bazer & Rollin, 2011, p. 112).

Pigs in Tinfoil Diapers: R&D Courtesy of the Taxpayers

We've seen that meat and dairy producers benefit tremendously from federal handouts aimed at protecting them from the normal winds of market forces. There's also another form of corporate welfare they enjoy. While most businesses have to factor in line items for their own research and development budgets, that tab is conveniently often picked up by the US taxpayers when it comes to the meat industry. Looking inside this hidden world of USDA, meat industry research offers a glimpse behind a curtain into one of the most bizarre worlds imaginable. In this case, reality truly is odder than fiction.

With USDA-funded research titles such as *The Effect of Scrotal Insulation on Testicular Gene Expression* (NIFA, 2009), it's not hard to see just how free the federal government feels to use taxpayer dollars for even the most peculiar purposes. Below are more examples.

Many sows in the pork industry—locked inside their crates—are artificially inseminated by hand. There are workers who help facilitate boars' ejaculation, and workers who deliver that fresh semen to the sows. Regrettably, from the pork producers' perspective, there can be little reliance on cryogenically preserved boar semen, as it simply doesn't remain potent. Animal factory operators must get the goods in the right place in a very small window of time. Instead of allowing pork companies to attempt to figure their problem out on their own, USDA doles out cash to researchers who want to lend these pig producers a hand.

One such researcher at the University of Wisconsin, fueled by federal grants, devised a method of putting tinfoil diapers on pigs' scrotums and then collecting their warmed semen for testing. While USDA has funded at least three large experiments in which researchers warm the scrotums of boars with foil diapers for semen collection studies, they've yet to crack the code on how to move away from reliance on fresh semen use (NIFA, 2009, 2010, 2013).

In a similar vein, at Mississippi State University, the USDA has funded research in which boars are subjected to heat stress, and then various data collections such as "scrotal surface gradient temperature" are measured before, during, and after the boar is stimulated to ejaculate (NIFA, 2014a). Then the collection specimen is tested and frozen, after which they rinse, lather, and repeat (NIFA, 2014a).

Also it's not solely pigs whose semen USDA is particularly fascinated by. The list of rooster semen experiments is long as well, with taxpayers handing their dollars over to researchers trying desperately to help poultry producers preserve their birds' genetic material. In just one 2014-2015 example, the government gave US$410,067 to researchers trying to extend the shelf life of rooster semen in order to boost poultry profits (NIFA, 2014b).

In related USDA-funded research on extending the viability of turkey semen, the experimenters were uncharacteristically open about the goal of their research: "Current methodology for storing turkey semen does not provide the fertility rates necessary for profitability," (NIFA, 2006, p. 1; Pelaz & Long, 2008, p. 1). In other words, rather than allowing turkey producers to conduct their own R&D for profitability purposes, the primary goal of the research was to grow profits for the poultry industry.

Not all federally-funded ag research deals in the semen world, however. A side effect of intensive genetic selection for rapid growth in broiler chickens is that an unusually high rate of birds suffer from tibial dyschondroplasia (TD), an illness in the growth plate of the tibia that can lead to painful lameness in the bird. In addition to such lameness, TD can cause necrosis, spontaneous fracture, and in some cases the inability even to stand. Some birds with severe TD can move only by crawling on their hocks. This condition is a major cause of mortality in broiler factories, leading to decreased industry revenue. As a result, the USDA stepped in to fund research on the problem. In it, researchers used federal dollars to feed pesticides to baby chicks to induce TD. Despite the obvious animal suffering caused by the experimentation, the end results were lackluster and didn't provide any options for prevention or treatment to the birds (NIFA, 2004).

Alas, due to lobbying efforts of the meat industry, the federal Animal Welfare Act—designed in part to protect animals in labs—exempts all agricultural research. In other words, chickens and pigs used in experimentation labs have no federal protection if the research is ag-related. (Chickens wouldn't have protection either way since the Act exempts all birds, along with rats and mice.)

Such research raises serious questions about what involvement the federal government should have in funding research that aids the production of animals for food, and whether it should instead focus such dollars on aiding industries which could better improve human health, such as the fruit and vegetable industries.

CONCLUSION: LIBERTARIANS BY CONVENIENCE, SOCIALISTS BY HABIT

The meat, egg, and dairy industries occupy an astoundingly coveted position among American businesses: They get bailouts when they overproduce, have their most costly business expense subsidized, get free money to market their products, and even get generous gifts for research and development for which they don't pay. At the same time as their hand remains firmly in the federal cookie jar, they oppose any regulations that would result in them being more socially responsible, while supporting regulations that would criminalize whistleblowers in their midst.

The farm animal industries may like to proclaim a libertarian mantra when asked to curb their abuses, but when they suffer from lack of demand, their clamor toward socialism is stark. They consistently come to Congress and USDA with outstretched arms and cupped palms, seeking to defy the normal laws of economics that other businesses must navigate.

As troubling as this is, the fact that these industries contribute so massively to many social ills—institutionalized animal cruelty, environmental destruction, and public health crises, to name a few—the fact that there are no strings attached to their corporate welfare receipt is truly perplexing. That these industries get such regal treatment from the US government is a testament to the grip they hold over many lawmakers and regulators. In fact, while USDA is charged with regulating the nation's meat, egg, and dairy producers, more often it acts not as a regulatory body, but as a trade association.

At a time when deficit reduction and federal spending loom so large in America's political life, this kind of agricultural welfare should be the first to go on the butcher block.

ACKNOWLEDGMENT

Heartfelt gratitude to Lucinda Wheeler and Thomas Goodman for their research aid.

REFERENCES

Agricultural Marketing Service (AMS). (2013). USDA announces purchase program for chicken products. Retrieved from http://www.ams.usda.gov/AMSv1.0/getfile?dDocName=STELPRDC5104998

Agricultural Marketing Service (AMS). (2014). USDA chicken products purchase program announced. Retrieved from http://www.ams.usda.gov/AMSv1.0/getfile?dDocName=STELPRDC5108873

Bittman, M. (2012, January 10). We're eating less meat. Why? *New York Times. Opinianator.* Retrieved from http://opinionator.blogs.nytimes.com/2012/01/10/were-eating-less-meat-why/

Cheeke, P. R. (1999). *Contemporary issues in animal agriculture* (2nd ed.). Danville, IL: Interstate Publishers.

Coburn, T. (2014). Wastebook 2014: what Washington doesn't want you to read. Washington, DC: United States Senate. Retrieved from http://fulltextreports.com/2014/10/24/wastebook-2014-what-washington-doesnt-want-you-to-read/

Colvin, J. (2014). Christie vetoes politically charged pig crate bill. Retrieved from http://news.yahoo.com/christies-decision-pig-crates-down-wire-191903239--election.html

Editorial Board. (2013, April 26). Cruelty to farm animals demands exposure. *The Washington Post*. Retrieved from http://www.washingtonpost.com/opinions/cruelty-to-farm-animals-demands-exposure/2013/04/26/9a972c8e-a6bf-11e2-a8e2-5b98cb59187f_story.html

Foley, J. (2013). It's time to rethink America's corn system. *Scientific American*. Retrieved from http://www.scientificamerican.com/article/time-to-rethink-corn/

Food Marketing Initiative (FMI). (n.d.). Gestation crates: problems and solutions. Retrieved from https://www.fmi.org/docs/animal-welfare/hsus-gestation-crate-factsheet.pdf?sfvrsn=3

Friedrich, B. (2012). National Pork Producers Council: anti-science and anti-animal. Retrieved from http://www.commondreams.org/views/2012/09/08/national-pork-producers-council-anti-science-anti-animal

Gerke, R. (2012, March 20). Herbert signs so-called 'ag-gag' bill. *The Salt Lake Tribune*. Retrieved from http://www.sltrib.com/53758916

Global Development and Environment Institute (GDAE). (2013). Feeding the factory farm. Tufts University. Retrieved from http://www.ase.tufts.edu/gdae/policy_research/BroilerGains.htm

Grandin, T. (Ed.). (2010). *Improving animal welfare: a practical approach*. Cambridge, MA: CAB International.

Grier, P. (2010). Michelle Obama says 'Let's move' on obesity in American kids. *The Christian Science Monitor*. Retrieved from http://www.csmonitor.com/USA/2010/0209/Michelle-Obama-says-Let-s-Move-on-obesity-in-American-kids

Harish (2013). Factory farming and the price of meat. *Counting Animals: A place for people who love animals and numbers*. Retrieved from http://www.countinganimals.com/factory-farming-and-the-price-of-meat/

Iowa Pork Producers Association (IPPA). (2012). USDA announces major pork purchase. Retrieved from http://www.iowapork.org/usda-announces-major-pork-purchase/

Kluger, J. (2009). Pork gets a swine flu bailout. *Time Magazine*, Retrieved from http://content.time.com/time/health/article/0,8599,1921649,00.html

Meat & Poultry Staff (MPS). (2013). USDA plan for dark meat chicken. *Meat + Poultry*. Retrieved from http://www.meatpoultry.com/articles/news_home/Regulatory/2013/09/USDA_plan_for_dark_meat_chicke.aspx?ID=%7bCD131736-E7BE-4303-819B-7827E73C1332%7d&cck=1

Muirhead, S. (2011). Ag's go-to message not resonating. *FoodLink*. Retrieved from http://feedstuffs-foodlink.com/story-ags-goto-messages-not-resonating-71-66622

National Institute of Food and Agriculture (NIFA). (2004). Decreasing production related bone diseases in poultry. Report from the Current Research Information System (CRIS), Accession No. 0404529. Washington, DC: United States Department of Agriculture, Research, Education and Economic Information System, NIFA. Retrieved from http://portal.nifa.usda.gov/web/crisprojectpages/0404529-decreasing-production-related-bone-diseases-in-poultry.html

National Institute of Food and Agriculture (NIFA). (2006). Analysis of sperm storage mechanisms in poultry. Report from the Current Research Information System (CRIS), Accession No. 0405837. Washington, DC: United States Department of Agriculture, Research, Education and Economic Information System, NIFA. Retrieved from https://portal.nifa.usda.gov/web/crisprojectpages/0405837-analysis-of-sperm-storage-mechanisms-in-poultry.html

National Institute of Food and Agriculture (NIFA). (2009). The effect of scrotal insulation on oxidative damage to sperm and testicular gene expression. Report from the Current Research Information Systems (CRIS), Accession No. 0208789. Washington DC, USA: United States Department of Agriculture, Research, Education and Economic Information System, NIFA. Retrieved from http://www.reeis.usda.gov/web/crisprojectpages/0208789-the-effect-of-scrotal-insulation-on-oxidative-damage-to-sperm-and-testicular-gene-expression.html

National Institute of Food and Agriculture (NIFA). (2010). The effect of scrotal insulation on testicular gene expression. Report from the Current Research information Systems (CRIS), Accession No. 0220407. Washington DC, USA: United States Department of Agriculture, Research Education and Economic Information System, NIFA. Retrieved from http://www.reeis.usda.gov/web/crisprojectpages/0220407-the-effect-of-scrotal-insulation-on-testicular-gene-expression.html

National Institute of Food and Agriculture (NIFA). (2013). Understanding and control of male fertility in swine. Report from the Current Research Information System (CRIS), Accession No. 0229466. Washington DC, USA: United States Department of Agriculture, Research, Education and Economic Information System, NIFA. Retrieved from http://reeis.usda.gov/web/crisprojectpages/0229466-understanding-and-control-of-male-fertility-in-swine.html

National Institute of Food and Agriculture (NIFA). (2014a). Molecular strategies to circumvent or mitigate seasonal infertility in boars. Report from the Current Research Information System (CRIS), Accession No. 1002384. Washington DC, USA: United States Department of Agriculture, Research, Education and Economic Information System, NIFA. Retrieved from http://portal.nifa.usda.gov/web/crisprojectpages/1002384-molecular-strategies-to-circumvent-or-mitigate-seasonal-infertility-in-boars.html

National Institute of Food and Agriculture (NIFA). (2014b). Poultry germplasm preservation: modulatingmembrane lipids for successful cryopreservation of semen from valuable genetic stocks. Report from the Current Research Information System (CRIS), Accession No. 1000438. Washington DC, USA: United States Department of Agriculture, Research, Education and Economic Information System, NIFA. Retrieved from http://www.reeis.usda.gov/web/crisprojectpages/1000438-poultry-germplasm-preservation-modulating-membrane-lipids-for-successful-cryopreservation-of-semen-from-valuable-genetic-stocks.html

Olson, L. (2013). Commentary by Linden Olson: words. *Pork Network*, Retrieved from http://www.porknetwork.com/pork-news/latest/Commentary-by-Linden-Olson-Words-204219211.html

Pelaz, J., & Long, J. (2008). Characterizing the glycocalyx of poultry spermatoza: Ii low temperature storage of turkey semen and sperm mobility phenotype impact the cabohydrate component of membrane glycoconjugates. *Andrology*, *29*, 431–439. doi:10.2164/jandrol.107.004259

Plumber, B. (2014). How the U.S. government spends millions to get people to eat more pizza. *The Washington Post*. Retrieved from http://www.washingtonpost.com/blogs/wonkblog/wp/2014/02/10/13-percent-of-americans-are-eating-pizza-on-any-given-day/

Pond, W. G., Bazer, F. W., & Rollin, B. E. (Eds.). (2011). *Animal welfare in animal agriculture: husbrandry, stewardship and sustainability in animal production*. Boca Raton, FL: CRC Press.

Shapiro, P. (2013). The Chicken Industry Loves Federal Handouts. *The Huffington Post*. Retrieved from http://www.huffingtonpost.com/paul-shapiro/the-chicken-industry_b_3947857.html

Simon, D. (2013). *Meatonomics: how the rigged economics of meat and dairy make you consume too much - and how to eat better, live longer, and spend smarter*. San Francisco, CA: Conari Press.

The Week Staff (WS). (2013). Farm subsidies: a welfare program for agribusiness. *The Week*, Retrieved from http://theweek.com/articles/461227/farm-subsidies-welfare-program-agribusiness

Wheat, D. (2015). Push for national chicken cage standards stalls. *Capital Press*. Retrieved from http://www.capitalpress.com/Nation_World/Nation/20150102/push-for-national-chicken-cage-standards-stalls

KEY TERMS AND DEFINITIONS

Ag-Gag: An effort by agribusiness to suppress whistleblowing at their facilities.

Subsidy: Financial aid or support extended to an economic sector generally with the aim of promoting economic and social policy.

Surplus: Something left over when requirements have been met; an excess of production or supply over demand.

USDA: United States Department of Agriculture.

Chapter 14
Sustainable Food Consumption:
A Mission Almost Impossible Because of the West

Amzad Hossain
Curtin University, Australia & Rajshahi University, Bangladesh

ABSTRACT

It is difficult to separate western consumerism from excessive meat consumption and through globalization this culture is spreading through the planet to traditional places, such as Bangladesh and the Indian subcontinent. The chapter argues that the socio-economic and planetary cost of increasing meat consumption is clearly untenable and initiating a process that restores natural resources is imperative. A major objective of this chapter is to raise awareness about the consequences from unsustainable meat production and consumption and the negative implication from a Western type of diet. Drawing on the spiritual messages from the Baul philosophers, it makes the case that preserving traditional flexitarianism, defined here as meat in the absence of any other food options or rare ceremonial meat consumption, is essential for the health of the planet and its inhabitants.

INTRODUCTION

The meaning of sustainable food consumption should derive from the most commonly used Brundtland 1987 definition of sustainable development: "the development that meets the needs of the present without compromising the ability of future generations to meet their own needs" (WCED, 1987, p. 16). Hence the meat we consume today should not compromise food for present and future people. There are however vast differences in the food we eat depending on culture and place and nowhere are these as evident as in relation to meat consumption.

Table 1 shows a comparison between countries based on FAO data. Bangladesh is the nation with the lowest meat consumption of 4.1 kg per person per year. The people of this country maintain a very low individual meat intake because of the prevalence of traditional culture which encourages plant-based meals. Influential spiritual leaders, such as the mystic Baul philosophers recognized by UNESCO (2008) as part of the intangible heritage of humanity, consume zero meat. This traditional culture however is

DOI: 10.4018/978-1-4666-9553-5.ch014

Table 1. Per capita meat supply, 2011 [kg] (FAOSTAT, 2015)

Country	Kg/Capita	Country	Kg/Capita	Country	Kg/Capita
New Zealand	126.9	Norway	65.9	Botswana	29.7
Australia	121.2	Panama	64.4	El Salvador	28.1
USA	117.6	Korea	62.2	Egypt	28
Oceania	116.2	Saudi Arabia	62	Thailand	27.8
Austria	106.4	Mexico	61	Nicaragua	27.4
Israel	102	South Africa	59.3	Lesotho	22.7
Spain	93.1	Turkmenistan	57.8	Peru	21.2
Brazil	93	Viet Nam	57.6	Zimbabwe	20.7
Canada	92.2	China	57.5	Syria	20.6
Portugal	90.3	Bulgaria	57.1	Lao	19.9
France	88.7	UAE	55.6	Iraq	18.9
Germany	87.9	Ecuador	54.4	Yemen	18.6
Iceland	87	Romania	53.4	**Africa**	*18.6*
Kuwait	87	Malaysia	53.3	Haiti	18.2
Italy	86.7	Paraguay	51.4	Niger	17.9
Americas	86.5	Cuba	49.7	Ghana	16.9
EU	82.6	Costa Rica	49	Kenya	16
UK	82.5	Ukraine	49	Pakistan	15.5
Sweden	81.9	Japan	48.8	Liberia	15
Greece	80.6	Colombia	47.6	Cambodia	14.9
Ireland	80.5	Lebanon	43.8	Senegal	14.5
Chile	80.1	Albania	43.4	**Least Developed Countries**	**14.1**
Montenegro	77.3	**World**	**42.2**	Tajikistan	14
Belgium	76.8	Myanmar	39.3	Afghanistan	13.4
Europe	76	Congo	37.6	Zambia	13.4
Poland	75.6	Moldova	37.5	Indonesia	12.9
Switzerland	74.7	Kyrgyzstan	36.8	Uganda	12.6
Finland	74.4	FYR Macedonia	35.7	Nepal	12.2
Hungary	73.7	Timor-Leste	35.6	Togo	12.2
Mongolia	73.4	Honduras	35.5	Nigeria	9.5
Lithuania	72.9	Iran	35.4	Tanzania	8.9
Netherlands	72.7	Philippines	34.4	Mozambique	7.9
Kazakhstan	69.7	Morocco	33.7	Sri Lanka	6.3
Uruguay	69.5	Turkey	33.4	Rwanda	6.2
Latvia	68.5	Libya	31.3	India	4.2
Russia	66.9	**Asia**	*31.2*	Bangladesh	4.1

being eroded with the processes of globalization and the increasing influence of the West, including western style meat-rich diets.

Using the example of Bangladesh, the chapter argues that eastern traditional diets long ago embraced the notion of flexitarianism as part of a harmonious way of living with nature. These traditions need not only to be preserved but unless they are also spread to the western world, sustainable development will continue to remain a mirage in a hot and exhausted planet. We first examine the roots of the Bangladeshi flexitarian diet and then outline why, due to western influences, sustainable food consumption currently appears to be a mission impossible. The final section argues that meat reduction is an easy one-stop solution that the world needs for avoiding the existential threats of climate change and environmental deterioration.

SUSTAINABLE FOOD CONSUMPTION

There are many definitions of flexitarianism (Raphaely & Marinova, 2014; Marinova & Raphaely, 2015; Dagevos, 2015) which essentially stress the importance of meat reduction in western countries. In the context of a developing and transitioning country like Bangladesh however flexitarianism may be defined completely differently. Flexitarians are people who eat animal protein, especially meat, only when there is no other alternative and participate in occasional meat intake because of religious or social ceremonial reasons.

The sustainability philosopher-practitioners in the Baul tradition of Bangladesh maintain: "Eating meat destroys human longevity. If this is disregarded, nature of course can retaliate"[1]. The Bauls – saintly mendicants, are mostly unlettered, yet full of poetic, musical, and philosophical talent (Hossain, 1990). They come mainly from Muslim and Hindu backgrounds. The Bauls are naturalist by belief and practice and live simple lives singing as they go along from place to place and as feelings come to them (Hossain & Marinova, 2009). Ordinary people regard them as supreme Pundits (scholars) as the messages they convey in their spontaneously composed songs are of great relevance (Sen, 1929). In 2005 the Baul tradition was included in UNESCO's list of Masterpieces of the Oral and Intangible Heritage of Humanity.

Baul philosopher-practitioner Pramil Sing asserts that: "a lifestyle pattern that includes a very low meat intake is associated with greater longevity" (Singer & Mason, 2007, p. 224). According to the vegan Baul guru Aziz Fakir[2], ceremonial meat consumption, which is inherently infrequent, is socially needed and ecologically sustainable. Some traditions in the Indian sub-continent observe ceremonial meat consumption because of various socio-religious needs.

The Baul philosophers of Bangladesh and the Ayurvedic and spiritual healers of the East see culture as being at the centre of dietary habits and consider the western consumption of meat as self-destruction. Bangladesh's Baul philosophers say that ecological foodprint (rather than footprint) should be the basis for sustainable development (Hossain, Marinova & Hossain Rhaman, 2014). Globalization however is inflicting powerful and aggressive influence on other cultures, mainly seeking profit making. According to Brenkert (2008, p. 2), "international marketers have been charged with imposing the values of their home countries on the countries in which they do business, destroying local businesses, and manipulating people". The powerful West is corrupting traditional flexitarianism in Bangladesh by increasing the amount of livestock production and unfairly promoting meat consumption. Livestock marketers see meat as a means of profiteering and promote its consumption as a "healthy" diet and a sign of civilizational advancement. Little attention is given to the social, cultural and environmental costs that this involves.

SUSTAINABILITY COST OF INCREASING MEAT CONSUMPTION

As the figures in Table 1 show, rich countries (together with Brazil) are responsible for excessively large quantities of meat consumption. World supply of meat is also growing globally at close to twice the rate of population growth (Marinova & Raphaely, 2015). Within a national economy, production and supply of goods are aimed at satisfying people's need-based wants but globalization's objective is for supply to precede and often exceed demand. This artificially creates demand for consumables which are often not needed (Ellis et al., 2011, p. 20) and the related advertising may be not only false but also grievous in nature (Schlegelmilch, 1998, p. 109). In the case of meat, globalization rarely delivers increased supply of livestock or meat to meet existing human needs in starving populations. Instead, it creates superfluous demand for increased meat consumption in places like Bangladesh where flexitarian traditional diets have prevailed.

One of the problems with this is the socio-ecological costs of the ever-increasing supply of meat as the planet's boundaries are being pushed to the limit. This is manifested through the depleting or degrading of resources, challenges for human health and morality, biodiversity loss and serious concern about climate change. While the world's population grew twofold from 3 to 6 billion between 1961 and 2011, the supply of meat increased fourfold – from 81 to 320 million of tonnes, and the number of animals slaughtered each year for human consumption during this period grew from 7 to 62.5 billion (calculated from FAOSTAT, 2014). A further doubling of meat production is projected for the period 2000-2050 (Steinfeld et al., 2006) during which population is projected to increase by 50% (calculated from UNDESA, 2013). Were everybody to consume as much meat as in the current diet of the average New Zealander, Australian or American, we would need at least four additional planets not only to provide the necessary resources but also to absorb the resulting waste (Environmental Careers Organisation, 2004, p. 357).

Currently "(t)he livestock sector represents the world's single largest human use of land and largest source of water pollutants" (Jackson, 2009, p. 274). Hoekstra (2013, p. 51) explains that livestock puts a large claim on the Earth's natural resources that accounts for 70 percent of all agricultural land and 30 percent of the land surface of the planet. Estimates of livestock's contribution to greenhouse gases (GHG) range between 20 (Eshel, Shepon, Makov & Milo, 2014) and 51 (Goodland & Anhang, 2009) percent of the global total. With respect to water use, Hoekstra (2013, p. 52) points out: "the global footprint of animal production amounts to 2,422 billion m^3/yr. One-third of this total is related to beef cattle, another 19 percent to dairy cattle." Farming of livestock is inefficient in relation to calorie conversion – it takes 38 calories of feed to produce 1 calorie of beef (Eshel et al., 2014); requires an average of 25 kcal of fossil energy to produce 1 kcal of animal protein which is 10 times greater than the fossil energy required per kcal of plant protein (Hoekstra, 2013). The Australian study of Doran-Browne, Eckard, Behrendt, Rose and Kingwell (2015) shows that per unit of nutrient density (covering protein, fibre, vitamins A, C and E, calcium, iron, magnesium and potassium) and per protein equivalent, untrimmed beef produces respectively 79 and 54 times more GHG compared to flour.

From an economic point of view, in 2015 livestock accounts for 40 percent of the global value of agricultural production, being more than 50 percent in the industrialized countries (FAOSTAT, 2015). However, the environmental impacts of the pollution created by livestock, deforestation and biodiversity loss due to land conversion into pastures or for growing feed are usually unaccounted. For example, livestock is the largest source of water pollutants which severely limits the scope for the future productivity of natural systems (Jackson, 2009, p. 274). An example at a local scale is the city of Delmarva (with a population of 4 million) which struggles to absorb the manure generated by the 600 million chickens raised annually (Singer & Mason, 2007). The excess manure washes off into the rivers and streams or gets into the underground water (Singer & Mason, 2007).

All scientific evidence points to the fact that meat-based diets are detrimental to the environment. It is however difficult to preserve and encourage flexitarianism in a globalized world.

A MISSION ALMOST IMPOSSIBLE: WESTERN ATTITUDES GENERATING A MISSION IMPOSSIBLE

The spread of western culture in the form of globalization appears as a new type of colonization (Thai, Rahm & Coggburn, 2007, p. 451). This strives to transform the culture of the East into consumerism, including its eating habits. In Bangladesh for example, frequent meat consumption by people is considered a carnivorous animal behavior irrespective of the religious background. By many consumerism is viewed as a new demonic religion in the globalized world (Esposito & Watson, 2000). Western presence and marketing, including meat intensive fast food outlets in the cities, are encouraging behavioral changes, imposing commercial values and corrupting other societies.

Eating meat is slowly increasing in Bangladesh with the westernization of the food habits starting from the elite classes who accept the consumption culture of the West as superior to their own. As rich people generally tend to consume higher on the food chain than the poor (Lemons, Westra & Goodland, 1998), when people's incomes increase they start to eat more meat. Hence we see the trend of ever-increasing meat consumption and this is a major factor for the world hurtling away from sustainability. Flexitarianism is a most needed eating behaviour.

The globalization of western culture in the name of development and progress has a serious negative impact on the planet. Westerners "express concern about the environment yet live materialistic lifestyles that result in high levels of waste" (Hamilton, Dennis & Baker, 2005, p. ix) and this includes meat intensive dietary habits. As in Bangladish, a flexitarian should endorse the ecological value of reverence to Mother Nature from which follow values of responsibility, frugality, diversity and justice for all living creatures (Skolimowski, 1993, p. 35).

Western consumerist culture is self-destructing as it is using up natural resources and polluting the land, water and air. Based on individualism, it also destroys social bondage (Crocker & Linden, 1998; Casey, 1995). There is a lot for Bangladesh to loose be it to succumb to western influences in food habits.

On the contrary, there is a lot to be gained were flexitarianism, or lessening of meat consumption to the absence of any alternatives or to a ceremonial level, to be adopted across the world. The Indian subcontinent is a good example for this and there is need not only to maintain its food culture but also for it to be spread across the globe. The accomplishment of such a level of sustainable meat consumption culture in the West however appears as an almost impossible mission.

Making the Impossible Possible

Sustainable food consumption will remain a mission impossible both in the West and around the globe unless there is a significant cultural change towards reduced meat intake. This seems like a very easy way to restore hope for the future. Many argue that returning to the spiritual values of eastern religions, including Buddhism, Hinduism and the teaching of the Bauls, could hold the solution for reducing meat consumption. The new religion of consumerism needs to be rejected in favour of a collective mindset which endorses harmonious and non-abusive interaction with nature and all living species together with modesty, respectfulness and responsibility as interfaith values. It is difficult to understand why the highly (officially) educated western civilization shows such ignorance or a willful defiance of the values which support sustainability.

There are ways to transform the culture of overconsumption, including meat eating, towards a sustainable culture. The living traditions of the Indian subcontinent are a source of hope for sustainability with respect to the moral and practical aspects of meat consumption.

CONCLUSION: A ONE-STOP SOLUTION

Sustainability is a race against time for the common good. Meat reduction is an easy one-stop solution for the current global environmental problems, including climate change. Were the world to adopt the flexitarian diet of the Indian subcontinent, the immediate ecological benefits would allow sustainable food production not only for current and future generations, it would free up the atmosphere, waterways, land and resources for other more meaningful human endeavours. The obsession with eating meat is dangerous but it can easily be stopped if people return to the core values and meaning of life as espoused by the spiritual leaders of the East who strive for perfection as a way of gaining longevity. Embracing and venerating nature can allow meeting human materialistic needs in a most modest way where meat is rarely present, if at all, and the future of life on earth is sustainable.

Bangladesh vibrates with songs sung by thousands of Bauls around the clock which direct people's consumption, restrain them from consuming meat and encourage them to live on a grain-based diet with fruits and vegetables in order to achieve longevity and avoid nature's retaliation to those who do not live a life in harmony with it. This ecstatic song of Harun Baul[3] who dedicated his life to the mission of making the Earth ever sustainable is a joyous pledge to people not just in Bangladesh but all around the globe: "This earth has to remain as it is; leave this beautiful earth beautiful"[4].

REFERENCES

Brenkert, G. C. (2008). *Marketing ethics*. Oxford, UK: Blackwell Publishing.

Casey, C. (1995). *Work, self and society*. London, UK: Routledge.

Crocker, D., & Linden, T. (1998). *Ethics of consumption: the good life, and global; stewardship*. Oxford, UK: Rowman and Littlefield Publishers.

Dagevos, H. (2015). Exploring flexitarianism: meat r*eduction in a meat-centred food culture*. In T. Raphaely & D. Marinova (Eds.), *Impacts of meat consumption on health and environmental sustainability*. Hershey, PA: IGI Global.

Doran-Browne, N. A., Eckard, R. J., Behrendt, R., Rose, S., & Kingwell, R. S. (2015). Nutrient density as a metric for comparing greenhouse gas emissions from food production. *Climatic Change, 129*(1), 73–87. doi:10.1007/s10584-014-1316-8

Ellis, N., Fitchett, J., Higgins, M., Jack, G., Lim, M., Sare, M., & Tadajewski, M. (2011). *Marketing: a critical textbook*. New York, NY: Sage Publications.

Environmental Careers Organisation. (2004). *The eco guide to careers that make a difference: environmental work for a sustainable world*. Washington, DC: Island Press.

Eshel, G., Shepon, A., Makov, T., & Milo, R. (2014). Land, irrigation water, greenhouse gas and reactive nitrogen burdens of meat, eggs and dairy production in the United States. *Proceedings of the National Academy of Sciences of the United States of America*, *111*(33), 11996–12001. doi:10.1073/pnas.1402183111 PMID:25049416

Esposito, J. L., & Watson, M. (2000). *Religion and global order*. Cardiff, UK: University of Wales Press.

Food and Agriculture Organization of the United Nations. Statistics Division (FAOSTAT). (2014). Production: livestock primary and livestock processed. Retrieved from http://faostat.fao.org/site/569/DesktopDefault.aspx?PageID=569#ancor

Food and Agriculture Organization of the United Nations. Statistics Division (FAOSTAT). (2015). Food balance: food supply, livestock and fish primary equivalent. Retrieved from http://faostat3.fao.org/browse/FB/CL/E

Goodland, R., & Anhang, J. (2009, November/December). Livestock and climate change: What if the key actors in climate change are... cows, pigs and chickens? *World Watch Magazine*, *22*(6), 10–19.

Hamilton, C., Dennis, R., & Baker, D. (2005). Wasteful consumption in Australia. Discussion Paper No. 77. Canberra, Australia. Retrieved from http://www.tai.org.au/documents/downloads/DP77.pdf

Hoekstra, A. Y. (2013). *The water footprint of modern consumer society*. London, UK: Earthscan from Routledge.

Hossain, A. (1990). *Religion of songs: the beliefs and practices of the Bauls of Bangladesh* [Master of Arts Thesis]. Deakin University, Melbourne, Australia.

Hossain, A., & Marinova, D. (2009). Biosystems management: Muslim/Hindu applications in Bangladesh. *Man in India*, *89*(4), 557–566.

Hossain, A., Marinova, D., & Hossain Rhaman, P. (2014). Islamic insights on sustainability. In D. Humphreys & S. S. Stober (Eds.), *Transitions to sustainability: theoretical debates for a changing planet* (pp. 50–62). Champaign, IL: Common Ground Publishing.

Jackson, T. (2009). *Prosperity without growth: economics for a finite planet*. London, UK: Earthscan.

Lemons, J., Westra, L., & Goodland, R. (1998). *Ecological sustainability and integrity: concepts and approaches*. London, UK: Kluwer Academic Publishers. doi:10.1007/978-94-017-1337-5

Marinova, D., & Raphaely, T. (2015). Preface. In T. Raphaely & D. Marinova (Eds.), *Impacts of meat consumption on health and environmental sustainability*. Hershey, PA: IGI Global.

Raphaely, T., & Marinova, D. (2014). Flexitarianism: Decarbonising through flexible vegetarianism. *Renewable Energy*, *67*, 90–96. doi:10.1016/j.renene.2013.11.030

Schlegelmilch, B. (1998). *Marketing ethics: an international perspective*. London, UK: International Thomson Business Press.

Sen, K. M. (1929). *Medieval mysticism of India*. London, UK: Luzac and Company.

Singer, P., & Mason, J. (2007). *The ethics of what we eat: why our food choices matter*. Emmaus, PA: Rodale.

Skolimowski, H. (1993). *A sacred place to dowell: living with reverence upon the Earth*. London, UK: Element Book Ltd.

Steinfeld, H., Gerber, P., Wassenaar, T., Castel, V., Rosales, M., & de Hann, C. (2006). Livestock's long shadow: environmental issues and options. Rome, Italy: Food and Agriculture Organisation (FAO).

Thai, K. V., Rahm, D., & Coggburn, J. D. (2007). *Handbook of globalisation and the environment*. London, UK: CRC Press. doi:10.1201/9781420016932

United Nations Department of Economic and Social Affairs (UNDESA). (2013). Population Division, Population Estimates and Projections Section. Retrieved from http://esa.un.org/wpp/unpp/p2k0data.asp

United Nations Educational Scientific and Cultural Organization (UNESCO). (2008). Baul songs. Retrieved from http://www.unesco.org/culture/ich/rL/00107

World Commission on Environment and Development (WCED). (1987). *Our common future*. Oxford, UK: Oxford University Press.

KEY TERMS AND DEFINITIONS

Baul: A mystic spiritual minstrel on the Indian subcontinent, including Bangladesh.

Consumerism: A culture which encourages the acquisition and use of increasing amounts of goods and services, including food related.

Globalization: A process associated with the spread of western products and lifestyle for profit making.

Intangible Heritage of Humanity List: A worldwide list of oral and other intangible cultural treasures established by UNESCO.

Traditional Diet: Diets in pre-industrial cultures comprising foods that have been eaten for centuries.

(Traditional) Flexitarianism: Understood here as people who eat animal protein only when there is no other alternative and participate in occasional meat intake because of religious or social ceremonial reasons.

Western Culture: Refers to cultures in Europe, United States of America, Canada, Australia, New Zealand and parts of Latin America.

ENDNOTES

[1]   Khele mansho, jiboni shakti hoy dhongso; na manile ai kalam, prakriti hoy tomar bam, obossho (in Bangla).

[2]   Aziz Shah Fakir (105) is one of the oldest as well as most renowned Baul gurus. He lives with his wife Laily at the tomb of his guru Darvish Kalu Shah Fakir (d. 1971) in the village of Choraikole of Bangladesh's Kushtia district. Aziz Shah Fakir has hundreds of disciples all over Bangladesh who visit him regularly to receive his reflective teachings on doing good deeds, supporting good people, opposing and encountering bad deeds and people, spirituality, naturalism as well as about longevity through vegetarian dietary habits and controlled sensuality.

[3]   Harun Baul (who died in 2013 at the age of 75) of Choraikole, Kumarkhali, Kushtia was one of the disciples of Baul guru Aziz Shah Fakir. He never went to school, but learnt to read and write simple Bangla. Harun was a poet-philosopher-singer. Amzad Hossain published articles on the Baul's views on sustainability management on behalf of Harun (I'm Harun Baul Speaking) in "The News from Bangladesh" – a daily news monitoring service.

[4]   "Ai prithibi jemon ase temon e thik robe; sundar a prithibi sere tomai chole jete hobe" (in Bangla)

Chapter 15
Meat Myths and Marketing

Diana Bogueva
Curtin University, Australia

Ian Phau
Curtin University, Australia

ABSTRACT

This chapter explores how marketing uses the creation and perpetuation of myths to reinforce demand for meat amongst mainstream consumers. It explores advertising misinformation including with regards the place of meat in our culture, its nutritional value, its association with affluence, masculinity and the benefits of small-scale production. The power of marketing is within the context of whether marketing has a role to play in decreasing rather than perpetuating meat-consumption.

INTRODUCTION

Myths in society, whether related to food or any other aspects of everyday life, convey particular messages. Associated with notions of morality, heroes or natural phenomena, they are stories enlivened by people who accept them as valid and meaningful guides of behaviour. Myths can be used to justify and guide beliefs and cultural practices. This acceptance may be society-wide or limited to particular groups.

Consumer behaviour is as influenced by myths as any other aspects of life and this produces a variety of conscious or unconscious responses and decisions often expressing particular worldviews, values and lifestyles. According to Geertz (in Chernus, 2012, n.p.), a myth says: "because the world is the way it is, living as we do (or ought to) is uniquely satisfying and fulfilling".

Myths often contain elements that are both true and false blending fiction with empirical facts. The more truth they hold, the more convincing they are, the harder they are to contest, and therefore the more influence they have (Chernus, 2012). Marketing often contributes to both the establishing and reaffirming of myths and in so doing, encourage certain actions that become part of accepted normative thoughts and behaviours. Consumers may think that advertising is harmless and appealing only to some. However encouragement to perform certain actions, misinformation and myths repeated often enough become part of the accepted norm and steer certain behaviours.

DOI: 10.4018/978-1-4666-9553-5.ch015

This chapter looks at marketing myths related to meat consumption. Dispelling such messages is important not only for the benefit of the individual consumer but also for the collective good because of the heavy social and ecological footprint meat consumption and production have. Through consumer behaviour, marketing myths assume a physical reality that impacts human health, social wellbeing and the natural environment. This chapter analyses several popular meat myths created or exploited by the marketing industry, including: "We were meant to eat meat"; "Meat is good for you"; "Real men eat meat"; "Meat is part of our culture"; "Meat is affordable" and "Organic meat is much better".

MYTH 1: "WE WERE MEANT TO EAT MEAT"

Hunting, domestication, killing animals, socialising and eating meat in many forms have been important components of human progress (Smil, 2002; Pobiner cited in Kasper, 2013) including our intellectual and physical growth. Scientific evidence suggests that meat consumption may have contributed to our evolutionary heritage and is linked to key characteristics that have made us human mammals with larger brains, smaller guts and developed language (Smil, 2013b; Choi, 2012; Dominguez-Rodrigo et al., 2012). Smil (2013a, n.p.) explains: "Larger brains benefited from consuming high-quality proteins in meat-containing diets, and, in turn, hunting and killing of large animals, butchering of carcasses and sharing of meat have inevitably contributed to the evolution of human intelligence... and socializing". The ability to secure meat played a major role in human evolution. With the domestication of livestock, hunting was gradually replaced by the planned slaughter of livestock (Burket, 1983). Cooking allowed humans to develop more sophisticated tastes for meat. Nowadays 70 billion animals are slaughtered each year to be consumed by 7 billion humans.

In the contemporary industrialised world we have a very simple reason to eat meat – because this is what we've been taught to do. As Joy explains: "We do not need meat to survive or even to be healthy... We eat animals simply because it is what we have always done, and because we like the way they taste" (Joy, 2011, p. 29).

The large numbers of animals raised to support these dietary habits are putting enormous stress on the environmental limits of the planet and changing its ecological balance. Researchers are increasingly calling for reduction of meat consumption, promotion of more rational meat eating (Smil, 2013a, 2013b) and flexitarian diets (Raphaely & Marinova, 2014; Verain, Dagevos & Antonides, 2015). This is not a simple task as, in addition to habit and socialisation, marketers use the interrelatedness of human evolution with eating animals to further lure and guide consumers in the direction of high meat intake.

Meat-eating behaviour is a habit developed throughout the years passed to children not only by socialisation – including parents, friends and schooling - but also reinforced by advertising. Meat myths are prominent among the cluttered messages of contemporary food marketing whether it tries to build up, change or expand present or future habits and consumer preferences. It relies on our willingness to adopt, or reject, certain behaviours. Samuel Johnson is credited with the words: "The chains of habit are too weak to be felt until they are too strong to be broken" (Esar, 1968, p. 363). The historical truth that humans had to eat meat for their survival no longer applies but the message continues to be used in marketing and it will take tremendous effort for the chains of the habit to be broken.

MYTH 2: "MEAT IS GOOD FOR YOU"

For centuries meat was seen as essential for human health but recent scientific evidence shows that a plant-based diet is "good for us and for the planet" (Stanton, 2012, p. 5). Marketing meat is not necessarily health- or nutrition-driven but mainly oriented towards selling the products. Debunking the myth that an excessively meat-based diet is good for you is a challenge as shown in the highly disputed processes surrounding the development of dietary guidelines in the EU, US and Australia where such efforts were strongly contested by proponents of the livestock and related industries.

While the medical community seems to agree that reducing "meat consumption in the human diet dovetails with dietary guidelines for increased consumption of vegetables, legumes, fruits, nuts and wholegrain products" (Stanton, 2012, p. 5), the marketing of meat continues with advertising messages in the press, on TV, radio, billboards, Internet and elsewhere portraying it as essential to human health. Diets recommending high intake of animal-based proteins, such as the Atkins, Pier Dukan or the paleo diets, are being promoted by celebrities and movie stars without their legitimacy, safety or scientific nutritional validity confirmed.

Consumption of red meat however is credibly associated with increased risk of heart disease and stroke (HSPH, 2015). Amongst others, the World Health Organization (2003), the American Institute for Cancer Research and the World Cancer Research Fund support the evidence that high intakes of meat are associated with colorectal cancer. Excessive meat consumption has also been found to be a risk factor for developing type 2 diabetes (Barnard, Levin & Trapp, 2014). Campbell and Campbell (2006) disclose many health hazards associated with meat production and consumption including the dangers presented to humans by meat-borne pathogens, bacteria and use of hormones. According to the Pew Commission (2008), the meat industry itself increases the potential for pathogen and infectious disease transfer from animals to humans, raises the risk of food-borne infections and antimicrobial resistance, and exposes workers to a number of adverse health conditions, including bronchitis and asthma. To add to the growing list, excessive meat consumption is also associated with obesity (Wang & Beydoun, 2009).

Reputable health organisations recommend limiting the intake of red meat in the human diet – to a maximum of 500 g per week (WCRF & AICR, 2007) or a maximum of 65 g per day (NHMRC, 2013). These organisations send clear messages about restricting red meat consumption because of its negative health impacts and the recommended amounts are vastly different from the consumption quantities portrayed by the myth that copious quantities of meat is good for you.

The rising popularity of more ecological diets, such as flexitarian, vegetarian and vegan diets, is linked to perceptions of these as being a more moral, ethical option (Berley & Singer, 2007; de Bakker & Dagevos, 2012) with health benefits being an added bonus. Such ethical moral motivations for meat avoidance include environmental awareness, concern about use of resources and planetary health as well as sensitivity to the suffering of other sentient beings. Food choices are thus no longer value neutral or driven only by necessity and availability (in the developed world) but have a distinctive moral component (Rozin, Homes, Faith & Wansink, 2012).

MYTH 3: "REAL MEN EAT MEAT"

The high value placed on eating meat is related to its relevance in western society as a symbol of achievement, power and domination (Fiddles, 1991). According to Ruby and Heine (2011, p. 448), throughout much of European history meat was "a staple for gentry and a rare treat for the peasants". As the middle class emerged within society the impression of growing affluence was supported by eating of meat. According to Smil (2002), carnivorousness continues to evoke strong emotions, being a nearly universal symbol of affluence, wellbeing, satiety and contentment. Considered the tip of the food pyramid (Twigg, 1983), meat also defines social status. A study by Allen (2005) shows that people aspiring to improve their social position choose high-status foods such as meat and reject the low-status fruit and vegetables.

Along with social status, different foods are also linked to gender and sexuality. Medium-rare steak is the food most associated with men and maleness (Rozin et al., 2012). Historically meat, and especially red meat, is linked to masculinity (Fiddles, 1991). It carries the image of strength, force and aggression – qualities valued in a competitive world (Twigg, 1983). Men are able to show their dominance and virility by conquering beasts (Lupton, 1996); killing and butchering is associated with positively aggressive male behavior; meat production and consumption are positively depicted as human control over nature (Baker, Thompson & Palmer-Barnes, 2002), "power-over" (Warren, 2000) and the "master identity" (Plumwood, 1993; Rogers, 2008) whose self-centred biological drives require controlled rational expression (Kheel, 2008).

According to Kimmel (1996), eating meat is a vehicle of resurrecting manhood also when traditional gender roles are challenged, such as with the broader involvement of women in the workplace. The idea of manhood continues to be overemphasized by advertising and marketing campaigns in the fast-food culture. Domino's Pizza, Burger King, El Taco and McDonalds have produced TV and billboard advertisements repeatedly suggesting that real men eat more meat and that compromised masculinity can be regained through meat consumption (Rogers, 2008). Many advertisements present another macho idea – the bigger the meat, the better. In a Burger King advertisement, a man is served a small portion in a fancy restaurant after which he leaves his date singing "I am man..." conveying the message that a real man needs more meat. In recent years marketing is specifically targeting the Millennial Generation (or Generation Y), whose spending power is expected to soon overtake that of the Baby Boomers (Moss, 2014), attempting to reach them with advertisements such as "Beef. It's what's for dinner" and "Cool 2B Real". The image of the "real man" however is not always as attractive as the (marketing) myths suggest.

Some feminist scholars (e.g. Gruen, 2014) draw parallels between the oppressive structures and values in society that affect both animals and women. According to Adams (2010), the connection between meat and masculinity goes far beyond typical sexist advertising and legitimises violence and oppression on all levels.

In contrast to the dominant belief of power, meat also symbolises death, violence, cruelty and the denial of empathy (Twigg, 1983; Adams, 2010). A study of Allen, Wilson, Ng and Dunne (2000) shows that people who rate themselves as omnivores tend to be higher in authoritarianism, prefer hierarchical structures and place more importance on power. By comparison, people who identify as vegetarians tend to value more social justice, peace and equality (Allen et al., 2000). A preference for meat or vegetarian options seems to be dependent on whether the values symbolised by the food choice fit the individual's personal principles (Allen, Gupta & Monnier, 2008).

Alongside the masculinization of meat is the feminization of vegetarianism. Rogers (2008) however points out that vegetarianism is a threat to hegemonic masculinity from people who are motivated by concerns about the environment and/or animal rights. Many vegetarians also identify themselves as feminists because they share a similar interest in rejecting patriarchal structures and hegemonic practices in society (Kwan & Roth, 2011). Positively for some, vegetarian men purposefully abstaining from meat are perceived by others as more principled and more virtuous (Ruby & Heine, 2011) for whom emotions are also more important (Allen et al., 2000). By refusing to eat meat such groups and individuals are participating in a physical resistance against institutional power and dominance (Kwan & Roth, 2011).

In particular, food consumption is a social marker for constructing social identities and lifestyles for both meat eaters and vegetarians.

MYTH 4: "MEAT IS PART OF OUR CULTURE"

Meat eating is part of Western culture with the typical family meal and social gatherings centred on it. Sausage sizzles and meat BBQs are the norm not only at home but also for parties, school fundraisings, sporting events, concerts, birthday celebrations, weddings, elections and others. On top of meat consumption being a deeply engraved social norm and habit, various marketing techniques are constantly used to reinforce these norms and to continue to make the customer want, need and buy it. The challenge for marketers is to make the public feel comfortable with what they are purchasing – pretty pictures of grazing happy cows and freely running chickens is one of the ways to convince the consumers.

In Australia the two main players contributing to the advertising of meat are Meat and Livestock Australia (MLA), which represents the beef and lamb industry, and Australian Pork Limited, the producers owned industry supporting the Australian pork sector. To attract viewers, entertained them and make them remember the messages, everyday situations, humour, irony, music, local jargon and parodies are used in the advertisements launched by these organisations. Examples include the MLA's Australia Day lamb commercials with a "lambassador" and "Put another steak on a barbie" as well as Australian Pork's "Get some pork on your fork". Lamb in particular is being targeted at young people and promoted as representing the Australian identity, lifestyle and way of socialising (see Figure 1 and Figure 2). Chicken is also marketed as a symbol of locally produced quality food (Figure 3).

Figure 1. Fueling Australia's love for lamb: Targeting the young generation
(Authors' Own Image).

Figure 2. Fueling Australia's love for lamb: Using lamb as equivalent to eat
(Authors' Own Image).

Figure 3. Using public transport for advertising meat: a public transport bus advertising chicken meat in Perth, Western Australia
(Authors' Own Image).

The same approach is used in the TV series "The Dinner Project" and Master Chef programs which aim to inspire consumers about cooking healthy meals using beef and lamb recipes. Hundreds of cooking websites, books, special TV shows, cooking competitions, advertisements on billboards, public transport, print and digital media are an unbreakable part of the big meat propaganda machine and its chains. They not only shape Australia's culture but also flawlessly serve the purpose to engage and create loyal consumers.

The presence of meat in the Western diet is taken as a given, whereas vegetarianism and veganism are treated as practices outside of the norm and requiring explanation (Wilson, Weatherall & Butler, 2004). These social expectations are one of the main barriers to adopting a meat-free diet (Taylor, 2012) or having a meat-free meal. As Kheel (2008, p. 236) states: "Despite some evidence of increasing tolerance for vegetarians today, the pressure to comply with the norm of meat eating still operates as a powerful cultural and economic force".

In reality, both omnivorism and vegetarianism represent particular dietary choices, and the preference to include meat in someone's diet should require as much explanation as the decision to exclude it (Fiddles, 1991). Both, meat eating and plant-based options are choices based on a set of assumptions about animals, humans and the world (Joy, 2011). It is a myth that because we have eaten meat in the past we have to continue to do so in the future. Health related and environmental concerns supported by scientific evidence are creating a new culture in which meat should have an increasingly modest place.

MYTH 5: "MEAT IS AFFORDABLE"

Throughout the years the meat industry transformed itself from small and medium scale livestock farms to big industrial production enterprises (PwC, 2011). According to the United Nations report (Steinfeld et al., 2006), approximately 80% of the meat sector growth in the world is occurring through industrial livestock production. One of the main results of this industrial livestock production is that meat has become readily available and inexpensive. With this availability of cheap meat demand has grown nationally and internationally. The global transition to industrialized production resulted in the growing power of the livestock industry with consolidation continuing to increase with the demand for more meat. A handful of large corporations dominate the meat industry (in Australia, United States, Europe) with enormous facilities, feedlots and slaughtering houses.

Beef industry is Australia's largest agricultural commercial activity and the country is the second largest exporter of beef (after Brazil) and the eighth largest producer of beef in the world. In 2011, there were about 26.6 million cattle in Australia, with 91% of the total herd used for beef production and the remaining 9% for milk (ABARES, 2011). Most consumers enjoying the "affordable" price of meat are unaware that the majority of animals slaughtered for food produced in the world are raised and fattened in feedlots and increasingly smaller spaces. Similar to the US, the use of feedlots in Australia has grown significantly since the 1980s and is expected to continue to increase in the future (PwC, 2011).

Popular works (e.g. Foer, 2009) and investigative journalism (e.g. Lynn White from Animals Australia) have lifted the veil on the meat industry's treatment of animals. Whether it be overcrowding, unnatural feed which promotes physical suffering, inhumane (and sometimes ineffective) slaughtering, or many other questionable practices, contemporary factory farming has been criticized on the grounds of animal cruelty. The denial of animal suffering (an attitude more common in men) and dissociating meat products with animals (more common among women) (Rothgerber, 2013) allows ongoing widespread promotion of meat consumption. Numerous related strategies to justify eating meat are consistently used by advertisers; advertisements about beef, pork, lamb or chicken do not show the real picture about how the animals are raised but instead include happy pastoral scenes or mouth-watering images or ready to eat foods. Companies apportion vast budgets to promote meat – Cargill, JBS and McDonalds spent respectively $1,792 million, $1,594 million and $768 million in 2011. Much smaller amounts are available for advocacy for farm animals – the largest campaigner to date being the Humane Society of the United States with $126 million (Harish, 2012).

People are being spared the truth (or deceived) not only by the extensive marketing, through use of myths, of meat products but also through the price of meat which externalises the costs of animal suffering, public health, environmental deterioration and the health of the people working in factory farms and slaughterhouses. Ongoing perpetuation of common meat myths through marketing vehicles comes with a very high price tag attached.

MYTH 6: "ORGANIC MEAT IS MUCH BETTER"

Alternative small scale and organic livestock production systems, including their marketing, lie outside the conventional mainstream way of raising animals for food production and rely on local and niche markets. They produce meat by adopting ostensibly sustainable practices that allow production without polluting the environment or depleting the planet's resources including pasture-based organic farming with small flocks and herds, minimal antibiotic use and preserving the ecosystem. Irrespective of the better conditions for raising animals in these farms, many practices in the business, including breed selection, artificial insemination, weaning of calves and ultimately slaughter, remain the same. Although the small-scale farms provide better quality meat, they face fierce competition from the mega livestock industry with cheaper prices and abundant quantities (McWilliams, 2009). The ecological footprint of these small-scale producers is also larger. For example, grass-grazing cows emit considerably more methane than grain-fed and require more land converted to pasture (McWilliams, 2009).

Many of these livestock products started to carry labels identifying them as different from the mainstream. Marketing uses such sustainability, environmental and ethical labelling as an instrument to generate demand for "good" food products (Binnekamp & Ingenbleek, 2008). On the Australian market consumers witnessed a surge of "Free-range", "Hormone free", "Humane Certified", "Organic", "Grass-fed", "Local produce" and other "better" meat options offered generally at a higher price which conscious consumers are ready to accept.

Big meat producing corporations are also using compassionate and organic ideas about humanely produced meat and animal products deploying some of these methods and labels into components of their brand. This "sustainability" trend however does not reduce meat consumption - it just makes the consumer feel more peaceful with their choice and obfuscates them from having to take any other action in relation to meat choices. According to Boyle (2012), marketing should ask people to eat less and better quality, more sustainable meat. Organic meat and sustainability labels should not lure consumers to create another meat myth perpetuating excessive an justifiable meat consumption.

CONCLUSION

Meat consumption choices are very complex and influenced by an array of factors, including stage of life, friends, entertainment, perceptions, attitudes, beliefs, knowledge, norms and values. Marketing strategies identify and understand meat consumers' psychology. Marketing messages are targeted at providing exactly what is wanted – in quantity, place and time, in order to make a profit. For many consumers, marketing created or perpetuated myths allow them to justify or explain their own behaviours and offer an excuse not to change their consumption habits. If humans do care about health, morality, environmental sustainability and the welfare of animals, a change in excessive meat consumption as well as a change in the marketing, myths and practices that underpin such high intake (especially in the western world), is urgently needed.

The chapter examined myths and marketing around eating meat. Although meat consumption has been commonplace for millennia, the relatively new imperatives of climate change, environmental deterioration, resources limits, animal wellbeing, pollution from animal agriculture and importantly, issues related to human morality are making humans reassess their dietary habits. While the meat industry pushes for consumers all around the world to eat more of its products, change is slowly staring to appear. Meat-free Mondays and Veggie Thursday campaigns endorsed by doctors and environmentalists are active in 36 countries. Foer (2009, p. 24) writes: "We have the burden and the opportunity of living in the moment when the critique of factory farming broke into the popular consciousness. We are the ones of whom it will be fairly asked, 'What did you do when you learned the truth about eating animals?' A different kind of marketing that can help a transition away from meat is needed but ultimately it remains up to us to make the personal choice whether to stop accepting the myths promoted by the livestock industry and see the bigger truth about eating animals.

REFERENCES

Adams, C. J. (2010). The sexual politics of meat: a feminist-vegetarian critical theory (20th Anniversary Ed.). New York, NY/London, UK: Continuum.

Allen, M. W. (2005, December 5-7). Social structures, status seeking and the basic food groups. Paper presented at the Australian and New Zealand Market Academy (ANZMAC) 2005 Conference, University of Western Australia, Peerth, Australia. Retrieved from http://www.anzmac.org/conference_archive/2005/cd-site/pdfs/3-Consumer-Beh/3-Allen.pdf

Allen, M. W., Gupta, R., & Monnier, A. (2008). The Interactive effect of cultural symbols and human values on taste evaluation. *The Journal of Consumer Research*, *35*(2), 294–308. doi:10.1086/590319

Allen, M. W., Wilson, M., Ng, S. H., & Dunne, M. (2000). Values and beliefs of vegetarians and omnivores. *The Journal of Social Psychology*, *140*(4), 405–422. doi:10.1080/00224540009600481 PMID:10981371

Australian Bureau of Agriculture and Resource Economics and Science (ABARES). (2011). Australian commodity statistics 2011. Canberra, Australia: ABARES. Retrieved from http://data.daff.gov.au/brs/data/warehouse/agcstd9abcc002/agcstd9abcc0022011/ACS_2011_1.0.3.pdf

Baker, S., Thompson, K. E., & Palmer-Barnes, D. (2002). Crisis in the meat industry: A values-based approach to communications strategy. *Marketing Communications, 8*(1), 19–30. doi:10.1080/13527260110108319

Barnard, N., Levin, S., & Trapp, C. (2014). Meat consumption as a risk factor for type 2 diabetes. *Nutrients, 6*(2), 897–910. doi:10.3390/nu6020897 PMID:24566443

Berley, P., & Singer, Z. (2007). *The flexitarian table: Inspired, flexible meals for vegetarians, meat lovers and everyone in between.* New York, NY: Houghton Mifflin Harcourt.

Binnekamp, M., & Ingenbleek, P. (2008). Do 'good' food products make others look 'bad'? Spin-off effects of labels for sustainable food production in the consumer perception. *British Food Journal, 110*(9), 843–864. doi:10.1108/00070700810900576

Boyle, E. (2012). *High steaks: why and how to eat less meat.* Gabriola Island, Canada: New Society Publishers.

Burket, W. (1983). *Homo necans: the anthropology of ancient Greek sacrificial ritual and myth* [translated]. Berkley, CA: University of California.

Campbell, T. C., & Campbell, T. M. (2006). *The China study.* Dallas, TX: BenBella Books.

Chernus, I. (2012). Mythic America: essays about America's national myths in the past, present and future. Retrieved from https://mythicamerica.wordpress.com/

Choi, C. Q. (2012). Eating meat made us human, suggest new skull fossil. Retrieved from http://www.livescience.com/23671-eating-meat-made-us-human.html

de Bakker, E., & Dagevos, H. (2012). Reducing meat consumption in today's consumer society: Questioning the citizen-consumer gap. *Journal of Agricultural & Environmental Ethics, 25*(6), 877–894. doi:10.1007/s10806-011-9345-z

Dominguez-Rodrigo, M., Pickering, T. R., Diez-Martin, F., Mabulla, A., Musiba, C., Trancho, G., & Arriaza, C. S. et al. (2012). Earliest porotic hyperostosis on a 1.5-million-year-old hominin, Olduvai Gorge, Tanzania (oldest evidence of anemia in human evolution. *Public Library of Science (PLoS). One, 7*(10), 1–6. doi:10.1371/journal.pone.0046414

Esar, E. (1968). *20,000 quips and quotes.* New York, NY: Doubleday.

Fiddles, N. (1991). *Meat: a natural symbol.* London, UK: Routledge.

Foer, J. S. (2009). *Eating animals.* New York, NY: Little, Brown and Company.

Gruen, L. (2014). *Ecofemenism: women, animals and nature.* London, UK: Bloomsbury Academic.

Harish. (2012). Meat industry advertising. Retrieved from http://www.countinganimals.com/meat-industry-advertising/

Harvard, T. H. Chan School of Public Health (HSPH). (2015). The nutrition source: protein. Retrieved from http://www.hsph.harvard.edu/nutritionsource/what-should-you-eat/protein/

Joy, M. (2011). *Why we love dogs, eat pigs and wear cows: an introduction to carnism.* San Francisco, CA: Conari Press.

Kasper, L. R. (2013). The splendid table: why do we eat meat? Tracing the evolutionary history. Retrieved from http://www.splendidtable.org/story/why-do-we-eat-meat-tracing-the-evolutionary-history

Kheel, M. (2008). *Nature ethics: an ecofemenist perspective.* Lanham, MD: Rowman and Littlefield.

Kimmel, M. (1996). *Manhood in America: a cultural history.* New York, NY: Simon and Schuster.

Kwan, S., & Roth, L. M. (2011). The everyday resistance of vegetarianism. In C. Bobel & S. Kwan (Eds.), *Embodied resistance: challenging the norms, breaking the rules* (pp. 186–196). Nashville, TN: Vanderbilt University Press.

Lupton, D. (1996). *Food, the body and the self.* London, UK: Sage Publications.

McWilliams, J. E. (2009). *Just Food: where locavores get it wrong and how we can truly eat responibly.* New York, NY: Little, Brown and Company.

Moss, L. (2014). The marketing of meat: why beef and pork producers are so focused on millennials. Retrieved from http://www.mnn.com/money/sustainable-business-practices/stories/the-marketing-of-meat-why-beef-and-pork-producers-are

National Health and Medical Research Council (NHMRC). (2013). Australian Dietary Guidelines. Canberra: Australian Government, NHMRC, Department of Health and Ageing. Retrieved from https://www.nhmrc.gov.au/guidelines-publications/n55

Pew Comission on Industrial Farm Animal Production (Pew Commission). (2008). Putting meat on the table: industrial farm animal production in America. Baltimore, MD: Pew Charitable Trusts and Johns Hopkins Bloomberg School of Public Health. Retrieved from http://www.ncifap.org/_images/pcifapsmry.pdf

Plumwood, V. (1993). *Femenism and the mastery of nature.* New York, NY: Routledge.

PricewaterhouseCoopers (PwC). (2011). The Australian beef industry: the basics. Sydney, Australia: PwC. Retrieved from http://www.pwc.com.au/industry/agribusiness/assets/Australian-Beef-Industry-Nov11.pdf

Raphaely, T., & Marinova, D. (2014). Flexitarianism: A more moral dietary option. *International Journal of Sustainable Society, 6*(1-2), 189–211. doi:10.1504/IJSSOC.2014.057846

Rogers, R. (2008). Beasts, burgers and hummers: meat and the crisis of masculinity in contemporary television advertising. *Environmental Communication: A Journal of Nature and Culture, 2*(3), 281-301.

Rothgerber, H. (2013). Real men don't eat (vegetable) quiche: Masculinity and the justification of meat consumption. *Psychology of Men & Masculinity, 14*(4), 363–375. doi:10.1037/a0030379

Rozin, P., Homes, J., Faith, M. S., & Wansink, B. (2012). Is meat male? a quantitative multimethod framework to establish metaphoric relationships. *The Journal of Consumer Research, 39*(3), 629–643. doi:10.1086/664970

Ruby, M. B., & Heine, S. J. (2011). Meat, morals and masculinity. *Appetite, 56*(2), 447–450. doi:10.1016/j.appet.2011.01.018 PMID:21256169

Smil, V. (2002). Eating meat: Evolution, patterns and consequences. *Population and Development Review*, *28*(4), 599–639. doi:10.1111/j.1728-4457.2002.00599.x

Smil, V. (2013a). *Should we eat meat?: Evolution and consequence of modern carnivory*. Chichester, UK: Wiley-Blackwell. doi:10.1002/9781118278710

Smil, V. (2013b, July 19). Should humans eat meat [Excerpt]? *Scientific American*, Retrieved from http://www.scientificamerican.com/article/should-humans-eat-meat-excerpt/http://www.scientificamerican.com/article/should-humans-eat-meat-excerpt/

Stanton, R. (2012). A plant based diet - good for us and for the planet. *The Medical Journal of Australia*, *1*(2), 5–6. doi:10.5694/mjao11.11508 PMID:25369929

Steinfeld, H., Gerber, P., Wassenaar, T., Castel, V., Rosales, M., & de Hann, C. (2006). Livestock's long shadow: environmental issues and options. Rome, Italy: Food Agriculture Organisation (FAO). Retrieved from ftp://ftp.fao.org/docrep/fao/010/a0701e/a0701e00.pdf

Taylor, N. (2012). Reversing meat-eating culture to combat climate change. Haslemere, UK: World Preservation Foundation (WPF).

Twigg, J. (1983). Vegetarianism and the meanings of meat. In A. Murcott (Ed.), *The sociology of food and eating: essays on the sociological significance of food* (pp. 18–30). Aldershot, UK: Gower.

Verain, M., Dagevos, H., & Antonides, G. (2015). Flexitarianism a range of sustainable food styles. In L. A. Reisch & J. Thøgersen (Eds.), *Handbook of research on sustainable consumption* (pp. 209–223). Cheltenham, UK: Edward Elgar Publishing. doi:10.4337/9781783471270.00023

Wang, Y., & Beydoun, M. A. (2009). Meat consumption is associated with obesity and central obesity among US adults. *International Journal of Obesity*, *33*(6), 621–628. doi:10.1038/ijo.2009.45 PMID:19308071

Warren, K. (2000). *Ecofemenist philosophy: a western perspective on what it is and why it matters*. Lanham, MD: Rowman and Littlefield.

Wilson, M. S., Weatherall, A., & Butler, C. (2004). A rhetorical approach to discussions about health and vegetarianism. *Journal of Health Psychology*, *9*(4), 567–581. doi:10.1177/1359105304044040 PMID:15231057

World Cancer Research Fund (WCRF), & American Institute for Cancer Research (AICR). (2007). Food, nutrition, physical activity and the prevention of cancer: a global perspective. Washington, DC: AICR. Retrieved from http://www.aicr.org/assets/docs/pdf/reports/Second_Expert_Report.pdf

KEY TERMS AND DEFINITIONS

Advertising: Communication promoting, encouraging or manipulating the use of particular products, services or particular behaviors.

Flexitarian: A person who consumes red meat in quantities less than 455 g per week and reduces other animal protein intakes.

Myth: A story or belief that guides behavior and cultural practices with or without factual basis.

Vegan: A person who abstains from eating any animal-based product.

Vegetarian: A person who eats mainly a plant-based diet and abstains from meat and fish consumption.

Chapter 16
Intensive Speciesism Regions in Brazil:
Ethical and Social Stains in Territories Supplying Global Meat Markets

Luciano Florit
University of Blumenau, Brazil

Cristiane Sbardelati
University of Blumenau, Brazil

ABSTRACT

This chapter brings together ethical, social and territorial implications of the meat industry in the regions of Brazil where animal husbandry is the main economic activity. This article thus articulates concepts elaborated in the field of environmental ethics applying them to sociological analysis of territorial development using the notion of 'regions of intensive speciesism'. The notion is elaborated as a conceptual tool to highlight territories whose socio-economic development pattern accepts the production of meat as a supposed "regional vocation". In such territories, non-human sentient beings are unquestionably put into the same category as things. The identification of this pattern enunciates interrelated implications of meat production that are not usually recognized. These implications include the high rates of health problems affecting workers in slaughterhouses; the symbolic and economic domination over territories and people by the agroindustry; and the drastic moral inconsideration of sentient beings. The article is based on the case study of Concórdia, a micro region located in the state of Santa Catarina in the south of Brazil.

INTRODUCTION

Food consumption patterns describing how most people meet their needs are largely defined by producers according to the logic globalized capitalism imposes on productive systems. This logic works, fundamentally, without great moral considerations, unless rules are clearly defined through regulatory and politically legitimate processes.

DOI: 10.4018/978-1-4666-9553-5.ch016

On the consumer side, according to the specific socio-economic and cultural conditions, there are higher or lower possibilities for interfering with the established patterns through individual choices. However, the expression "consumption pattern" refers to a situation in which practices observe regularities that are somehow not dependant on individual choices. From a probabilistic point of view, our food consumption practices are highly predictable and persistent. This is because individual decisions tend to be repeated routinely in established patterns, without a very conscious decision making process.

In this chapter we show how a certain part of the planet, in this case Concórdia situated in the south of Brazil, has established a productive pattern specialized in supplying pork and poultry to other regions in the same way that other regions supply goods, that is, with no moral considerations whatsoever in relation to these activities. Why should a region that is specialized in the production of pork and poultry have to submit to moral considerations that are different from a region that produces other ordinary goods? The answer is that, although similar to other commodities in the sense that living species are cultivated by farmers and processed industrially, the raw material that is part of the meat production line is made of bodies of individuals who have feelings, preferences and experience pain. Moreover, although similar to other manufactured goods, due to the embedded precision work involved, body parts of pigs and chickens are not mere pieces. They are however treated as such.

The ethical need to distinguish meat production from other types of industrial products and to explain that animals are not mere things is certainly not new. However, in the last four decades, this issue has gained new dimensions due to the development of industrial production and consumption standards linked to the distribution of roles in the current international division of labour. This has influenced the specialization of certain regions as providers of animal products to the international market for use by very distant consumers.

Concórdia – the subject of this case study, is a micro region situated in the state of Santa Catarina in the south of Brazil. Santa Catarina is one of the most important states in Brazil in terms of agro-industrial meat production (poultry, pork and derivatives), supplying both the domestic and international markets. Some of the most well-known agro-industries in Brazil, such as Brasil Foods (currently a merge of Sadia and Perdigão), originated in Concórdia and currently occupy prominent positions in the international meat market.

Animal agriculture is the main economic activity in the Concórdia micro-region, making up 70% of the GDP in some of its municipalities which in turn provide raw materials and labour to large slaughterhouses. This socio-economic and territorial configuration results from a historical process that enables the articulation of interests of various social groups with the agro-industries being the dominant players. Such configuration produces a distinctive development pattern with defined characteristics, shaped with no moral considerations around the radical instrumentalization of the lives and bodies of non-human sentient beings.

Economic, political, environmental and cultural dimensions constitute this scheme and link productive arrangements to socio-cultural values supported by normative and symbolic parameters. According to Florit, de Oliveira, Fleuri and Wartha (2016, p. 239), "(f)rom an environmental point of view this has resulted in the transformation of the landscape and the domesticated animal species both supported by socially defined conceptions of nature". Giddens (2009) describes 'patterns' as a set of practices that generally routinely reproduced. They are associated with a territory which consequently is also a product of these practices (Florit et al., 2016). The economic, political, cultural and environmental patterns connected with a 'region' indicate specific geographic areas where they operate and from where the means of their reproduction are attained (Theis, 2008).

Two sections are presented below. First, a theoretical discussion explaining the relationship between environmental ethics and territorial development is offered. This is followed by an empirical analysis of the case study which covers aspects of intensive speciesism and the economic, political and historical conditions shaping its appearance.

FROM ENVIRONMENTAL ETHICS TO REGIONAL ANALYSIS

The last decades have seen an increasing number of discussions questioning the ethical and moral implications of our relationships with non-human beings, which affect, or should affect, our notion of sustainability. These questions, which challenge the norms and morals in regard to the treatment of non-human beings, are essentially formulated in the field of environmental and animal ethics. However questions concerning the moral consideration of non-human beings, in particular the question of interests of animals, are usually contained within the context of environmental ethics. As such they are treated at a philosophical and normative level without systematically exploring the net of social and political relations that causes and supports these problems. It is thus important to place this type of question in an analysis that also captures the social and political relations. They sustain, naturalize and reproduce the circumstances under which environmental ethics formulates its objections.

Environmental ethics is the academic field in which the moral principles of our actions towards nature and non-human beings are critically considered. Within this context, environmental ethics helps in answering questions such as: Can we justify an activity that will create jobs but causes species to become endangered? Should we protect a national park or should we produce energy which will improve the life quality of certain people? Under what conditions could human activities that cause the death of animals be justified? The answers to such questions cannot be established only on factual information such as the amount of energy expected to be generated by one hydroelectric power plant or the feed conversion efficiency of one animal species. Even if such information contributes to the decision making process the issues being analysed require value judgment supported by preferences, aspirations and moral conceptions socially accepted by certain groups or by society in general.

The critical reflection around the ethical validity of these value judgments constitutes the essence of environmental ethics. This field formulates questions such as: Do we have a moral obligation concerning nature as a whole, species or individual living beings? Is the individual of a wild species more important than the individual of a domesticated species? Does a living being have a value in itself or is it only valuable if it is useful to satisfy human needs? The second range of issues refers to abstract and specific philosophical analysis. However, the first one clearly refers to circumstances that often arise in the course of concrete socio-economic development processes. This implies that the concrete development processes are subject to a critical ethical reflection, although the decisive factors that determine these processes are usually solely economic.

There are several branches of environmental ethics and various ways of classifying them. Robert Elliot categorizes the different branches of environmental ethics following four approaches centred on: a) Human beings; b) Animals; c) Life; d) Ethics of the whole or ecological holism (Elliot, 2004). This classification is based on environmental ethics' fundamental problem, namely how far to expand the scope of moral consideration. To what extent and for what reasons would we have to take moral responsibility for non-human beings? Does this responsibility include animals or other living beings such as plants or landscapes

When discussing sustainable development, it is common to find reasons to deal with nature considering only human needs. However, many scholars in this field share the belief that one of the most problematic traits of Western civilization is exacerbated anthropocentrism when dealing with nature. Such human-centred belief, which is defined and supported by religious traditions and moral philosophies, is reflected in political, economic and developmental conceptions that historically implemented the morality which justified the urge to transform nature for human domination (Sterba, 2010; Singer, 2002, Chp. 10).

The relevance of this reflection in regard to sustainable development becomes more evident if another distinction between the two branches of environmental ethics – holism and individualism, is established. Holists claim that since environmentalism focuses primarily in entities that represent totalities, such as ecosystems and species, environmental ethics should engage in supporting the moral status and intrinsic value of these entities. Therefore, it is necessary to adopt some form of holism, for which the classic paradigms are ill-prepared. The entities which are at the core of holistic concerns are composed of both biotic elements (animals, plants etc.) and abiotic components (rivers, rocks etc.). This form of justification explores the need to systematize the ethical implications of the Darwinian legacy which, in turn, informs preservationist thoughts and biological sciences' worldviews (Callicott, 2001).

On the other hand, the individualistic branch of thought seeks ethical foundations to defend the moral value of non-human individuals. The reasoning behind this is that these foundations could be more consistent and avoid new problems that normally arise when defending total entities[1]. Paul Taylor, supported by Kantian principles, defends a biocentric perspective in which one would have moral responsibilities towards "each individual organism… conceived of as a teleological centre of life, pursuing its own good in its own way" (Taylor, 2011, pp. 44-45). From another perspective, we have a moral responsibility towards beings that have the ability to experience sensations, feel pain and suffer. These are referred to as "sentient beings". Along these lines, through a consequentialist argumentation formulated in the utilitarian tradition, Peter Singer argues that the quality of sentience is sufficient to ensure that these beings have interests which should be considered, regardless of the species to which they belong. From a Kantian perspective, Tom Regan argues that these beings should be considered "subjects-of-a-life" which makes them worthy of rights. As right bearers, under no circumstances should these beings have their dignity and interests violated, regardless of the benefits that could eventually be generated from their hypothetical instrumental use.

The sentience-based arguments include all animals that have a complex nervous system, leaving doubts as to whether other animal categories should be morally considered[2]. However the interests of other animal categories could be preserved if based on the fact that they are subjects-of-a-life; thus making the latter a more inclusive category.

Although these lines of thought are often called "animalistic", they can also be applied in defence of broad landscapes that include plants, abiotic elements etc. This is true to the extent that wild sentient beings would not be able to ensure their interests without the preservation of the ecosystems to which they belong. Thus these perspectives provide foundations for the ethical reflection about the human treatment of both domesticated and wild animals. This is not true about the holistic perspectives which focus on "natural" or wild landscapes without taking domesticated animals into consideration. Therefore the individualistic perspectives focused on sentience enable analysis that crosses the boundaries between wild and domesticated environments once the foundation of their moral consideration is based on being sentient beings and is regardless of their relationship with humans.

The moral disregard for sentient beings due to their species is considered by many authors as *speciesism*. The Australian philosopher Peter Singer is one of the most influential authors discussing speciesism in contemporary society[3]. Singer is a contemporary author of utilitarianism - a consequentialist ethics that assesses actions according to their consequences. Utilitarian reasoning generally prescribes the maximization of wellbeing to individuals affected by an action as the criterion to establish whether this action is ethically justifiable or not. The distinctiveness of Singer's argument is his emphasis that the scope of moral consideration, in other words, the extension of categories of beings to which humans have moral obligation to care for their wellbeing, must be extended to include all creatures that can be considered sentient - beings capable of experiencing suffering and pain. Considering these beings have the capability to feel pain means that they also have a spontaneous tendency to seek satisfaction through what brings them well-being, and a sense of mental identity that implies appreciation for their own life. According to Singer sentient beings are those who, due to their nervous system, possess the ability to experience pleasure and pain, characterizing the fact that they have interests and preferences. These preferences cause them to seek satisfaction in accordance with their nature and avoid threatening feelings. In this sense sentience is what favours animals' physical and mental self-protection (Singer, 2004).

Speciesism is therefore a category that expresses the moral disregard of sentient beings based on their species. Hence to restrict our moral consideration towards humans is to practice speciesism. Thus, speciesism can be understood as a prejudice or biased attitude in favour of the interests of members of one's own species and against the interests of members of other species.

This is therefore the criterion endorsed by Singer to establish the boundary of moral consideration: sentient beings, including humans and non-humans. For Singer, " the capacity for suffering and enjoyment is…not only necessary, but also sufficient for us to say that a being has interests - at an absolute minimum, an interest in not suffering" (Singer, 2004, p. 9). Singer adds:

If a being suffers, there can be no moral justification for refusing to take that suffering into consideration. No matter what the nature of the being, the principle of equality requires that its suffering be counted equally with the like suffering - insofar as rough comparisons can be made - of any other being. If a being is not capable of suffering, or of experiencing enjoyment or happiness, there is nothing to be taken into account. So the limit of sentience (using the term as a convenient if not strictly accurate shorthand for the capacity to suffer and/ or experience enjoyment) is the only defensible boundary of concern for the interests of others (Singer, 2004, p. 10).

Although sometimes accused of radicalizing his arguments Singer sets himself a very conservative criterion to define the boundary of moral consideration. Unlike for example Taylor, Singer does not propose a model of biocentric ethics that has any form of life as a criterion of moral consideration.

So Singer is not criticized for the boldness of his position but rather, for its limitations. His attitude is deliberate; he chooses arguments that are highly reasonable and not counter-intuitive and which can be supported by science. He knows it is very hard to argue from an ethical point of view that causing suffering to a being who feels pain is innocuous.

Singer also chooses criteria that are feasible in practice if specific political attitudes are adopted – a consideration which is not so obvious with other perspectives that defend the intrinsic value of life for all and every being. For Singer there is a clear opportunity to consider alternatives to at least reduce the suffering inflicted on animals by industry in today's society. In other words, what is sometimes pointed out as limitations in Singer's arguments is also its strength[4].

Singer's contribution and the debate around his arguments have been fundamental in reaching a certain consensus on this issue. In fact, most authors who study this subject, including some Singer's critics, conclude that contemporary society needs to urgently correct the way animals are treated, especially those raised in the context of industrial systems. In an article that summarizes the state of the art of this discussion, Lori Gruen (2004) concludes:

While there are different philosophical principles that can help to decide how we should treat animals, they all share something that is beyond dispute: we should not treat animals the way our society treats them today. Independently of how lines are drawn, there are no plausible reasons to treat animals other than as beings worthy of moral consideration. (p. 480)

Non-human animals are highly discounted by industry considered only for their instrumental value. The instrumental value of an entity refers to the value that it has merely to the extent that it serves certain purposes. In other words, animals do not have any recognized value in themselves but only as means for human purposes. This allows industry to consider animals merely as raw materials. Philosophically speaking, animals are seen as things assessed on the value of being useful constituents for the production of other goods.

This symbolic reduction has enabled particular production patterns and territorial development which have experienced increased growth over the last few decades. However the ethical analysis of this pattern by itself cannot fully explain why the interests of certain types of non-human animals are systematically disregarded. A full explanation requires inclusion, as part of the theoretical approach and analysis model, of the social constraints (political, economic and cultural) which shape the societies where these practices are carried out

In globalized capitalism this issue is certainly related to consumption patterns that tend to acquire a global character. From a production point of view, however, the problem has a territorial dimension – in countries, regions and micro regions, where certain social relations determine the concentration of activities of breeding and slaughter of animals.

We define regions of intensive speciesism (RIS) as those in which the process of social construction of territory is associated with animal exploitation, mediated by specific relationships between humans. In these regions the conceptions of the nature of non-humans have been created in a history (Florit, 2004; Gudynas, 1999) that became entrenched through a regional way of life or "regional vocations" (Florit, 2012; Grava, 2013). These so-called vocations lead these regions to establish specific inter-regional relationships, both at a national and international scale, providing other regions with animal products while absorbing the associated environmental and ethical liabilities. The development pattern in these regions is based on reducing the conception of sentient beings to the category of ordinary things, denying them moral consideration. This happens to the extent that productive specialization requires animals to be considered as raw materials in a manufacture process. These regions have their own dynamics in which inherent aspects of production modernization and speciesism are especially intense, largely because the territorial segregation makes them immune from the contemporary debate about the treatment of animals.

This dynamic is the outcome of a social-political and economic process which occurs both at the subjective individual level and at the objective dimension of social life. From the subjective point of view, the symbolic conversion of animals to a merely instrumental status is combined with cultural values that facilitate political domination of subordinated social groups. This symbolic radical reduction contrasts with the selective moralization of animals, which characterizes contemporary practices with pets (Eder, 1996)[5] and cultural reflexivity around eating patterns, animal rights, global solidarity etc.

From an objective point of view, political legitimacy is given to an economic model that enables the livelihood of farmers and low-skilled industrial workers. This political order becomes a robust system, since it relies on the referred "regional vocations" (supposedly natural and self-evident) that lead to the denial of other production possibilities. These alleged vocations are backed up by social and political relationships linked to specific players who benefit from these so-called vocations. These players include economic agents, especially big slaughterhouses, and political agents at the municipal, state and national, levels.

Before the constitution of the RIS, farmers traditionally met their subsistence needs in their local communities through small-scale animal production with very limited integration with markets. However now, in RIS, animal production is oriented exclusively to meet big markets outside the region by large-scale production systems in which the logic of continuous increase of productivity prevails.

The transformation that led to this process was not just economic and technological but was also symbolic and required adjustments in subjectivity and identity of the farmers involved. These adjustments hide the fact that the former traditional farmer and the current industrial worker (sometimes the same person at different life stages) work in conditions with fundamental differences (both qualitatively and quantitatively) when considering the moral justifications for the activity. On the one hand, traditional producers use animals on a small scale to take care of their own livelihood and subsistence needs, following their own rules and rituals, with care and emotional recognition typical of their lifestyles. On the other hand, industrial workers are employed in capitalist relationships where production is motivated and designed to reproduce and expand capital and non-human animals are viewed as mere commodities.

Due to the strategies adopted by companies in the RISs, this activity tends to lead to an oligopolistic labour market, characterized by very few buyers of the labour force (the large companies) and a large number of suppliers. Thus this activity becomes the sole source of income for a significant portion of the population who do not have any other option but to repeatedly kill and cause suffering to sentient beings.

Regions of Intensive Speciesism are therefore regions whose development patterns are based on the premise of denigrating non-human sentient beings to the status of mere things. This reduction is constitutive of RIS as a regime of accumulation in which animals are raw materials in a highly competitive market whose dynamic pushes for a continuous increase in productivity and cost reduction.

In order to establish a framework to enable the identification of regions that fit this pattern, we use the number of slaughtered animals *per capita* in a given territory. This indicator shows the density of sentient animals being slaughtered in a specific territory with respect to the size of the human population in that territory. This method differs from the usual calculations which take into consideration the total weight of the carcasses or their monetary value and do not reveal the number of slaughtered beings. Thus the *per capita* slaughter ratio of a territory can be considered as an indicator of the degree of speciesism in the prevailing development pattern of this territory.

INTENSIVE SPECIESISM IN BRAZILIAN REGIONS

Although the breeding and slaughter of animals historically played a key role in shaping the Brazilian territory, today this influence seems to have entered a new phase. Cattle breeding is the activity that takes up the largest share of land in the country – about 158.7 million hectares in 2006, compared to 59.8 million hectares dedicated to food crops in the same year (Schlesinger, 2010). According to estimates, pastures occupy 20% of the Brazilian territory (Rosa, 2006).

Brazil has the largest commercial cattle herd in the world (Schlesinger, 2010, p. 5) and is also the world's largest meat exporter. National policies are strongly directed at increasing Brazilian participation in the international meat trade. According to the Ministry of Agriculture it is expected that by 2020 Brazilian meat production will supply 44.5% of the world market. Poultry will account for 48.1% of world exports and pork will account for 14.2%, which would mean that in 2020 Brazil would hold the position of world's leading exporter of beef and chicken (MAPA, 2012). Currently, over 50% of the world's beef market is dominated by Brazilian companies due to the strong internationalization of the sector and large overseas acquisitions made by Brazilian slaughterhouses (Schlesinger, 2010, pp. 25-26). In recent years the Brazilian cattle industry has acquired a new relevance in the territorial configuration of Brazil due to its participation in the global meat market and international division of production.

This growth comes despite livestock breeding, particularly cattle and pig farming, having been linked to a number of negative environmental impacts in recent years, such as the deforestation of diverse Brazilian biomes (da Motta, Hargrave, Luedemann & Sarmiento Gutierrez, 2011), greenhouse gas emissions (GHG) (FAO, 2009), the pollution of water resources (Denardin & Sulzbach, 2005) and soil degradation (de Almedia, 2011). Therefore, this new relevance has implications that are economic, environmental, and as emphasized here, implies a new stage in which speciesism constitutes one of the foundations of development patterns.

From an empirical point of view, the number of animals that make up the Brazilian productive system is very substantial (5.4 billion) if considered from the perspective of speciesism critics who attach importance to the experience of all sentient beings. More than 95% (birds and pigs) of all farmed animals live in confined conditions (Table 1). If we take into consideration the animals slaughtered in the same year, the number is somewhat smaller (5 billion), due to the life cycle of cattle (Table 2).

The number of slaughtered animals distributed by state reveals that the intensity of slaughtered sentient beings does not necessarily correspond to the states that have higher recent environmental impact due to livestock, or whose impact is more widespread because of deforestation and other environmental degradation processes. The Amazonian states are not included amongst the six states with the largest number of slaughters; even though today they are considered "villains" of deforestation and as a consequence strong agents of current environmental degradation associated with livestock. The reason is that the prevalent figures in terms of number of slaughtered animals are those stemming from the farming of birds and pigs.

Table 1. Animal population (cattle, pigs and birds*)

Species	Population
Cattle	209.541.109
Birds	5.197.861.850
Pigs	38.956.758
Total animal population	5.446.359.717

Brazil (IGBE, n.d.).
* Roosters, hens, chickens, chicks.

Table 2. Total slaughtered animals (cattle, pigs and birds*)

Species	Slaughters
Cattle	29.278.095
Birds	4.988.320.741
Pigs	32.510.569
Total slaughtered animals	5.050.109.405

Brazil (IGBE, n.d.).
* Roosters, hens, chickens, chicks.

The largest concentration of slaughtered animals is situated in the southern states: Paraná, Santa Catarina and Rio Grande do Sul respectively because of the pig and poultry sectors which operate predominantly with confined animals. In terms of slaughters *per capita,* the highest number is in the state of Santa Catarina, followed by Paraná, and Rio Grande do Sul.

Within the state of Santa Catarina the *per capita* concentration is more pronounced in some micro regions. According to data from 2010 (IGBE, n.d.; Florit, Sbardelati, Grava, Refosco & Pinto, 2014), while in Brazil the average is 26.47 slaughters *per capita*, in the micro region of Concórdia this number reaches 1,162 annual slaughters per person. In other words, Brazil, which occupies a prominent place worldwide in terms of livestock farming, witnesses a large internal regional concentration, considering the amount of sentient beings slaughtered in one year compared to the population of a given territory.

Employment offered by the meat industry is also concentrated in Concordia. The figures in Table 3 show the municipality of Concórdia with 26.21% of formal jobs being in the sectors of slaughter and manufacture of meat products. In the micro region, this percentage is 24.87%, while in the whole state of Santa Catarina, the percentage is down to 3.7%. In other words, in the RIS of Concórdia, 1 in 4 people

Figure 1. Slaughter per capita in Brazil, Santa Catarina and the micro region of Concórdia E.
Data source: IBGE/SIDRA data, Pesquisa Trimestral do Abate de Animais (Trimestral Research on Animal Slaughter, no date,).

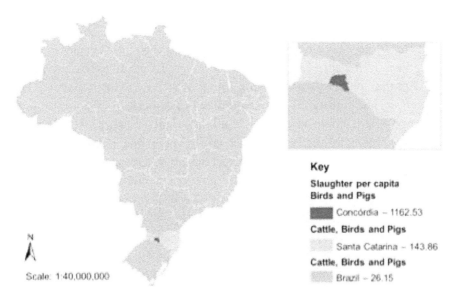

Table 3. Formal jobs in the slaughter industry and meat products manufacture in Santa Catarina, micro region of Concórdia and Municipality of Concórdia in 2011

	Total Number of Jobs	Jobs in the Meat Sector	Representation
Santa Catarina	1.504.114	55.143	3.66%
Micro Region of Concórdia *	42.357	10.533	24.87%
Municipality of Concórdia	21.341	5.594	26.21%

(SEBRAE, 2013; Grava, 2013).

* Data from micro region of Concórdia refer to the year of 2010

looking for a job will end up employed by the meat industry. If we consider the significant portion of population that has low professional skills, then for these people job opportunities outside the meat sector are really small.

From a historical point of view, the social structure in which the meat industry is the core component arises from a process that took place in the twentieth century. The existing local population prior to this time, such as the mestizos and native indigenous, were expelled from the territory due to conflicts with settlers and the construction of a railway that would connect the West of the state with the Southeast of the country. The conflicts were marked by numerous deaths which resulted in the local inhabitants fleeing to other regions.

Before colonization this region was inhabited by semi-nomadic indigenous groups, mostly belonging to the Laklãnõ Xoklengs and Kaigangs tribes (Ferreira, 1992), and by a population made of Brazilian mestizos, also known as caboclos (Nodari, 2009). The caboclo population lived in extreme poverty and land occupation was their dominant way of accessing land. They lived from subsistence agriculture and had a strong practice of collectivism. Initially, these lands were of no interest to livestock owners, but as the railway construction progressed in the West of Santa Catarina and land became more valuable, the caboclo population was forced to move to other regions in search of shelter and food. The railway construction began in 1911 and was undertaken by Companhia Brasil Railway (Company Brazil Railway), which helped farmers to expel the local population in order to vacate land to be sold to settlers.

The colonization occurred in the first half of the twentieth century as a result of a campaign to attract Italian and German farmers from the state of Rio Grande do Sul. From 1920 onwards the land occupation process was conducted by private companies that made colonization their business, buying state-owned lands or acquiring them in exchange for work. The lands were then sold to settlers (Nodari, 2009) who would cultivate them through family agriculture. The productive activity was developed through the work of family members and sometimes with the help of neighbours.

The predominance of small properties directed all upcoming economic development. It was common for farmers to breed pigs for their own private consumption and sell the surplus in the local and neighbouring markets. Some merchants would connect this productive region, especially the pigs, to the markets in the Southeast and return with manufactured products (Goularti Filho, 2007).

Due to the urbanization process and the emergence of the agro-industry sector the farmer has become a businessperson growing economically whilst still remaining linked to earlier agricultural roots. In the cities, families driven out of the countryside due to the modernization of agriculture, are employed directly by companies. In the rural areas the smallholders who remain in the agriculture sector are either indirect workers for the private companies or workers in agricultural cooperatives through the "integration" system (Schreineir, 1994, p. 16)[6].

In this way the small and medium landowners had their socioeconomic conditions changed when agricultural modernization started in the 1970s. These social changes resulted in the fragmentation of the small producers' cultivation, followed by reorganization to meet the new requirements of waged labour. In this way agriculture became increasingly more reliant on industry.

The modernization of agriculture integrated Concórdia's agricultural production with the international market, homogenizing the geographical area, while specializing and concentrating production. This homogenization established a new commercial relationship between the countryside and the city through production cooperatives, banks and financial incentives and the resellers of agricultural equipment.

This process also reframed the social conception of work in the region. Until the 1970s, the agriculture in Concórdia was primarily based in small farms owned by colonizers. Farmers were depicted as orderly, hard-working and honourable people, self-sufficient through the noble work on their land. This representation was built on a rather severe religious foundation. When the socio-economic changes happened in the 1970s and 1980s, the term "work" was revised and resized, and is now linked to disciplining and the normalization of people to factory labour.

New principles have arisen reframing traditional values that were once followed by rural workers who now work in slaughterhouses. As a result, new rules of conduct also appeared. These rules play an important role in the agro-industrial chain, both for slaughterhouse workers and integrated farmers. Today, many of the integrated farmers who produce pork and poultry say that they cannot imagine themselves performing another activity or changing professions because "breeding beasts" is the only skill they have. Thus, this employment is one more extension of the regional "vocation" which normalises killing sentient beings.

The most powerful actor of this process in Concórdia is Brasil Foods (formerly Sadia Company). Created in 1944, it soon gained strength and became the municipality "boss". Currently Brasil Foods (BRF) has 105,240 employees of which 5,300 work in Concórdia factories (Brazil Foods, 2013). This represents 25% of the total jobs in the municipality and around 95% of the jobs in the municipality meat sector characterizing the oligopsonistic nature of the labour market.

The high number of employees, the habit of supervising the unions, the ability to determine the election results and the participation in public posts at a municipal, state and federal level give BRF (and Sadia) power of decisive influence on the municipality's political life. The company executives always occupy strategic political posts beneficial to both the company and the industry. For instance, Sadia's founder, Attilio Fontana (1900 - 1989), was the Mayor of Concórdia, Congressman, Deputy Governor of Santa Catarina and President of the Bank of the State of Santa Catarina. His nephew, Victor Fontana (born 1916) was a Congressman, Deputy Governor of Santa Catarina and President of the Bank of the State of Santa Catarina. His grandson, Luiz Fernando Furlan (born 1946), was Chairman of the Board at BNDES (National Bank for Economic and Social Development) and Minister for Development, Industry and International Trade under the Lula government.

One of the consequences of this social formation is reflected in the health of the personnel working in slaughterhouses. In addition to physical and mental illnesses, they often suffer appalling work related accidents.

The large amount of repetitive tasks performed by workers has led to an increasing number of Repetitive Strain Injuries (RSI), also known as Work-related Musculoskeletal Disorders (WMSDs). There is also a high occurrence of psychological illnesses and severe mutilation of limbs and body parts. The kind of work done in slaughterhouses can bring irreversible damage for workers, including permanent disability. The fast-paced work, repetitive movements and higher productivity requirements in increasingly shorter amounts of time directly affect the staff and their work.

According to the research 'Perfil de Agravos à Saúde em Trabalhadores de Santa Catarina' (Disease Profile of Workers in Santa Catarina, 2013), there are approximately 2,700 leaves of absence each year in the meat sector. Between 2005 and 2011, there were 19,374 workers with work-related illnesses, which represent approximately 39% of the industry workers. Workers on disability leave are on average 37 years old (IFH, 2013, p. 33).

These figures show inadequate working conditions, which are reflected in the significant number of granted pension benefits. From 2005 to 2011, there were 1607 workers on leave who were diagnosed with depression; 1271 workers on leave due to back pain and 1238 workers on leave due to shoulder injuries. Diagnosed depression among slaughterhouse workers is 341% higher than among workers in other economic activities. The occurrence of soft tissue disorders among slaughterhouse workers is 426% higher than among workers in other sectors. Hence, the meat industry, especially pork and poultry, has been recognized as a crippling industry that destroys the worthy life of many of its workers.

The suffering inflicted on animals in this context does not cause any concern among the relevant social powers. Importantly, this issue is also almost ignored by the animal rights movement in Santa Catarina. According to a survey completed by Cunha (2015), there are twenty animal rights entities in the state of Santa Catarina, one of which, founded in 2006, is located in the municipality of Concórdia. These institutions carry out important work generally aimed at animal rights within urban issues. None of these groups openly acknowledge the problem of systematically breeding and slaughtering animals (Cunha, 2015). Thus indirectly and unintentionally, they also contribute to the lack of public discussion about the experienced reality in these territories.

This panorama reveals a scenario of economic and political domination and socio-cultural hegemony over the RIS of Concórdia. Its practices go beyond the region.

In a nutshell, the RIS of Concórdia is an example of a territorial development pattern shaped with no moral considerations or objections around the radical instrumentalization of lives and bodies of non-human sentient beings. The current practices stems from a process of political, economic and symbolic domination which produces high rates of health related problems in addition to the systematic and customary death of sentient beings (Figure 2).

CONCLUSION

Development is a process that in theory aims to enable the realization of human aspirations perceived as legitimate. These aspirations are not static. On the contrary, they change as they become progressively informed by social reflections around values considered fundamental, and the realization of the consequences resulting from practices adopted during the development process. Accordingly, notions related to desired development models have gone through recurrent and beneficial revisions in the contemporary world.

Since regional development patterns result from social and political creations, it is also the duty of these creations to deliberate on the values that they provide and reproduce. Accordingly, the relationship between regional development and values deserves to be considered in the context of regional planning and also when reflecting about our consumption patterns.

In recent decades, environmental ethics has advanced significantly in relation to the implicit values in the use of nature and domesticated animals. However, some territories which are specialized in activities, such as animal husbandry, deepen their supposed "vocations" without a consistent assimilation of these objections. The speciesism category, as well as others that were shaped in the field of environmental ethics, tends to be used as a normative parameter to evaluate individual behaviour with insufficient concern about the social conditions in which these practices are structured and socially reproduced. Nevertheless it is critical to place this kind of discussion in the context of the social and political relations that sustain, disseminate and reproduce the circumstances in which environmental and animal ethics are formulated.

Figure 2. Regions of intensive speciesism

The demand for ethics in relation to the treatment of animals is an issue that will be considered as a relevant dimension of development. It questions some of the fundamental values that underpin the "regional vocations" in various parts of Brazil. Therefore, for those regions whose development models are strongly grounded in animal husbandry, the ethical imperatives could represent either a threat or an opportunity.

However the demand for ethical conduct evident among the population in some consumer markets is tenuous and ambivalent in Brazil. This may deepen the current disparity in terms of territorialisation of values, making the Brazilian regions amplify their "vocation" for meat production and assume the ethical burden that is increasingly rejected by the population of importing countries. This could possibly leave such regions in the role of mere executors of tasks that are ethically challenged by consumer markets.

With regard to the production of pork and poultry, Brazil observes development patterns that although located in specific territories are global in scale, since their dynamics and reproduction stems from integration with global markets. The spatial segregation of these patterns is the hallmark of this phenomenon.

Such segregation, due to its invisibility to most consumers, supports this resistance to global discussions around the moral consideration of animals. Its political isolation also facilitates the violation of rights including those of human beings who suffer through degrading conditions. So while in some large consumer centres these activities become the subject of incisive and controversial questions, peripheral regions are willing to absorb this critique by building specialized territories whose spatial segregation protects them from extra-economic questions. This reflects a specific pattern of territory development:

the industrial labour market is an oligopsony; relationships between integrated family farmers and agro-industries are unbalanced; and political and symbolic dynamics designate the meat sector as a "regional vocation". That is, the region has a necessary so-called inclination for such activity.

However, challenging the practices in the regions mentioned above can result in positive effects. The fact that many human beings can only succeed economically through the suffering and systematic killing of animals is an issue that, if revisited, could bring dignity to the human beings involved.

The history of development is in a way the history of things that are normal at a certain point in time, but that under certain circumstances become inadmissible. To deliberate about this issue does not mean to start placing animals first at the expense of human beings who depend on this economic activity to survive. Instead, thinking about the unnecessary suffering of animals also means thinking about the unnecessary suffering of human beings who deserve opportunities that are up to the same moral considerations to which humanity is starting to aspire.

The logic of necessity that underlies speciesist practices follows a plurality of situations that makes the use and killing of animals happen under different motivations. Basically these motivations can be to meet subsistence needs driven by the use value of the meat, or, may be to make a profit, in which case the animals are appreciated only by their trade value. Both motivations can be considered speciesist. Nonetheless, they are profoundly different practices in terms of ethical content. The first practice tends to be imbued with moral rules which do not completely rule out all forms of compassion. It tends to be associated with certain emotions (for example, when children of a rural family flee to avoid witnessing the slaughter) and does not ignore the fundamental ethical dilemma of the situation. Meanwhile, the latter practice tends to simplify the operation merely to its productive aspect by considering animals to be ordinary things or raw materials. This is a purely instrumental rationality as seen in Concórdia – the Region of Intensive Speciesism discussed here. The first logic is associated with use value and integrated with the lifestyle of small farmers, indigenous populations and other traditional cultures. Conversely, the second logic is associated with the trade value of meat. It drives the meat industry chain whose results are deleterious and unacceptable from an ethical, environmental and social point of view.

The relationship that human beings establish with each other affects the relationship between human beings and other living species. In turn, the relationship that humans establish with non-human living beings also affects the relationship that humans have with each other. Usually sociological theses tend to see the relationship with nature (and therefore with non-human living beings) as a result of social relations. This is in fact a valid perception which is not only based on "objective" observations but also carries humanistic conceptions about human qualities. Nevertheless, once we start to consider the moral value of non-human living beings, it is imperative that we make efforts to gasp the other side of the process, which is certainly recursive. This implies an understanding that if on the one hand we are hardly able to achieve ethical relationships with animals in the context of exploitation between humans, on the other hand we will also not be able to achieve properly dignified relationships between humans while we continue to consider animals as mere things.

ACKNOWLEDGMENT

This article is a result of research project that has financial support from CNPq.

The authors are grateful to Christine Moura Eon for the translation to English and to Marina Beatriz Borgmann for the revision.

REFERENCES

Brazil Foods. (2013). Relatório anual e de sustentabilidade 2013 [Annual sustainability report]. Retrieved from http://www.brf-global.com/brasil/

Callicott, J. B. (2001). The land ethic. In D. Jamieson (Ed.), *A companion to environmental philosophy* (pp. 204–217). Malden, MA: Blackwell Publishers. doi:10.1002/9780470751664.ch14

Cunha, L. C. (2015). If natural entities have intrinsic value, should we then abstain from helpining animals who are victims of natural processes. *Relations, 3.*1. https://www.academia.edu/12378106/ If_Natural_Entities_Have_Intrinsic_Value_Should_We_Then_Abstain_from_Helping_Animals_Who_ Are_Victims_of_Natural_Processes

da Motta, R. S., Hargrave, J., Luedemann, G., & Sarmiento Gutierrez, M. B. (2011) Climate change in Brazil: Economic, social and regulatory aspects. Institute for Applied Economic Research. Retrieved from http://www.ipea.gov.br/agencia/images/stories/PDFs/livros/livros/livro_climatechange.pdf

de Almedia, R. G. (2011). Pastagens: um desafio nacional [Pasture: a national challenge]. Retrieved from http://www.portaldbo.com.br/Portal/Artigos/Pastagens-um-desafio-nacional/1048

Denardin, V. F., & Sulzbach, M. T. (2005). Os possíveis caminhos da sustentabilidade para a agropecuária da região Oeste de Santa Catarina [The possible sustainability paths for agriculture in the West region of Santa Catarina]. *Desenvolvimento em Questão, 3*(6), 87–115.

Eder, K. (1996). *The social construction of nature: a sociology of ecological enlightenment.* London, UK: Sage.

Elliot, R. (2004). La ética ambiental. [Environmental Ethics] In P. Singer (Ed.), Compendio de ética [Compendium of ethics] (pp. 391–404). Madrid, Spain: Alianza Editorial.

Felipe, S. T. (2003). *Por uma questão de princípios: alcance e limites de ética de Peter Singer em defesa dos animais* [A matter of principles: the power and limits of Peter Singer ethics for animal rights]. Florianópolis, Brazil: Boiteux.

Felipe, S. T. (2007). Etica e experimentação animal: fundamentosabolicionistas [Ethics and animal experimentation: abolitionist foundations]. Florianópolis, Brazil: Universidade Federal de Santa Catarina (UFSC).

Ferreira, A. G. Z. (1992). *Concórdia: o rastro de sua história* [Concordia: its history tracks]. Concordia, Brazil: Fundação Municipal de Cultura.

Florit, L. F. (2004). *A reinvenção social do natural: natureza e agricultura no mundo contemporâneo* [The social reinvention of the natural: nature and agriculture in the contemporary world]. Blumenau, Brazil: Edifurb.

Florit, L. F. (2012). Dilemas éticos e politicos, humanos e não humanos na gestão pública do desenvolvimento territorial sustentável. [Ethical and political, human and non-human dilemas in sustainable territorial development] In A. Philippi, C. A. Sampaio, & V. Fernandes (Eds.), *Gestão de natureza pública e sustentabilidade* [*Management of public nature and sustainability*]. (pp. 247–268). Barueri, Brazil: Editora Manole.

Florit, L. F., de Oliveira, L. B., Fleuri, R. M., & Wartha, R. (2016). A 'European Valley' in South America: regionalisation, colonisation and environmental inequalities in Santa Catarina, Brazil. In J. Dessein, E. Battaglini, & L. Horlings (Eds) cultural sustainability and regional development: theories and practices of territorialisation (pp. 235-248). Abington, UK: Routledge.

Florit, L. F., Sbardelati, C., Grava, D., Refosco, J. C., & Pinto, A. C. (2014, July 9). *Implicações éticas e sociais da "vocação regional" pela suinocultura e avicultura na microrregião de Concórdia/SC [Ethical and social implications of "regional vocation" for pork and poultry in the micro region of Concórdia].* Paper presented at the 2nd Seminário de Desenvolvimento Regional, Estado e Sociedade (SEDRES): A diversidade regional brasileira em perspectiva [Regional Development Seminar, State and Society: The Brazilian regional diversity put in perspective], Universidade Estudual de Paraíba, Campina Grande, Brazil.

Food and Agriculture Organization (FAO). (2009). Escritório Regional para a América Latina e o Caribe [Regional Office for Latina America and the Caribbean]. Retrieved from http://www.rlc.fao.org/pt/temas/pecuaria/

Giddens, A. (2009). *Sociology* (6th ed.). Cambridge, UK: Polity Press.

Goularti Filho, A. (2007). *Formação econômica de Santa Catarina* [Economic formation of Santa Catarina]. 2nd ed.). Florianópolis, Brazil: Editora da UFSC.

Grava, D. S. (2013). *A construção social da pecuária como 'vocação regional' em Santa Catarina: notas críticas sobre suas impliações socioeconômicas, ambientais e éticas [The social construction of animal husbandry as 'regional vocation' in Santa Catarina].* (Masters in Regional Development Dissertation), Universidade de Blumenau, Brazil.

Gruen, L. (2004). Los animals. [The animals] In P. Singer (Ed.), *Compendio de ética* [Ethics compendium]. (pp. 469–482). Madrid, Spain: Alianza Editorial.

Gudynas, E. (1999). Concepciones de la naturaleza y desarrollo en América Latina [Conceptions of nature and development in Latin America]. *Persona y Sociedad, 13*(1), 101–125.

Instituto Brasileiro de Geografia e Estatistice (IGBE). (n.d.). Sistema IGBE de Recuperação Automática - SIDRA: pesquisa trimestral do abate de animais [IGBE system of Automatic Recovery - ISOAR: Quarterly survey about the slaughter of animals. Retrieved from http://www.sidra.igbegov.br/

Instituto Fator Humano (IFH). (2013). Perfil de agravos à saúde em trabalhadores de Santa Catarina [Disease profile of workers in Santa Catarina]. Santa Catarina, Brazil: Relatório de Pesquisa, Instituto Fator Humano em parceria com UFSC, Univali apoio MPT, FECESC, FETIESC, SINTIARC e DIEESE.

Ministerio da Agricultura Pecuària e Abastecimento (MAPA). (2012). Exportação [Exports]. Retrieved from http://www.agricultura.gov.br/animal/exportacao

Nodari, E. S. (2009). *Ethnicidades renegociadas: práticas socioculturais no oeste de Santa Catarina* [Ethnicities renegotiated: socio-cultural practices in Western Santa Catarina]. Florianópolis, Brazil: Editora da UFSC.

Oliveira, E. M., & Goldim, J. R. (2014). Legislação de proteção animal para fins cientificos e a não inclusão dos invertebrados - análise bioética [Animal protection legislation for scientific purposes and non-inclusion of invertebrates - bioethical analysis]. *Revista Bioética, 22*(1), 45–56. doi:10.1590/S1983-80422014000100006

PDDR-Concórdia. (2014). *Plano diretor de desenvolvimento rural de Concórdia* [Plan for rural development of Concórdia]. Concórdia, Brazil: Secretaria de Agricultura. [Agriculture Department]

Rosa, F. R. T. (2006). *Areas de pastagem versus agricultura: o que aconteceu em 2005* [Grazing areas versus agriculture: what happened in 2005]. São Paulo, Brazil: Scot Consultoria.

Schlesinger, S. (2010). *Onde pastar? O gado bovino no Brasil* [Where to graze? The cattle in Brazil] (1st ed.). Rio de Janeiro, Brazil: Fase.

Schreineir, D. F. (1994). *A formação de uma cultura do trabalho: cotidiano, trabalho e poder. Extremo Oeste do Paraná [The formation of a labour culture: daily life, work and power. Extreme West of Paraná] - 1970/1988.* [Masters in History Dissertation]. Universidade Federal de Santa Catarina (UFSC), Florianópolis, Brazil.

Servico de Apoio às Micro e Pequenas Empresas (SEBRAE). (2013). Santa Catarina em números: Concórdia [Santa Catarina in numbers: Concórdia]. Florianópolis, Brazil: SEBRAE.

Singer, P. (2002). *Ethica prática* [Practical ethics]. São Paulo, Brazil: Martins Fontes.

Singer, P. (2004). *Libertação animal* [Animal liberation]. Rio de Janeiro, Brazil: Lugano Editora.

Sterba, J. (2010). Kantians and utilitarians and the moral status of nonhuman life. In D. R. Keller (Ed.), *Environmental ethics: the big question* (pp. 182–192). Chichester, UK: Wiley Blackwell.

Taylor, P. W. (2011). *Respect to nature: a theory of environmental ethics.* Princeton, NJ: Princeton University Press.

Theis, I. M. (2008). *Desenvolvimento-e-territótio: questões teóricas, evidências empíricas* [Development and territory: theoretical issues, empirical evidences]. Santa Cruz, Brazil: Edunsic.

Thomas, K. (1996). *O homem e o mundo natural: mudanças de atitude em relação as plantas e aos animais (1500-1800)* [Human and the natural world: attitude changes towards plants and animals]. Sao Paulo, Brazil: Companhia das Letras.

Varner, G. (2001). Sentientism. In D. Jamieson (Ed.), *A companion to environmental philosophy* (pp. 192–203). Malden, MA: Blackwell. doi:10.1002/9780470751664.ch13

KEY TERMS AND DEFINITIONS

Environmental Ethics: A philosophical discipline which analyses the moral relationships between people and the ecological environment.

Nonhuman Sentient Beings: Animals other than humans.

Regions of Intensive Speciesism (RIS): Regions in which the process of social construction of territory is associated with animal exploitation, mediated by relationships between humans.

Speciesism: A philosophical approach which endorses the supremacy of the human being in relation to other animals.

ENDNOTES

[1] . To deepen the debate between individualistic and holistic refer to (Varner, 2001) and Callicot's response (Callicott, 2001).

[2] Regarding the state of scientific knowledge concerning sentience between invertebrates see Felipe, S. T. (2003).

[3] It is worth noting that although this expression obtained wider dissemination with Singer, the word "speciesism" was coined by Richard Ryder, in 1970.

[4] For a detailed study of the scope and limits of Peter Singer's work in regard to animals rights, refer to Felipe, S. T. (2003).

[5] For the critical analysis of these justifications refer to among others, Thomas, 1996; Felipe, 2007; Singer, 2004.

[6] The "integration" system is basically a contract between the agro-industry and the farmers. The integrative agro-industry provides animals to the farmers (pigs and chicks), along with all necessary inputs, such as feed, medication and technical assistance for raising and termination (i.e. preparation for slaughter) of the animals. When the animals – pigs or birds – reach the desired age and weight, the integrated farmer hands over the live animals to the agro-industry, who is in charge of removing the herd from the property. The price is established by the agro-industry in function of the animal weight. The payment is delivered later on a pre-scheduled time (PDDR-Concórdia, 2014). The contracts are unbalanced, based on exclusivity and dependency, and all the production risks and necessary investments for production are with the "integrated".

Chapter 17
Meat Production and Consumption:
An Ethical Educational Approach

Paula Brügger
Federal University of Santa Catarina, Brazil

Dora Marinova
Curtin University, Australia

Talia Raphaely
Curtin University, Australia

ABSTRACT

This chapter presents the results of studies that unveil how meat and other animal derived products are causing severe environmental impacts, social problems and ethical concerns regarding both human and non-human animals. Although there are many ways to tackle the issue a critical non-anthropocentric education that encompasses ethics as a dimension of sustainability, is proposed. Traditional non environmental education often legitimizes values that are averse to an ethic that could be described as correct regarding the relationship between humans and the other animal species and even many educational currents that call themselves "environmental" are guided by a shallow conservationist point of view. Although welfarist practices may in some contexts be of help, the authors propose the animal abolitionist perspective as the unique genuine foundation for education to build this new paradigm.

INTRODUCTION

There is now a consistent and increasing body of evidence that shows that meat and other animal production chains are destroying the foundations upon which life on Earth depends (e.g. the overview by (Raphaely & Marinova, 2014). This economic sector is dramatically threatening the cycle of renewable and non-renewable natural resources, such as water and topsoil, by polluting, damaging or guzzling resources at a rate with which biogeochemical cycles cannot cope. It is also causing and accelerating

DOI: 10.4018/978-1-4666-9553-5.ch017

climate change, either directly by releasing greenhouse gases along the production chain or indirectly by triggering deforestation and substantial land use changes. Raising livestock, including for dairy and egg production, is also causing habitat loss and impoverishment, and is thus responsible for ravaging the planet´s biodiversity. Excessive meat consumption is further directly and indirectly negatively affecting human health (Raphaely & Marinova, 2014).

There is ample scientific evidence about the devastating impacts livestock has on the planet and its inhabitants, including humans. Despite this, consumption of meat and other animal products is considered a dietary norm across the globe. Is there a way to stop the destruction of the planetary and public health by overcoming people's addiction to meat or ignorance about its impacts? Considered a very private act of consumption, external social pressure (such as dietary recommendations, perceived expectations by others, attempts to alter social norms or rewards) may not be very effective in changing what people eat; knowledge and beliefs however can alter and transform their diets (Patterson, Kristal & White, 1996). In order to provide a knowledge foundation for dietary change, this chapter starts with an overview of negative impacts of animal production and consumption on the planet and public health and then exposes the power of the livestock sector interested in maintaining the current meat trajectory. The second part of the chapter tackles the issue about the separation between human and non-human sentient beings and argues for a change in the beliefs[1] people commonly hold about animals. Such changes could be achieved through an ethical educational approach. In fact, education has been found time and again to be the only socio-demographic variable exerting a consistent and significant influence on dietary changes (Srinivasan, 2006).

NEGATIVE IMPACTS OF MEAT PRODUCTION AND CONSUMPTION

The monumental negative impacts of animal production and consumption affect both – the planet's ecological environment and the health of human population. They are discussed below.

Livestock's Monumental Ecological Footprint

Many important studies emerged in the last decade detailing livestock's monumental ecological footprint and one of the most influential is the 2006 analysis by the Food and Agriculture Organization of the United Nations – *Livestock's long shadow* (Steinfeld et al., 2006). This study reveals many serious negative externalities from the livestock sector, including impacts on climate change, water and biodiversity. Its contribution to climate change, estimated at 18 percent of all greenhouse gas emissions, is higher than that of transport (Steinfeld et al., 2006). Taking into account the direct effects along with the impacts of feed crop agriculture required for animal production and energy consumption, the study pointed out that the livestock sector is one of the top two or three main contributors to the most severe environmental problems at a local and global scale. It was identified as a key factor in deforestation, especially in Latin America, and the single largest anthropogenic user of land, accounting for 70 percent of all agricultural land and 30 percent of the land surface of the planet (Steinfeld et al., 2006). Most of the Red List threatened species suffer habitat loss and degradation in proximity to livestock (Steinfeld et al., 2006).

Several studies published since FAO's analysis prove that there is no reason to believe that this iconic study overestimated livestock´s negative impacts. On the contrary, many studies indicate that FAO's analysis of impacts is conservative (such as (Goodland & Anhang, 2009; Pelletier & Tyedmers, 2010). Although it is thus possible to find some variation in the data related to different impacts, livestock´s externalities remain at the frontline of the most significant damaging anthropogenic activities on the planet. For example, according to FAO's further assessments (Gerber et al., 2013; FAO, 2013), a quarter of the earth's terrestrial surface is used for ruminant grazing and a third of the global arable land is employed to grow feed for livestock, accounting for 40 percent of the total cereal production. The growing demand for animal products lead by the wealthy countries is the main driver of disease emergence, spread and persistence in extensive and intensive livestock systems and in food supply chains as well as of the overall endemic disease burdens in human and animal populations threatening global biodiversity and biosafety (FAO, 2013).

Latin America (together with the Caribbean) is the region of the planet with greatest biodiversity hosting "almost one half of the world's tropical forests, 33 percent of its total mammals, 35 percent of its reptilian species, 41 percent of its birds and 50 percent of its amphibians" and representing a source of abundant genetic resources (UNEP, 2010b, p. 1). This region is now severely threatened by the impacts of its livestock sector.

The South American specialized beef industry (excluding dairy herds) is responsible for about 1 billion tonnes CO_2-eq of greenhouse gases (GHG) per year contributing 54 percent to the total emissions from the global specialized beef production and 15 percent to the emissions from the entire global livestock sector (Gerber et al., 2013). Brazil is of special interest as the country has the largest commercial bovine herd on the planet (ABIEC, n.d.). When greenhouse gases originating from fires and deforestation in Brazil are excluded, livestock (beef and dairy) becomes the main source of GHG, contributing more than 42 percent of the country's emissions (de Zen, Barioni, Bonato, de Almedia & Rittl, 2008, p. 4). From an economic point of view, the livestock´s greenhouse gases emissions are approximately forty times higher than those of agriculture per unit of generated GDP (de Zen et al.). Even if nationally appropriate government policies for the mitigation of greenhouse gases emissions from the livestock sector are put in place for the ten-year period from 2011 to 2020 (Gerber et al., 2013, p. 95), the livestock sector in Brazil and worldwide would be directly responsible for uncountable other negative externalities, including deforestation and biodiversity loss.

The expansion of feed crops, predominantly soybean production, is a significant threat to biodiversity. As soybean is traded internationally, GHG emissions from soybean expansion in Latin America are attributed to the respective production units around the world using the imported soybean cakes; pasture expansion and deforestation however are happening right in the Latin American region (Gerber et al., 2013, p. 41). For example, the Brazilian State of Mato Grosso is the center of the country's soybean industry, with production spreading into savanna and rainforest biomes (Lathuillière, Johnson, Galford & Couto, 2014). A study which analyzed the environmental footprints of soybean production and resource flows accompanying exports to China and Europe for the 2000-2010 period, showed that soybean production was associated with 65% of the State's deforestation and 14–17% of total Brazilian land-use related carbon emissions (Lathuillière et al., 2014). The livestock industry is primarily responsible for the fast rate of deforestation in the Amazon – 191 percent increase in August and September 2014 compared to the same period in 2013 (Ambiente Brasil, 2015a). Mato Grosso is the state with the highest number of deforestation alerts (Ambiente Brasil, 2015b). The Brazilian tropical savannah ecoregion, the Cerrado, is also severely affected by the replacement of native vegetation with soybean which has shallow roots and does not allow water to properly infiltrate the soil and reach the artesian sheets. Consequently groundwater levels are diminishing leading to the end of rivers and water reservoirs (Barbosa in Dias, 2015).

According to a 2014 World Wide Fund for Nature report (WWF, 2014), the state of the world's biodiversity appears worse than ever. The Living Planet Index (LPI), which measures trends in thousands of vertebrate species populations, shows a decline of 52 percent between 1970 and 2010: "the number of mammals, birds, reptiles, amphibians and fish across the globe is, on average, about half the size it was 40 years ago" (WWF, 2014, p. 8). Habitat loss and degradation, exploitation and climate change are threatening globally declining biodiversity with Latin America showing a staggering fall of 83 percent since 1970 (WWF, 2014, p. 2). Thirteen well-known conservationist biologists argued that half of the planet should be put aside in permanently protected areas to allow the survival of 10 million other species with which we share the Earth's habitat (Noss et al., 2012). While the eminent scientists argue that focus solely on climate change may obscure other pressing problems, such as land use change, they do not point the finger at the damage caused by livestock raising. Although the 2010 *Global Biodiversity Outlook 3* (CBD, 2010) recommended a shift to a moderate consumption of meat, new reports from influential and prestigious international bodies have failed to bring this issue at the forefront of any discussions or policy making. This is particularly surprising given the categorical evidence that excessive animal consumption has detrimental consequences for human health (Marsh, Saunders & Zeuschner, 2015; Stanton, 2015).

Livestock's Heavy Public Health Footprint

Other chapters of this book provide more detailed information about the threats the livestock sector poses to public health. However what we want to stress here is the link between humans and farm animals through new diseases and antibiotic use.

The FAO Report *World Livestock 2013: Changing disease landscapes* (FAO, 2013) focuses on global health, which includes human health as well as planetary health and safety in food production. According to the report, "most of the new diseases that have emerged in humans over recent decades are of animal origin and are related to the human quest for more animal-source food" (FAO, 2013, p. 2). This includes new influenza viruses (e.g. H5N1 highly pathogenic avian influenza, swine influenza A H1N1 and H3N2v virus) as well as the human immunodeficiency virus 1 (HIV-1), bovine spongiform encephalopathy (BSE commonly known as mad cow disease), severe acute respiratory syndrome (SARS), Ebola viruses and many other veterinary public health risks. Although animal-based products are part of global food security, the new agro-ecological and socio-economic conditions for raising animals are "creating public health threats associated with an animal-to-human pathogen shift, which implies pandemic risks, food safety hazards and high burdens of zoonotic diseases" (FAO, 2013, p. 3). The intensification of livestock production, referred to as mass production as well as industrial or factory farming, and associated supply practices not only "drive disease emergence, spread and persistence" (FAO, 2013, p. 3) but are also causing instability and reduced resilience in the planet's ecosystems through animal waste generation, pollution and chemical contamination of surface and groundwater, land and air with risks of spreading pathogens.

Of large concern is the use of antibiotics to prevent disease in factory farms (as prophylaxis), as feed additive to stimulate growth (FAO, 2013) and the unintended wide release into the environment through animal sewage and runoff water from agricultural sites (WHO, 2015). These drugs are from all major classes of antibiotics prescribed in human medicine (Food and Water Watch, 2012). Their use for animals is in subtherapeutic or non-therapeutic doses that can cause antimicrobial resistance, enhanced environmental survival of pathogens and increased pathogenicity in both humans and animals. There is already evidence that antibiotics are loosing their effectiveness for human medication because of routine use in industrial farms (FWW, 2012). In fact, agriculture uses the largest share of the globally produced antibiotics (WHO, 2015; FWW, 2012). The Food and Drug Administration (FDA) reports that 80 percent of antibiotics in US are sold for consumption (through food or water) by food-producing animals (Bowers, 2013). Even when ionophores (a type of antibiotics not used in humans) are taken out of the sales data, the proportion of antibiotics destined for factory farming drops only to 70 percent (Bowers, 2013). Because of the link between antibiotic resistant infections and foodborne pathogens (Landers, Cohen, Wittum & Larson, 2012), reputable organizations, such as the World Health Organization (WHO, 2001), call for termination or restriction of nontherapeutic use of antibiotics for livestock animals, shared responsibility and one health approach in clinical practice, veterinary practice and agriculture

BEEF POWER: SOCIAL, POLITICAL, ECONOMIC, AND ETHICAL ISSUES

In 2015, livestock accounts for 40 percent of the global value of agricultural production, with this share being more than half in industrial countries (FAO, 2015). Despite the serious concerns about the environmental and pubic health consequences this sector has, demand for animal products is constantly on the increase and expected to grow by 70 percent by 2050 (Gerber et al., 2013, p. 83). Intensive factory farming is seen as the modern answer to this growing demand and there seems to be a widespread unjustifiable acceptance that the world needs to consume animal products. According to some (Fraser, 2005), consumers just want cheap meat. This gives enormous power to the livestock industry to claim that it is responding to people's demands, ask for protection from market forces through subsidies or weather events through disaster reliefs as well as to spread its influence and vested interests across political and government circles. Below are some examples from Brazil. It is not difficult to find similar cases with a slightly different socio-economic context in almost any other country where intensive farming exists. Even where intensive farming isn't a predominant economic component, other industries closely aligned or related to the livestock sector such as industrial feed producers and plant gene banks, grain processors and traders, fertilizer companies, the pharmaceutical and biotech industries, food retailers and animal genetic industries, exert enormous power, influence and control (ETC Group, 2011).

The livestock industry is linked to vested interests which bond economic and political sectors. (Schlesinger, 2010) In Brazil, influential groups form "clusters" to fund this unsustainable industry diverting financial resources from public funds. According to one investigation (Motoki, Broggi, Falcão, Favoretto & Casteli, 2013), between 2007 and 2012 the National Bank for Economic and Social Development (BNDES) of Brazil – the main vehicle of the country's government to stimulate economic growth, invested BRL 9.5 billion (around USD 3.7 billion) in the three largest beef companies in the country – JBS, Marfrig and Brazil Foods. Instead of providing loans to be repaid by the beef industry, BNDES became a shareholder in these companies diverting funds from the public sector (about 40 percent of BNDES' budget comes from the Workers´ Support Fund raised through taxes paid by all workers to fund salary bonuses and unemployment benefits). Out of the BRL 3.5 billion bonds issued by JBS between 2009 and 2010, BNDES purchased 99.9% and the total support received by JBS from the government is estimated at BRL 13.3 billion (Leitão & Gribel, 2012). When investigated by the Court of Auditors about its dealings with JSB, BNDES obstructed the process, refusing access to information and claiming "banking secrecy" (Amora, 2014).

The economic power of the meat industries in Brazil is linked directly to social and political influence. For instance, one of the main directors of Brazil Foods was the Minister for Development, Industry and Trade in the government of Luiz Inacio Lula da Silva, between 2002 and 2007 (Motoki et al., 2013). In the 2012 municipal elections, JBS was a major donor to the campaigns of candidates across the country (Motoki et al., 2013) and in the 2014 elections, the company was listed as the biggest donor (Balza, 2014) without any ideological discrimination clearly revealing its determination to preserve its political and economic power. Although the donations were legal and made according to the requirements of the law, they show the power of the beef industry to influence political life.

Originating in the 1950s (José Batista Sobrinho started to supply meat to the workers during the construction of Brasília), JBS is now the world's biggest producer of beef after acquiring meat companies in US, Europe, Australia and Brazil (HBF & FOEE, 2014). It is also among the world's top ten international food and beverage companies and responsible for the slaughter of 85,000 heads of cattle, 70,000 pigs and 12 million birds per day; its power tentacles spread across 150 countries (HBF & FOEE, 2014). It is not surprising then that any calls for animal rights are threatening to the existing power structures (Tuttle, 2005).

The cruel reality of factory farms, slaughterhouses, meat-processing plants and packaging assembly lines are shattering not only to animals but these jobs often jeopardizes the physical (e.g. repetitive strain injuries) and mental (e.g. psychic disorders) health of the workers. Those who make the meat products out of the livestock animals endure poor working conditions, earn low wages and have low social status (Motoki et al., 2013; HBF & FOEE, 2014) making this employment a modern form of slavery.

Ethical concerns are present in all dimensions of the issue about animal production and consumption – from anthropogenic impacts on biodiversity and climate change, to hazards in both human and non-human health. The most serious ethical aspect however is the pitiless exploitation and brutal slaughter of more than 60 billions of sentient beings every year. Anthropocentrism, the dominance of an instrumental rationality and a mechanistic point of view are at the foundation of the pragmatic value we attribute to nature, the understanding that all natural components are "objects" and instruments for human use, including other sentient beings. This makes it easy to justify animal exploitation and slaughter for human consumption. By comparison, a holistic ecological worldview sees the world as an integrated whole and recognizes the fundamental interdependence in the web of life (Capra, 1996).

THE ANIMAL HEART OF THE MATTER: A CRITICAL ETHICAL ASPECT

The word animal is used in this chapter to refer to non-human animals. Dichotomies, such as human and non-human animals, are a discontinuous and fragmented way of thinking. Far from being just a mode of expression, terms like these lead to the very essence of thought because words represent particular paradigms and ways of seeing the world (Brügger, 2008). Even with this limitation, perceiving ourselves as animals may be of help to understand how we treat livestock.

These sentient beings (formally recognized as such in Article 13 of Title II of the Treaty of Lisbon of the European Union, 2008 which came into effect in 2009) are routinely confined in minuscule spaces, deprived of the most basic experiences that nature would offer if they were in their natural environments and suffer horrific treatment and mutilation, such as debeaking, mulesing, castration, limb amputations, transportation for slaughter at extreme temperatures, lack of space, food and water supply. In most, if not all, Western countries, farm animals are not protected by law and denied any *locus standi* to avoid such mistreatment (other animals, including pets are). Such an injustice is even more abominable when we know that animal protein and other animal derived products are not essential to human health (Craig & Mangels, 2009), even in quantities that would not cause environmental degradation, social and political problems or trigger human diseases. This fact shifts the discussion beyond the ecological balance perspective. Although necessary, the environmental standpoint is deficient when these ethical issues are at stake.

Ethical discussions in Western culture on this subject are not new. From Pythagoras (Greek philosopher and mathematician) and Plutarch (Greek historian) to twentieth century authors such as Ruth Harrison (English activist), Carol Adams (feminist), Tom Regan (philosopher), Peter Singer (philosopher), Steven Wise (American legal scholar) and Gary Francione (professor in law), to name a few, have argued that non-human animals are sensitive and intelligent living beings who should not be exploited or killed.

There is a variety of positions and arguments concerning the question of animal "rights" *versus* welfarism (e.g. Phelps, 2007). The most heated debates are between those who preach abolitionism, that is the immediate abolishing of all use of animals in any circumstances as the only safe way to set free non-human animals from all tyranny and exploitation, and those who believe in welfarism, that is taking incremental steps to improve the living conditions of animals and reducing their suffering as the most secure way in achieving the same objective, depending on the context and on how such tactics are implemented. The first – who claim to profess the only genuine abolitionism – believe that tactics which do not focus on abolition in the short term, such as welfare laws, only delay the recognition of animal rights instead of promoting them, because such tactics perpetuate the idea that animals are properties devoid of inherent value. On the other hand, welfarist reforms become counterproductive when they are seen as the ultimate goal rather than only a milestone (Phelps, 2007).

It is reasonable to argue that welfarist practices – similar to recycling and the monetization of environmental externalities – are examples of relationships between society and nature that are marked by the absence of an ethical/epistemological revolutionary bias, in the sense of building a counter-hegemonic ideology. In the case of wefarist propositions, non-human animals are still regarded as property and in the other two examples, the dominant production processes remain something that only needs minor adjustments, perpetuating the vision of nature as a mere part of an entire production sector. What defines the potential nature of a set of practices, regulations or laws – whether revolutionary or solely reformist - is their purpose, objective or end, that is, their teleological account. There will be no revolutionary plea if our institutions and society in general treat such proposals as ultimate goals, not palliatives or paths towards a true paradigm shift. The teleological component along with the epistemological foundations has a dialectical relationship with the premises of a cause or its ethical stances (Brügger, 2009a). Although those who advocate abolitionist tactics *tout court* disagree precisely on the existence of such possibilities and frontiers, it is defended here that some isolated steps might help in achieving the final goal of abolitionism, depending on their dialectical relationship with other teleological constituents. Transcending reformist perspectives requires an ethical educational process based on an abolitionist paradigm.

An eloquent foundation when it comes to animal abolitionism derives from the concept of speciesism, proposed by Richard Ryder in the 1970s. He postulated that: "like racism or sexism – speciesism is a prejudice based upon morally irrelevant physical differences – we treat the other animals not as relatives but as unfeeling things. Our concern for the pain and distress of others should be extended to any ´painient´- pain-feeling - being regardless of his or her sex, class, race, religion, nationality or species" (Ryder, 2005). Ryder (2005) argues that all beings that feel pain deserve human rights.

Animals' emotional lives and sentience have been systematically studied for decades (e.g. Masson & McCarthy, 1996; Bekoff, 2000, 2007). In 2012, a group of outstanding neuroscientists, neuropharmacologists, neurophysiologists, neuroanatomists and computational neuroscientists assessed the available substantial data on conscious experience and related behaviours in human and non-human animals and affirmed that: "The absence of a neocortex does not appear to preclude an organism from experiencing affective states. Convergent evidence indicates that non-human animals have the neuroanatomical, neurochemical, and neurophysiological substrates of conscious states along with the capacity to exhibit intentional behaviours. Consequently, the weight of evidence indicates that humans are not unique in possessing the neurological substrates that generate consciousness. Non-human animals, including all mammals and birds, and many other creatures, including octopuses, also possess these neurological substrates" (The Cambridge Declaration on Consciousness, 2012, n.p.). This agreement based on evidence and principles shows that there are no more scientific arguments left to support the idea that humans have the right to use other animals as their property (Francione, 2004). Although only a fraction of the animal kingdom is covered by the Cambridge Declaration of Consciousness, namely *subphylum Vertebrata*, *Mammalia* and *Aves*, mammals and birds constitute the majority of the 60 billion of animals raised and slaughtered for food every year and many more exploited and killed for other animal products. Irrespective as to whether this is considered a legal, philosophical or ethical issue, such a starting point provides sufficient basis for the route towards abolitionism starting with education.

TOWARDS AN ETHICAL EDUCATIONAL APPROACH: THE ABOLITIONIST PERSPECTIVE

Livestock raised for human consumption is a grave anthropocentrically induced malady and the kaleidoscope of its negative externalities is its syndrome – a collection of signs and symptoms that point to the health threats from this condition. Many feasible solutions exist to tackle the "livestock syndrome". Public policies are urgently needed to tax the negative externalities that arise from this sector and make visible its true public costs. It is also theoretically viable to improve laws and tighten inspections on factory farms and other animal-processing plants to stop, denounce and punish animal abuses and human rights violations. This would be fair, since the costs of these externalities are distributed to society in general, while the profits are concentrated in the hands of a few. However as argued earlier, because of the dominance in our Western culture of an instrumental rationality coupled with the power of the livestock sector, problems are likely to arise in this pragmatic field of possible solutions.

For example, Gerber et al. (2013) believes that with further research and development together with improved measurement methodologies and evolved market-based instruments, it would be possible to resolve or at least minimize these externalities. Besides being a one-dimensional view of addressing livestock´s negative impacts, carbon markets including the Kyoto Clean Development Mechanism (CDM) and other sources of mitigation finance, such as the UN REDD (Reducing Emissions from Deforestation and Forest Degradation) program, as well as UN reports aimed at achieving practical actions (e.g. ten Brink et al., 2009), may be of little help in a world scenario where responsibilities are migrated from governments to markets, and social, economic and political inequalities abound (e.g. Packer, 2012).

Many benefits, both for the environment and human health, can be achieved directly by reducing animal protein products and intake. In a EU study, it is argued that "halving the consumption of meat, dairy products and eggs in the European Union would achieve a 40% reduction in nitrogen emissions, 25–40% reduction in greenhouse gas emissions and 23% per capita less use of cropland for food production. In addition, the dietary changes would also lower health risks" leading to a reduction in cardiovascular mortality (Westhoek et al., 2014, p. 196). Another study which assessed the impact of diet changes on the blue (freshwater in rivers, lakes, reservoirs and aquifers extracted for irrigation) and green (naturally infiltrated rain, attached to soil particles and accessible by plant roots) water footprints of food consumption found that "reducing animal products in the human diet offers the potential to save water resources, up to the amount currently required to feed 1.8 billion additional people globally" (Jalava, Kummu, Porkka, Siebert & Varis, 2014, p. 14).

Although these studies do not focus on animal ethics, these data are very elucidative because they describe a rich set of positive effects from reduction in animal product consumption for the human species but also on the biosphere. One can think of the term biosphere only within a selfish and strictly human perspective and this term together with others, such as biodiversity, are woven into an anthropocentric context. This is where the challenge begins: to transform the way we feel, envisage and deal with our biophysical environment. Here, we leave the dominion of coercive actions or pragmatic data to enter a realm of movement from necessity to liberty through the path of education (Gramsci, 1971; Freire, 1971).

From a genuine non-anthropocentric educational perspective, the concept of biosphere acquires "thickness" in space and time in order to encompass historical, political and ethical facets. Merely in this context can we consider the interests of non-human animals not only because this new attitude would positively enhance sustainability, but also because they are bearers of intrinsic value, "subjects of a life" (Regan, 2001), and for their capacity to suffer and feel pain (Ryder, 2005; The Cambridge Declaration on Consciousness, 2012). This ethical dimension needs to encompass the humanity's relationships with nature and its attempts to achieve sustainability. A way to accomplish this is to examine the link between environmental education and animal abolitionism as a key for building the new epistemological approach (Brügger, 2009b).

This is important because traditional non-environmental education often legitimizes values that are averse to an ethic relationship between humans and the other animal species. Even many educational approaches that call themselves "environmental" are guided by a shallow conservationist point of view that tolerates animal exploitation and trivializes the slaughter of animals (predators, for instance) whenever they are seen a menace to human activities. No consideration is given as to whether these human enterprises are necessary or are no more than profitable activities designed to satisfy consumerist whims. A wide and profound picture may unveil how fragile some of our taken for granted necessities can be. An interdisciplinary approach can show how critical environmental education is for considering the whole picture.

It could be argued that environmental education has been failing in its task and propose alternative terms, such as ecopedagogy, green or ecological education or even vegan or abolitionist education. Adjectives however matter only to some extent and they may turn out to be mere labels. What is crucial is to observe and analyze the binomial society-education in the perspective of its underlying knowledge and paradigms: "Epistemology is in itself ethics, and ethics is epistemology" (Marcuse, 1966, p. 125).

The clock is ticking with billions more animals being slaughtered and education, with or without any adjective, is struggling to deliver any critical thinking. Although it is possible to agree that many positive changes have been taking place in some contexts, the most glaring traits of the paradigm that has been responsible for the current environmental crisis, including the way we treat non-human animals, are still dominant, namely anthropocentrism, individualism, consumerism, supremacy of technology, productivity as an icon, competitiveness, efficiency and the culture-nature dichotomy. These values are inconspicuously anchored at the bottom of our minds; they are regarded as indisputable values, good in themselves. It is uncertain whether we are to enter a new era concerning the relationship between society and nature and between humans and non-human animals. Decades of environmental education have not succeeded in changing the way we relate to nature. However, has it really been a critical environmental teaching? When hegemonic values such as speciesism are maintained, the teaching process transmutes into a "cheating" tactic (Brügger, 2004a).

Considering the theme discussed in this chapter, the animal abolitionist perspective is the only authentically ethical way to relate to non-humans. While welfarist practices may in some cases and contexts help cleaning the way to abolitionism, welfarist foundations can in no case constitute a solid ground for a critical education. Education as a whole is not an assemblage of concepts which hover above social or ethical values. A deeper and more comprehensive look at graduate curricula of areas such as Biological Sciences, Pharmacology, Economics or Industrial Engineering reveals values and practices that are inextricably associated with the dominant paradigm. There are important counter-hegemonic studies that lead to animal abolitionism which emerge from these areas but they are still regarded either as peripheral or as mere curiosities, assigned to the so-called elective subjects. They constitute no more than "islands" of counter-hegemonic paradigms, not basic knowledge (Brügger, 2004b, pp. 167-170).

Fundamental questions here are: what kind of a professional will emerge from such curricula, what are we educating for and for whom? The technicist nature of modern education is designed to accomplish the demands of a market whose views are crystallized in hedonistic and anthropocentric values (Raphaely & Marinova, 2013) that disregard ethical, social and natural limits of the planet and the suffering of animals, among many other issues.

An ethical educational approach requires putting the abolitionist perspective at the core of any discipline, subject or program being taught. Moreover, the educational process has to transform itself. To transcend the existing hegemonic paradigm the following five principles need to be applied to achieve ethical education (Brügger, 2004b, pp. 164-170; Raphaely & Marinova, 2013):

1. **Inter- and Transdisciplinary Approach:** The first and most comprehensive principle is to recognize that the mere choice of conventional topics (e.g. deforestation or biodiversity loss) will not modify the foundations of education per se unless these topics are regarded as "generating themes" in order to encompass historical, ethical, cultural and political aspects, to quote a few. The topics need to be explored looking for knowledge within, between and across all disciplines (Gadotti, 2008; Raphaely, Marinova & Todorov, 2010; Raphaely & Marinova, 2013). This requires a systemic view that addresses the topics in their spatial and temporal "thickness" building new epistemic fields and not condemning this "new" approach to remain an island with a different rationality;

2. **Non-Anthropocentric Paradigms:** The second pillar is the adoption of non-anthropocentric (zoo-bio-ecocentric) paradigms. Peace requires refraining from exerting power over others regardless of its form: speciesism, racism, sexism or imperialism. Diversity similarly demands freedom and equality, not hierarchies or ego-actions. Values that support the dominant anthropocentric paradigm need to be challenged and rejected (Spretnak, 1999);

3. **Metaphorical Nature of Knowledge:** An important principle is to recognize the metaphorical nature of knowledge. Science is a human creation devoid of neutrality both in its epistemological and ethical aspects. This means that we must understand the limits of science (and technology) in providing solutions and reliable answers in many contexts and subjects. As Werner Heisenberg brilliantly affirmed in his 1958 book *Physics and Philosophy: The Revolution in Modern Science*,

"what we observe is not nature itself but nature exposed to our method of questioning" (Heisenberg, 1999, p. 58);

4. **Multi-Dimensional Way of Thinking:** Another point is that a one-dimensional way of thinking prevails in our techno-scientificist culture in which debates on conflictive perspectives are rarely stimulated. A critical education cannot be based on a monolithic thought because its transformative potential would be depleted. It requires multi-dimensionality, the acceptance of mixed realities and imagining of different possible realities (Raphaely & Marinova, 2013); and

5. **Multiple Intelligences:** A major problem of traditional education is that it does not encourage an ethics that allows personal attributes to flourish. This could be accomplished by diversifying and encouraging other modalities of constructing "knowledge" and by establishing equilibrium between the cognitive and the affective dimensions of the educational process. Zohar and Marshall (2001), for example, point out that psychologists have for some time recognized that in addition to the commonly identified intelligence quotient (IQ), there also exist emotional and spiritual intelligences. A multiple intelligence approach would favour the development of other matrices of knowledge, such as based on intuitive reason, spiritual and emotional intelligence.

Many of us carried as children strong convictions about animal sentience without the need for any evidence. Maybe not all children have feelings and creeds about animal sentience, but the truth is that – when it is the case – society takes the charge to give children untruthful answers to their questions, responses that challenge their own internal beliefs in order to promote adaptation. This modification of our inner essence gradually intensifies to the extent that our primeval matrix of rationality is erased or severely blurred. This acquisition of new "lenses" - through which we perceive and experience our surroundings – finally blinkers the instruction that comes from within, which is part of our essence. Awakening the kid inside us would help promote altruistic values and reconnect us to nature. Adopting an ethical educational approach will allow children and grown-ups to feel self-empowered to also perform the right action.

CONCLUSION

A substantial global diet change away from animal products is needed to contain climate change, improve the state of the planet's ecosystems and reduce humans' impact on the environment (UNEP, 2010a). It is also needed to improve and avoid further threats to human health. There are however big forces at play which hold the reins of such a process of transformation. Two of them are the power and vested interests of the livestock industry which transcend national boundaries, and an educational system that encourages maintaining the *status quo*.

Is it legitimate to continue to educate people without challenging the ethical foundation of knowledge? Against the clear evidence that non-human animals are sentient beings who feel pain as human animals do, what right do we have to continue to exploit beings for consumption? In the quest to counter the hegemonic forces and domination of old paradigms and principles that profess the supremacy of some and condemn others to the slaughter, animal abolitionism is a privileged way of education and freedom to change the *status quo* and bridge the human-nature dichotomy.

REFERENCES

Ambiente Brasil. (2015a). Desmatamento tem alta na Amazônia em Agosto e Setembro, diz Imazon [Deforestation in the Amazon high in August and September, said Imazon]. Retrieved from http://noticias. ambientebrasil.com.br/clipping/2014/10/20/109734-desmatamento-tem-alta-na-amazonia-em-agosto-e-setembro-diz-imazon.html

Ambiente Brasil. (2015b). Inpe detecta aumento do desmate da Amazônia entre Agosto e Outubro [INPE detects increases in Amazon Deforestation between August and October]. Retrieved from http://noticias. ambientebrasil.com.br/clipping/2014/11/29/110876-inpe-detecta-aumento-do-desmate-da-amazonia-entre-agosto-e-outubro.html

Amora, D. (2014, September 17). BNDES obstrui auditoria sobre empréstimos de R$ 8 bi ao JSB/Friboi, sez TCU [BNDES blocks audit of R$ 8 billion in loans to the JBS/Friboi, says TCU]. *Folha De S. Paulo.* Retrieved from http://www1.folha.uol.com.br/mercado/2014/09/1517673-bndes-obstrui-auditoria-sobre-emprestimos-de-r-8-bi-ao-jbsfriboi-diz-tcu.shtml

Associação Brasileira das Indústrias Exportadoras de Carnes (ABIEC). (n.d.). Rebanho Bovino Brasileiro [Brazilian Beef Herd]. Retrieved from http://www.abiec.com.br/3_rebanho.asp

Balza, G. (2014). Campeã em doações, Friboi virou gigante da carne com R$ 10 bi do BNDES [Champion in donations, Friboi turned meat giant with US$ 10 billion from BNDES]. Retrieved from http://eleicoes.uol.com.br/2014/noticias/2014/08/10/campea-em-doacoes-friboi-virou-gigante-da-carne-com-r-10-bi-do-bndes.htm

Bekoff, M. (2000). *The smile of a dolphin: remarkable accounts of animal emotions.* Washington, DC: Random House/Discovery Books.

Bekoff, M. (2007). *The emotional lives of animals: a leading scientist explores animal joy, sorrow and empathy - and why they matter.* Novato, CA: New World Library.

Bowers, B. (2013). Rep. Louise Slaughter says 80% of antibiotics are fed to livestock. Retrieved from http://www.politifact.com/truth-o-meter/statements/2013/oct/15/louise-slaughter/rep-louise-slaughter-says-80-antibiotics-are-fed-l/

Brügger, P. (2004a). 25 years past Tbilisi: environmental teaching or teaching? In W. Leal Filho & M. Littledyke (Eds.), *International perspectives in environmental education* (pp. 129–138). Frankfurt, Germany: Peter Lang International Academic Publishers.

Brügger, P. (2004b). *Educação ou adestramento ambiental* (3rd ed.). Florianópolis, Brazil: Letras Contemporâneas.

Brügger, P. (2008). Vivissecção fé cega faca amolada? [Vivisection: blind faith, sharpened knife?] In C. A. Molinaro, F. L. F. de Medeiros, I. F. Sarlet, & T. Fensterseifer (Eds.), *A dignidade de vida e os direitos fundamentais para além dos humanos: uma discussão necessária* [The dignity of life and the fundamental rights beyond the human: a needed discussion]. (pp. 145–146). Belo Horizonte, Brazil: Fórum.

Brügger, P. (2009a). Para além da dicotomia abolicionismo versus bem-estarismo [Apart from the abolition dichotomy versus welfarism]. Retrieved from http://www.anda.jor.br/26/11/2009/para-alem-da-dicotomia-abolicionismo-versus-bem-estarismo

Brügger, P. (2009b). Nosotros y los otros animales: Especismo veganismo y educación ambiental [We and the other animals: speciesism, veganism and environmental education]. *Linhas Criticas Brasillia, 15*(29), 197–214.

Capra, F. (1996). *The web of life: a new scientific understanding of living systems.* New York, NY: Anchor Books.

Craig, W. J., & Mangels, A. R. (2009). Position of the American Dietetic Association: Vegetarian diets. *Journal of the American Dietetic Association, 109*(7), 1266–1282. doi:10.1016/j.jada.2009.05.027 PMID:19562864

De Zen, S., Barioni, L. G., Bonato, D. B. B., de Almedia, M. H. S. P., & Rittl, T. F. (2008). Pecuária de corte brasileira, impactos ambientais e emissões de gases efeito estufa (GEE): Sumário Executivo [Brazilian beef cattle, environmental impacts and emissions of greenhouse gases (GHG)]. Sao Paulo, Brazil: Universidade de Sao Paulo. Retrieved from http://www.cepea.esalq.usp.br/pdf/Cepea_Carbono_pecuaria_SumExec.pdf

Dias, E. (2015). O Cerrado está extinto e isso leva ao fim dos rios e dos reservatórios de água [The Cerrado is extinct and this leads to the end of rivers and water reservoirs]. *Jornal Opcão,* 2048. Retrieved from http://www.jornalopcao.com.br/entrevistas/o-cerrado-esta-extinto-e-isso-leva-ao-fim-dos-rios-e-dos-reservatorios-de-agua-16970/

European Union. (2008). Treaty of Lisbon. Consolidated versions of the Treaty on European Union and the Treaty on the Functioning of the European Union. Retrieved from http://eur-lex.europa.eu/legal-content/EN/TXT/?uri=uriserv:OJ.C_.2008.115.01.0001.01.ENG

Food and Agriculture Organisation of the United Nations (FAO). (2015). World agriculture: towards 2015/2030. Livestock production. Rome, Italy: FAO, Corporate Document Repository. Retrieved from http://www.fao.org/docrep/005/y4252E/y4252e07.htm

Food and Agriculture Organisation of the United Nations (FAO). (2013). World livestock 2013: changing disease landscapes. Rome, Italy: FAO. Retrieved from http://www.fao.org/docrep/019/i3440e/i3440e.pdf

Food and Water Watch (FWW). (2012). Antibiotic Resistance 101: How antibiotic misuse in factory farms can make you sick. Washington, DC: FWW. Retrieved from http://documents.foodandwaterwatch.org/doc/AntibioticResistance.pdf

Francione, G. L. (2004). Animals: property of persons? In C. R. Sunstein & M. C. Nussbaum (Eds.), *Animal rights: current debates and new directions* (pp. 108–142). New York, NY: Oxford University Press.

Fraser, D. (2005). Animal welfare and the intensification of animal production: an alternative interpretation. Rome, Italy: Food and Agriculture Organisation of the United Nations (FAO). Retrieved from ftp://ftp.fao.org/docrep/fao/009/a0158e/a0158e00.pdf

Freire, P. (1971). *Pedagogy of the oppressed.* New York, NY: Herder and Herder.

Gadotti, M. (2008). Education for sustainability: A critical contribution to the Decade of Education for Sustainable Development. *Green Theory and Praxis: The Journal of Ecopedagogy, 4*(1), 15–64. doi:10.3903/gtp.2008.1.3

Gerber, P. J., Steinfield, H., Henderson, B., Mottet, A., Opio, C., & Dijkman, J. ... Tempio, G. (2013). Tackling climate change through livestock: a global assessment of emissions and mitigation opportunities. Rome, Italy: Food and Agriculture Organisation of the United Nations (FAO). Retrieved from http://www.fao.org/docrep/018/i3437e/i3437e.pdf

Goodland, R., & Anhang, J. (2009, November/December). Livestock and climate change: What if the key actors in climate change are... cows, pigs and chickens? *World Watch Magazine, 22*(6), 10–19.

Gramsci, A. (1971). *Selections from the prison notebooks.* London, UK: Lawrence and Wishart.

Group, E. T. C. (2011). Who will control the green economy? *Corporate concentration in the Life Industries: Communique,* 107. Retrieved from http://www.etcgroup.org/content/who-will-control-green-economy-0

Heinrich Böll Foundation (HBF), & Friends of the Earth Europe (FOEE) (2014). Meat Atlas: Facts and figures about the animals we eat. Ahrensfelde, Germany: HBF and FOEE. Retrieved from http://www.boell.de/sites/default/files/meat_atlas2014_kommentierbar.pdf

Heisenberg, W. (1999). *Physics and philosophy: the revolution in modern science.* Amherst, NY: Prometheus Books.

Jalava, M., Kummu, M., Porkka, M., Siebert, S., & Varis, O. (2014). Diet change - a solution to reduce water use? *Environmental Research Letters, 9*(7). doi:10.1088/1748-9326/9/7/074016

Landers, T. F., Cohen, B., Wittum, T. E., & Larson, E. L. (2012). A review of antibiotics use in food animals: Perspective policy and potential. *Public Health Reports, 127*(1), 4–22. PMID:22298919

Lathuillière, M. J., Johnson, M. S., Galford, G. L., & Couto, E. G. (2014). Environmental footprints show China and Europe's evolving resource appropriation for soybean production in Mato Grosso, Brazil. *Environmental Research Letters, 9*(7). doi:10.1088/1748-9326/9/7/074001

Leitão, M., & Gribel, A. (2012). Coluna no Globo: Nas asas oficiais [Column on the Globe: The official wings]. Retrieved from http://oglobo.globo.com/economia/miriam/posts/2012/05/10/nas-asas-oficiais-444362.asp

Marcuse, H. (1966). *One-dimensional man: studies in the ideology of advanced industrial societies.* Boston, MA: Beacon Press.

Marsh, K., Saunders, A., & Zeuschner, C. (2015). Red meat and health: evidence regarding red meat, health and chronic disease risk. In D. Marinova & T. Raphaely (Eds.), *Impacts of Meat Consumption on Health and Enviuronmental Sustainability.* Hershey, PA: IGI Global.

Masson, J. F., & McCarthy, S. (1996). *When elephants weep: the emotional lives of animals.* New York, NY: Dell Publishing.

Motoki, C., Broggi, F., Falcão, S. N., Favoretto, T., & Casteli, T. (2013). Moendo gente: a situação do trabalho nos frigoríficos [Grinding people: The labour situation in refrigerators]. Sao Paulo, Brazil: ONG Repórter Brazil. Retrieved from http://escravonempensar.org.br/upfilesfolder/materiais/arquivos/moendo_gente_final.pdf

Noss, R. F., Dobson, A. P., Baldwin, R., Beier, P., Davis, C. R., Dellasala, D. A., & Tabor, G. et al. (2012). Bolder thinking for conservation. *Conservation Biology*, *26*(1), 1–4. doi:10.1111/j.1523-1739.2011.01738.x PMID:22280321

Packer, I. (2012). A engenharia legal e o papel do estado na transição para um 'capitalismo verde' [Legal engineering and the role of the state in the transition to 'green capitalism']. *Proposta*, *36*(125), 26–31.

Patterson, R. E., Kristal, A. R., & White, E. (1996). Do beliefs, knowledge and perceived norms about diet and cancer predict dietary change? *American Journal of Public Health*, *86*(10), 1394–1400. doi:10.2105/AJPH.86.10.1394 PMID:8876507

Pelletier, N., & Tyedmers, P. (2010). Forecasting potential global environmental costs of livestock production 2000-2050. *Proceedings of the National Academy of Sciences of the United States of America (PNAS)*, *107*(43), 18371-18374.

Phelps, N. (2007). *The longest struggle: animal advocacy from Pythagoras to PETA*. New York, NY: Lantern Books.

Raphaely, T., & Marinova, D. (2013). Sustainability humanistic education: A new pedagogy for a better world. *International Journal of Education. Economics and Development*, *4*(2), 170–189.

Raphaely, T., & Marinova, D. (2014). Flexitarianism: A more moral dietary option. *International Journal of Sustainable Society*, *6*(1-2), 189–211. doi:10.1504/IJSSOC.2014.057846

Raphaely, T., Marinova, D., & Todorov, V. (2010). Sustainability education: What on earth are we doing? *Management and Sustainable Development*, *26*(2), 49–60.

Regan, T. (2001). *Defending animal rights*. Chicago, IL: University of Illinois Press.

Ryder, R. (2005, August 6). All beings that feel pain deserve human rights. *The Guardian*. Retrieved September 6, 2015, from http://www.theguardian.com/uk/2005/aug/06/animalwelfare

Schlesinger, S. (2010). Onde pastar? O gado bovino no Brasil [Where to graze? The cattle in Brazil] (1st ed.). Rio de Janeiro, Brazil: Federação de Órgãos para Assistência Social e Educacional (FASE).

Secretariat of the Convention of Biological Diversity (CBD). (2010). Global biodiversity outlook 3. Montréal, Canada: CBD. Retrieved from http://www.cbd.int/doc/publications/gbo/gbo3-final-en.pdf

Spretnak, C. (1999). *The resurgence of the real: body, nature and place in a hypermodern world*. New York, NY: Routledge.

Srinivasan, C. S. (2006, August 12-16). *WHO dietary norms: a quantitative evaluation of potential consumption impacts in the United States*. Paper presented at the 26th International Association of Agricultural Economists Conference, Gold Coast, Queensland, Australia. Retrieved from http://purl.umn.edu/25292

Stanton, R. (2015). Meat and dietary guidelines. In D. Marinova & T. Raphaely (Eds.), *Impacts of meat consumption on health and environmental sustainability*. Hershey, PA: IGI Global.

Steinfeld, H., Gerber, P., Wassenaar, T., Castel, V., Rosales, M., & de Hann, C. (2006). Livestock's long shadow: environmental issues and options. Rome, Italy: Food Agriculture Organisation (FAO). Retrieved from ftp://ftp.fao.org/docrep/fao/010/a0701e/a0701e00.pdf

ten Brink, P., Berghöfer, A., Schröter-Schlaack, C., Sukhdev, P., Vakrou, A., White, S., & Wittmer, H. (2009). The economics of ecosystems and biodiversity for national and international policy makers. Executive summary: responding to the value of nature. United Nations Environment Programme (UNEP). Retrieved from http://www.unep.org/pdf/TEEB_D1_Summary.pdf

The Cambridge Declaration on Consciousness. (2012). Retrieved from http://yourbrainandyou. com/2012/08/24/the-cambridge-declaration-on-consciousness/

Tuttle, W. (2005). *The world peace diet: eating for spiritual health and social harmony*. New York, NY: Lantern Books.

United Nations Environment Programme (UNEP). (2010a). Assessing the environmental impacts of consumption and production, priority products and materials. Paris, France: UNEP, International Panel for Sustainable Resource Management. Retrieved from http://www.greeningtheblue.org/sites/default/ files/Assessing%20the%20environmental%20impacts%20of%20consumption%20and%20production.pdf

United Nations Environment Programme (UNEP). (2010b). State of biodiversity in Latin America and the Caribbean. Retrieved from http://www.unep.org/delc/Portals/119/LatinAmerica_StateofBiodiv.pdf

Westhoek, H., Lesschen, J. P., Rood, T., Wagner, S., De Marco, A., Murphy-Bokern, D., & Oenema, O. et al. (2014). Food choices, health and environment: Effects of cutting Europe's meat and dairy intake. *Global Environmental Change*, *26*, 196–205. doi:10.1016/j.gloenvcha.2014.02.004

World Health Organisation (WHO). (2001). WHO global strategy for containment of antimicrobial resistance. Geneva, Switzerland: WHO. Retrieved from http://www.who.int/drugresistance/WHO_Global_Strategy_English.pdf

World Health Organisation (WHO). (2015). Drug resistance: antimicrobial use. Retrieved from http:// www.who.int/drugresistance/use/en/

World Wildlife Fund (WWF). (2014). Living planet report 2014: summary. Gland, Switzerland: WWF. Retrieved from http://assets.wwf.org.uk/downloads/living_planet_report_2014_summary.pdf?_ga=1.3 8233059.1148581730.1413300010

Zohar, D., & Marshall, I. (2001). *Spiritual intelligence: the ultimate intelligence*. London, UK: Bloomsbury Publishing.

KEY TERMS AND DEFINITIONS

Animal Abolitionism: Animal rights advocacy opposing the use of animals in any capacity by humans.

Beef Power: The influence of the livestock sector.

Ethical Education: Education that endorses animal abolitionism.

Speciesism: A philosophical approach which endorses the supremacy of the human being in relation to other animals.

Welfarism: Advocacy aimed at improving the living conditions of animals.

ENDNOTES

[1] These mental constructions and attitudes towards the animal world might be underpinned by religious, secular values or might be based on scientific evidence. Irrespective of what the basis of the beliefs is, education plays an important role in their formation.

Section 4
Conclusion

Chapter 18
After Meat

Jonathan Balcombe
Author, USA

ABSTRACT

This chapter describes a future scenario in which humankind abruptly stops eating animals. Its purpose is to explore the impacts of animal agriculture using a fictional device. Several ways, in which the end of meat might happen are suggested, including a virulent global human epidemic originating from the meat supply. The ensuing state of the world is examined along six axes: animals, the environment, human health, the economy, food, and society. Because of the enormous scale of animal agriculture today, a sudden end to meat would cause massive upheavals across all of these domains. However, it is predicted that positive gains would be significant, outweighing negative outcomes.

INTRODUCTION

Seven billion humans consume a lot of meat. Most of the animals killed by humans today are killed to be eaten. The intensive factory farming systems in which most of these animals are reared are surely the largest-scale source of animal suffering today. Animal agriculture is also increasingly documented as a major factor in serious/potentially life-threatening environmental and human health concerns (Akhtar, Greger, Ferdowsian, & Frank, 2009). Perhaps because they are awakening to these implications, Americans have reduced their meat consumption by ten percent over the last decade (Larson, 2012a) and an increasing number of them are choosing meat and other animal products sourced from animals raised in more humane conditions.

As yet, however, the US decline in meat consumption is not reflected in overall global patterns. Meat consumption has been rising in some developing nations, and meteorically in China, with a nine-fold increase between 1978 and 2012 (Larson, 2012b). Even though the average Chinese citizen today eats only half as much meat as the average American, China's substantially larger human population translates to double the total volume of meat consumed today than in the United States (Larson, 2012b), accounting for a quarter of the global total (Larson, 2012b).

Nevertheless, to the extent that the United States may be seen as a bellwether to the rest of the world, we may expect meat's American decline to reverberate elsewhere. An August 2014 study from Data

DOI: 10.4018/978-1-4666-9553-5.ch018

monitor Consumer involving 25,000 respondents from 25 countries reported that a third of global consumers are cutting back on meat, confirming an April 2014 study by FGI Research that found a growing dietary trend toward flexitarianism (Guyton, 2014; Meatless Monday, 2014) . The alternative meat food sector is booming, and Meatless Monday campaigns are established in 35 countries worldwide (Meatless Monday, 2014).

BACKGROUND

Quite aside from how unlikely, or for many how undesirable, it would be for humans to stop eating animals entirely, there is no restriction on asking what might transpire if we did. For this essay I have removed the reins of uncertainty to speculate on such an audacious future. I examine six domains: the animals, the environment, human health, the economy, food, and society.

A SUDDEN END TO MEAT

Possible Scenario(s)

It helps to consider first how an end to meat might come about, because this pertains to how a post-meat world might function. For instance, a gradual fade-out of meat consumption would imply that we stopped eating animals as a matter of choice, whereas a virulent meat-borne zoonosis that forced an end to meat consumption would not be something of our choosing. For this essay, I have chosen the latter scenario, which assumes a more or less decisive end to meat, in which the transition to a vegetarian culture happens almost instantaneously. Here, briefly, is one fictional scenario, one that assumes the viability of an agriculture system totally devoid of animals

The year is 2019. World human population is approaching eight billion. Animal agriculture has become more concentrated and centralized than ever and is now a near monopoly, with just two companies controlling 98% of the American meat supply from hoof to hamburger.

In early October, the United States is slammed by a grave food-borne epidemic. Symptoms are severe and the resulting illness uncompromising. Fever rapidly develops, accompanied by violent nausea. Death usually follows within 24 to 48 hours.

The Centers for Disease Control issues a nation-wide advisory to avoid eating chicken. International embargos are placed on US meat. However it all happens too late. A new strain of bird flu has been incubating in the bodies of consumers for days. And it is extremely contagious in the meat supply, spread by rendered animal tissues used widely in livestock feed. Fish feed derived from terrestrial rendering products has found its way into aquaculture, and from there to wild fish populations. The commercial fishing and aquaculture industries have been exposed.

Within a week, the epidemic has metastasized. It has gone global, infecting 90% of the global meat supply. By November, over 980 million humans have died worldwide, dwarfing the Spanish influenza of 1918 and prompting a new term: 1BP—the 1 Billion Plague. Eating flesh becomes tantamount to playing Russian roulette. Within a week, human meat consumption effectively drops to zero.

Could it happen? Could we be faced with an apocalyptic pandemic transmitted to humans through animal tissues? About 75% of the new infectious diseases affecting humans over the past 10 years were caused by bacteria, viruses and other pathogens that started in animals and animal products, according to the (WHO, n.d.). Outbreaks of lethal or potentially lethal pathogens are among the leading headline grabbers of the current era. At the time of writing (November 2014), newscasts are saturated with the largest Ebola virus outbreak in history. The epidemic has officially infected over 14,000, so far killing over 5,000 Africans (Barber, 2014)—although the World Health Organization believes the actual case load is probably about three times greater (Miles, 2014). Not making headlines, a concomitant outbreak of the bird flu variant H5N8 led to the immediate slaughter of 150,000 chickens on a farm in the Netherlands, and millions more in South Korea (Escritt, 2014).The swine flu first identified in Veracruz, Mexico in April 2009, had killed at least 18,000 humans on five continents within the first year (WHO, 2009).

So far, no zoonosis has led to a massive human die-off, but experts don't take the risk lightly. Mad Cow Disease and its connection to the brain-wasting Creutzfeldt-Jakob disease in humans made headlines in the late 1980s, leading to international trade embargos and widespread fears that eating beef was unsafe. Mortality rates would soar if we suddenly found ourselves unable to treat, say, an outbreak of a virulent new antibiotic-resistant strain of E. coli 0157:H7 bacteria.

Other pathways to the end of meat can be imagined, however improbable. Some brain parasites alter the behaviour of their hosts, so an infectious agent could corrupt our minds against meat. For example, Toxoplasmosis deviously manipulates the brains of one of its hosts—rats or mice—to lose their natural fear of its final host, cats, thereby making them more likely to become cat food (Ingam, Goodrich, Robey, & Eisen, 2013). Also, red meat allergies have been associated with the lone star tick bite (Goetz, 2012). In a different twist, perhaps a "compassion gene" sweeps through the human population, kindling our innate childhood empathy for animals. Or maybe sentient animals are granted the right to life, and killing them to eat becomes an act of murder. The abolition of slavery, suffrage for women, and the civil rights movement are reminders that epochal, morally driven social change happens.

After Meat

Predicting the future is an imperfect art. A thought experiment of such magnitude as the end of human consumption of animals would yield unexpected outcomes impossible to foresee, especially as we peer further into the future. Prognostications rely heavily on an understanding of the past, and much of what follows references meat's recent and current place in the world. Beyond that, I openly confess to taking liberties with an unknown future.

The Animals

At the time we stopped, humans were eating more meat than ever before. Between 1961 and 2007, the world's total meat supply had quadrupled to over 280 million tons. Approximately 70 billion land animals were raised around the world for human consumption in 2010, according to Humane Society International (2013). Every hour in the United States, more than 1 million land animals were killed for food (Akhtar et al., 2009). Three hundred chickens died every second.

Had meat consumption ended gradually, then perhaps we would have found a way to care for the enormous short-term surplus of animals otherwise destined for slaughter. However because it ended abruptly when a virulent, ineradicable virus made it too risky to consume, the animals weren't so lucky. Raising animals is costly, so the collapse of the consumer base forced producers to stop breeding and raising more of them. Releasing them to their own devices was unthinkable for ecological and human safety reasons. A massive culling operation commenced, with tens of millions of caged chickens, pigs and cattle gassed in their sheds. It resembled a larger-scale version of the 2001 outbreak of foot-and-mouth disease in Britain, which had led to the extermination and immolation of ten million cattle and sheep at a cost of $12 billion (CNN, 2014).

Nevertheless, far fewer animals were killed that first year than in a typical year during the age of meat, when most factory farmed animals were being sent to slaughter at less than one year of age: six months for pigs, and just 5-7 weeks for chickens. The Great Cull only terminated those living at the time, whereas more than one generation of pigs and chickens (and turkeys, lambs and calves) were going to the knife in a typical pre-Cull year.

Cows, chickens, and pigs however have not gone extinct. During the initial destruction of livestock, animal protection organizations mobilized large-scale rescue campaigns. Funded by a sympathetic and nostalgic segment of the public, pre-existing farm animal sanctuaries—of which several dozen were scattered across the United States—quickly expanded their capacities. Donations flooded in and new sanctuaries sprang up. Today, relic populations of cows wander pastures and nurse calves, hand-fed pigs wallow in mud and sleep on thick straw, and small flocks of chickens and turkeys forage, brood and bathe in the dust and sun.

The sum total of animal suffering in the world has plummeted. Of every 100 animals killed by humans, 98 had died to be eaten, and more than half were being reared in the notoriously inhumane conditions of factory farms. Abuses of animals in the animal production and slaughter milieu were well-documented (Eisnitz, 1997; Singer, 2010). Tail amputations, beak-searing, horn disbudding, castration, teeth-clipping, and ear punching—all without anaesthesia or veterinary treatment—are now gone. Gestation crates, battery cages, veal crates, broiler sheds, and other forms of intensive confinement are now taught in history classes, and cattle prods, kill-cones, neck slicers, scalding tanks, pluckers, and eviscerators have become museum exhibits. Bottom trawlers, long-lines and purse-seines no longer subject fishes and other marine life to death by crushing, decompression, or asphyxiation.

With a large grant from the National Endowment for the Humanities, the National Museum of Animals and Society has developed a sprawling interpretive park themed on the age of meat. It features animatronic animals, videos, a walk-through battery cage barn and—for adults only—The Processor, a virtual reality ride through the stages of slaughter from loading and transport to shackling and rendering. Veal crates, gestation crates, and ceiling conveyors have been repurposed for seating, handrails, and people movers.

Inevitably, some livestock escaped to survive in the wild. Natural selection has favoured the ancestral traits these animals originally had in their wild state. Cattle are gradually coming to resemble the European aurochs, chickens the Red jungle fowl of Asia, and pigs the European boar. In North America, feral turkeys have interbred with native wild birds, hastening their return to ancestral traits.

Marine and freshwater life has been affected by an end to human consumption of fishes and other aquatic animals. At the turn of the third millennium, man's quest for fish flesh had accounted for the deaths of between 300 billion and over two trillion fishes per year (Cooke & Cowx, 2004; Mood, 2010). Those numbers didn't include bycatch—non-target species that were discarded back into the water, mostly dead. Populations of targeted "food fish" had been collapsing under the weight of human appetite. Global fish catches had been declining since the late 1980s (Watson & Pauly, 2001). The number of collapsed fisheries increased exponentially after 1950 (FAO, 2007), contributing to over 100 confirmed cases of marine population extinctions by the year 2000 (Dulvy, Sadovy, & Reynolds, 2003).

Some populations, such as Northern Cod and Orange Roughy, have not rebounded even with a total end to their exploitation, owing to the systematic removal of older, more knowledgeable individuals.

Otherwise, positive effects are being recorded across the oceans. Within ten years of fishing fleets and canning operations grinding to a halt, marine surveys are documenting the return of several formerly "commercially extinct" species, including all three species of Bluefin tuna. Clean-up operations organized in an unprecedented ocean partnership between Australia and Japan have removed nearly half of the 160,000 tons of fishing gear that were being left in the oceans each year (WAPI, 2014).

The Environment

Meat's popularity belied its inefficiency as a food source. It is a common misconception that feeding a planet's worth of vegetarian humans would require more grains and vegetables than feeding a planet's worth of omnivores. Most of the corn and soy grown in the world was fed to cattle, pigs and chickens. Worldwide, we were feeding 7 billion people on 70 billion land animals. A Stanford University economist calculated that two to five times more grain was required to produce the same amount of calories through livestock as through direct grain consumption, and as much as 10 times more in the case of American grain-fed beef (Bittman, 2008). A Cornell University ecologist calculated that 800 million people could be fed with grain eaten by U.S. livestock (Pimentel & Pimentel, 1997)

Land use and fossil fuel consumption were commensurate with these excesses. Meat production (grazing and feed-crop production) accounted for 70 percent of all agricultural land (FAO, 2006) and about 45 percent of the total land surface of the planet (Thornton, Herrero, & Ericksen, 2011). In addition to being a major contributor to land degradation, air and water pollution, water shortage and biodiversity loss, animal agriculture was a major driver of climate change, producing nearly a fifth of the world's greenhouse gases, more than the transportation sector (FAO, 2006). Some land grazed by cattle was not arable, but the end of meat has still freed up millions of square miles to nature.

The upshot is that less food is being grown and transported in the post-meat world. The consequences include less land needed and used for agriculture, lower air and water pollution, and a slowing of greenhouse gas production.

Resigned to the lost domestic and international demand for its former beef supply, the Brazilian government announces the first ever reversal of a century-long trend in rainforest declines. Emboldened by this turn-around, the Brazilian National Congress enacts a new plan to establish the remaining 1.5 million square miles of its now expanding rainforest as a national park system, including a network of interpretive centers connected by 300 miles of trails and boardwalks. Conservation International teams with National Geographic to lead the establishment of new national parks on land reclaimed from former livestock grazing and feed-crop production. The US Midwest breadbasket, once devoted mostly to feed-crops, now features Prairie National Park, an extensive patchwork of reclaimed grasslands connected by natural corridors and with a full biodiversity complement including bison, grizzly, and wolf populations.

The Mississippi River is no longer flushing manure-sourced phosphorus and nitrogen into the oceans, though it continues to deliver toxics from a constellation of petrochemical plants and oil refineries along Louisiana's "chemical corridor". Within a decade, the former "dead zone" in the Gulf of Mexico has shrunk, as happened in the Black Sea when fertilizer use plunged following the 1991 collapse of the Soviet Union (Mee, 2006).

Projections for rising global temperatures have fallen short, and by mid-century climate conditions have stabilized. This is due to a diversity of coordinated measures by nations around the world—curbing human overpopulation, converting to green energy production, streamlining mass transportation, reducing industrial pollution—but the shift to plant-based agriculture is hailed as a keystone to stemming climate change. It all happened too late to avoid several famines and mass human migrations from floods or parched equatorial regions, but there is newfound hope that humanity will be able to steer the planet towards pre-20th Century climate conditions.

Human Health

The link between meat consumption and the leading killers in the developed nations was well established when we stopped eating animals. Worldwide, ischaemic heart disease and stroke were the first and the second leading causes of human deaths (WHO, 2014). Animal products were the primary source of saturated fats that promote cardiovascular disease, and dozens of studies had linked animal protein consumption to colorectal, breast, endometrial, and prostate cancers (Akhtar et al., 2009). When meat consumption stopped, people who avoided animal products had significantly lower body mass indexes than meat eaters (Spencer, Appleby, Davey, & Key, 2003). Globally, obesity rates had more than doubled between 1980 and 2008, and more than 1.4 billion adults were either overweight or obese (WHO, 2008). A 2010 United Nations report calling "a global shift towards a vegan diet vital to saving the world from hunger, fuel poverty and the worst impacts of climate change" could just as easily have been a clarion call for improving human health (Carus, 2010).

Within a year, the CDC was announcing declining levels of deaths from heart disease, strokes, and cancer, and in rates of obesity and diabetes. Ten years on, the National Cancer Institute announces a 50% drop in cancers of the bowel and stomach, a 30% drop in breast, prostate, pancreatic, and liver cancers, and 25% declines in leukaemia and brain cancers.

Overseas, the pattern is similar. Across Europe, Asia and urban Africa, an "epidemic" of health gains is being reported by public health agencies. Morbidity and mortality from heart disease, strokes, obesity, diabetes, and some cancers are showing a pattern not seen since meat was rationed during World War II. People are living longer. 2025 sees the launch of a Centenarian Olympics. The U.S. President, herself a spring chicken at 83, is now sending birthday congratulations only to 101 year olds.

In a long-overdue move, the NIH invests in prevention by forming INR: the Institute of Nutrition Research, focusing on innovative ways to address the new health landscape post-meat. Plant-based nutrition booklets are distributed to under-served populations, farmers markets are established in former urban "food deserts," and a little-known cookbook titled Eat Vegan on $4 a Day becomes a best seller. Many hospitals, operating at well below capacity, are pressed into new services: research for breeding and recovery of critically endangered species; R&D labs for new sources of alternative meats, including algal and fungal (e.g., Quorn) and dairy products.

The food supply is safer. Meat recalls don't happen, and outbreaks of *E. coli* from tainted spinach, tomatoes, and other plant crops are virtually unheard of now that they are no longer fertilized by manure (E. coli is named for the colon where it grows, and plants don't have colons). There remains the risk of contamination of vegetable and fruit crops by the ubiquitous norovirus—transmitted in human feces and vomit—due to unsafe food handling practices.

The Economy

The end of meat was, of course, a catastrophe for a very large industrial complex. According to a 1992 estimate, 700,000 full- and part-time workers were employed in animal agriculture in the United States alone (Iowa State University, 1992). Globally, animal agriculture employed 1.3 billion people and created livelihoods for one billion of the world's poor (FAO, 2006).

Nature however abhors a vacuum. Just as a huge tree falling in the jungle spawns a spasm of new growth in the gap, so too has the end of animal agriculture stimulated new opportunities for the production, distribution and sale of food. Food production has decentralized as former meat giants Tyson, Smithfield, JBS, Shuanghui Group, and others were forced to scale back operations and seek new channels of food production. The cessation of livestock feeding has also shrunk grain production below 1990 levels. While geopolitical factors continue to hamper efforts to address regional hunger, more humans are being fed with less.

Post-meat, the U.S. Congress quickly enacted a farm bill bearing no resemblance to its predecessors. Between 1995 and 2010, roughly two-thirds of the $200 billion in government crop subsidies had gone to animal-feed crops, tobacco and cotton, while farmers who grew fruits, vegetables and tree nuts received practically no regular direct subsidies (Allen, 2011). Established plant food crop growers are now getting a portion, and seed monies are being distributed to ranchers and other small-scale farmers whose businesses relied wholly or partly on animal production.

New fast-tracked grant offerings from the National Academy of Sciences, the Small Business Innovation Research grants program, the Pew Charitable Trusts, the International Monetary Fund, and myriad private foundations have been established to drive the Mission Transition forward. A network of advisors provides free consulting to growers. Hog farmers in the eastern seaboard states have followed the lead of tobacco farmers, who began switching to chickpeas back in 2012 to satiate demand for hummus, whose yearly market share grew 18% to $315 million that year (Kesmodel & Fletcher, 2013). In Texas, an enterprising cattle baron has transformed twenty 75-acre feedlots and a glut of fertilizer into a successful organic spinach and asparagus production facility, which by year's end is the major supplier to the desert states.

Worldwide, pre-transition animal agriculture subsidies that the UN had estimated were accounting for 31% of the global farm income now support food grown to be eaten by humans and not livestock. Import/export tariffs are relaxed on shipments of vegetables, fruit, and grains to better ensure that no one goes hungry.

In the only analysis of its kind, the Physicians Committee for Responsible Medicine had estimated in 1992 that the yearly healthcare costs of meat consumption in the U.S. were between $29 billion and $61 billion (Barnard, Nicholson, & Howard, 1995). By one estimate, the United States health care bill for obesity was $190 billion in 2005 (Cawley & Meyerhoefer, 2010). After meat, health insurance premiums have dropped as the population loses weight and morbidity rates decline. Medicare costs for a longer-lived population of the elderly are offset by lower morbidity.

Food

Ironically, initial fears that the bite, chew, juiciness, and flavour of animal flesh would be forever lost turned out to be illusory. All flesh ultimately comes from plants, and the lines between meat and faux meat were already becoming blurred. By 2012, US Department of Agriculture (USDA) regulations were permitting up to 30 percent soy products in school lunch meats (Bittman, 2012a), and a blind taste test conducted by a faux meat developer had led New York Times food critic Mark Bittman to proclaim: *Really: Would I rather eat cruelly raised, polluting, unhealthful chicken, or a plant product that's nutri-*

tionally similar or superior, good enough to fool me and requires no antibiotics, cutting off of heads or other nasty things? (Bittman, 2012b). By that time, while the global meat consumption was still rising, it had fallen by some twelve percent in a decade in the United States (Bittman, 2012b). These trends were widely thought to be driven by omnivores, not vegetarians; a Harris poll commissioned by the Vegetarian Resource Group reported that a third of Americans were eating meatless meals "a significant amount of the time" (Stahler, n.d.).

The meatless age has precipitated a flourishing of companies formerly huddled at the edge of the dinner plate. Meat-lovers, desperate for a fix, are also reaching for a wide range of already-available non-animal meat products. In three years, Tofurky, Gardein, SmartLife, Field Roast, Morningstar, and Boca have become household brands sprawling across the supermarket meat freezers.

Rerouted meat subsidies, R&D investments, and economies of scale mean that meat alternatives are no longer a niche market. Before meat ended, venture capital firms were funnelling $350 million into food start-ups, some of which were specializing in non-animal meat and dairy products. Within 6 months, sales of meat substitutes have blossomed from 0.3% to 3% of total food market share, held back only by production rates. Bill Gates, who in 2013 had publically extolled the taste of meat alternatives from Beyond Meat, Hampton Creek Foods, and Lyrical, announces a large grant to facilitate mass production and distribution of these products (Aubrey, 2013). A vacant meat packing plant in Missouri is retrofitted as the center of operations.

Fueled by investments from Smithfield and Tysons, Modern Meadow perfects the use of 3D printing technology to produce meat (and leather) from clean laboratory animal cell lines. Former meat giants leverage their infrastructure towards full-scale production of in vitro pork, beef, veal, turkey and chicken. Fast food chains McDonalds, Burger King, and Wendy's have all launched new menus. McDonalds' "MacFuture" menu features burgers and nuggets made from in vitro meats, soy, wheat gluten, cassava, and quinoa. KFQ (Kentucky Fried Quorn) is available from Snack Box to Mega Bucket. Within 18 months, faux meats cost less than pre-transition animal equivalents, and are rated indistinguishable from "the real thing" by 80% of market survey participants.

Society

The causes of human violence are broad and complex. In the centuries leading up to the end of meat eating by humans, significant long- and short-term declines in violence had been occurring in many domains, including military conflict, homicide, genocide, torture, and the treatment of racial and ethnic minorities (Pinker, 2011). In his sweeping book *The Better Angels of our Nature: Why Violence has Declined,* the psychologist Steven Pinker attributes these trends to a series of civilizing processes, including: the rise of statehoods with judiciary processes, a rise in respect for and empowerment of women, technological advances expanding the exchange of goods over long distances (thus making other peoples more valuable alive than dead), and enormous advances in literacy and reason (Pinker, 2011).

While the total number of animals being killed by humans was still rising at the time of the pandemic, nevertheless humankind's moral concern for animals had risen unprecedentedly in the decades leading up to it. A number of scholars had begun to emphasize what they referred to as "the link"—the connection between cruelty to animals and human violence (Ascione & Arkow, 1999; Lockwood & Ascione, 1997). Since Upton Sinclair described growing violence among slaughter workers in his 1906 novel *The Jungle*, much speculation but little science had been applied to a possible link between slaughtering and eating animals, and societal violence. Social geographer Michael Broadway described rising crime rates in two rural districts in Kansas and Nebraska, after slaughterhouses were established there (Broadway, 1990). Perhaps the most in-depth study of this possible link examined rates of violent crime across several industries: animal slaughtering (excluding poultry), iron and steel forging, truck trailer manufacturing, motor vehicle metal stamping, sign manufacturing, and industrial laundering). They found that "slaughterhouse employment increases total arrest rates, arrests for violent crimes, arrests for rape, and arrests for other sex offenses in comparison with other industries." (Fitzgerald, Kalof, & Dietz, 2009, p. 158).

In the wake of meat, wholesale rates of societal violence have continued to drop, although attributing this to dietary shift is tenuous. There are however indicators that a cultural shift away from consuming animal products has a liberating effect on compassion. No longer needing to justify a lifestyle choice that is intrinsically violent, the new plant-based humanity is becoming a kinder and gentler one. Public protests against wide-scale animal abusive spectacles boomed in the wake of the pandemic, leading to the end of ritual animal sacrifices, such as that of the annual Islamic Hajj to Mecca. At the Gadhimai Festival in Nepal—where every five years half a million animals were crudely slaughtered to please the Hindu goddess of power—animal-themed paper balloons are now sent skyward.

The frequency and fierceness of many territorial disputes are declining, and some experts are attributing this, in part, to the fact that less land is now required to feed human populations (see The Environment, above). Elsewhere, conflicts are generally fewer and less bloody. It is nothing new that most people simply want to live in peace.

CONCLUSION

Is this a rose-tinted future? It is inescapable that a sudden end to meat would be a huge upheaval given the sheer magnitude and reach of the industry. However there is no getting around it: animal agriculture, in its current state, foists an enormous, multifaceted burden on the planet, including us. The hamburger is vastly more costly than its deceptively low sticker price. Our history shows that we would deal with the initial destabilization of an end to meat, and science suggests that we would emerge better off. An end of factory farms would be a massive ethical advance for humankind.

What if humans reverted to meat consumption? Would we see a return of the factory farm, massive animal agriculture subsidies, and pre-transition rates of heart disease and strokes? Depending on how long human societies had been meatless, the new era might look quite different. Following a few years of Meatless Mondays-to-Sundays, demand for animal flesh could be a fraction of its former scale. Humankind might come to look on industrial-scale meat production with revulsion, choosing instead products from sources that adhere to higher ethical standards. In this revised food culture, small family farms would raise animals in pastoral settings where they would live longer, more contented lives. Meat supply would be localized, precluding the need for transport to slaughterhouses.

REFERENCES

Akhtar, A. Z., Greger, M., Ferdowsian, H., & Frank, E. (2009). Health professionals' roles in animal agriculture, climate change, and human health. *American Journal of Preventive Medicine*, *36*(2), 182–187.

Allen, A. (2011, October 3). U.S. touts fruit and vegetables while subsidizing animals that become meat. The Washington Post. Retrieved September 8, 2015, from http://www.washingtonpost.com/national/ health-science/us-touts-fruit-and-vegetables-while-subsidizing-animals-that-become-meat/2011/08/22/ gIQATFG5IL_story.html

Ascione, F., & Arkow, P. E. (1999). *Child abuse, domestic violence, and animal abuse: linking the circles of compassion for prevention and intervention.* West Lafayette, IN: Purdee University Press.

Aubrey, A. (2013). Why Bill Gates is investing in chicken-less eggs. Retrieved September 8, 2015, from http://www.npr.org/sections/thesalt/2013/06/13/191029875/why-bill-gates-is-investing-in-chicken-less-eggs

Barber, E. (2014). Ebola death toll surpasses 5,000 worldwide. Time. Retrieved September 8, 2015, from http://time.com/3582804/ebola-death-toll-surpasses-5000-worldwide

Barnard, N. D., Nicholson, A., & Howard, J. L. (1995). The medicinal costs attributable to meat consumption. *Preventive Medicine, 24*(6), 646–655.

Bittman, M. (2008, January 27). Rethinking the meat guzzler. New York Times. Retrieved September 8, 2015, from http://www.nytimes.com/2008/01/27/weekinreview/27bittman.html?pagewanted=all&_r=0

Bittman, M. (2012a, January 10). We're eating less meat why? The New York Times. Retrieved September 8, 2015, from http://opinionator.blogs.nytimes.com/2012/01/10/were-eating-less-meat-why/

Bittman, M. (2012b, March 9). A chicken without guilt. The New York Times. Retrieved September 8, 2015, from http://www.nytimes.com/2012/03/11/opinion/sunday/finally-fake-chicken-worth-eating. html?_r=0

Cable News Network (CNN). (2014). Foot and mouth disease: fast facts. Retrieved December 1, 2014, from http://www.cnn.com/2013/09/02/health/foot-and-mouth-disease-fast-facts/

Carus, F. (2010, June 3). UN urges global move to meat and dairy free-diet. The Guardian. Retrieved September 8, 2015, from http://www.theguardian.com/environment/2010/jun/02/un-report-meat-free-diet

Cawley, J., & Meyerhoefer, C. (2010). The medical care costs of obesity: an instrumental variables approach. NBER Working Paper 16467. New York, NY: Cornell University, Department of Policy Analysis and Management. Retrieved September 8, 2015, from http://www.nber.org/papers/w16467

Cooke, S. J., & Cowx, I. G. (2004). The role of recreational fisheries in global fish crisis. *Bioscience, 54*(9), 857–859.

Dulvy, N., Sadovy, Y., & Reynolds, J. (2003). Extinction vaulnerability in marine populations. *Fish and Fisheries, 4*(1), 25–64.

Eisnitz, G. (1997). *Slaughterhouse: the shocking story of greed, neglect and inhumane treatment inside the U.S. meat industry.* Amherst, NY: Prometheus Books.

Escritt, T. (2014). Authorities identify highly contagious bird flu strain. Scientific American. Retrieved November 17, 2014, from http://www.scientificamerican.com/article/authorities-identify-highly-contagious-bird-flu-strain/?WT.mc_id=SA_DD_20141117

Fitzgerald, A. J., Kalof, L., & Dietz, T. (2009). Slaughterhouse and increased crime rates: An empirical analysis of the spillover from 'The Jungle' into the surrounding community. *Organization & Environment*, *22*, 158–184.

Food and Agriculture Organization of the United Nations (FAO). (2006). *Livestock's long shadow: environmental issues and options*. Rome, Italy: FAO, Animal Health and Production Division.

Food and Agriculture Organization of the United Nations (FAO). (2007). *The state of world fisheries and aquaculture 2006*. Rome, Italy: FAO, Fisheries and Aquaculture Division.

Goetz, G. (2012). Red meat allergy likely caused by tick bites. Food Safety News. Retrieved January 12, 2015, from http://www.foodsafetynews.com/2012/06/red-meat-allergy-likely-caused-by-tick-bites/

Guyton, G. (2014). Spotlight on flexitarianism. GlobalMeat News.com. Retrieved November 17, 2014, from http://www.globalmeatnews.com/Analysis/Spotlight-on-flexitarianism-are-consumers-cutting-their-meat-intake

Humane Society International (HSI). (2013). Animal agriculture one of the largest contributers to global warming, UN body reaffirms. Humane Society International. Retrieved December 1, 2014, from http://www.hsi.org/news/press_releases/2013/09/fao_report_climate_change_092713.html

Ingam, W. M., Goodrich, L. M., Robey, E. A., & Eisen, M. B. (2013). Mice infected with low-virulence strains of Toxoplasma gondii lose their inate aversion to cat urine, even after extensive parasite clearance. *PLoS ONE*, *8*(9), e75246. doi:10.1371/journal.pone.0075246

Iowa State University. (1992). Livestock confinement dusts and gases. University Extension. Retrieved December 1, 2014, from http://nasdonline.org/static_content/documents/1627/d001501.pdf

Kesmodel, D., & Fletcher, O. (2013). Hummus is conquering America: tobacco farmers open fields to chickpeas; a bumper crop. The Wall Street Journal. Retrieved March 3, 2015, from http://www.wsj.com/articles/SB10001424127887323798104578453174022015956

Larson, J. (2012a). Meat consumption in China now double that in the United States. Permaculture Research Institute. Retrieved March 27, 2015, from http://permaculturenews.org/2012/05/02/meat-consumption-in-china-now-double-that-in-the-united-states/

Larson, J. (2012b). Peak meat: US meat consumption falling. Earth Policy Institute. Retrieved November 21, 2014, from http://www.earthpolicy.org/data_highlights/2012/highlights25

Lockwood, R., & Ascione, F. E. (1997). *Cruelty to animals and interpersonal violence: readings in research and application*. West Lafayette, IN: Purdee University Press.

Meatless Monday. (2014). New study confirms continued rise of the flexitarian diet. Meatless Monday. Retrieved November 17, 2014, from http://www.meatlessmonday.com/articles/new-study-confirms-continued-rise-flexitarian-diet/

Mee, L. (2006). Reviving dead zones. Scientific American. Retrieved December 1, 2014, from http://www.scientificamerican.com/article/reviving-dead-zones/

Miles, T. (2014). Official WHO Ebola toll near 5,000 with the true number nearer 15,000. Reuters. Retrieved October 23, 2014, from http://www.reuters.com/article/2014/10/22/us-health-ebola-who-idUSKCN0IB23220141022

Mood, A. (2010). Worst things happen at sea: the welfare of wild-caught fish. Fishcount.org.uk. Retrieved May 23, 2013, from http://www.fishcount.org.uk/published/standard/fishcountfullrptSR.pdf

Pimentel, D., & Pimentel, M. (1997). Sustainability of meat-based and plant-based diets and the environment. *The American Journal of Clinical Nutrition, 78*(3), 660S–663S.

Pinker, S. (2011). *The better angels of our nature*. New York, NY: Viking.

Singer. (2010, September 14). Fish the forgotten victims of our plate. The Guardian. Retrieved from http://www.theguardian.com/commentisfree/cif-green/2010/sep/14/fish-forgotten-victims

Spencer, E. A., Appleby, P. N., Davey, G. K., & Key, T. J. (2003). Diet and body mass index in 38 000 EPIC-Oxford meat-eaters, fish-eaters, vegetarians and vegans. *International Journal of Obesity, 27,* 728–734.

Stahler, C. (n.d.). How often do americans eat vegetarian meals and how many adults in the U.S. are vegans? The Vegetarian Resource Group. Retrieved December 1, 2014, from http://www.vrg.org/journal/vj2011issue4/vj2011issue4poll.php

Thornton, P., Herrero, M., & Ericksen, P. (2011). Livestock and climate change. Nairobi, Kenya: International Livestock Research Institute. Retrieved September 8, 2015, from https://cgspace.cgiar.org/bitstream/handle/10568/10601/IssueBrief3.pdf

Watson, R., & Pauly, D. (2001). Systematic distortions in world fisheries catch trends. *Nature, 414,* 534–536.

World Animal Protection (WAP). (2014). Fishing's phantom menace. Retrieved September 8, 2015, from https://www.worldanimalprotection.us.org/sites/default/files/us_files/summary_sea-change-tackling-ghost-fishing-gear-summary_us.pdf

World Health Organization (WHO). (2008). Obesity and overweight. Media centre. Retrieved September 8, 2015, from http://www.who.int/mediacentre/factsheets/fs311/en

World Health Organization (WHO). (2009). Pandemic (H1N1) 2009 - update 100. Emergencies preparedness, response. Retrieved September 8, 2015, from http://www.who.int/csr/don/2010_05_14/en/

World Health Organization (WHO). (2014). The top 10 causes of death. Media Centre. Retrieved March 27, 2015, from http://www.who.int/mediacentre/factsheets/fs310/en/index2.html

World Health Organization (WHO). (n.d.). Ten facts on food safety. Fact File. Retrieved March 2, 2015, from http://www.who.int/features/factfiles/food_safety/facts/en/index4.html

KEY TERMS AND DEFINITIONS

Dead Zone: A large area of water devoid of life due to pollution.

Farm Bill: The primary piece of legislation affecting United States agriculture and food policy, passed approximately every five years.

Meat Recall: A command, usually government-issued, that a meat product be removed from the food supply, usually due to its being declared unfit to eat.

Meatless Mondays: A campaign to reduce meat consumption and promote vegetarianism by urging consumers to shun meat one day per week.

One Billion Plague (1BP): A [fictional] endemic that claims the lives of approximately one billion people.

Pandemic: An infectious epidemic that has spread across large geographic regions and/or segments of the human populace.

Toxoplasmosis: A parasitic disease of mammals, especially cats, caused by a protozoan.

Zoonosis: A disease transmissible from animals to humans.

Conclusion

The negative health and environmental impacts of meat consumption are far-reaching as presented in the chapters of this book. Simple solutions through changing human diet are within our reach. There are a lot of alternatives to animal proteins which offer exciting opportunities. It is time to participate, by reducing meat consumption, in creating better prospects for the human race and all other species on this planet.

As we were finishing this book the G7 summit in Bonn produced a commitment to phase out fossil fuels by the end of this century in order to cut greenhouse gas emissions. Despite their large global warming impact and huge opportunities for reduction in dangerous anthropogenic emissions, changes in meat production and consumption were again not on the agenda of this and, in fact, most previous climate change international meetings. However, the latest Intergovernmental Panel on Climate Change (IPCC, 2014) in its assessment reports for the first time highlights that different policy directions are needed when looking at a 20- or 100-year time horizon. Over twenty years, from all sectors of the economy, the largest contributor to global warming potential is agriculture – 22% compared to 20% for industry, 17% for electricity and heat production, 9.8% for transport and 5.7 for buildings. A major reason behind agriculture's high contribution is the livestock sector and the methane emissions associated with enteric fermentation and manure management. Methane is 84 times more powerful than CO_2 (IPCC, 2014). Nitrous oxide – another greenhouse gas linked to farming and agriculture is 264 times more powerful than CO_2 (IPCC, 2014). It is obvious that the climate change imperatives are much broader than the currently emphasized alternative fuels and renewable energy. If we are to avert global warming's worst case scenarios meat reduction is an imperative.

Commitments for 2100 are close to the hundred-year scale and, although essential for changing the technological trajectories and use of natural resources, we need a much quicker response to stabilize climate change at a 2oC temperature increase. Being less than one generation long, the twenty-year horizon is a timeframe most people are comfortable with. They can relate to events in the past and can plan for the future on this scale. It is within this timeframe that meat reduction can have an immediate impact.

Were we to drastically cut the consumption of animal proteins now, the emitted methane will remain in the earth's atmosphere for 20 years before gradually vanishing. By limiting further addition of food animal-related methane and nitrous oxide we can free up the air from these powerful gases to pave the road for a better future and allow the atmospheric space for renewable energy and other new climate friendly technologies to phase in.

By drastically reducing meat consumption, we will also achieve better personal and public health goals including reductions in obesity rates, cancers, cardiovascular diseases, saving antibiotics and diminishing pressures on land, forests, water and food security. There will be more opportunities to reduce global poverty. We will not have to condemn billions of food animals to short lives of exploitation. There will be new business opportunities around plant-based proteins...

The scientific evidence about meat consumption is clear. We hope we managed to communicate this to you. Eating less meat is the best option we have to safeguard human and environmental health.

Dora Marinova
Curtin University, Australia

Talia Raphaely
Curtin University, Australia

REFERENCE

Intergovernmental Panel on Climate Change (IPCC). (2014). [Synthesis Report. Geneva, Switzerland: IPCC.]. *Climatic Change*, 2014.

Compilation of References

Aarestrup, F. M., & Wegener, H. (2009). Hearing on H.R. 1549, the Preservation of Antibiotics for Medical Treatment Act of 2009. Washington, DC: Government of the United States of America, US House of Representatives Committee on Rules. Retrieved from http://goo.gl/4YI8mv

Aarestrup, F. M., Wegener, H. C., & Collignon, P. (2008). Resistance in bacteria of the food chain: Epidemiology and control strategies. *Expert Review of Anti-Infective Therapy, 6*(5), 733–750. doi:10.1586/14787210.6.5.733 PMID:18847409

Abete, I., Romaguera, D., Vieira, A. R., Lopez de Munain, A., & Norat, T. (2014). Association between total, processed, red and white meat consumption and all-cause, CVD and IHD mortality: A meta-analysis of cohort studies. *The British Journal of Nutrition, 112*(5), 762–775.

Abid, Z., Cross, A. J., & Sinha, R. (2014). Meat, dairy, and cancer. *The American Journal of Clinical Nutrition, 100*(Supplement 1), 386S–393S.

Adams, C. J. (2010). The sexual politics of meat: a feminist-vegetarian critical theory (20th Anniversary Ed.). New York, NY/London, UK: Continuum.

Adams, C. J. (2013). *The sexual politics of meat*. New York, NY: Bloomsbury Academic.

Adeyemo, O. K., Ayodeji, I. O., & Aiki-Raji, C. O. (2002). The water quality and sanitary conditions in a major abattoir (Bodija) in Ibadan, Nigeria. *African Journal of Biomedical Research, 5*, 51–55.

Agapakis, C. (2012). Steak of the art: the fatal flaws of in vitro meat. Retrieved from http://blogs.discovermagazine.com/crux/2012/04/24/steak-of-the-art-the-fatal-flaws-of-in-vitro-meat/UgAx51JQ0qt

Agricultural Marketing Service (AMS). (2013). USDA announces purchase program for chicken products. Retrieved from http://www.ams.usda.gov/AMSv1.0/getfile?dDocName=STELPRDC5104998

Agricultural Marketing Service (AMS). (2014). USDA chicken products purchase program announced. Retrieved from http://www.ams.usda.gov/AMSv1.0/getfile?dDocName=STELPRDC5108873

Agwa, O. K., Sito, E., & Ogugbue, C. J. (2013). A spatial assessment of the microbiological and physicochemical quality of a stream receiving raw abattoir waste. *Middle-East Journal of Scientific Research, 14*(7), 879–886.

Ajzen, I. (1991). The theory of planned behaviour. *Organizational Behavior and Human Decision Processes, 50*(2), 179–211. doi:10.1016/0749-5978(91)90020-T

Akhtar, A. Z., Greger, M., Ferdowsian, H., & Frank, E. (2009). Health professionals' roles in animal agriculture, climate change, and human health. *American Journal of Preventive Medicine, 36*(2), 182–187.

Alexander, D. D., & Cushing, C. A. (2009). Quantitative assessment of red meat or processed meat consumption and kidney cancer. *Cancer Detection and Prevention, 32*(5-6), 340–351.

Alexander, D. D., Mink, P. J., Cushing, C. A., & Sceurman, B. (2010). A review and meta-analysis of prospective studies of red and processed meat intake and prostate cancer. *Nutrition Journal*, *9*(50), 1–17.

Alexander, D. D., Morimoto, L. M., Mink, P. J., & Cushing, C. A. (2010). A review and meta-analysis of red and processed meat consumption and breast cancer. *Nutrition Research Reviews*, *23*(2), 349–365.

Alexandratos, N., & Bruinsma, J. (2012). World agriculture towards 2030/50 (no. 12-03). Food and Agriculture Organization of the United Nations (FAO). Retrieved from http://www.fao.org/fileadmin/templates/esa/Global_persepctives/world_ag_2030_50_2012_rev.pdf

Alfaia, C. P. M., Alves, S. P., Martins, S. I. V., Costa, A. S. H., Fontes, C. M. G. A., Lemos, J. P. C., & Prates, J. A. M. et al. (2009). Effect of the feeding system on intramuscular fatty acids and conjugated linoleic acid isomers of beef cattle, with emphasis on their nutritional value and discriminatory ability. *Food Chemistry*, *114*(3), 939–946. doi:10.1016/j.foodchem.2008.10.041

Algers, B., Blokhuis, H. J., Botner, A., Broom, D. M., Costa, P., Domingo, M., & Wierup, M. et al. (2009). Scientific opinion on the overall effects of farming systems on dairy cow welfare and disease. Scientific Opinion of the Panel on Animal Health and Animal Welfare (Question No EFSA-Q- 2006-113). *European Food Safety Authority*, *1143*, 1–38.

Allen, A. (2011, October 3). U.S. touts fruit and vegetables while subsidizing animals that become meat. The Washington Post. Retrieved September 8, 2015, from http://www.washingtonpost.com/national/health-science/us-touts-fruit-and-vegetables-while-subsidizing-animals-that-become-meat/2011/08/22/gIQATFG5IL_story.html

Allen, M. W. (2005, December 5-7). Social structures, status seeking and the basic food groups. Paper presented at the Australian and New Zealand Market Academy (ANZMAC) 2005 Conference, University of Western Australia, Peerth, Australia. Retrieved from http://www.anzmac.org/conference_archive/2005/cd-site/pdfs/3-Consumer-Beh/3-Allen.pdf

Allen, M. W., Gupta, R., & Monnier, A. (2008). The Interactive effect of cultural symbols and human values on taste evaluation. *The Journal of Consumer Research*, *35*(2), 294–308. doi:10.1086/590319

Allen, M. W., Wilson, M., Ng, S. H., & Dunne, M. (2000). Values and beliefs of vegetarians and omnivores. *The Journal of Social Psychology*, *140*(4), 405–422. doi:10.1080/00224540009600481 PMID:10981371

Alliance to Save Our Antibiotics. (2014). Antimicrobial resistance - why the irresponsible use of antibiotics in agriculture must stop. The Soil Association. Retrieved from http://www.soilassociation.org/LinkClick.aspx?fileticket=G9q4uEb5deI%3D&tabid=1841

Allison, I., Bindoff, N. L., Bindschadler, R. A., Cox, P. M., de Noblet, N., & England, M. H. … Weaver, A. J. (2009). The Copenhagen Diagnosis: updating the world on the latest climate science. Retrieved June 10, 2015, from http://www.ccrc.unsw.edu.au/sites/default/files/Copenhagen_Diagnosis_HIGH.pdf

Almiron, N., & Zoppeddu, M. (2014). Eating meat and climate change: The media blind spot - a study of Spanish and Italian press coverage. *Environmental Communication*, *9*(3), 307–325. doi:10.1080/17524032.2014.953968

Altorf-van der Kuil, W., Engberink, M. F., Vedder, M. M., Boer, J. M., Verschuren, W. M., & Geleijnse, J. M. (2012). Sources of dietary protein in relation to blood pressure in a general dutch population. *Public Library of Science (Plos)*. *One*, *7*(2), 1–8.

Ambiente Brasil. (2015a). Desmatamento tem alta na Amazônia em Agosto e Setembro, diz Imazon [Deforestation in the Amazon high in August and September, said Imazon]. Retrieved from http://noticias.ambientebrasil.com.br/clipping/2014/10/20/109734-desmatamento-tem-alta-na-amazonia-em-agosto-e-setembro-diz-imazon.html

Ambiente Brasil. (2015b). Inpe detecta aumento do desmate da Amazônia entre Agosto e Outubro [INPE detects increases in Amazon Deforestation between August and October]. Retrieved from http://noticias.ambientebrasil.com.br/clipping/2014/11/29/110876-inpe-detecta-aumento-do-desmate-da-amazonia-entre-agosto-e-outubro.html

American Broadcasting Company (ABC). (2011, August 19). Bill Clinton's vegan diet inspiring others. *ABC News*. Retrieved from http://abcnews.go.com/WNT/video/bill-clintons-vegan-diet-inspiring-14344849

American Broadcasting Company (ABC). (2013, December 8). Jay-Z and Beyonce's three week vegan diet. *ABC News*. Retrieved from http://abcnews.go.com/GMA/video/jay-beyonces-week-vegan-diet-21139910

American Heart Association (AHA). (2015). Saturated fats. Retrieved April 2, 2015, from http://www.heart.org/HEARTORG/GettingHealthy/NutritionCenter/HealthyEating/Saturated-Fats_UCM_301110_Article.jsp

American Veterinary Medical Association. (2015). Judicious therapeutic use of antimicrobials: position statement. Retrieved from https://www.avma.org/KB/Policies/Pages/Judicious-Therapeutic-Use-of-Antimicrobials.aspx

Aminov, R. I. (2010). A brief history of the antibiotic era: lessons learned and challenges for the future (Article 134). Frontiers in Microbiology, 1, 1-7. doi:10.3389/fmicb.2010.00134

Amora, D. (2014, September 17). BNDES obstrui auditoria sobre empréstimos de R$ 8 bi ao JSB/Friboi, sez TCU [BNDES blocks audit of R$ 8 billion in loans to the JBS/Friboi, says TCU]. *Folha De S. Paulo*. Retrieved from http://www1.folha.uol.com.br/mercado/2014/09/1517673-bndes-obstrui-auditoria-sobre-emprestimos-de-r-8-bi-ao-jbsfriboi-diz-tcu.shtml

Ananthakrishnan, A. N., Du, M., Berndt, S. I., Brenner, H., Caan, B. J., Casey, G., & Chan, A. T. et al. (2015). Red meat intake, NAT2 and risk of colorectal cancer: A pooled analysis of 11 studies. *Cancer Epidemiology, Biomarkers & Prevention*, 24(1), 198–205.

Andreoletti, O., Budka, H., Buncic, S., Colin, P., Collins, J. D., & de Koijer, A. ... Vanopdenbosch, E. (2008). Foodborne antimicrobial resistance as a biological hazard. Scientific Opinion of the Panel on Biological Hazards (Question No EFSA-Q-2007-089). Agreement by the BIOHAZ Panel for public consultation 5-6 March 2008. Public consultation 17 April - 27 May 2008. *European Food Safety Authority, 765*, 1-87. Retrieved September 8, 2015, from http://www.efsa.europa.eu/sites/default/files/scientific_output/files/main_documents/biohaz_op_ej765_antimicrobial_resistance_en,3.pdf?ssbinary=true

Andrews, N. C. (1999). Disorders of iron metabolism. *The New England Journal of Medicine, 341*(26), 1986–1995.

Animal Liberation Victoria (ALV). (2012). Why vegan? Retrieved from http://veganeasy.org/Why-Vegan

Animals Australia. (2014). Exposing live export cruelty. Retrieved April 8, 2015, from http://www.animalsaustralia.org/investigations/live-export/

Antibiotics Alliance. (2014). Antimicrobial resistance - why the irresponsible use of antibiotics in agriculture must stop. A briefing from the Alliance to Save Our Antibiotics. London, UK: Antibiotics Alliance. Retrieved September 8, 2015, from http://www.soilassociation.org/LinkClick.aspx?fileticket=G9q4uEb5deI%3D&tabid=1841

Antibiotics Alliance. (2014). *Antimicrobial resistance - why the irresponsible use of antibiotics in agriculture must stop: A briefing from the Alliance to Save Our Antibiotics*. London, UK: Antibiotics Alliance.

Antilla, L. (2005). Climate of skepticism: US newspaper coverage of the science of climate change. *Global Environmental Change, 15*(4), 338–352. doi:10.1016/j.gloenvcha.2005.08.003

Appel, L. J., Moore, T. J., Obarzanek, E., Vollmer, W. M., Svetkey, L. P., Sacks, F. M., & Karanja, N. et al. (1997). A clinical trial of the effects of dietary patterns on blood pressure. DASH Collaborative Research Group. *The New England Journal of Medicine, 336*(16), 1117–1124.

Appleby, P. N., Davey, G. K., & Key, T. J. (2002). Hypertension and blood pressure among meat eaters, fish eaters, vegetarians and vegans in EPIC-Oxford. *Public Health Nutrition*, *5*(5), 645–654.

Appleby, P. N., Thorogood, M., Mann, J. I., & Key, T. J. (1999). The Oxford Vegetarian Study: An overview. *The American Journal of Clinical Nutrition*, *70*(3Suppl), 525S–531S.

Appleby, P. N., Thorogood, M., McPherson, K., & Mann, J. I. (1995). Associations between plasma lipid concentrations and dietary, lifestyle and physical factors in the Oxford Vegetarian Study. *Journal of Human Nutrition and Dietetics*, *8*(5), 305–314.

Armelagos, G. J. (1969). Disease in ancient Nubia. *Science*, *163*(3864), 225–258. doi:10.1126/science.163.3864.255 PMID:5762604

Ascherio, A., Katan, M. B., Zock, P. L., Stampfer, M. J., & Willett, W. C. (1999). Trans fatty acids and coronary heart disease. *The New England Journal of Medicine*, *340*(25), 1994–1998.

Ascione, F., & Arkow, P. E. (1999). *Child abuse, domestic violence, and animal abuse: linking the circles of compassion for prevention and intervention*. West Lafayette, IN: Purdee University Press.

Asplund, T., Hjerpe, M., & Wibeck, V. (2013). Framings and coverage of climate change in Swedish specialized farming magazines. *Climatic Change*, *117*(1), 197–209. doi:10.1007/s10584-012-0535-0

Associação Brasileira das Indústrias Exportadoras de Carnes (ABIEC). (n.d.). Rebanho Bovino Brasileiro [Brazilian Beef Herd]. Retrieved from http://www.abiec.com.br/3_rebanho.asp

Aston, L. M., Smith, J. N., & Powles, J. W. (2012). Impact of a reduced red and processed meat dietary pattern on disease risks and greenhouse gas emissions in the UK: A modelling study. *British Medical Journal*, *2*(5), 1–9.

Astrup, A., Dyerberg, J., Elwood, P., Hermansen, K., Hu, F. B., Jakobsen, M. U., & Willett, W. C. et al. (2011). The role of reducing intakes of saturated fat in the prevention of cardiovascular disease: Where does the evidence stand in 2010? *The American Journal of Clinical Nutrition*, *93*(4), 684–688.

Aubrey, A. (2013). Why Bill Gates is investing in chicken-less eggs. Retrieved September 8, 2015, from http://www.npr.org/sections/thesalt/2013/06/13/191029875/why-bill-gates-is-investing-in-chicken-less-eggs

Aune, D., Chan, D. S., Vieira, A. R., Navarro Rosenblatt, D. A., Vieira, R., Greenwood, D. C., & Norat, T. et al. (2013). Red and processed meat intake and risk of colorectal adenomas: A systematic review and meta-analysis of epidemiological studies. *Cancer Causes & Control*, *24*(4), 611–627.

Aune, D., Ursin, G., & Veierod, M. B. (2009). Meat consumption and the risk of type 2 diabetes: A systematic review and meta-analysis of cohort studies. *Diabetologia*, *52*(11), 2277–2287.

Aust, M. O., Godlinski, F., Travis, G. R., Hao, X., McAllister, T. A., Leinweber, P., & Thiele-Bruhn, S. (2008). Distribution of sulfamethazine, chlorotetracycline and tylosin in manure and soil of Canadian feedlots after subtherapeutic use in cattle. *Environmental Pollution*, *156*(3), 1243–1251. doi:10.1016/j.envpol.2008.03.011 PMID:18440678

Australia China Connections (ACC). (2014). Food safety in China opens doors for Australia's agri sector. Retrieved from http://www.chinaconnections.com.au/en/magazine/current-issue/1940-food-safety-in-china-opens-doors-for-australia%E2%80%99s-agri-sector

Australian Bureau of Agriculture and Resource Economics and Science (ABARES). (2011). Australian commodity statistics 2011. Canberra, Australia: ABARES. Retrieved from http://data.daff.gov.au/brs/data/warehouse/agcstd9abcc002/agcstd9abcc0022011/ACS_2011_1.0.3.pdf

Australian Bureau of Statistics (ABS). (1999). National nutrition survey: foods eaten, Australia, 1995. Retrieved February 1, 2015, from http://www.abs.gov.au/AUSSTATS/abs@.nsf/0/9A125034802F94CECA2568A9001393CE Australian Bureau of Statistics (ABS). (2014). Australian health survey: nutrition first results - food and nutrients, 2011-12. Retrieved February 1, 2015, from http://www.abs.gov.au/ausstats/abs@.nsf/PrimaryMainFeatures/4364.0.55.007?OpenDocument

Australian Bureau of Statistics (ABS). (2013). Energy Accounts Australia 2011-12 (No. 4604.0). Retrieved June 10, 2015, from http://www.abs.gov.au/ausstats/abs@.nsf/mf/4604.0

Australian Commission on Safety and Quality in Health Care. (2013). Australian one health antimicrobial resistance colloquium. Background paper. Retrieved from http://www.safetyandquality.gov.au/wp-content/uploads/2013/07/Briefing-paper-for-One-Health-AMR-Colloquium-participants-Final-Jul-2013.pdf

Australian Government. (2012). Australia in the Asian Century. White paper. Canberra, Australia: Department of the Prime Minister and Cabinet. Retrieved from http://www.murdoch.edu.au/ALTC-Fellowship/_document/Resources/australia-in-the-asian-century-white-paper.pdf

Australian Government. (2013). China country strategy. Australia in the Asian Century: towards 2025. Canberra, Australia: Department of Foreign Affairs and Trade. Retrieved from http://apo.org.au/files/Resource/dfat_chinacountrys-trategy_aug_2013.pdf

Australian Institute of Health and Welfare (AIWH). (2008). Australia's health 2008. The eleventh biennial health report of the Australian Institute of Health and Welfare. Canberra, Australia: AIWH. Retrieved September 8, 2015, from http://www.aihw.gov.au/WorkArea/DownloadAsset.aspx?id=6442453674

Azadbakht, L., & Esmaillzadeh, A. (2009). Red meat intake is associated with metabolic syndrome and the plasma C-reactive protein concentration in women. *The Journal of Nutrition*, *139*(2), 335–339.

Babio, N., Sorli, M., Bullo, M., Basora, J., Ibarrola-Jurado, N., Fernandez-Ballart, J., & Salas-Salvado, J. et al. (2012). Association between red meat consumption and metabolic syndrome in a Mediterranean population at high cardiovascular risk: Cross-sectional and 1-year follow-up assessment. *Nutrition, Metabolism, and Cardiovascular Diseases*, *22*(3), 200–207.

Babio, N., Toledo, E., Estruch, R., Ros, E., Martinez-Gonzalez, M. A., Castaner, O., & Salas-Salvado, J. et al. (2014). Mediterranean diets and metabolic syndrome status in the PREDIMED randomized trial. *Canadian Medical Association Journal*, *186*(17), E649–E657.

Baer, D. J., Judd, J. T., Clevidence, B. A., & Tracy, R. P. (2004). Dietary fatty acids affect plasma markers of inflammation in healthy men fed controlled diets: A randomized crossover study. *The American Journal of Clinical Nutrition*, *79*(6), 969–973.

Baghurst, K., Record, S., & Leppard, P. (2000). Red meat consumption in Australia: Intakes, nutrient contribution and changes over time. *Australian Journal of Nutrition and Dietetics*, *57*(4), S3–S37.

Bailey, R., Froggatt, A., & Wellesley, L. (2014). Livestock - climate change's forgotten sector: global public opinion on meat and dairy consumption. London, UK: Chatham House, The Royal Institute of International Affairs. Retrieved from http://www.chathamhouse.org/sites/files/chathamhouse/field/field_document/20141203LivestockClimateChangeBaileyFroggattWellesley.pdf

Bailey, R., Froggatt, A., & Wellesley, L. (2014). *Livestock - climate change's forgotten sector: global public opinion on meat and dairy consumption*. London, UK: Chatham House, The Royal Institute of International Affairs. Retrieved September 7, 2015, from http://www.chathamhouse.org/sites/files/chathamhouse/field/field_document/20141203LivestockClimateChangeBaileyFroggattWellesley.pdf

Bajželj, B., Richards, K. S., Allwood, J. M., Smith, P., Dennis, J. S., Curmi, E., & Gilligan, C. A. (2014). Importance of food-demand management for climate mitigation. *Nature Climate Change*, 4(10), 924–929. doi:10.1038/nclimate2353

Baker, S., Thompson, K. E., & Palmer-Barnes, D. (2002). Crisis in the meat industry: A values-based approach to communications strategy. *Marketing Communications*, 8(1), 19–30. doi:10.1080/13527260110108319

Balza, G. (2014). Campeã em doações, Friboi virou gigante da carne com R$ 10 bi do BNDES [Champion in donations, Friboi turned meat giant with US$ 10 billion from BNDES]. Retrieved from http://eleicoes.uol.com.br/2014/noticias/2014/08/10/campea-em-doacoes-friboi-virou-gigante-da-carne-com-r-10-bi-do-bndes.htm

Bandera, E. V., Kushi, L. H., Moore, D. F., Gifkins, D. M., & McCullough, M. L. (2007). Consumption of animal foods and endometrial cancer risk: A systematic literature review and meta-analysis. *Cancer Causes & Control*, 18(9), 967–988.

Bao, W., Bowers, K., Tobias, D. K., Hu, F. B., & Zhang, C. (2013). Prepregnancy dietary protein intake, major dietary protein sources, and the risk of gestational diabetes mellitus: A prospective cohort study. *Diabetes Care*, 36(7), 2001–2008.

Bao, W., Rong, Y., Rong, S., & Liu, L. (2012). Dietary iron intake, body iron stores, and the risk of type 2 diabetes: A systematic review and meta-analysis. *BioMed Central Medicine*, 10, 1–13.

Baragona, S. (2010). Use of antibiotics in livestock debated. Retrieved from http://www.voanews.com/content/use-of-antbiotics-in-livestock-debated-93683324/169529.html

Barber, E. (2014). Ebola death toll surpasses 5,000 worldwide. Time. Retrieved September 8, 2015, from http://time.com/3582804/ebola-death-toll-surpasses-5000-worldwide

Barlow, S., Chesson, A., Collins, J. D., Flynn, A., Hardy, A., & Jany, K.-D. … Vannier, P. (2008). Scientific opinion of the Scientific Committee. Food safety, animal health and welfare and environmental impact of animals derived from cloning by somatic cell nucleus transfer (scnt) and their offspring and products obtained from those animals (Question No EFSA-Q-2007-092). *European Food Safety Authority, 767*, 1-49. Retrieved September 8, 2015, from http://www.efsa.europa.eu/sites/default/files/scientific_output/files/main_documents/sc_op_ej767_animal_cloning_en.pdf

Barnard, N. D., Cohen, J., Jenkins, D. J., Turner-McGrievy, G., Gloede, L., Jaster, B., & Talpers, S. et al. (2006). A low-fat vegan diet improves glycemic control and cardiovascular risk factors in a randomized clinical trial in individuals with type 2 diabetes. *Diabetes Care*, 29(8), 1777–1783.

Barnard, N. D., Gloede, L., Cohen, J., Jenkins, D. J., Turner-McGrievy, G., Green, A. A., & Ferdowsian, H. (2009). A low-fat vegan diet elicits greater macronutrient changes, but is comparable in adherence and acceptability, compared with a more conventional diabetes diet among individuals with type 2 diabetes. *Journal of the American Dietetic Association*, 109(2), 263–272.

Barnard, N. D., Nicholson, A., & Howard, J. L. (1995). The medicinal costs attributable to meat consumption. *Preventive Medicine*, 24(6), 646–655.

Barnard, N. D., Scialli, A. R., Turner-Mcgrievy, G., & Lanou, A. J. (2004). Acceptability of a low-fat vegan diet compares favourably to a step II diet in a randomized, controlled trial. *Journal of Cardiopulmonary Rehabilitation*, 24(4), 229–235.

Barnard, N., Levin, S., & Trapp, C. (2014). Meat consumption as a risk factor for type 2 diabetes. *Nutrients*, 6(2), 897–910. doi:10.3390/nu6020897 PMID:24566443

Barnard, R. J., Jung, T., & Inkeles, S. B. (1994). Diet and exercise in the treatment of NIDDM. The need for early emphasis. *Diabetes Care*, 17(12), 1469–1472.

Baron, S., Bedarida, T., Cottart, C. H., Vilbert, F., Vessieres, E., Ayer, A.,... Nivet-Antoine, V. (2014). Dual effects of resveratrol on arterial damage induced by insulin resistance in aged mice. *The Journal of Gerontology: Series A Biological Sciences and Medical Sciences, 69*(3), 260-269. doi: 2.10.1093/gerona/glt1081

Barton, D. (2013). The rise of the middle class in China and its impact on the Chinese and world economies. In China-United States Exchange Foundation, *US-China economic relations in the next ten years: towards deeper engagement and mutual benefit* (Ch. 7, pp. 1-12). Retrieved from http://www.chinausfocus.com/us-china-2022/

Bassett, E. J., Keith, M. S., Armelagos, G. J., Martin, D. L., & Villanueva, A. R. (1980). Tetracycline labelled human bone from ancient Sudanese Nubia (A.D. 350). *Science, 209*(4464), 1532–1534. doi:10.1126/science.7001623 PMID:7001623

Bastide, N. M., Pierre, F. H., & Corpet, D. E. (2011). Heme iron from meat and risk of colorectal cancer: A meta-analysis and a review of the mechanisms involved. *Cancer Prevention Research (Philadelphia, Pa.), 4*(2), 177–184.

Bazzano, L. A., He, J., Ogden, L. G., Loria, C. M., Vupputuri, S., Myers, L., & Whelton, P. K. (2002). Fruit and vegetable intake and risk of cardiovascular disease in US adults. *The American Journal of Clinical Nutrition, 76*, 93–99. PMID:12081821

Beasley, J. M., & Wylie-Rosett, J. (2013). The role of dietary proteins among persons with diabetes. *Current Atherosclerosis Reports, 15*(9), 1–8.

Bedford, J. L., & Barr, S. I. (2005). Diets and selected lifestyle practices of self-defined adult vegetarians from a population-based sample suggest they are more 'health conscious'. *The International Journal of Behavioral Nutrition and Physical Activity, 2*(1), 1–11.

Bee, G., Guex, G., & Herzog, W. (2004). Free range rearing of pigs during the winter: Adaptations in muscle fiber characteristics and the effects on adipose tissue composition and meat quality traits. *Journal of Animal Science, 82*(4), 1206–1218. PMID:15080344

Bekoff, M. (2000). *The smile of a dolphin: remarkable accounts of animal emotions*. Washington, DC: Random House/Discovery Books.

Bekoff, M. (2007). *The emotional lives of animals: a leading scientist explores animal joy, sorrow and empathy - and why they matter*. Novato, CA: New World Library.

Bellarby, J., Tirado, R., Leip, A., Weiss, F., Lesschen, J. P., & Smith, P. (2013). Livestock greenhouse gas emissions and mitigation potential in Europe. *Global Change Biology, 19*(1), 3–18. doi:10.1111/j.1365-2486.2012.02786.x PMID:23504717

Benker-Coker, M. O., & Ojior, O. O. (1995). Effect of slaughterhouse wastes on the water quality of Ikpoba River, Nigeria. *Bioresource Technology, 52*(1), 5–12.

Bennett, B. J., de Aguiar Vallim, T. Q., Wang, Z., Shih, D. M., Meng, Y., Gregory, J., & Lusis, A. J. et al. (2013). Trimethylamine-N-oxide, a metabolite associated with atherosclerosis, exhibits complex genetic and dietary regulation. *Journal of Cell Metabolism, 17*(1), 49–60.

Bennett, W. L. (1996). *News: the politics of illusion* (3rd ed.). White Plains, NY: Longman.

Berkow, S. E., & Barnard, N. (2006). Vegetarian diets and weight status. *Nutrition Reviews, 64*(4), 175–188.

Berkow, S. E., & Barnard, N. D. (2005). Blood pressure regulation and vegetarian diets. *Nutrition Reviews, 63*(1), 1–8.

Berkow, S. E., Bernard, N., Katcher, H., & Eckart, J. (2010). Four Theraputic Diets: Adherance and Acceptanility. *Canadian Journal of Dietetic Practice and Researc, 71*(4), 199–204.

Berley, P., & Singer, Z. (2007). *The flexitarian table: Inspired, flexible meals for vegetarians, meat lovers and everyone in between.* New York, NY: Houghton Mifflin Harcourt.

Berners-Lee, M., Hoolohan, C., Cammack, H., & Hewitt, C. N. (2012). The relative greenhouse gas impacts of realistic dietary choices. *Energy Policy, 43,* 184–190. doi:10.1016/j.enpol.2011.12.054

Bernstein, A. M., Pan, A., Rexrode, K. M., Stampfer, M., Hu, F. B., Mozaffarian, D., & Willett, W. C. (2012). Dietary protein sources and the risk of stroke in men and women. *Stroke, 43*(3), 637–644.

Bernstein, A. M., Sun, Q., Hu, F. B., Stampfer, M. J., Manson, J. E., & Willett, W. C. (2010). Major dietary protein sources and risk of coronary heart disease in women. *Circulation, 122*(9), 876–883.

Beverland, M. B. (2014). Sustainable eating: Mainstreaming plant-based diets in developed economies. *Journal of Macromarketing, 34*(3), 369–382. doi:10.1177/0276146714526410

Bianco, A. (2005). The future of The New York Times. *Bloomberg Businessweek,* Retrieved from http://www.bloomberg.com/bw/stories/2005-01-16/the-future-of-the-new-york-times

Billings, M. (1997). The influenza pandemic of 1918. Retrieved April 8, 2015, from https://virus.stanford.edu/uda/

Binnekamp, M., & Ingenbleek, P. (2008). Do 'good' food products make others look 'bad'? Spin-off effects of labels for sustainable food production in the consumer perception. *British Food Journal, 110*(9), 843–864. doi:10.1108/00070700810900576

Bittman, M. (2008, January 27). Rethinking the meat guzzler. New York Times. Retrieved September 8, 2015, from http://www.nytimes.com/2008/01/27/weekinreview/27bittman.html?pagewanted=all&_r=0

Bittman, M. (2012, January 10). We're eating less meat. Why? *New York Times. Opinianator.* Retrieved from http://opinionator.blogs.nytimes.com/2012/01/10/were-eating-less-meat-why/

Bittman, M. (2012a, January 10). We're eating less meat why? The New York Times. Retrieved September 8, 2015, from http://opinionator.blogs.nytimes.com/2012/01/10/were-eating-less-meat-why/

Bittman, M. (2012b, March 9). A chicken without guilt. The New York Times. Retrieved September 8, 2015, from http://www.nytimes.com/2012/03/11/opinion/sunday/finally-fake-chicken-worth-eating.html?_r=0

Blomqvist, L., Brook, B. W., Ellis, E., Kereiva, P., & Nordhaus, T. (2013). Does the shoe fit? Real versus imagined ecological footprints. *Public Library of Science (PLoS). Biology, 11*(11), e1001700. doi:10.1371/journal.pbio.1001700

Bolton, T. (2006). *News on the Net: a critical analysis of the potential of online alternative journalism to challenge the dominance of mainstream news media.* Lilydale, Australia: Swinburne University of Technology.

Bond, L., & Butler, H. (2009). The Gatehouse Project: a multi-level integrated approach to promoting wellbeing in schools. In A. Killoran & M. Kelley (Eds.), *Evidence-based public health: effectiveness and efficiency* (pp. 250–269). Oxford, UK: Oxford University Press. doi:10.1093/acprof:oso/9780199563623.003.016

Bonequi, P., Meneses-Gonzalez, F., Correa, P., Rabkin, C. S., & Camargo, M. C. (2013). Risk factors for gastric cancer in Latin America: A meta-analysis. *Cancer Causes & Control, 24*(2), 217–231.

Bøtner, A., Broom, D. M., Doherr, M. G., Domingo, M., Hartung, J., & Keeling, L. … Wierup, M. (2010). Scientific opinion on the influence of genetic parameters on the welfare and the resistance to stress of commercial broilers. EFSA Panel on Animal Health and Welfare. *European Food Safety Authority, 8*(7), 1-82. Retrieved September 8, 2015, from http://www.efsa.europa.eu/sites/default/files/scientific_output/files/main_documents/1666.pdf

Bowers, B. (2013). Rep. Louise Slaughter says 80% of antibiotics are fed to livestock. Retrieved from http://www.politifact.com/truth-o-meter/statements/2013/oct/15/louise-slaughter/rep-louise-slaughter-says-80-antibiotics-are-fed-l/

Bowers, K., Yeung, E., Williams, M. A., Qi, L., Tobias, D. K., Hu, F. B., & Zhang, C. (2011). A prospective study of prepregnancy dietary iron intake and risk for gestational diabetes mellitus. *Diabetes Care, 34*(7), 1557–1563.

Boyd-Barrett, O. (2006). Alternative reframing of mainstream media frames. In D. K. Thussu (Ed.), *Media on the move: global flow and contra-flow* (pp. 178–194). London, UK: Routledge.

Boyle, E. (2012). *High steaks: why and how to eat less meat.* Gabriola Island, Canada: New Society Publishers.

Bradbear, C., & Friel, S. (2011). Food systems and environmental sustainability: a review of the Australian evidence. Working Paper October 2011. Canberra, Australia: Australian National University (ANU), College of Medicine, National Centre for Epidemiology and Population Health. Retrieved September 8, 2015, from http://nceph.anu.edu.au/files/Food%20systems%20and%20Environmental%20Sustainability%20A%20review%20of%20the%20Evidence%20NCEPHWorkingPaperOctober2011x%20(3).pdf

Bray, G. A., Vollmer, W. M., Sacks, F. M., Obarzanek, E., Svetkey, L. P., & Appel, L. J. (2004). A further subgroup analysis of the effects of the DASH diet and three dietary sodium levels on blood pressure: Results of the DASH-Sodium Trial. *The American Journal of Cardiology, 94*(2), 222–227.

Brazil Foods. (2013). Relatório anual e de sustentabilidade 2013 [Annual sustainability report]. Retrieved from http://www.brf-global.com/brasil/

Brenkert, G. C. (2008). *Marketing ethics.* Oxford, UK: Blackwell Publishing.

Breslow, J. M. (2014, January 8). Illinois "nightmare bacteria" outbreak raises alarms. *Frontline.* Retrieved September 6, 2015, from http://www.pbs.org/wgbh/pages/frontline/health-science-technology/hunting-the-nightmare-bacteria/illinois-nightmare-bacteria-outbreak-raises-alarms/

Brian, P. W. (1957). Effects of antibiotics on plants. *Annual Review of Plant Physiology, 81*(1), 413–426. doi:10.1146/annurev.pp.08.060157.002213

Briske, D. D., Ash, A. J., Dernerc, J. J., & Huntsinger, L. (2014). Commentary: A critical assessment of the policy endorsement for holistic management. *Agricultural Systems, 125*, 50–53. doi:10.1016/j.agsy.2013.12.001

Bristow, E., & Fitzgerald, A. J. (2011). Global climate change and the industrial animal agriculture link: The construction of risk. *Society & Animals, 19*(3), 205–224. doi:10.1163/156853011X578893

Brockington, D. (2009). *Celebrity and the environment: fame, wealth and power in conservation.* New York, NY: Zed Books.

Broedbaek, K., Siersma, V., Andersen, J. T., Petersen, M., Afzal, S., & Hjelvang, B., cPoulsen, H.E. (2011). The association between low-grade inflammation, iron status and nucleic acid oxidation in the elderly. *Free Radical Research, 45*(4), 409–416.

Brown, C. (2014, November 11). Chinese demand for Australian beef drives $75 million Townsville abattoir plan. *ABC News.* Retrieved from http://www.abc.net.au/news/2014-11-10/second-abattoir-proposed-for-north-queensland-port-town/5877062

Brown, E. (2011, August 18). Bill Clinton talks about being a vegan. *Los Angeles Times.* Retrieved from http://articles.latimes.com/2011/aug/18/news/la-heb-bill-clinton-vegan-20110818http://articles.latimes.com/2011/aug/18/news/la-heb-bill-clinton-vegan-20110818

Brown, L. (2011). *World on the edge: how to prevent environmental and economic collapse.* New York, NY: Norton.

Brugère, J. F., Borrel, G., Gaci, N., Tottey, W., O'Toole, P. W., & Malpuech-Brugere, C. (2014). Archaebiotics: Proposed therapeutic use of archaea to prevent trimethylaminuria and cardiovascular disease. *Gut Microbes, 5*(1), 5–10.

Brügger, P. (2009a). Para além da dicotomia abolicionismo versus bem-estarismo [Apart from the abolition dichotomy versus welfarism]. Retrieved from http://www.anda.jor.br/26/11/2009/para-alem-da-dicotomia-abolicionismo-versus-bem-estarismo

Brügger, P. (2004a). 25 years past Tbilisi: environmental teaching or teaching? In W. Leal Filho & M. Littledyke (Eds.), *International perspectives in environmental education* (pp. 129–138). Frankfurt, Germany: Peter Lang International Academic Publishers.

Brügger, P. (2004b). *Educação ou adestramento ambiental* (3rd ed.). Florianópolis, Brazil: Letras Contemporâneas.

Brügger, P. (2008). Vivissecção fé cega faca amolada? [Vivisection: blind faith, sharpened knife?] In C. A. Molinaro, F. L. F. de Medeiros, I. F. Sarlet, & T. Fensterseifer (Eds.), *A dignidade de vida e os direitos fundamentais para além dos humanos: uma discussão necessária* [The dignity of life and the fundamental rights beyond the human: a needed discussion]. (pp. 145–146). Belo Horizonte, Brazil: Fórum.

Brügger, P. (2009b). Nosotros y los otros animales: Especismo veganismo y educación ambiental[We and the other animals: speciesism, veganism and environmental education]. *Linhas Criticas Brasillia, 15*(29), 197–214.

Buchman, T. G., Dushoff, J., Effron, M. B., Ehrlich, P. R., Fitzpatrick, S., Laxminarayan, R., & Solomkin, J. S. et al. (2008). Antibiotic overuse: The influence of social norms. *Journal of the American College of Surgeons, 207*(2), 265–275. doi:10.1016/j.jamcollsurg.2008.02.035 PMID:18656057

Burket, W. (1983). *Homo necans: the anthropology of ancient Greek sacrificial ritual and myth* [translated]. Berkley, CA: University of California.

Burr, M. L., & Butland, B. K. (1988). Heart disease in British vegetarians. *The American Journal of Clinical Nutrition, 48*(3Suppl), 830–832.

Butler, M., & Cooper, M. (2012, December 7). New antibiotics: what's in the pipeline? The Conversation, Retrieved from https://theconversation.com/new-antibiotics-whats-in-the-pipeline-10724

Cable News Network (CNN). (2014). Foot and mouth disease: fast facts. Retrieved December 1, 2014, from http://www.cnn.com/2013/09/02/health/foot-and-mouth-disease-fast-facts/

Callicott, J. B. (2001). The land ethic. In D. Jamieson (Ed.), *A companion to environmental philosophy* (pp. 204–217). Malden, MA: Blackwell Publishers. doi:10.1002/9780470751664.ch14

Calverd, A. (2005, July). A radical approach to Kyoto. *Physics World.* Retrieved from http://physicsworldarchive.iop.org/pdf?site=pwa&bkdir=18/7&pdf=phwv18i7a46&pdfhash=77DB6720F54408CAD26A0A7B4DBB5D05&doctime=Fri%2C24 Apr 2015 03%3A35%3A32 GMT

Cameron, E. (2014). *Climate change: implications for agriculture: key findings from the Intergovernmental Panel on Climate Change Fifth Assessment Report.* Cambridge, UK: University of Cambridge, Institute for Sustainability Leadership.

Campbell, D. (2009). Animal welfare and market access: a dairy industry perspective. Paper presented at the Proceedings of the Society of Dairy Cattle Veterinarians of the NZVA Annual Conference, Rotorua, NZ.

Campbell, T. C., & Campbell, T. M. (2005). *The China study.* Dallas, TX: BenBella Books.

Cao, L., Tian, W., Wang, J., Malcom, B., Liu, H., & Zhou, Z. (2013). Recent food consumption trends in China and trade implications to 2020. *Australasian Agribusiness Review*, *21*, 15–44.

Capra, F. (1996). *The web of life: a new scientific understanding of living systems*. New York, NY: Anchor Books.

Carlet, J., Jarlier, V., Harbarth, S., Voss, A., Goossens, H., & Pittet, D. (2012). Ready for a world without antibiotics? The Pensières Antibotic Risistance Call to Action. *Antimicrobial Resistance and Infection Control*, *1*(11), 1–13. doi:10.1186/2047-2994-1-11 PMID:22958725

Carpenter, K. J. (1986). The history of enthusiasm for protein. *Nutrition (Burbank, Los Angeles County, Calif.)*, *116*(7), 1364–1370.

Carter, J., Jones, A., O'Brien, M., Ratner, J., & Wuerthner, G. (2014). Holistic management: Misinformation on the science of grazed ecosystems. International Journal of Biodiversity, *2014*. Article ID, *163431*, 1–10. doi:10.1155/2014/163431

Carus, F. (2010, June 3). UN urges global move to meat and dairy free-diet. The Guardian. Retrieved September 8, 2015, from http://www.theguardian.com/environment/2010/jun/02/un-report-meat-free-diet

Carus, F. (2010, June 3). UN urges global move to meat and dairy-free diet. *The Guardian*. Retrieved from http://www.theguardian.com/environment/2010/jun/02/un-report-meat-free-diet

Carvalho, A., & Burgess, J. (2005). Cultural circuits of climate change in U.K. broadsheet newspapers, 1985-2003. *Risk Analysis*, *25*(6), 1457–1469. doi:10.1111/j.1539-6924.2005.00692.x PMID:16506975

Casey, C. (1995). *Work, self and society*. London, UK: Routledge.

Cassidy, E. S., West, P. C., Gerber, J. S., & Foley, J. A. (2013). Redefining agricultural yields: From tonnes to people nourished per hectare. *Environmental Research Letters*, *8*(3), 1–8. doi:10.1088/1748-9326/8/3/034015

Cawley, J., & Meyerhoefer, C. (2010). The medical care costs of obesity: an instrumental variables approach. NBER Working Paper 16467. New York, NY: Cornell University, Department of Policy Analysis and Management. Retrieved September 8, 2015, from http://www.nber.org/papers/w16467

Centers for Disease Control and Prevention (CDC). (2014). CDC and zoonotic disease. Retrieved December 17, 2014, from http://www.cdc.gov/about/facts/cdcfastfacts/zoonotic.html

Centre for Science in the Public Interest (CSPI). (2014). Antibiotic Resistance Project. Retrieved February 21, 2015, from http://www.cspinet.org/ar/

Centrers for Disease Control and Prevention (CDC). (2013). Antbiotic resistance threat in the United States, 2013. Atlanta, GA: CDC, Division of Healthcare Quality Promotion, National Center for Emerging and Zoonotic Infectious Diseases. Retrieved September 8, 2015, from http://www.cdc.gov/drugresistance/pdf/ar-threats-2013-508.pdf

Centres for Disease Control and Prevention (CDC). (2013a, March 5). Action needed now to halt spread of deadly bacteria: data show more inpatients suffering infections from bacteria resistant to all or nearly all antibiotics. CDC Newsroom. Retrieved from http://www.cdc.gov/media/releases/2013/p0305_deadly_bacteria.html

Centres for Disease Control and Prevention (CDC). (2013b). Antbiotic resistance threats in the United States 2013. Atlanta, GA: CDC, Division of Healthcare Quality Promotion, National Center for Emerging and Zoonotic Infectious Diseases. Retrieved from http://www.cdc.gov/drugresistance/pdf/ar-threats-2013-508.pdf

Centres for Disease Control and Prevention (CDC). (2013c). Vital signs: Carbapenem-resistant enterobacteriaceae. *Morbidity and Mortality Weekly Report*, *62*(9), 165–170. PMID:23466435

Chan, D. S., Lau, R., Aune, D., Vieira, R., Greenwood, D. C., Kampman, E., & Norat, T. (2011). Red and processed meat and colorectal cancer incidence: Meta-analysis of prospective studies. *Public Library of Science (PLoS). One, 6*(6), 1–11.

Chang-Claude, J., Hermann, S., Eilber, U., & Steindorf, K. (2005). Lifestyle determinants and mortality in German vegetarians and health-conscious persons: Results of a 21-year follow-up. *Cancer Epidemiology, Biomarkers & Prevention, 14*(4), 963–968.

Chang, Q., Wang, W., Regev-Yochay, G., Lipstich, M., & Hanage, W. P. (2015). Antibiotics in agriculture and the risk to human health: How worried should we be? *Evolutionary Applications, 8*(3), 240–247. doi:10.1111/eva.12185 PMID:25861382

Cheeke, P. R. (1999). *Contemporary issues in animal agriculture* (2nd ed.). Danville, IL: Interstate Publishers.

Chen, C., Hare, B., Hagemann, M., Höhne, N., Moltmann, S., & Schaeffer, M. (2011). Cancun climate talks: keeping options open to close the gap. Climate Action Briefing Paper. New York, NY: Climate Analytics, PIK and ECOFYS. Retrieved from http://climateactiontracker.org/assets/publications/briefing_papers/briefing_paper_cancun.pdf

Chen, G. C., Lv, D. B., Pang, Z., & Liu, Q. F. (2013). Red and processed meat consumption and risk of stroke: A meta-analysis of prospective cohort studies. *European Journal of Clinical Nutrition, 67*(1), 91–95.

Chen, Y. H., Xia, E. Q., Xu, X. R., Li, S., Ling, W. H., Wu, S., & Li, H. B. et al. (2012). Evaluation of benzo[a]pyrene in food from China by high-performance liquid chromatography-fluorescence detection. *International Journal of Environmental Research and Public Health, 9*(11), 4159–4169.

Chen, Y., Pan, L., Liu, N., Troy, F. A., & Wang, B. (2014). LC-MS/MS quantification of N-acetylneuraminic acid, N-glycolylneuraminic acid and ketodeoxynonulosonic acid levels in the urine and potential relationship with dietary sialic acid intake and disease in 3- to 5-year-old children. *The British Journal of Nutrition, 111*(2), 332–341.

Chernus, I. (2012). Mythic America: essays about America's national myths in the past, present and future. Retrieved from https://mythicamerica.wordpress.com/

China Meat. (2014). Statistical data on the pork, beef and lamb imports in the first half year. Retrieved from http://www.chinameat.cn/html/n1/1_1_0_0/2014-08-05/2014080509363507.shtml

Chiu, B. C.-H., Ji, B.-T., Dai, Q., Gridley, G., McLaughlin, J. K., Gao, Y.-T., & Chow, W.-H. et al. (2003). Dietary factors and risks of colon cancer in Shanghai, China. *Cancer Epidemiology, Biomarkers & Prevention, 12*(3), 201–208. PMID:12646508

Choi, C. Q. (2012). Eating meat made us human, suggest new skull fossil. Retrieved from http://www.livescience.com/23671-eating-meat-made-us-human.html

Choices, N. H. S. (2013). Meat in your diet. Retrieved December 17, 2014, from http://www.nhs.uk/Livewell/Goodfood/Pages/meat.aspx

Chomsky, N. (1989). *Necessary illusions: thought control in democratic societies*. Boston, MA: South End Press.

Chopra, I., Hesse, L., & O'Neill, A. (2002). Discovery and development of new anti-bacterial drugs. In H. vand der Goot (Ed.), Pharmacochemistry library (Vol. 32) (pp. 213-225). Amsterdam, the Netherlands: Elsevier. doi:10.1016/S0165-7208(02)80022-8

Chukwu, O. (2008). Analysis of groundwater pollution from abattoir waste in Minna, Nigeria. *Research Journal of Dairy Sciences, 2*(4), 74–77.

Chukwu, O., Adeoye, P. A., & Chidiebere, I. (2011). Abattoir wastes generation, management and the environment: A case of Minna, North Central Nigeria. *International Journal of Biosciences*, *1*(6), 100–109.

Claire, J. S. (2012, July 16). Ellen no longer vegan. *Vegans Health*. Retrieved from http://www.veganshealth.com/ellen-no-longer-vegan/

Clayman, S. E., & Reisner, A. (1998). Gatekeeping in action: Editorial conferences and assessments of newsworthiness. *American Sociological Review*, *63*(2), 178–199. doi:10.2307/2657322

Cleugh, H., Smith, M. S., Battaglia, M., & Graham, P. (Eds.). (2011). Climate change: science and solutions for Australia. Collingwood, Australia: Commonwealth Scientific and Industrial Research Organisation (CSIRO).

Clifton, P. M. (2011). Protein and coronary heart disease: The role of different protein sources. *Current Atherosclerosis Reports*, *13*(6), 493–498.

Clover, C. (2007, May 30). Go vegan to help climate, says Government. *The Telegraph*. Retrieved from http://www.telegraph.co.uk/news/earth/earthnews/3295795/Go-vegan-to-help-climate-says-Government.html

Clucas, I. (1997). A study of the options for utilization of bycatch and discards from marine capture fisheries. Fisheries Circular No. 928. Rome: Italy: Food and Agriculture Organization of the United Nations (FAO), Fishery Industries Division, Fisheries Department. Retrieved from http://www.fao.org/docrep/w6602e/w6602e00.HTM

Coburn, T. (2014). Wastebook 2014: what Washington doesn't want you to read. Washington, DC: United States Senate. Retrieved from http://fulltextreports.com/2014/10/24/wastebook-2014-what-washington-doesnt-want-you-to-read/

Colditz, G. A., Manson, J. E., Stampfer, M. J., Rosner, B., Willett, W. C., & Speizer, F. E. (1992). Diet and risk of clinical diabetes in women. *The American Journal of Clinical Nutrition*, *55*(5), 1018–1023.

Cole, J. R., & McCoskey, S. (2013). Does global meat consumption follow an environmental kuznets curve? *Sustainability: Science, Practice, &. Policy*, *9*(2), 26–36.

Cole, M., & Morgan, K. (2011). Vegaphobia: Derogatory discourses of veganism and the reproduction of speciesism in the UK national newspapers. *The British Journal of Sociology*, *62*(1), 134–153. doi:10.1111/j.1468-4446.2010.01348.x PMID:21361905

Collingon, P. (2005). Fluroquinolone use in food animals. *Emerging Infectious Diseases*, *11*(11), 1789–1792. doi:10.3201/eid1111.040630

Colvin, J. (2014). Christie vetoes politically charged pig crate bill. Retrieved from http://news.yahoo.com/christies-decision-pig-crates-down-wire-191903239--election.html

Compassion in World Farming (CIWF). (2011). Antibiotics in animal farming: public health and animal welfare. Surrey, UK: CIWF. Retrieved from https://www.ciwf.org.uk/media/3758863/Antibiotics-in-Animal-Farming-Public-Health-and-Animal-Welfare.pdf

Compassion in World Farming (CIWF). (2013). Statistics: Broiler chickens. Retrieved May 10, 2015, from https://www.ciwf.org.uk/media/5235303/Statistics-Broiler-chickens.pdf

Compassion in World Farming (CIWF). (2014a). Victory: notorious slaughterhouse is shutdown. Retrieved November 20, 2014, from http://www.philiplymbery.com/2014/11/victory-notorious-slaughterhouse-is-shutdown/

Compassion in World Farming (CIWF). (2014b). Welfare issues for egg laying hens. Retrieved April 8, 2015, from http://www.ciwf.org.uk/farm-animals/chickens/egg-laying-hens/welfare-issues/

Cooke, S. J., & Cowx, I. G. (2004). The role of recreational fisheries in global fish crisis. *Bioscience*, *54*(9), 857–859.

Cook, M., Molto, E., & Anderson, C. (1989). Flurochrome labelling in Roman period skeletons from Dakhleh Oasis, Egypt. *American Journal of Physical Anthropology*, *80*(2), 137–143. doi:10.1002/ajpa.1330800202 PMID:2679120

Cordell, D., Drangert, J., & White, S. (2009). The story of phosphorus: Global food security and food for thought. *Global Environmental Change*, *19*(2), 292–305. doi:10.1016/j.gloenvcha.2008.10.009

Cordts, A., Nitzko, S., & Spiller, A. (2014). Consumer response to negative information on meat consumption in Germany. *International Food and Agribusiness Management Review, 17*(A), 83-106.

Cornwall, J., Collie, G., & Ashton, P. (2000). *Sustaining a nation, celebrating 100 years of agriculture in Australia*. Sydney, Australia: Focus Publishing.

Craig, W. J., & Mangels, A. R. (2009). Position of the American Dietetic Association: Vegetarian diets. *Journal of the American Dietetic Association*, *109*(7), 1266–1282.

Cranston, M. (2015, May 4). Chinese agri giant wants $100m of Australian farmland. *Financial review*. Retrieved from http://www.afr.com/real-estate/chinese-agri-giant-wants-100m-of-australian-farmland-20150430-1mwvtw

Crocker, D., & Linden, T. (1998). *Ethics of consumption: the good life, and global; stewardship*. Oxford, UK: Rowman and Littlefield Publishers.

Cross, L. (1951). Veganism defined. Retrieved from http://www.ivu.org/history/world-forum/1951vegan.html

Cross, A. J., Ferrucci, L. M., Risch, A., Graubard, B. I., Ward, M. H., Park, Y., & Sinha, R. et al. (2010). A large prospective study of meat consumption and colorectal cancer risk: An investigation of potential mechanisms underlying this association. *Cancer Research and Clinical Oncology*, *70*(6), 2406–2414.

Cross, A. J., Freedman, N. D., Ren, J., Ward, M. H., Hollenbeck, A. R., Schatzkin, A., & Abnet, C. C. et al. (2011). Meat consumption and risk of esophageal and gastric cancer in a large prospective study. *The American Journal of Gastroenterology*, *106*(3), 432–442.

Cross, A. J., & Sinha, R. (2004). Meat-related mutagens/carcinogens in the etiology of colorectal cancer. *Environmental and Molecular Mutagenesis*, *44*(1), 44–55.

Crous-Bou, M., Fung, T. T., Prescott, J., Julin, B., Du, M., Sun, Q., & de Vivo, I. et al. (2014). Mediterranean diet and telomere length in Nurse's Health Study: Population based cohort study. *British Medical Journal*, *349*(7986), 1–11. PMID:25467028

Cudworth, E. (2011). *Social lives with other animals*. Basingstroke, UK: Palgrave Macmillan. doi:10.1057/9780230302488

Cui, L., & Su, X. Z. (2009). Discovery, mechanisms of action and combination theraphy of artemisinin. *Expert Review of Anti-Infective Therapy*, *7*(8), 999–1013. doi:10.1586/eri.09.68 PMID:19803708

Cullen, S. (2014, November 7). Beef deal: Australia, China close to finalising $1 billion agreement that would see 1 million cattle exported. *Abc News*. Retrieved from http://www.abc.net.au/news/2014-11-07/billion-dollar-beef-deal-china/5873496

Cunha, L. C. (2015). If natural entities have intrinsic value, should we then abstain from helpining animals who are victims of natural processes. *Relations, 3*.1. https://www.academia.edu/12378106/If_Natural_Entities_Have_Intrinsic_Value_Should_We_Then_Abstain_from_Helping_Animals_Who_Are_Victims_of_Natural_Processes

Curickshank, G. (2003). Cobb focuses on bottom line performance. *Poultry World*, *157*(7), 22–25.

da Motta, R. S., Hargrave, J., Luedemann, G., & Sarmiento Gutierrez, M. B. (2011) Climate change in Brazil: Economic, social and regulatory aspects. Institute for Applied Economic Research. Retrieved from http://www.ipea.gov.br/agencia/images/stories/PDFs/livros/livros/livro_climatechange.pdf

Daemmrich, A. (2013). The political economy of healthcare reform in China: Negotiating public and private. *SpringerPlus*, *2*(448), 1–13. PMID:24052932

Dagevos, H. (2015). Exploring flexitarianism: meat re*duction in a meat-centred food culture*. In T. Raphaely & D. Marinova (Eds.), *Impacts of meat consumption on health and environmental sustainability*. Hershey, PA: IGI Global.

Dagevos, H., & Voordouw, J. (2013). Sustainability and meat consumption: is reduction realistic. Sustainability: Science, Practice, &. *Policy*, *9*(2), 60–69.

Dahlquist, G. G., Blom, L. G., Persson, L. A., Sandstrom, A. I., & Wall, S. G. (1990). Dietary factors and the risk of developing insulin dependent diabetes in childhood. *British Medical Journal*, *300*(6735), 1302–1306.

DairyCo. (2014). Average UK milk yield. Retrieved April 8, 2015, from http://www.dairyco.org.uk/market-information/farming-data/milk-yield/average-milk-yield/ -. VU8cvGBi6Ct

Daniel, C. R., Cross, A. J., Graubard, B. I., Park, Y., Ward, M. H., Rothman, N., & Sinha, R. et al. (2012). Large prospective investigation of meat intake, related mutagens, and risk of renal cell carcinoma. *The American Journal of Clinical Nutrition*, *95*(1), 155–162.

Davis, K. E., Prasad, C., Vijayagopal, P., Juma, S., & Imrhan, V. (2014). Advanced glycation end products, inflammation, and chronic metabolic diseases: Links in a chain?[Epub ahead of print]. *Critical Reviews in Food Science and Nutrition*, (Sep): 26.

Dawkins, M. S. (1998). *Through our eyes only? The search for animal consciousness*. Oxford, UK: Oxford University Press. doi:10.1093/acprof:oso/9780198503200.001.0001

D'Costa, V. M., King, C. E., Kalan, L., Morar, M., Sung, W. W., Schwarz, C., & Wright, G. D. et al. (2011). Antibiotic resistance is ancient. *Nature*, *477*(7365), 457–461. doi:10.1038/nature10388 PMID:21881561

de Almedia, R. G. (2011). Pastagens: um desafio nacional [Pasture: a national challenge]. Retrieved from http://www.portaldbo.com.br/Portal/Artigos/Pastagens-um-desafio-nacional/1048

de Backer, C. J. S., & Hudders, L. (2014). From Meatless Mondays to Meatless Sundays: Motivations for meat reduction among vegetarians and semi-vegetarians who mildly or significantly reduce their meat intake. *Ecology of Food and Nutrition*, *53*(6), 639–657.

de Backer, C. J. S., & Hudders, L. (2015). Meat morals: Relationship between meat consumption consumer attitudes towards human and animal welfare and moral behaviour. *Meat Science*, *99*, 68–74.

de Bakker, E., & Dagevos, H. (2012). Reducing meat consumption in today's consumer society: Questioning the citizen-consumer gap. *Journal of Agricultural & Environmental Ethics*, *25*(6), 877–894.

de Biase, S. G., Fernandes, S. F., Gianini, R. J., & Duarte, J. L. (2007). Vegetarian diet and cholesterol and triglycerides levels. *Arquivos Brasileiros de Cardiologia*, *88*(1), 35–39.

de Boer, J., Schösler, H., & Aiking, H. (2014). 'Meatless days' or 'less but better?': Exploring strategies to adapt Western meat consumption to health and sustainability challenges. *Appetite*, *76*, 120–128.

De Hann, C., Schillhorn van Veen, T., Brandenburg, B., Gauthier, J., Le Gall, F., Mearns, R., & Simeon, M. (2001). *Livestock development: Implications for rural poverty, the environment and global food security*. Washington, DC: The World Bank. doi:10.1596/0-8213-4988-0 de Keyzer, W., van Caneghem, S., Heath, A. L., Vanaelst, B., Verschraegen, M., de Henauw, S., & Huybrechts, I. (2012). Nutritional quality and acceptability of a weekly vegetarian lunch in primary-school canteens in Ghent, Belgium: 'Thursday Veggie Day'. *Public Health Nutrition, 15*(12), 2326–2330.

de la Hamaide, S. (2012, January 11). Antibiotics for livestock vital to feed world: OIE. Reuters. Retrieved from http://www.reuters.com/article/2012/01/11/us-antibiotics-livestock-idUSTRE80A1JF20120111

de Oliveira Otto, M. C., Alonso, A., Lee, D. H., Delclos, G. L., Bertoni, A. G., Jiang, R., & Nettleton, J. A. et al. (2012). Dietary intakes of zinc and heme iron from red meat, but not from other sources, are associated with greater risk of metabolic syndrome and cardiovascular disease. *The Journal of Nutrition, 142*(3), 526–533.

de Savigny, D., & Adam, T. (2009). Systems thinking for health systems strengthening. Geneva, Switzerland: World Health Organization (WHO) & Alliance for Healthy Policy and Systems Research. Retrieved September 8, 2015, from http://www.who.int/alliance-hpsr/resources/9789241563895/en/

de Vries, M., & de Boer, I. J. M. (2010). Comparingenvironmental impacts for livestock products: A review of life cycle assessments. *Livestock Science, 128*(1), 1–11. doi:10.1016/j.livsci.2009.11.007

De Zen, S., Barioni, L. G., Bonato, D. B. B., de Almedia, M. H. S. P., & Rittl, T. F. (2008). Pecuária de corte brasileira, impactos ambientais e emissões de gases efeito estufa (GEE): Sumário Executivo [Brazilian beef cattle, environmental impacts and emissions of greenhouse gases (GHG)]. Sao Paulo, Brazil: Universidade de Sao Paulo. Retrieved from http://www.cepea.esalq.usp.br/pdf/Cepea_Carbono_pecuaria_SumExec.pdf

DeCoux, E. L. (2009). Speaking for the modern Prometheus: The legal and ideological significance of animal sufferring to the abolition movement. *Animal Law, 16*(9), 9–64.

DeGeneres, E. (2011). Change your life by going vegan with Ellen! Retrieved from http://www.ellentv.com/2011/08/02/change-your-life-by-going-vegan-with-ellen/

Delgado, C., Rosegrant, M., Steinfield, H., Ehui, S., & Courbois, C. (1999). Livestock to 2020: the next food revolution. Food, Agriculture and the Environment Discussion Paper 28. Washington, DC: International Food Policy Research Institute (IFPRI), Food Agriculture Organization (FAO) and International Livestock Research Institute (ILRI).

Dellavalle, C. T., Xiao, Q., Yang, G., Shu, X. O., Aschebrook-Kilfoy, B., Zheng, W., & Ward, M. H. et al. (2014). Dietary nitrate and nitrite intake and risk of colorectal cancer in the Shanghai Women's Health Study. *International Journal of Cancer, 134*(12), 2917–2926.

Denardin, V. F., & Sulzbach, M. T. (2005). Os possíveis caminhos da sustentabilidade para a agropecuária da região Oeste de Santa Catarina[The possible sustainability paths for agriculture in the West region of Santa Catarina]. *Desenvolvimento em Questão, 3*(6), 87–115.

Department for Environment. Food & Rural Affairs (DEFRA). (2013). Sustainable consumption report: follow up to the green food project. London, UK: DEFRA. Retrieved September 7, 2015, from https://www.gov.uk/government/publications/sustainable-consumption-report-follow-up-to-the-green-food-project

Department of Agriculture (DOA). (2013). Australian agriculture: reducing emissions and adapting to a changing climate. Key findings of the Climate Change Research Program. Canberra, Australia: DOA. Retrieved from http://www.agriculture.gov.au/ag-farm-food/climatechange/australias-farming-future/climate-change-and-productivity-research

Department of Agriculture and Rural Development (DARD). (2009). Guideline manual for the management of abattoirs and other waste of animal origin. Gauteng, Southern Africa: DARD, Gauteng Provincial Government. Retrieved September 8, 2015, from http://www.gdard.gpg.gov.za/Services1/Guideline%20Manual%20for%20the%20Management%20of%20Abatooirs.pdf

Department of the Environment (DOE). (2014). About the Carbon Farming Initiative. Retrieved December 13, 2014, from http://www.environment.gov.au/climate-change/emissions-reduction-fund/cfi/about

Deutsch, C. H. (2007). Trying to connect the dinner plate to climate change. *The New York Times*. Retrieved from http://www.nytimes.com/2007/08/29/business/media/29adco.html?_r=0

Devine, M. (2008a, September 11). Our dietary freedom is at steak. *The Sydney Morning Herald*. Retrieved from http://www.smh.com.au/news/miranda-devine/our-dietary-freedoms-at-steak/2008/09/10/1220857635063.html

Devine, M. (2008b, October 2). Act hastily, roo the scare tactics. *The Sydney Morning Herald*. Retrieved from http://www.smh.com.au/news/opinion/miranda-devine/act-hastily-roo-the-scare-tactics/2008/10/01/1222651169023.html?page=fullpage

Dias, E. (2015). O Cerrado está extinto e isso leva ao fim dos rios e dos reservatórios de água [The Cerrado is extinct and this leads to the end of rivers and water reservoirs]. *Jornal Opcão,* 2048. Retrieved from http://www.jornalopcao.com.br/entrevistas/o-cerrado-esta-extinto-e-isso-leva-ao-fim-dos-rios-e-dos-reservatorios-de-agua-16970/

Djoussé, L., & Gaziano, J. M. (2009). Dietary cholesterol and coronary artery disease: A systematic review. *Current Atherosclerosis Reports*, *11*(6), 418–422.

Dominguez-Rodrigo, M., Pickering, T. R., Diez-Martin, F., Mabulla, A., Musiba, C., Trancho, G., & Arriaza, C. S. et al. (2012). Earliest porotic hyperostosis on a 1.5-million-year-old hominin, Olduvai Gorge, Tanzania (oldest evidence of anemia in human evolution. *Public Library of Science (PLoS). One*, *7*(10), 1–6. doi:10.1371/journal.pone.0046414

Dong, H., Bogg, L., Rehnberg, C., & Diwan, V. (1999). Association between health insurance and antibiotics prescribing in four counties in rural China. *Health Policy (Amsterdam)*, *48*(1), 29–45. doi:10.1016/S0168-8510(99)00026-3 PMID:10539584

Doran-Browne, N. A., Eckard, R. J., Behrendt, R., Rose, S., & Kingwell, R. S. (2015). Nutrient density as a metric for comparing greenhouse gas emissions from food production. *Climatic Change*, *129*(1), 73–87. doi:10.1007/s10584-014-1316-8

Dow, S. (2007, July 3). Into the meat of the issue. *The Sydney Morning Herald*. Retrieved from http://www.smh.com.au/news/environment/into-the-meat-of-the-issue/2007/07/02/1183351125939.html

Doyle, J. (2011). Sustainable consumption? Reframing meat and dairy consumption in the politics of climate change. In J. Doyle (Ed.), *Mediating climate change* (pp. 123–144). Burlington, VT: Ashgate.

Drøyvold, W. B., Midthjell, K., Nilsen, T. I. L., & Holmen, J. (2005). Change in body mass index and its impact on blood pressure: A prospective population study. *Obesity (Silver Spring, Md.)*, *29*, 650–655. PMID:15809666

Dulvy, N., Sadovy, Y., & Reynolds, J. (2003). Extinction vaulnerability in marine populations. *Fish and Fisheries*, *4*(1), 25–64.

Easy, V. (2012). Eating out guide. Retrieved from http://www.veganeasy.org/Eating-Out-Guide

Economou, V., & Gousia, P. (2015). Agriculture and food animals as a source of antimicrobial-resistant bacteria. *Infection and Drug Resistance*, *8*, 49–61. doi:10.2147/IDR.S55778 PMID:25878509

Eder, K. (1996). *The social construction of nature: a sociology of ecological enlightenment*. London, UK: Sage.

Editorial Board. (2013, April 26). Cruelty to farm animals demands exposure. *The Washington Post*. Retrieved from http://www.washingtonpost.com/opinions/cruelty-to-farm-animals-demands-exposure/2013/04/26/9a972c8e-a6bf-11e2-a8e2-5b98cb59187f_story.html

Eisler, P. (2013, March 6). Deadly superbugs invade U.S. health care facilities. USA Today. Retrieved from http://www.usatoday.com/story/news/nation/2012/11/29/bacteria-deadly-hospital-infection/1727667/

Eisnitz, G. (1997). *Slaughterhouse: the shocking story of greed, neglect and inhumane treatment inside the U.S. meat industry.* Amherst, NY: Prometheus Books.

Elferink, E., Nonhebel, S., & Moll, H. (2008). Feeding livestock food residue and the consequences for the environmental impact of meat. *Journal of Cleaner Production, 16*(12), 1227–1233. doi:10.1016/j.jclepro.2007.06.008

Elliot, R. (2004). La ética ambiental. [Environmental Ethics] In P. Singer (Ed.), Compendio de ética [Compendium of ethics] (pp. 391–404). Madrid, Spain: Alianza Editorial.

Elliott, P., Stamler, J., Dyer, A. R., Appel, L., Dennis, B., Kesteloot, H., & Zhou, B. et al. (2006). Association between protein intake and blood pressure: The INTERMAP Study. *Journal of the American Medical Association Internal Medicine, 166*(1), 79–87.

Ellis, N., Fitchett, J., Higgins, M., Jack, G., Lim, M., Sare, M., & Tadajewski, M. (2011). *Marketing: a critical textbook.* New York, NY: Sage Publications.

Environmental Careers Organisation. (2004). *The eco guide to careers that make a difference: environmental work for a sustainable world.* Washington, DC: Island Press.

Erb, K., Mayer, A., Kastner, T., Sallet, K., & Haberl, H. (2012). The impact of industrial grain fed livestock production on food security: an extended literature review. Vienna, Austria: Alpen-Adria University Klagenfurt, Institute of Social Ecology. Retrieved from http://www.fao.org/fileadmin/user_upload/animalwelfare/the_impact_of_industrial_grain_fed_livestock_production_on_food_security_2012.pdf

Ericksen, P. J., Ingram, J. S. I., & Liverman, D. M. (2009). Food security and global environmental change: Emerging challenges. *Environmental Science & Policy, 12*(4), 372–377. doi:10.1016/j.envsci.2009.04.007

Erickson, J. (2014, September 5). Dietary recommendations may be tied to increased greenhouse gas emissions. *Michigan News.* Retrieved from http://ns.umich.edu/new/releases/22359-dietary-recommendations-may-be-tied-to-increased-greenhouse-gas-emissions

Esar, E. (1968). *20,000 quips and quotes.* New York, NY: Doubleday.

Escritt, T. (2014). Authorities identify highly contagious bird flu strain. Scientific American. Retrieved November 17, 2014, from http://www.scientificamerican.com/article/authorities-identify-highly-contagious-bird-flu-strain/?WT.mc_id=SA_DD_20141117

Eshel, G. (2010). Grass-fed beef packs a punch to the environment. Retrieved from http://blogs.reuters.com/environment/2010/04/08/grass-fed-beef-packs-a-punch-to-environment/

Eshel, G., & Martin, P. (2006). Diet, energy and global warming. *Earth Interactions, 10*(9), 1–17. doi:10.1175/EI167.1

Eshel, G., Shepon, A., Makov, T., & Milo, R. (2014). Land, irrigation water, greenhouse gas and reactive nitrogen burdens of meat, eggs and dairy production in the United States.[PNAS]. *Proceedings of the National Academy of Sciences of the United States of America, 111*(33), 11996–12001. doi:10.1073/pnas.1402183111 PMID:25049416

Esposito, J. L., & Watson, M. (2000). *Religion and global order.* Cardiff, UK: University of Wales Press.

Estruch, R., Martinez-Gonzalez, M. A., Corella, D., Salas-Salvado, J., Ruiz-Gutierrez, V., Covas, M. I., & Ros, E. et al. (2006). Effects of a Mediterranean-style diet on cardiovascular risk factors: A randomized trial. *Annals of Internal Medicine, 145*(1), 1–11.

Estruch, R., Ros, E., Salas-Salvado, J., Covas, M. I., Corella, D., Aros, F., & Martinez-Gonzalez, M. A. et al. (2013). Primary prevention of cardiovascular disease with a Mediterranean diet. *The New England Journal of Medicine, 368*(14), 1279–1290.

European Commission (EC). (1997). *Council Directive 97/2/EC of 20 January 1997 amending Directive 91/629/EEC laying down minimum standards for the protection of calves.* Brussels, Belgium: EC. Retrieved September 8, 2015, from http://ec.europa.eu/food/fs/aw/aw_legislation/calves/97-2-ec_en.pdf

European Commission (EC). (1999). *Council Directive 1999/74/EC laying down minimum standards for the protection of laying hens.* Brussels, Belgium: EC. Retrieved from http://eur-lex.europa.eu/LexUriServ/LexUriServ.do?uri=OJ:L: 1999:203:0053:0057:EN:PDF

European Commission (EC). (2001). *Commission directive 2001/93/EC of 9 November 2001 amending Directive 91/630/ EEC laying down minimum standards for the protection of pigs.* Brussels, Belgium: EC. Retrieved September 8, 2015, from http://faolex.fao.org/cgi-bin/faolex.exe?rec_id=029442&database=faolex&search_type=link&table=result&lang =eng&format_name=@ERALL

European Environment Agency (EEA). (2014). *Environmental indicator report 2014: environmental impacts of production-consumption systems in Europe.* Luxembourg: Publications Office of the European Union. Retrieved September 7, 2015, from http://www.eea.europa.eu/publications/environmental-indicator-report-2014

European Union. (2008). Treaty of Lisbon. Consolidated versions of the Treaty on European Union and the Treaty on the Functioning of the European Union. Retrieved from http://eur-lex.europa.eu/legal-content/EN/TXT/?uri=uriserv:O J.C_.2008.115.01.0001.01.ENG

Eurostat. (2015). *Meat production and foreign trade - head - monthly data.* Luxembourg: Eurostat. Retrieved September 8, 2015, from http://appsso.eurostat.ec.europa.eu/nui/show.do?dataset=apro_mt_pheadm&lang=en

Factiva. (2014). Vegan. Retrieved from http://new.dowjones.com/products/factiva/

Farmer, B., Larson, B. T., Fulgoni, V. L. III, Rainville, A. J., & Liepa, G. U. (2011). A vegetarian dietary pattern as a nutrient-dense approach to weight management: An analysis of the national health and nutrition examination survey 1999-2004. *Journal of the American Dietetic Association, 111*(6), 819–827.

Farvid, M. S., Ding, M., Pan, A., Sun, Q., Chiuve, S. E., Steffen, L. M., & Hu, F. B. et al. (2014). Dietary linoleic acid and risk of coronary heart disease: A systematic review and meta-analysis of prospective cohort studies. *Circulation, 130*(18), 1568–1578.

Felipe, S. T. (2007). Etica e experimentação animal: fundamentosabolicionistas [Ethics and animal experimentation: abolitionist foundations]. Florianópolis, Brazil: Universidade Federal de Santa Catarina (UFSC).

Felipe, S. T. (2003). *Por uma questão de princípios: alcance e limites de ética de Peter Singer em defesa dos animais* [A matter of principles: the power and limits of Peter Singer ethics for animal rights]. Florianópolis, Brazil: Boiteux.

Fernandez-Cao, J. C., Arija, V., Aranda, N., Bullo, M., Basora, J., Martinez-Gonzalez, M. A., & Salas-Salvado, J. et al. (2013). Heme iron intake and risk of new-onset diabetes in a Mediterranean population at high risk of cardiovascular disease: An observational cohort analysis. *BioMed Central Public Health, 13*, 1042–1048.

Ferrante, J. (2011). *Sociology: a global perspective* (8th ed.). Belmont, CA: Wadsworth.

Ferreira, A. G. Z. (1992). *Concórdia: o rastro de sua história* [Concordia: its history tracks]. Concordia, Brazil: Fundação Municipal de Cultura.

Ferrucci, L. M., Sinha, R., Ward, M. H., Graubard, B. I., Hollenbeck, A. R., Kilfoy, B. A., & Cross, A. J. et al. (2010). Meat and components of meat and the risk of bladder cancer in the NIH-AARP Diet and Health Study. *Cancer, 116*(18), 4345–4353.

Feskens, E. J., & Kromhout, D. (1990). Habitual dietary intake and glucose tolerance in euglycaemic men: The Zutphen Study. *International Journal of Epidemiology, 19*(4), 953–959.

Feskens, E. J., Sluik, D., & van Woudenbergh, G. J. (2013). Meat consumption, diabetes, and its complications. *Current Diabetes Reports, 13*(2), 298–306.

Fiddles, N. (1991). *Meat: a natural symbol*. London, UK: Routledge.

Fitzgerald, A. J., Kalof, L., & Dietz, T. (2009). Slaughterhouse and increased crime rates: An empirical analysis of the spillover from 'The Jungle' into the surrounding community. *Organization & Environment, 22*, 158–184.

Flew, T. (2008). Not yet the internet election: Online media, political commentary and the 2007 Australian Federal Election. *Media International Australia Incorporating Culture and Policy, 126*, 5–13.

Florit, L. F., de Oliveira, L. B., Fleuri, R. M., & Wartha, R. (2016). A 'European Valley' in South America: regionalisation, colonisation and environmental inequalities in Santa Catarina, Brazil. In J. Dessein, E. Battaglini, & L. Horlings (Eds) cultural sustainability and regional development: theories and practices of territorialisation (pp. 235-248). Abington, UK: Routledge.

Florit, L. F., Sbardelati, C., Grava, D., Refosco, J. C., & Pinto, A. C. (2014, July 9). *Implicações éticas e sociais da "vocação regional" pela suinocultura e avicultura na microrregião de Concórdia/SC [Ethical and social implications of "regional vocation" for pork and poultry in the micro region of Concórdia].* Paper presented at the 2nd Seminário de Desenvolvimento Regional, Estado e Sociedade (SEDRES): A diversidade regional brasileira em perspectiva [Regional Development Seminar, State and Society: The Brazilian regional diversity put in perspective], Universidade Estudual de Paraíba, Campina Grande, Brazil.

Florit, L. F. (2004). *A reinvenção social do natural: natureza e agricultura no mundo contemporâneo* [The social reinvention of the natural: nature and agriculture in the contemporary world]. Blumenau, Brazil: Edifurb.

Florit, L. F. (2012). Dilemas éticos e politicos, humanos e não humanos na gestão pública do desenvolvimento territorial sustentável. [Ethical and political, human and non-human dilemas in sustainable territorial development] In A. Philippi, C. A. Sampaio, & V. Fernandes (Eds.), *Gestão de natureza pública e sustentabilidade* [*Management of public nature and sustainability*]. (pp. 247–268). Barueri, Brazil: Editora Manole.

Foer, J. S. (2009, October 28). Eating animals is making us sick. CNN. Retrieved from http://edition.cnn.com/2009/OPINION/10/28/opinion.jonathan.foer/index.html?iref=allsearch

Foer, J. S. (2009). *Eating animals*. New York, NY: Little, Brown and Company.

Foley, J. (2013). It's time to rethink America's corn system. *Scientific American*. Retrieved from http://www.scientificamerican.com/article/time-to-rethink-corn/

Fonseca-Nunes, A., Jakszyn, P., & Agudo, A. (2014). Iron and cancer risk--a systematic review and meta-analysis of the epidemiological evidence. *Cancer Epidemiology, Biomarkers & Prevention, 23*(1), 12–31.

Foo, S. Y., Heller, E. R., Wykrzykowska, J., Sullivan, C. J., Manning-Tobin, J. J., & Moore, K. J., … Rosenzweig, A. (2009). Vascular effects of a low-carbohydrate high-protein diet. *Proceedings of the National Academy of Sciences of the United States (PNAS), 106*(36), 15418-15423.

Food Agriculture Organization (FAO). (2010). *Global forest resource assessment*. Rome, Italy: FAO.

Food and Agriculture Organisation of the United Nations (FAO). (2013). World livestock 2013: changing disease landscapes. Rome, Italy: FAO. Retrieved from http://www.fao.org/docrep/019/i3440e/i3440e.pdf

Food and Agriculture Organisation of the United Nations (FAO). (2015). World agriculture: towards 2015/2030. Livestock production. Rome, Italy: FAO, Corporate Document Repository. Retrieved from http://www.fao.org/docrep/005/y4252E/y4252e07.htm

Food and Agriculture Organization (FAO). (2009). Escritório Regional para a América Latina e o Caribe [Regional Office for Latina America and the Caribbean]. Retrieved from http://www.rlc.fao.org/pt/temas/pecuaria/

Food and Agriculture Organization of the United Nations (FAO). (2006). *Livestock's long shadow: environmental issues and options*. Rome, Italy: FAO, Animal Health and Production Division.

Food and Agriculture Organization of the United Nations (FAO). (2006). Spotlight/2006: Livestock Impacts on the Environment. *FAO Magazine*. Retrieved from http://www.fao.org/ag/magazine/0612sp1.htm

Food and Agriculture Organization of the United Nations (FAO). (2007). *The state of world fisheries and aquaculture 2006*. Rome, Italy: FAO, Fisheries and Aquaculture Division.

Food and Agriculture Organization of the United Nations (FAO). (2014). FAOSTAT. Retrieved from http://faostat.fao.org/site/610/DesktopDefault.aspx?PageID=610#ancor

Food and Agriculture Organization of the United Nations (FAO). (2014a). 2013 food balance sheets. Retrieved June 10, 2015, from http://faostat3.fao.org/download/FB/FBS/E

Food and Agriculture Organization of the United Nations (FAO). (2014b). Production/ livestock, primary. Retrieved June 10, 2015, from http://faostat3.fao.org/download/Q/QL/E

Food and Agriculture Organization of the United Nations Statistics Division (FAOSTAT). (2015). Livestock Primary. Production: Meat Indigenous Poultry. Retrieved May 10, 2015, from http://faostat3.fao.org/browse/Q/QL/E

Food and Agriculture Organization of the United Nations. Statistics Division (FAOSTAT). (2014). Food Security Indicators. Retrieved from http://faostat3.fao.org/download/D/FS/E

Food and Agriculture Organization of the United Nations. Statistics Division (FAOSTAT). (2014). Production: livestock primary and livestock processed. Retrieved from http://faostat.fao.org/site/569/DesktopDefault.aspx?PageID=569#ancor

Food and Agriculture Organization of the United Nations. Statistics Division (FAOSTAT). (2015). Food balance: food supply, livestock and fish primary equivalent. Retrieved from http://faostat3.fao.org/browse/FB/CL/E

Food and Drug Administration (FDA). (2000). The human health impact of fluroquinolone resistant Campylobacter attributed to the consumption of chicken. Rockville, MD: FDA, Center for Veterinary Medicine. Retrieved from http://goo.gl/iZhFU7

Food and Drug Administration (FDA). (2010). CVM Updates - DVM Reports on Antimicrobials Sold or Distributed for Food-Producing Animals. Silver Spring, MD: FDA. Retrieved from http://www.fda.gov/AnimalVeterinary/NewsEvents/CVMUpdates/ucm236143.htm

Food and Drug Administration (FDA). (2013a). Guidance for industry, new animal drugs and new animal drug combination products administered in or on medicated feed or drinking water of food-producing animals: recommendations for drug sponsors for voluntarily aligning product use conditions with GFI#209. Rockville, MD: Center for Veterinary Medicine, FDA. Retrieved from http://goo.gl/iZhF

Food and Drug Administration (FDA). (2013b). Summary report on antimicrobials sold or distributed for use in food-producing animals. Silver Spring, MD: Government of the United States of America, Department of Health and Human Services, FDA. Retrieved February 21, 2015, from

Food and Water Watch (FWW). (2012). Antibiotic Resistance 101: How antibiotic misuse in factory farms can make you sick. Washington, DC: FWW. Retrieved from http://documents.foodandwaterwatch.org/doc/AntibioticResistance.pdf

Food Marketing Initiative (FMI). (n.d.). Gestation crates: problems and solutions. Retrieved from https://www.fmi.org/docs/animal-welfare/hsus-gestation-crate-factsheet.pdf?sfvrsn=3

Forestell, C. A., Spaeth, A. M., & Kane, S. A. (2012). To eat or not to eat red meat: A closer look at the relationship between restrained eating and vegetarianism in college females. *Appetite*, *58*(1), 319–325.

Foster-Fishman, P. G., Nowell, B., & Yang, H. (2007). Putting the system back into systems change: A framework for understanding and changing organizational and community systems. *American Journal of Community Psychology*, *39*(3-4), 197–215. doi:10.1007/s10464-007-9109-0 PMID:17510791

Fox, K. (2009, December 28). Call meat happy, but it is never humane. *The Sydney Morning Herald*. Retrieved from http://www.smh.com.au/it-pro/call-meat-happy-but-it-is-never-humane-20091227-lg61.html

Fox, N., & Ward, K. J. (2008a). Health ethics and environment: A qualitative study of vegetarian motivations. *Appetite*, *50*(2-3), 422–429. doi:10.1016/j.appet.2007.09.007 PMID:17980457

Fox, N., & Ward, K. J. (2008b). You are what you eat? Vegetarianism, health and identity. *Social Science & Medicine*, *66*(12), 2585–2595. doi:10.1016/j.socscimed.2008.02.011 PMID:18378056

Francione, G. L. (2004). Animals: property of persons? In C. R. Sunstein & M. C. Nussbaum (Eds.), *Animal rights: current debates and new directions* (pp. 108–142). New York, NY: Oxford University Press.

Franklin, T. L., Kolasa, K. M., Griffin, K., Mayo, C., & Badenhop, D. T. (1995). Adherance to very-low-fat diet by a group of cardiac rehabilitation patients in the rural southeastern United States. *Archives of Family Medicine*, *4*(6), 551–554.

Fraser, D. (2005). Animal welfare and the intensification of animal production: an alternative interpretation. Rome, Italy: Food and Agriculture Organisation of the United Nations (FAO). Retrieved from ftp://ftp.fao.org/docrep/fao/009/a0158e/a0158e00.pdf

Fraser, G. E. (1999). Associations between diet and cancer, ischemic heart disease, and all-cause mortality in non-Hispanic white California Seventh-day Adventists. *The American Journal of Clinical Nutrition*, *70*(3), 532S–538S.

Fraser, G. E. (2009). Vegetarian diets: What do we know of their effects on common chronic diseases? *The American Journal of Clinical Nutrition*, *89*(5), 1607S–1612S.

Fraser, G. E., & Shavlik, D. J. (2001). Ten years of life: Is it a matter of choice? *Journal of the American Medical Association Internal Medicine*, *161*(13), 1645–1652.

Freeman, C. P. (2013). Stepping up to the veggie plate: framing veganism as living for your values. In E. Plec (Ed.), *Perspectives on human-animal communication: internatural communication* (pp. 93–112). New York, NY: Taylor and Francis.

Freire, P. (1971). *Pedagogy of the oppressed*. New York, NY: Herder and Herder.

Freudberg, D. (2013, October 1). The diet-climate connection. The Huffington Post. Retrieved from http://www.huffingtonpost.com/david-freudberg/the-dietclimate-connectio_b_2446929.html

Friedlander, J., Riedy, C., & Bonfiglioli, C. (2014). A meaty discourse: What makes meat news? *Food Studies*, *3*(3), 27–43.

Friedrich, B. (2012). National Pork Producers Council: anti-science and anti-animal. Retrieved from http://www.commondreams.org/views/2012/09/08/national-pork-producers-council-anti-science-anti-animal

Friel, S. (2010). Climate change, food insecurity and chronic disease: Sustainable and healthy policy opportunities for Australia. *New South Wales Public Health Bulletin, 21*(6), 129–133. doi:10.1071/NB10019 PMID:20637169

Friel, S., Barosh, L. J., & Lawrence, M. (2013). Towards healthy and sustainable food consumption: An Australian case study. *Public Health Nutrition, 17*(5), 1156–1166. doi:10.1017/S1368980013001523 PMID:23759140

Friel, S., Dangour, A. D., Garnett, T., Lock, K., Chalabi, Z., Roberts, I., & Haines, A. et al. (2009). Public health benefits of strategies to reduce greenhouse-gas emissions: Food and agriculture. *Lancet, 374*(9706), 2016–2025. doi:10.1016/S0140-6736(09)61753-0 PMID:19942280

Friends of the Earth (FOE). (2008). What's feeding our food? The environmental and social impacts of the livestock sector. London, UK: FOE. Retrieved September 8, 2015, from http://www.foe.co.uk/sites/default/files/downloads/livestock_impacts.pdf

Fung, T. T., Hu, F. B., Hankinson, S. E., Willett, W. C., & Holmes, M. D. (2011). Low-carbohydrate diets, dietary approaches to stop hypertension-style diets, and the risk of postmenopausal breast cancer. *American Journal of Epidemiology, 174*(6), 652–660.

Fung, T. T., Hu, F. B., Wu, K., Chiuve, S. E., Fuchs, C. S., & Giovannucci, E. (2010). The Mediterranean and Dietary Approaches to Stop Hypertension (DASH) diets and colorectal cancer. *The American Journal of Clinical Nutrition, 92*(6), 1429–1435.

Fung, T. T., McCullough, M. L., Newby, P. K., Manson, J. E., Meigs, J. B., Rifai, N., & Hu, F. B. et al. (2005). Diet-quality scores and plasma concentrations of markers of inflammation and endothelial dysfunction. *The American Journal of Clinical Nutrition, 82*(1), 163–173.

Fung, T. T., McCullough, M., van Dam, R. M., & Hu, F. B. (2007). A prospective study of overall diet quality and risk of type 2 diabetes in women. *Diabetes Care, 30*(7), 1753–1757.

Fung, T. T., van Dam, R. M., Hankinson, S. E., Stampfer, M., Willett, W. C., & Hu, F. B. (2010). Low-carbohydrate diets and all-cause and cause-specific mortality: Two cohort studies. *Annals of Internal Medicine, 153*(5), 289–298.

Gadotti, M. (2008). Education for sustainability: A critical contribution to the Decade of Education for Sustainable Development. *Green Theory and Praxis: The Journal of Ecopedagogy, 4*(1), 15–64. doi:10.3903/gtp.2008.1.3

Galvin, N. (2008, December 16). Eat your greens, good living. *The Sydney Morning Herald.*

García-Fernández, E., Rico-Cabanas, L., Rosgaard, N., Estruch, R., & Bach-Faig, A. (2014). Mediterranean diet and cardiodiabesity: A review. *Nutrients, 6*(9), 3474–3500.

Garcia, P. T., Pensel, N. A., Sancho, A. M., Latimori, N. J., Kloster, A. M., Amigone, M. A., & Casal, J. J. (2008). Beef lipids in relation to animal breed and nutrition in Argentina. *Meat Science, 79*(3), 500–508. doi:10.1016/j.meatsci.2007.10.019 PMID:22062910

Garnaut, R. (2008). *The garnaut climate change review: final report.* New York, NY: Cambridge University Press.

Garnett, T. (2009). Livestock-related greenhouse gas emissions: Impacts and options for policy makers. *Environmental Science & Policy, 12*(4), 491–503. doi:10.1016/j.envsci.2009.01.006

Garnett, T. (2011). Where are the best opportunities for reducing greenhouse gas emissions in the food system (including the food chain)? *Food Policy, 36,* S23–S32. doi:10.1016/j.foodpol.2010.10.010

Garnett, T. (2013). Food sustainability: Problems, perspectives and solutions. *The Proceedings of the Nutrition Society*, *72*(1), 29–39.

Gartin, M., Brewis, A. A., & Schwartz, N. A. (2010). Nonprescription antibiotic therapy: Cultural models on both sides of the counter and both sides of the border. *Medical Anthropology Quarterly*, *24*(1), 85–107. doi:10.1111/j.1548-1387.2010.01086.x PMID:20420303

Gerber, P. J., Steinfield, H., Henderson, B., Mottet, A., Opio, C., & Dijkman, J. … Tempio, G. (2013). Tackling climate change through livestock - a global assessment of emissions and mitigation opportunities. Rome: Italy: Food and Agriculture Organisation of the United Nations (FAO). Retrieved from http://www.fao.org/docrep/018/i3437e/i3437e.pdf

Gerber, P. J., Steinfield, H., Henderson, B., Mottet, A., Opio, C., & Dijkman, J. … Tempio, G. (2013). Tackling climate change through livestock: a global assessment of emissions and mitigation opportunities. Rome, Italy: Food and Agriculture Organisation of the United Nations (FAO). Retrieved from http://www.fao.org/docrep/018/i3437e/i3437e.pdf

Gerke, R. (2012, March 20). Herbert signs so-called 'ag-gag' bill. *The Salt Lake Tribune*. Retrieved from http://www.sltrib.com/53758916

Gibbs, H. K., Ruesch, A., Achard, F., Clayton, M., Holmgren, P., Ramankutty, N., & Foley, J. (2010). Tropical forests were the primary sources of new agricultural land in the 1980s and 1990s.[PNAS]. *Proceedings of the National Academy of Sciences of the United States of America*, *107*(38), 16732–16737. doi:10.1073/pnas.0910275107 PMID:20807750

Giddens, A. (2009). *Sociology* (6th ed.). Cambridge, UK: Polity Press.

Gigleux, I., Jenkins, D. J., Kendall, C. W., Marchie, A., Faulkner, D. A., Wong, J. M., & Lamarche, B. et al. (2007). Comparison of a dietary portfolio diet of cholesterol-lowering foods and a statin on LDL particle size phenotype in hypercholesterolaemic participants. *The British Journal of Nutrition*, *98*(6), 1229–1236.

Gilsing, A. M., Fransen, F., de Kok, T. M., Goldbohm, A. R., Schouten, L. J., de Bruine, A. P., & Weijenberg, M. P. et al. (2013). Dietary heme iron and the risk of colorectal cancer with specific mutations in KRAS and APC. *Carcinogenesis*, *34*(12), 2757–2766.

Glasser, C. L. (2011). *Moderates and radicals under repression: the u.s. animal rights movement, 1990-2010. (Doctor of Philosophy in Sociology)*. Irvine, CA: University of California.

Glick-Bauer, M., & Yeh, M. C. (2014). The Health Advantage of a Vegan Diet: Exploring the Gut Microbiota Connection. *Nutrients*, *6*(11), 4822–4838.

Global Agriculture. (2015). Meat and animal feed. *Agriculture at a crossroads: findings and recommendations for future farming*. Retrieved from http://www.globalagriculture.org/report-topics/meat-and-animal-feed.html

Global Development and Environment Institute (GDAE). (2013). Feeding the factory farm. Tufts University. Retrieved from http://www.ase.tufts.edu/gdae/policy_research/BroilerGains.htm

Godfay, H. C. J., Beddington, J. R., Crute, I. R., Haddad, L., Lawrence, D., Muir, J. F., & Toulmin, C. et al. (2010). Food security: The challenge of feeding 9 billion people. *Science*, *327*(5967), 812–818.

Goetz, G. (2012). Red meat allergy likely caused by tick bites. Food Safety News. Retrieved January 12, 2015, from http://www.foodsafetynews.com/2012/06/red-meat-allergy-likely-caused-by-tick-bites/

Goforth, R., & Goforth, C. (2000). Appropriate Regulation of Antibiotics in Livestock Feed. *Boston College Environmental Affairs Review*, *28*(1), 39–77.

Goggin, G. (2006). The Internet, online and mobile cultures. In S. Cunningham & G. Turner (Eds.), *The media and communications in Australia* (2nd ed., pp. 259–278). Crows Nest, Australia: Allen & Unwin.

Goldenberg, S. (2013, July 29). Climate study predicts a watery future for New York, Boston and Miami. *The Guardian*. Retrieved from http://www.theguardian.com/environment/2013/jul/29/climate-new-york-boston-miami-sea-level

Goodland, R. (2012). FAO Yields to Meat Industry Pressure on Climate Change. Retrieved from bittman.blogs.nytimes.com/2012/07/11/fao-yields-to-meat-industry-pressure-on-climate-change/?_r=0

Goodland, R. (2013b). Meat, lies and videotape (A deeply flawed TED talk). Retrieved from http://www.freefromharm.org/agriculture-environment/meat-lies-videotape-a-deeply-flawed-ted-talk

Goodland, R., & Anhang, J. (2009, November/December). Livestock and climate change: what if the key actors in climate change are... cows, pigs and chickens? Washington, DC: Worldwatch Institute. *World Watch Magazine*, *22*(6), 10-19.

Goodland, R. (2010). Livestock and climate change: Critical comments and responses. *World Watch Magazine*, *23*(2), 7–9.

Goodland, R. (2013a). Lifting livestock's long shadow. *Nature Climate Change*, *3*(1), 2. doi:10.1038/nclimate1755

Goodland, R. (2014). A fresh look at livestock greenhouse gas emissions and mitigation potential in Europe. *Global Change Biology*, *20*(7), 2042–2044. doi:10.1111/gcb.12454 PMID:24166774

Goodland, R., & Anhang, J. (2009, November/December). Livestock and climate change: What if the key actors in climate change are cows, pigs and chickens? *World Watch Magazine*, *22*(6), 10–19.

Goodland, R., & Anhang, J. (2009, November/December). Livestock and climate change: What if the key actors in climate change are... cows, pigs and chickens? *World Watch Magazine*, *22*(6), 10–19.

Goodman, A., & Goodman, D. (2005). *The exception to the rulers: exposing oily politicians, war profiteers and the media that love them*. New York, NY: Hyperion.

Google. (2013). Google trends: vegan, vegetarian. Retrieved from https://www.google.com.au/trends/explore -q=Vegan%2C%20Vegetarian

Goossens, H., Ferech, M., Vander Stichele, R., & Elseviers, M.European Survelliance of Antimicrobial Consumption (ESAC) Project Group. (2005). Outpatient antibiotic use in Europe and association with resistance: A cross-national database study. *Lancet*, *365*(9459), 579–587. doi:10.1016/S0140-6736(05)70799-6 PMID:15708101

Gore, A. (Writer) & D. Guggenheim (Director). (2006). An inconvenient truth. Hollywood, CA: Paramount Pictures.

Goularti Filho, A. (2007). *Formação econômica de Santa Catarina* [Economic formation of Santa Catarina]. 2nd ed.). Florianópolis, Brazil: Editora da UFSC.

Government of New South Wales (NSW). (2008). Atlas of NSW: economy, agriculture-livestock. Land and property information. Retrieved from http://atlas.nsw.gov.au/public/nsw/home/topic/article/agriculture-livestock.html

Government of the United Kingdom. (2011). The future of food and farming: challenges and choices for global sustainability. London, UK: Government Office for Science. Retrieved September 8, 2015, from https://www.gov.uk/government/uploads/system/uploads/attachment_data/file/288329/11-546-future-of-food-and-farming-report.pdf

Government of the United Kingdom. (2013, May 12). A*ntimicrobial resistance poses 'catastrophic threat,' says Chief Medical Officer*. Press Release. Retrieved September 8, 2015, from https://www.gov.uk/government/news/antimicrobial-resistance-poses-catastrophic-threat-says-chief-medical-officer--2

Graca, J., Oliveira, A., & Calheiros, M. M. (2015). Meat beyond the plate: Data-driven hypotheses for understanding consumer willingness to adopt a more plant-based diet. *Appetite*, *90*, 80–90.

Gramsci, A. (1971). *Selections from the prison notebooks*. London, UK: Lawrence and Wishart.

Grandin, T. (Ed.). (2010). *Improving animal welfare: a practical approach*. Cambridge, MA: CAB International.

Grant, J., & Taylor, J. (2013, September 18). Who killed Mrs X? The global threat of antimicrobial resistance. The Rand Blog. Retrieved from http://www.rand.org/blog/2013/09/who-killed-mrs-x-the-global-threat-of-antimicrobial.html

Grant, W. B. (2013). A multicountry ecological study of cancer incidence rates in 2008 with respect to various risk-modyfying factors. *Nutrients*, *6*(1), 163–189. doi:10.3390/nu6010163 PMID:24379012

Grasgreen, A. (2011, August 18). Southern veganism. *Inside Higher Education*. Retrieved from https://www.insidehigh-ered.com/news/2011/08/18/university_of_north_texas_creates_vegan_dining_hall

Grava, D. S. (2013). *A construção social da pecuária como 'vocação regional' em Santa Catarina: notas críticas sobre suas impliações socioeconômicas, ambientais e éticas [The social construction of animal husbandry as 'regional voca-tion' in Santa Catarina]*. (Masters in Regional Development Dissertation), Universidade de Blumenau, Brazil.

Green, C. (2008). How to end 98% of animal abuse in the next 25 years. Retrieved from https://faunalytics.org/how-to-end-98-of-animal-abuse-in-the-next-25-years/

Grier, P. (2010). Michelle Obama says 'Let's move' on obesity in American kids. *The Christian Science Monitor*. Re-trieved from http://www.csmonitor.com/USA/2010/0209/Michelle-Obama-says-Let-s-Move-on-obesity-in-American-kids

Grigoryan, L., Burgerhof, J. G., Degener, J. E., Deschepper, R., Lundborg, C. S., & Monnett, D. L., … SAR Consortium. (2007). Attitudes, beliefs and knowledge concerning antibiotic use and self-medication: A comparative European Study. *Pharmacoepidemiology and Drug Safety*, *16*(11), 1234–1243. PMID:17879325

Grigoryan, L., Burgerhof, J. G., Degener, J. E., Deschepper, R., Lundborg, C., Monnet, D., & Haaijer-Ruskamp, F. et al. (2008). Determinants of self-medication with antibiotics in Europe: The impacts of beliefs, country wealth and the health-care system. *The Journal of Antimicrobial Chemotherapy*, *61*(5), 1172–1179. doi:10.1093/jac/dkn054 PMID:18296694

Grillenberger, M., Newman, C. G., Murphy, S. P., Bwibo, N. O., van't Veer, P., Hautvast, J. G., & West, C. E. (2003). Food supplements have a positive impact on weight gain and the addition of animal source foods increases lean body mass of Kenyan schoolchildren. *The Journal of Nutrition*, *133*(11), 3957S–3964S.

Group, E. T. C. (2011). Who will control the green economy? *Corporate concentration in the Life Industries: Commu-nique*, 107. Retrieved from http://www.etcgroup.org/content/who-will-control-green-economy-0

Gruen, L. (2004). Los animals. [The animals] In P. Singer (Ed.), *Compendio de ética* [Ethics compendium]. (pp. 469–482). Madrid, Spain: Alianza Editorial.

Gruen, L. (2014). *Ecofemenism: women, animals and nature*. London, UK: Bloomsbury Academic.

Gudynas, E. (1999). Concepciones de la naturaleza y desarrollo en América Latina[Conceptions of nature and develop-ment in Latin America]. *Persona y Sociedad*, *13*(1), 101–125.

Gunasekera, D., Tulloh, C., Ford, M., & Heyhoe, S. (2008, June 13). Climate change opportunities and challenges in Australian agriculture. Paper presented at the Faculty of Agriculture, Food and Natural Resources (FAFNR '08) An-nual Symposium, Sydney, Australia. Retrieved September 8, 2015, from http://agricoop.nic.in/Climatechange/ccr/files/Climate%20Change-Australian%20Agriculture.pdf

Gupta, S. (2011, August 29). Dr Sanjay Gupta reports: The last heart attack. *CNN*. Retrieved from http://sanjayguptamd. blogs.cnn.com/2011/08/29/sanjay-gupta-reports-the-last-heart-attack/

Gupta, A., Nelson, J. M., Barrett, T. J., Tauxe, R. V., Rossiter, S. P., Friedman, C. R., & Angulo, F. J. et al. (2004). Antimicrobial resistance among Campylobacter strains, United States, 1997-2001. *Emerging Infectious Diseases, 10*(6), 1102–1109. doi:10.3201/eid1006.030635 PMID:15207064

Gustavsson, J., Cederberg, C., Sonesson, U., Van Otterdiijk, R., & Meybeck, A. (2011). Global food losses and food waste. Rome, Italy: Food Agriculture Organisation (FAO).

Guyton, G. (2014). Spotlight on flexitarianism. GlobalMeat News.com. Retrieved November 17, 2014, from http://www. globalmeatnews.com/Analysis/Spotlight-on-flexitarianism-are-consumers-cutting-their-meat-intake

Habermeyer, M., Roth, A., Guth, S., Diel, P., Engel, K. H., Epe, B., & Eisenbrand, G. et al. (2014). Nitrate and nitrite in the diet: How to assess their benefit and risk for human health. *Molecular Nutrition & Food Research, 59*(1), 106–128.

Haddad, E. H., & Tanzman, J. S. (2003). What do vegetarians in the United States eat? *The American Journal of Clinical Nutrition, 78*(3), 626S–632S.

Halkjaer, J., Olsen, A., Overvad, K., Jakobsen, M. U., Boeing, H., Buijsse, B., & Tjonneland, A. et al. (2011). Intake of total, animal and plant protein and subsequent changes in weight or waist circumference in European men and women: The Diogenes project. *International Journal of Obesity, 35*(8), 1104–1113.

Hall, N. L., & Taplin, R. (2008). Room for climate advocates in a coal-focused economy? NGO influence on the Australian climate policy. *The Australian Journal of Social Issues, 43*(3), 359–379.

Hallström, E., Röös, E., & Börjesson, P. (2014). Sustainable meat consumption: A quantitative analysis of nutritional intake, greenhouse gas emissions and land use from a swedish perspective. *Food Policy, 47*, 81–90. doi:10.1016/j. foodpol.2014.04.002

Hamilton, C., Dennis, R., & Baker, D. (2005). Wasteful consumption in Australia. Discussion Paper No. 77. Canberra, Australia. Retrieved from http://www.tai.org.au/documents/downloads/DP77.pdf

Hamshere, P., Sheng, Y., Moir, B., Syed, F., & Gunning-Trant, C. (2014). *Analysis of China's food demand to 2050*. Paper presented at the 42nd Australian Bureau of Agriculture and Resource Economics and Science (ABARES) Outlook 2015 Conference, Canberra, Australia. Retrieved from http://data.daff.gov.au/data/warehouse/9aat/2014/WhatChinaWants/An alysisChinaFoodDemandTo2050_v.1.0.0.pdf

Hanan, E., & Davis, M. (2008, April 11). Age blighted by bias, selling its soul: so say the staff. *The Australian*. Retrieved from http://www.highbeam.com/doc/1G1-177695483.html

Hance, J. (2009). New report: boreal forests contain more carbon than tropical forests per hectare. *Mongabay*. Retrieved from http://news.mongabay.com/2009/1112-hance_boreal.html

Hansen, J. (2009). *Storms of my grandchildren: the truth about the coming climate catastrophe and our last chance to save humanity*. New York, NY: Bloomsbury.

Hansen, J., Sato, M., Kharecha, P., Beerling, D., Mason-Delmotte, V., Pagani, M., & Zachos, J. (2008). Target atmospheric CO_2: Where should humanity aim? *Open Atmospheric Science Journal, 2*, 217–231. doi:10.2174/1874282300802010217

Harcup, T. (2005). "I'm doing this to change the world": Journalism in alternative and mainstream media. *Journalism Studies, 6*(3), 361–374. doi:10.1080/14616700500132016

Harish (2013). Factory farming and the price of meat. *Counting Animals: A place for people who love animals and numbers*. Retrieved from http://www.countinganimals.com/factory-farming-and-the-price-of-meat/

Harish. (2012). Meat industry advertising. Retrieved from http://www.countinganimals.com/meat-industry-advertising/

Harper, L. A., Denmead, O. T., Freney, J. R., & Byers, F. M. (1999). Direct measurements of methane emissions from grazing and feedlot cattle. *Animal Science (Penicuik, Scotland)*, *77*(6), 1392–1401.

Harrison, J. W., & Svec, T. A. (1998). The beginning of the end of the antibiotic era? Part II Proposed solutions to antibiotic abuse. *Quintessence International*, *29*(4), 223–229. PMID:9643260

Harvard School of Public Health (HSPH). (2015). Fats and cholesterol: out with the bad, in with the good. Retrieved September 8, 2015, from http://www.hsph.harvard.edu/nutritionsource/what-should-you-eat/fats-and-cholesterol/

Harvard, T. H. Chan School of Public Health (HSPH). (2015). The nutrition source: protein. Retrieved from http://www.hsph.harvard.edu/nutritionsource/what-should-you-eat/protein/

Havlik, P., Valin, H., Herrero, M., Obersteiner, M., Schmid, E., & Rufino, M. C., …, Notenbaert, A. (2013). Climate change mitigation through livestock system transitions. *Proceedings of the National Academy of Sciences (PNAS)*, *111*(10), 3709-3714.

Havlík, P., Valin, H., Mosnier, A., Obersteiner, M., Baker, J. S., Herrero, M., & Schmid, E. et al. (2013). Crop productivity and the global livestock sector: Implications for land use changes and greenhouse gas emissions. *American Journal of Agricultural Economics*, *95*(2), 442–448. doi:10.1093/ajae/aas085

Hawe, P., Shiell, A., & Riley, T. (2009). Theorising interventions as events in systems. *American Journal of Community Psychology*, *43*(3-4), 267–276. doi:10.1007/s10464-009-9229-9 PMID:19390961

Hawkings, N. J., Butler, C. C., & Wood, F. (2008). Antibiotics in the community: A typology of user behaviours. *Patient Education and Counseling*, *73*(1), 146–152. doi:10.1016/j.pec.2008.05.025 PMID:18640805

Hayley, A., Zinkiewicz, L., & Hardiman, K. (2015). Values, attitudes and frequency of meat consumption: Predicting meat-reduced diets in Australia. *Appetite*, *84*, 98–106.

Health Action International Asia Pacific (HAIAP), & Third World Network (TWN) Penang in Association with Consumers Association of Penang. (2013). Antibiotic Use and Antibiotic Resistance in Food Animals in Malaysia: A Threat to Human and Animal Health. Penang, Malaysia: HAIAP and TWN.

Health Council of the Netherlands (HCN). (2011). Guidelines for a healthy diet: the ecological perspective. The Hague, the Netherlands: HCN. Retrieved September 7, 2015, from http://www.gezondheidsraad.nl/sites/default/files/201108E.pdf

Hedenus, F., Wirsenius, S., & Johansson, D. J. A. (2014). The importance of reduced meat and dairy consumption for meeting stringent climate change targets. *Climatic Change*, *124*(1-2), 79–91. doi:10.1007/s10584-014-1104-5

Heinrich Böll Foundation (HBF), & Friends of the Earth Europe (FOEE) (2014). Meat Atlas: Facts and figures about the animals we eat. Ahrensfelde, Germany: HBF and FOEE. Retrieved from http://www.boell.de/sites/default/files/meat_atlas2014_kommentierbar.pdf

Heinz, B., & Leeb, R. (1998). Getting down to the meat: The symbolic construction of meat consumption. *Communication Studies*, *49*(1), 86–99. doi:10.1080/10510979809368520

Heisenberg, W. (1999). *Physics and philosophy: the revolution in modern science*. Amherst, NY: Prometheus Books.

Helgason, T., Ewen, S. W., Ross, I. S., & Stowers, J. M. (1982). Diabetes produced in mice by smoked/cured mutton. *Lancet*, *320*(8306), 1017–1022.

Heller, M. C., & Keoleian, G. A. (2014). Greenhouse gas emission estimates of U.S. dietary choices and food loss. *Journal of Industrial Ecology, 19*(3), 391–401. doi:10.1111/jiec.12174

Hellmich, N. (2011). Bill Clinton declares vegan victory. *USA Today*. Retrieved September 9, 2015, from http://usatoday30.usatoday.com/yourlife/food/diet-nutrition/2011-08-24-Bill-Clinton-vegan_n.htm

Helms, M., Simonsen, J., Olsen, K. E., & Mølbak, K. (2005). Adverse halth events associated with antimicrobial drug resistance in Campylobacter species: A registry-based cohort study. *The Journal of Infectious Diseases, 191*(7), 1050–1055. doi:10.1086/428453 PMID:15747238

Henchion, M., McCarthy, M., Resconi, V. C., & Troy, D. (2014). Meat consumption: Trends and quality matters. *Meat Science, 98*, 561–568.

Henderson, G. (2013). Commentary: Savory's solution will change views of livestock. *Drovers Cattlenetwork*. Retrieved from http://www.cattlenetwork.com/search/site/Savory%2527s%2520herding%2520instinct

Hendrie, G. A., Ridoutt, B. G., Wiedmann, T. O., & Noakes, M. (2014). Greenhouse gas emissions and the Australian diet - Comparing dietary recommendations with average intakes. *Nutrients, 6*(1), 289–303. doi:10.3390/nu6010289 PMID:24406846

Henzell, T. (2007). *Australian agriculture: its history and challenges*. Collingwood, Australia: CSIRO Publishing.

Herrero, et al. (2011). Comment to the editor: Livestock and greenhouse gas emissions: the importance of getting the numbers right. In R. Goodland, & J. Anhang (Eds.), Animal Feed Science and Technology, 172(3-4), 166-167: 779-782. doi:10.1016/j.anifeedsci.2011.12.028

Herrero, M., Gerber, P., Vellinga, T., Garnett, T., Leip, A., & Opio, C. et al. (2011). Livestock and greenhouse gas emissions: The importance of getting the numbers right. *Animal Feed Science and Technology, 166–167*, 779–782. doi:10.1016/j.anifeedsci.2011.04.083

Hewitson, G. (2012, February 14). Will consumer horror undo the meat industry? *Food Magazine*, Retrieved from http://www.foodmag.com.au/News/Will-consumer-horror-undo-the-meat-industry-

Hill, M. (2011). Vegan diets becoming more popular, more mainstream. *Bloomberg Businessweek*, Retrieved from http://www.businessweek.com/ap/financialnews/D9KICAT00.htm

Hoekstra, A. Y. (2013). *The water footprint of modern consumer society*. London, UK: Earthscan from Routledge.

Hoffman, U. (2013). Agriculture at the crossroads: assuring food security in developing countries. In U. Hoffman (Ed.), *Trade and environment review 2013: Wake up before it is too late.* (pp. 1-8). Geneva, Switzerland: United Nations Conference on Trade and Development. Retrieved from http://unctad.org/en/PublicationsLibrary/ditcted2012d3_en.pdf

Hoffman, S., Stallings, S. F., Bessinger, R. C., & Brooks, G. T. (2013). Differences between health and ethical vegetarians: Strength of conviction, nutrition knowledge, diatry restriction and duration of adherence. *Appetite, 65*, 139–144. doi:10.1016/j.appet.2013.02.009 PMID:23416470

Holmberg, S. D., Solomon, S. L., & Blake, P. A. (1987). Health and economic impacts of antimicrobial resistance. *Reviews of Infectious Diseases, 9*(6), 1065–1078. doi:10.1093/clinids/9.6.1065 PMID:3321356

Holtrop, G., Johnstone, A. M., Fyfe, C., & Gratz, S. W. (2012). Diet composition is associated with endogenous formation of N-nitroso compounds in obese men. *The Journal of Nutrition, 142*(9), 1652–1658.

Hooda, J., Shah, A., & Zhang, L. (2014). Heme, an essential nutrient from dietary proteins, critically impacts diverse physiological and pathological processes. *Nutrients, 6*(3), 1080–1102.

Hoolohan, C., Berners-Lee, M., McKinstry-West, J., & Hewitt, C. N. (2013). Mitigating the greenhouse gas emissions embodied in food through realistic consumer choices. *Energy Policy*, *63*, 1065–1074. doi:10.1016/j.enpol.2013.09.046

Hossain, A. (1990). *Religion of songs: the beliefs and practices of the Bauls of Bangladesh* [Master of Arts Thesis]. Deakin University, Melbourne, Australia.

Hossain, A., & Marinova, D. (2009). Biosystems management: Muslim/Hindu applications in Bangladesh. *Man in India*, *89*(4), 557–566.

Hossain, A., Marinova, D., & Hossain Rhaman, P. (2014). Islamic insights on sustainability. In D. Humphreys & S. S. Stober (Eds.), *Transitions to sustainability: theoretical debates for a changing planet* (pp. 50–62). Champaign, IL: Common Ground Publishing.

Howat, P., Maycock, B., Cross, D., Collins, J. D., Jackson, L., Burns, S., & James, R. (2003). Towards a more unified definition of health promotion. *Health Promotion Journal of Australia*, *14*(2), 82–85.

Howat, P., Sleet, D., Elder, R., & Maycock, B. (2004). Preventing alcohol-related traffic injury: A health promotion approach. *Traffic Injury Prevention*, *5*(3), 208–219. doi:10.1080/15389580490465238 PMID:15276921

Hua, N. W., Stoohs, R. A., & Facchini, F. S. (2001). Low iron status and enhanced insulin sensitivity in lacto-ovo vegetarians. *The British Journal of Nutrition*, *86*(4), 515–519.

Huang, T., Yang, B., Zheng, J., Li, G., Wahlqvist, M. L., & Li, D. (2012). Cardiovascular disease mortality and cancer incidence in vegetarians: A meta-analysis and systematic review. *Annals of Nutrition & Metabolism*, *60*(4), 233–240.

Hughes, R., Cross, A. J., Pollock, J. R., & Bingham, S. (2001). Dose-dependent effect of dietary meat on endogenous colonic N-nitrosation. *Carcinogenesis*, *22*(1), 199–202.

Humane Society International (HSI). (2013). Animal agriculture one of the largest contributers to global warming, UN body reaffirms. Humane Society International. Retrieved December 1, 2014, from http://www.hsi.org/news/press_releases/2013/09/fao_report_climate_change_092713.html

Humane Society of the United States (HSUS). (2009). *The welfare of animals in the meat, egg and dairy industries. HSUS Reports: Farm industry impacts on animals Paper 2* (Vol. 370, pp. 1253–1263). Washington, DC: HSUS.

Hummel, R., Tschäpe, H., & Witte, W. (1986). Spread of plasmid-mediated nourseothricin Resistance due to antibiotic use in animal husbandry. *Journal of Basic Microbiology*, *26*(8), 461–466. doi:10.1002/jobm.3620260806 PMID:3033194

Hunnicutt, J., He, K., & Xun, P. (2014). Dietary iron intake and body iron stores are associated with risk of coronary heart disease in a meta-analysis of prospective cohort studies. *The Journal of Nutrition*, *144*(3), 359–366.

Idel, A., Fehlenberg, V., & Reichert, T. (2013). *Livestock production and food security in a context of climate change and environmental and health challenges. Background Paper*. Berlin, Germany: Germanwatch.

Ingam, W. M., Goodrich, L. M., Robey, E. A., & Eisen, M. B. (2013). Mice infected with low-virulence strains of Toxoplasma gondii lose their inate aversion to cat urine, even after extensive parasite clearance. *PLoS ONE*, *8*(9), e75246. doi:10.1371/journal.pone.0075246

Instituto Brasileiro de Geografia e Estatistice (IGBE). (n.d.). Sistema IGBE de Recuperação Automática - SIDRA: pesquisa trimestral do abate de animais [IGBE system of Automatic Recovery - ISOAR: Quarterly survey about the slaughter of animals. Retrieved from http://www.sidra.igbegov.br/

Instituto Fator Humano (IFH). (2013). Perfil de agravos à saúde em trabalhadores de Santa Catarina [Disease profile of workers in Santa Catarina]. Santa Catarina, Brazil: Relatório de Pesquisa, Instituto Fator Humano em parceria com UFSC, Univali apoio MPT, FECESC, FETIESC, SINTIARC e DIEESE.

Intergovermental Panel on Climate Change (IPCC). (2013). Summary for policy makers. In T. F. Stocker, D. Qin, G.-K. Plattner, M. Tignor, S. K. Allen, J. Boschung, & P. M Midgley et al. (Eds.), *Climate change 2013: the physical science basis. Contribution of Working Group 1 to the Fifth Assement Report of the Intergovernmental Panel on Climate Change (IPCC)* (pp. 3–30). Cambridge, UK and New York, NY: Cambridge University Press.

Intergovernmental Panel on Climate Change (IPCC). (2014). History. Retrieved from http://ipcc.ch/organization/organization_history.shtml

International Energy Agency (IEA). (2011a). IEA: time running out to limit earth's warming. *Inquirer.net*. Retrieved from http://newsinfo.inquirer.net/91179/iea-time-running-out-to-limit-earths-warming

International Energy Agency (IEA). (2011b). *World energy outlook 2011*. Paris, France: IEA.

International Energy Agency (IEA). (2013). Key world energy statistics. Retrieved June 10, 2015, from http://www.iea.org/publications/freepublications/publication/KeyWorld2014.pdf

Iowa Pork Producers Association (IPPA). (2012). USDA announces major pork purchase. Retrieved from http://www.iowapork.org/usda-announces-major-pork-purchase/

Iowa State University. (1992). Livestock confinement dusts and gases. University Extension. Retrieved December 1, 2014, from http://nasdonline.org/static_content/documents/1627/d001501.pdf

Itodo, I. N., Awulu, J. O., & Harrison, R. M. (1999). *Understanding our environment: An introduction to environmental chemistry and pollution* (3rd ed.). London, UK: The Royal Society of Chemistry.

Jackson, T. (2009). *Prosperity without growth: economics for a finite planet*. London, UK: Earthscan.

Jacobsen, M. (2013). Saying good riddance to the clean-plate club. *The New York Times*. Retrieved from http://parenting.blogs.nytimes.com//2013/08/02/saying-goodbye-to-the-clean-plate-club/

Jakob, M., & Hilaire, J. (2015). Climate science: Unburnable fossil-fuel reserves. *Nature, 517*(7533), 150–152. doi:10.1038/517150a PMID:25567276

Jakobsen, M. U., Dethlefsen, C., Joensen, A. M., Stegger, J., Tjønneland, A., Schmidt, E. B., & Overvad, K. (2010). Intake of carbohydrates compared with intake of saturated fatty acids and risk of myocardial infarction: Importance of the glycemic index. *The American Journal of Clinical Nutrition, 91*(6), 1764–1768.

Jakobsen, M. U., O'Reilly, E. J., Heitmann, B. L., Pereira, M. A., Bälter, K., Fraser, G. E., & Ascherio, A. et al. (2009). Major types of dietary fat and risk of coronary heart disease: A pooled analysis of 11 cohort studies. *The American Journal of Clinical Nutrition, 89*(5), 1425–1432.

Jakobsen, M. U., Overvad, K., Dyerberg, J., & Heitmann, B. L. (2008). Intake of ruminant trans fatty acids and risk of coronary heart disease. *International Journal of Epidemiology, 37*(1), 173–182.

Jalava, M., Kummu, M., Porkka, M., Siebert, S., & Varis, O. (2014). Diet change - a solution to reduce water use? *Environmental Research Letters, 9*(7). doi:10.1088/1748-9326/9/7/074016

James, S. J., & James, C. (2010). The food cold-chain and climate chain. *Food Research International, 43*(7), 1944–1956. doi:10.1016/j.foodres.2010.02.001

Jamin, E. L., Riu, A., Douki, T., Debrauwer, L., Cravedi, J. P., Zalko, D., & Audebert, M. (2013). Combined genotoxic effects of a polycyclic aromatic hydrocarbon (B(a)P) and an heterocyclic amine (PhIP) in relation to colorectal carcinogenesis. *Public Library of Science (PLoS). One, 8*(3), 1–11.

Jenkins, D. J., Kendall, C. W., Faulkner, D. A., Kemp, T., Marchie, A., Nguyen, T. H., & Singer, W. et al. (2008). Long-term effects of a plant-based dietary portfolio of cholesterol-lowering foods on blood pressure. *European Journal of Clinical Nutrition, 62*(6), 781–788.

Jenkins, D. J., Kendall, C. W., Faulkner, D., Vidgen, E., Trautwein, E. A., Parker, T. L., & Connelly, P. W. et al. (2002). A dietary portfolio approach to cholesterol reduction: Combined effects of plant sterols, vegetable proteins, and viscous fibers in hypercholesterolemia. *Metabolism: Clinical and Experimental, 51*(12), 1596–1604.

Jenkins, D. J., Kendall, C. W., Marchie, A., Faulkner, D. A., Josse, A. R., Wong, J. M., & Connelly, P. W. et al. (2005). Direct comparison of dietary portfolio vs statin on C-reactive protein. *European Journal of Clinical Nutrition, 59*(7), 851–860.

Jenkins, D. J., Kendall, C. W., Marchie, A., Faulkner, D. A., Wong, J. M., de Souza, R., & Connelly, P. W. et al. (2003). Effects of a dietary portfolio of cholesterol-lowering foods vs lovastatin on serum lipids and C-reactive protein. *Journal of the American Medical Association, 290*(4), 502–510.

Jenkins, D. J., Kendall, C. W., Marchie, A., Jenkins, A. L., Augustin, L. S., Ludwig, D. S., & Anderson, J. W. et al. (2003). Type 2 diabetes and the vegetarian diet. *The American Journal of Clinical Nutrition, 78*(3), 610S–616S.

Jenkins, D. J., Kendall, C. W., Nguyen, T. H., Marchie, A., Faulkner, D. A., Ireland, C., & Singer, W. et al. (2008). Effect of plant sterols in combination with other cholesterol-lowering foods. *Metabolism: Clinical and Experimental, 57*(1), 130–139.

Jensen, P., Blokhuis, H., Broom, D. M., Bubna-Littitz, H., Canali, E., & Dantzer, R. … Zygoyiannis, D. (1997). The Welfare of intensively kept pigs. Brussels, Belgium: European Commission, Scientific Veterinary Committee, Animal Welfare Section.

Jiang, T., Hanslow, K., & Pearce, D. (2009). On-farm impacts of Australian emissions trading scheme - an economic analysis. Barton, Australia: Australian Government (RIRDC).

Jiang, R., Manson, J. E., Meigs, J. B., Ma, J., Rifai, N., & Hu, F. B. (2004). Body iron stores in relation to risk of type 2 diabetes in apparently healthy women. *Journal of the American Medical Association, 291*(6), 711–717.

Jiao, L., Kramer, J. R., Chen, L., Rugge, M., Parente, P., Verstovsek, G., & El-Serag, H. B. et al. (2013). Dietary consumption of meat, fat, animal products and advanced glycation end-products and the risk of Barrett's oesophagus. *Alimentary Pharmacology & Therapeutics, 38*(7), 817–824.

Jiao, L., Stolzenberg-Solomon, R., Zimmerman, T. P., Duan, Z., Chen, L., Kahle, L., & Sinha, R. et al. (2015). Dietary consumption of advanced glycation end products and pancreatic cancer in the prospective NIH-AARP Diet and Health Study. *The American Journal of Clinical Nutrition, 101*(1), 126–134.

Johnson, C. M., Wei, C., Ensor, J. E., Smolenski, D. J., Amos, C. I., Levin, B., & Berry, D. A. (2013). Meta-analyses of colorectal cancer risk factors. *Cancer Causes & Control, 24*(6), 1207–1222.

Johnston, L. M., Matteson, C. L., & Finegood, D. T. (2014). Systems science and obesity policy: A novel framework for analyzing and rethinking population-level planning. *American Journal of Public Health, 104*(7), 1270–1278. doi:10.2105/AJPH.2014.301884 PMID:24832406

Joint Expert Technical Advisory Committee on Antibiotic Resistance (JETACAR). (1999). *The use of antibiotics in food-producing animals: antibiotic-resistant bacteria in animals and humans*. Canberra, Australia: Government of Australia, Commonwealth Department of Health and Aged Care and Commonwealth Department of Agriculture, Fisheries and Forestry.

Joyce, A., Dixon, S., Comfort, J., & Hallett, J. (2008). The cow in the room: Public knowledge of the links between dietary choices and health and environmental impacts. *Environmental Health Insights*, *1*, 31–34. PMID:21572845

Joyce, A., Dixon, S., Comfort, J., & Hallett, J. (2012). Reducing the environmental impact of dietary choice: Perspectives from a behavioural and social change approach. *Journal of Environmental and Public Health*, *2012*, 1–7. doi:10.1155/2012/978672 PMID:22754580

Joyce, A., Hallett, J., Hannelly, T., & Carey, G. (2014). The impact of nutritional choices on global warming and policy implications: Examining the link between dietary choices and greenhouse gas emissions. *Journal of Energy and Emission Control Technologies*, *2*, 33–43. doi:10.2147/EECT.S58518

Joy, M. (2011). *Why we love dogs, eat pigs and wear cows: an introduction to carnism*. San Francisco, CA: Conari Press.

Jungbluth, N., Tietje, O., & Scholz, R. W. (2000). Food purchases: Impacts from the consumers' point of view investigated with a modular LCA. *The International Journal of Life Cycle Assessment*, *5*(3), 134–142. doi:10.1007/BF02978609

Kahleova, H., Matoulek, M., Malinska, H., Oliyarnik, O., Kazdova, L., Neskudla, T., & Pelikanova, T. et al. (2011). Vegetarian diet improves insulin resistance and oxidative stress markers more than conventional diet in subjects with Type 2 diabetes. *Diabetic Medicine*, *28*(5), 549–559.

Kahn, S., & Hanjra, M. A. (2009). Footprints of water and energy inputs in food production - global perspectives. *Food Policy*, *34*(2), 140–150.

Kaluza, J., Akesson, A., & Wolk, A. (2014). Processed and unprocessed red meat consumption and risk of heart failure: Prospective study of men. *Circulation: Heart Failure*, *7*(4), 552–557.

Kaluza, J., Larsson, S. C., Hakansson, N., & Wolk, A. (2014). Heme iron intake and acute myocardial infarction: A prospective study of men. *International Journal of Cardiology*, *172*(1), 155–160.

Kaluza, J., Wolk, A., & Larsson, S. C. (2012). Red meat consumption and risk of stroke: A meta-analysis of prospective studies. *Stroke*, *43*(10), 2556–2560.

Kaluza, J., Wolk, A., & Larsson, S. C. (2013). Heme iron intake and risk of stroke: A prospective study of men. *Stroke*, *44*(2), 334–339.

Kasper, L. R. (2013). The splendid table: why do we eat meat? Tracing the evolutionary history. Retrieved from http://www.splendidtable.org/story/why-do-we-eat-meat-tracing-the-evolutionary-history

Kasser, T. (2011). Ecological Challenges, Materialistic Values and Social Change. In R. Biswas-Diener (Ed.), *Positive Psychology as Social Change* (pp. 89–108). New York, NY: Springer. doi:10.1007/978-90-481-9938-9_6

Kastorini, C. M., Milionis, H. J., Esposito, K., Giugliano, D., Goudevenos, J. A., & Panagiotakos, D. B. (2011). The effect of Mediterranean diet on metabolic syndrome and its components: A meta-analysis of 50 studies and 534,906 individuals. *Journal of the American College of Cardiology*, *57*(11), 1299–1313.

Katcher, H. I., Ferdowsian, H. R., Hoover, V. J., Cohen, J. L., & Barnard, N. D. (2010). A worksite vegan nutrition program is well-accepted and improves health-related quality of life and work productivity. *Annals of Nutrition & Metabolism*, *56*(4), 245–252.

Katz, D. L., & Meller, S. (2014). Can we say what diet is best for health? *Annual Review of Public Health, 35*(1), 83–103. doi:10.1146/annurev-publhealth-032013-182351 PMID:24641555

Kayange, N., Kamugisha, E., Mwizamholya, D. L. M., Jeremiah, S., & Mshana, S. E. (2010). Predictors of positive blood culture and deaths among neonates with suspected neonatal sepsis in a tertiary hospital, Mwanza-Tanzania. *BioMed Central Pediatrics, 10*(39), 1–9. PMID:20525358

Kay, P., Blackwell, P. A., & Boxall, A. B. A. (2004). Fate of veterinary antibiotics in a macroporous tile drained clay soil. *Environmental Toxicology and Chemistry, 23*(5), 1136–1144. doi:10.1897/03-374 PMID:15180364

Kayser, M., Nitzko, S., & Spiller, A. (2013). Analysis of differences in meat consumption patterns. *International Food and Agribusiness Management Review, 16*, 43–56.

Kellow, N. J., & Savige, G. S. (2013). Dietary advanced glycation end-product restriction for the attenuation of insulin resistance, oxidative stress and endothelial dysfunction: A systematic review. *European Journal of Clinical Nutrition, 67*(3), 239–248.

Kemper, N. (2008). Veterinary antibiotics in the aquatic and terrestrial environment. *Ecological Indicators, 8*(1), 1–13. doi:10.1016/j.ecolind.2007.06.002

Kenter, J. (2007, June 6). A vegetarian diet reduces the diner's carbon footprint. *The New York Times*. Retrieved from http://www.nytimes.com/2007/06/06/business/worldbusiness/06iht-greencol07.4.6029437.html?_r=0

Kesmodel, D., & Fletcher, O. (2013). Hummus is conquering America: tobacco farmers open fields to chickpeas; a bumper crop. The Wall Street Journal. Retrieved March 3, 2015, from http://www.wsj.com/articles/SB1000142412788 7323798104578453174022015956

Kessler, D. A. (2013, March 27). Antibiotics and the meat we eat. The New York Times. Retrieved from http://www.nytimes.com/2013/03/28/opinion/antibiotics-and-the-meat-we-eat.html

Keszei, A. P., Goldbohm, R. A., Schouten, L. J., Jakszyn, P., & van den Brandt, P. A. (2013). Dietary N-nitroso compounds, endogenous nitrosation, and the risk of esophageal and gastric cancer subtypes in the Netherlands Cohort Study. *The American Journal of Clinical Nutrition, 97*(1), 135–146.

Key, T. J., Appleby, P. N., Spencer, E. A., Travis, R. C., Roddam, A. W., & Allen, N. E. (2009). Mortality in British vegetarians: Results from the European Prospective Investigation into Cancer and Nutrition (EPIC-Oxford). *The American Journal of Clinical Nutrition, 89*(5), 1613S–1619S.

Key, T. J., Fraser, G. E., Thorogood, M., Appleby, P. N., Beral, V., Reeves, G., & McPherson, K. et al. (1998). Mortality in vegetarians and non-vegetarians: A collaborative analysis of 8300 deaths among 76,000 men and women in five prospective studies. *Public Health Nutrition, 1*(1), 33–41.

Key, T. J., Fraser, G. E., Thorogood, M., Appleby, P. N., Beral, V., Reeves, G., & McPherson, K. et al. (1999). Mortality in vegetarians and nonvegetarians: Detailed findings from a collaborative analysis of 5 prospective studies. *The American Journal of Clinical Nutrition, 70*(3Suppl), 516S–524S.

Kheel, M. (2008). *Nature ethics: an ecofemenist perspective*. Lanham, MD: Rowman and Littlefield.

Kickert, W. J. M., Klijn, E., & Koppenjan, J. F. M. (1997). Introduction: Management perspective on policy networks. In J. M. Walter, E.-H. Kickert, & J. F. M. Koppenjan (Eds.), *Managing complex networks: strategies for the public sector* (pp. 1–11). London, UK: Sage. doi:10.4135/9781446217658.n1

Kiesel, L. (2010, June 27-30). A comparative rhetorical analysis of US and UK newspaper coverage of the correlation between livestock production and climate change. *Proceedings of the 10th Biennial Conference on Communication and the Environment*, Portland, OR. Retrieved from https://theieca.org/contents-2009-conference-communication-and-environment-proceedings

Kilbourne, W. E., Beckmann, S. C., & Thelen, E. (2002). The role of the dominant social paradigm in environmental attitudes: A multinational examination. *Journal of Business Research, 55*(3), 193–204. doi:10.1016/S0148-2963(00)00141-7

Kilbourne, W. E., & Polonsky, M. J. (2005). Environmental attitutdes and their relation to the dominant social paradigm among university students in New Zealand and Australia. *Australasian Marketing Journal, 13*(2), 37–48. doi:10.1016/S1441-3582(05)70076-8

Kim, E., Coelho, D., & Blachier, F. (2013). Review of the association between meat consumption and risk of colorectal cancer. *Nutrition Research (New York, N.Y.), 33*(12), 983–994.

Kimmel, M. (1996). *Manhood in America: a cultural history*. New York, NY: Simon and Schuster.

Klein, N. (2014). *This changes everything: capatilism vs the climate*. New York, NY: Simon and Schuster.

Kluger, J. (2009). Pork gets a swine flu bailout. *Time Magazine*, Retrieved from http://content.time.com/time/health/article/0,8599,1921649,00.html

Knowler, W. C., Barrett-Connor, E., Fowler, S. E., Hamman, R. F., Lachin, J. M., Walker, E. A., & Nathan, D. M. (2002). Reduction in the incidence of type 2 diabetes with lifestyle intervention or metformin. *The New England Journal of Medicine, 346*(6), 393–403.

Knowles, T. G., Kreslin, S. C., Haslam, S. M., Brown, S. N., Green, L. E., Butterworth, A., & Nicol, C. J. et al. (2008). Leg disorders in broiler chickens: Prevalence, risk factors and prevention. *PLoS ONE, 3*(2), 1–7. doi:10.1371/journal.pone.0001545 PMID:18253493

Knox, R. (2012). How using antibiotics in animal feed creates superbugs. Retrieved from http://www.npr.org/sections/thesalt/2012/02/21/147190101/how-using-antibiotics-in-animal-feed-creates-superbugs

Koeth, R. A., Wang, Z., Levison, B. S., Buffa, J. A., Org, E., Sheehy, B. T., & Hazen, S. L. et al. (2013). Intestinal microbiota metabolism of L-carnitine, a nutrient in red meat, promotes atherosclerosis. *Nature Medicine, 19*(5), 576–585.

Kontogianni, M. D., Panagiotakos, D. B., Pitsavos, C., Chrysohoou, C., & Stefanadis, C. (2008). Relationship between meat intake and the development of acute coronary syndromes: The CARDIO2000 case-control study. *European Journal of Clinical Nutrition, 62*(2), 171–177.

Kosamu, B. M., Mawenda, J., & Mapoma, H. W. T. (2011). Water quality changes due to abattoir effluent: A case on Mchesa Stream in Blantyre, Malawi. *African Journal of Environmental Science and Technology, 5*(8), 589–594.

Krans, B. (2015). Few new drugs: why the antibiotic pipeline is running dry. Retrieved from http://www.healthline.com/health/antibiotics/why-pipeline-running-dry

Kratz, M. (2005). Dietary cholesterol, atherosclerosis and coronary heart disease. *Handbook of Experimental Pharmacology, 170*, 195–213.

Kretzer, M. (2013). A convenient truth: Al Gore goes vegan. Retrieved from http://www.peta.org/blog/al-gore-goes-vegan/

Kris-Etherton, P. M., Harris, W. S., & Appel, L. J.AHA Nutrition Committee. (2003). Omega-3 fatty acids and cardiovascular disease: New recommendations from the American Heart Association. *Arteriosclerosis, Thrombosis, and Vascular Biology, 23*(2), 151–152. doi:10.1161/01.ATV.0000057393.97337.AE PMID:12588750

Krishnasamy, V., Otte, J., & Silbergeld, E. (2015). Antimicrobial use in Chinese swine and broiler poultry production. *Antimicrobial Resistance and Infection Control*, *4*(17), 1–9. doi:10.1186/s13756-015-0050-y PMID:25922664

Kristensen, T., Mogensen, L., Hermansen, J. E., Knudsen, M. T., & Flysjö, A. (2006). Life cycle assessment of the livestock industry - Greenhouse gas emission. Tjele, Denmark: Aarhus University, Faculty of Science and Technology, Department of Agroecology. Retrieved September 7, 2015, from http://pure.au.dk/portal/files/44403731/Life_cycle_analysis_of_the_livestock_industry.doc

Kroenke, C. H., Kwan, M. L., Sweeney, C., Castillo, A., & Caan, B. J. (2013). High- and low-fat dairy intake, recurrence, and mortality after breast cancer diagnosis. *Journal of the National Cancer Institute*, *105*(9), 616–623. doi:10.1093/jnci/djt027 PMID:23492346

Kromhout, D., Geleijnse, J. M., Menotti, A., & Jacobs, D. R. Jr. (2011). The confusion about dietary fatty acids recommendations for CHD prevention. *The British Journal of Nutrition*, *106*(5), 627–632.

Kruska, R. L., Reid, R. S., Thornton, P. K., Henninger, N., & Kristjanson, P. M. (2003). Mapping livestock orientated agricultural production systems for the developing world. *Agriculture Systems*, *77*(1), 39–63. doi:10.1016/S0308-521X(02)00085-9

Kunutsor, S. K., Apekey, T. A., Walley, J., & Kain, K. (2013). Ferritin levels and risk of type 2 diabetes mellitus: An updated systematic review and meta-analysis of prospective evidence. *Diabetes/Metabolism Research and Reviews*, *29*(4), 308–318.

Kwan, S., & Roth, L. M. (2011). The everyday resistance of vegetarianism. In C. Bobel & S. Kwan (Eds.), *Embodied resistance: challenging the norms, breaking the rules* (pp. 186–196). Nashville, TN: Vanderbilt University Press.

Laestadius, L. I., Neff, R. A., Barry, C. L., & Frattaroli, S. (2013). Meat consumption and climate change: The role of non-governmental organizations. *Climatic Change*, *120*(1-2), 25–38. doi:10.1007/s10584-013-0807-3

Laestadius, L. I., Neff, R. A., Barry, C. L., & Frattaroli, S. (2014). "We don't tell people what to do": An examination of the factors influencing NGO decisions to campaign for reduced meat consumption in light of climate change. *Global Environmental Change*, *29*, 32–40.

Lal, R., & Pimentel, D. (2009). Biofuels: Beware crop residues. *Science*, *326*(5958), 1345–1346.

Landers, T. F., Cohen, B., Wittum, T. E., & Larson, E. L. (2012). A review of antibiotics use in food animals: Perspective policy and potential. *Public Health Reports*, *127*(1), 4–22. PMID:22298919

Larson, J. (2012a). Meat consumption in China now double that in the United States. Permaculture Research Institute. Retrieved March 27, 2015, from http://permaculturenews.org/2012/05/02/meat-consumption-in-china-now-double-that-in-the-united-states/

Larson, J. (2012b). Peak meat: US meat consumption falling. Earth Policy Institute. Retrieved November 21, 2014, from http://www.earthpolicy.org/data_highlights/2012/highlights25

Larson, L. (2013, September 5). Bill Clinton cheats - on his vegan diet. *Daily News*. Retrieved from http://www.nydailynews.com/news/politics/bill-clinton-cheats-vegan-diet-article-1.1446582

Larsson, S. C., Bergkvist, L., & Wolk, A. (2004). Milk and lactose intakes and ovarian cancer risk in the Swedish Mammography Cohort. *The American Journal of Clinical Nutrition*, *80*(5), 1353–1357. PMID:15531686

Larsson, S. C., & Orsini, N. (2014). Red meat and processed meat consumption and all-cause mortality: A meta-analysis. *American Journal of Epidemiology*, *179*(3), 282–289.

Larsson, S. C., & Wolk, A. (2012). Red and processed meat consumption and risk of pancreatic cancer: Meta-analysis of prospective studies. *British Journal of Cancer, 106*(3), 603–607.

Lasa, A., Miranda, J., Bullo, M., Casas, R., Salas-Salvado, J., Larretxi, I., & Portillo, M. P. et al. (2014). Comparative effect of two Mediterranean diets versus a low-fat diet on glycaemic control in individuals with type 2 diabetes. *European Journal of Clinical Nutrition, 68*(7), 767–772.

Lathuillière, M. J., Johnson, M. S., Galford, G. L., & Couto, E. G. (2014). Environmental footprints show China and Europe's evolving resource appropriation for soybean production in Mato Grosso, Brazil. *Environmental Research Letters, 9*(7). doi:10.1088/1748-9326/9/7/074001

Latvala, T., Niva, M., Mäkelä, J., Pouta, E., Heikkilä, J., Kotro, J., & Forsman-Hugg, S. (2012). Diversifying meat consumption patterns: Consumers' self-reported past behaviour and intentions for change. *Meat Science, 92*(1), 71–77.

Lauder, A., Enting, I., Carter, J., Clisby, N., Cowie, A., Henry, B., & Raupach, M. (2013). Offsetting methane emissions - an alternative to emission equivalance metrics. *International Journal of Greenhouse Gas Control, 12*(0), 419–429. doi:10.1016/j.ijggc.2012.11.028

Lawson, A. (2008, October 3). Environment turning middle-class vegan. *The Age*. Retrieved from http://www.stockand-land.com.au/news/agriculture/agribusiness/general-news/environment-turning-middleclass-vegan/1323831.aspx

Laxminarayan, R. (2015, April 3). Growth in livestock antibiotics raises risks for humans. Living on Earth. Retrieved from http://loe.org/shows/segments.html?programID=15-P13-00014&segmentID=2

Laxminarayan, R., Malani, A., Howard, D., & Smith, D. L. (2011). Extending the cure: policy response to the growing threat of antibiotic resistance. Washington, DC/New Delhi, India: Centre for Disease Dynamics, Economics and Policy (CDDEP).

Laxminarayan, R., van Boeckel, T., & Teillant, A. (2015). The economic costs of withdrawing antimicrobial growth promoters from the livestock sector. OECD Food, Agriculture and Fisheries Paper No. 78. Paris, France: Organisation for Economic Cooperation and Development (OECD). doi:10.1787/18156797

Laxminarayan, R., Duse, A., Wattal, C., Zaidi, A. K. M., Wertheim, H. F. L., Sumpradit, N., & Cars, O. et al. (2013). Antibiotic resistance - the need for global solutions. *The Lancet Infectious Diseases Commission, 13*(12), 1057–1098. doi:10.1016/S1473-3099(13)70318-9 PMID:24252483

Laxminarayan, R., Malani, A., Howard, D., & Smith, D. L. (2007). *Extending the cure: policy responses to the growing threat of antibiotic resistance*. Washington, DC: Resources for the Future.

Lea, E. J., Crawford, D., & Worsley, A. (2006). Public views of the benefits and barriers to the consumption of a plant based diet. *European Journal of Clinical Nutrition, 60*(7), 828–837. doi:10.1038/sj.ejcn.1602387 PMID:16452915

Lea, E. J., Worsley, A., & Crawford, D. (2005). Australian adult consumers' beliefs about plant foods: A qualitative study. *Health Education & Behavior, 32*(6), 795–808. doi:10.1177/1090198105277323 PMID:16267149

Lea, E., & Worsley, A. (2003). Benefits and barriers to the consumption of a vegetarian diet in Australia. *Public Health Nutrition, 6*(5), 505–511. doi:10.1079/PHN2002452 PMID:12943567

Lea, E., & Worsley, A. (2008). Australian consumers' food-related environmental beliefs and behaviours. *Appetitie, 50*(2-3), 207–214.

Lee, D. H., Folsom, A. R., & Jacobs, D. R. (2004). Dietary iron intake and Type 2 diabetes incidence in postmenopausal women: The Iowa Women s Health Study. *Diabetologia, 47*(2), 185–194.

Lee, J. E., Spiegelman, D., Hunter, D. J., Albanes, D., Bernstein, L., van den Brandt, P. A., & Smith-Warner, S. A. et al. (2008). Fat, protein, and meat consumption and renal cell cancer risk: A pooled analysis of 13 prospective studies. *Journal of the National Cancer Institute, 100*(23), 1695–1706.

Leitão, M., & Gribel, A. (2012). Coluna no Globo: Nas asas oficiais [Column on the Globe: The official wings]. Retrieved from http://oglobo.globo.com/economia/miriam/posts/2012/05/10/nas-asas-oficiais-444362.asp

Le, L. T., & Sabate, J. (2014). Beyond meatless, the health effects of vegan diets: Findings from the Adventist cohorts. *Nutrients, 6*(6), 2131–2147.

Lemons, J., Westra, L., & Goodland, R. (1998). *Ecological sustainability and integrity: concepts and approaches.* London, UK: Kluwer Academic Publishers. doi:10.1007/978-94-017-1337-5

Leneman, L. (1999). No animal food: The road to veganism in Britain, 1909-1944. *Society & Animals, 7*(3), 219–228. doi:10.1163/156853099X00095

Lenzen, S. (2008). Oxidative stress: The vulnerable beta-cell. *Biochemical Society Transactions, 36*(3), 343–347.

Leviston, Z., Price, J., Malkin, S., & McCrea, R. (2014). Fourth annual survey of Australian attitutes to climate change. Interim report. Perth, Australia: Commonwealth Scientific and Industry Research Organisation (CSIRO).

Levy, S. B. (2002). *The antibiotic paradox: how the misuse of antibiotics destroys their curative powers* (2nd ed.). Cambridge, MA: Perseus Publishing.

Lewis, J., & Boyce, T. (2009). Climate change and the media: the scale of the challenge. In T. Boyce & J. Lewis (Eds.), *Climate change and the media* (pp. 3–16). New York, NY: Peter Lang.

Lichtenstein, A. H., Appel, L. J., Brands, M., Carnethon, M., Daniels, S., Franch, H. A., & Wylie-Rosett, J. et al. (2006). Summary of American Heart Association Diet and Lifestyle Recommendations revision 2006. *Arteriosclerosis, Thrombosis, and Vascular Biology, 26*(10), 2186–2191.

Lindström, J., Peltonen, M., Eriksson, J. G., Louheranta, A., Fogelholm, M., Uusitupa, M., & Tuomilehto, J. (2006). High-fibre, low-fat diet predicts long-term weight loss and decreased type 2 diabetes risk: The Finnish Diabetes Prevention Study. *Diabetologia, 49*(5), 912–920.

Liu, H., Parton, K. A., Zhou, Z., & Cox, R. (2012). Away-from-home meat consumption in China. *Asian Journal of Agriculture and Development, 8*(2), 1–15.

Liu, J., & Savenije, H. H. G. (2008). Food consumption patterns and their effect on water requirement in China. *Hydrology and Earth System Sciences, 12*(3), 887–898. doi:10.5194/hess-12-887-2008

Locke, S. (2014a, November 17). China already swallows $2.3 billion worth of Australian meat before a Free Trade deal. *ABC News.* Retrieved from http://www.abc.net.au/news/2014-11-17/nrn-china-meat-consumption-prefta/5896162

Locke, S. (2014b, November 18). Beef and lamb exporters winners in the free trade agreement with China. *ABC News.* Retrieved from http://www.abc.net.au/news/2014-11-17/nrn-hold-beef-fta-china/5896494

Lockwood, R., & Ascione, F. E. (1997). *Cruelty to animals and interpersonal violence: readings in research and application.* West Lafayette, IN: Purdee University Press.

Lombard, M., Synder-Duch, J., & Bracken, C. C. (2002). Content analysis in mass communication: Assessment and reporting of intercoder reliability. *Human Communication Research, 28*(4), 587–604. doi:10.1111/j.1468-2958.2002.tb00826.x

Longmire, A., Taylor, C., & Wedderburn-Bisshop, G. (2014). Land use: agriculture and forestry. Discussion Paper, Zero Carbon Australia Land Use Report. Melbourne, Australia: The University of Melbourne, Melbourne Sustainable Society Institute.

Lorek, S., & Fuchs, D. (2013). Strong sustainable consumption governance - precondition for a degrowth path? *Journal of Cleaner Production, 38,* 36–43.

Lovejoy, J., & DiGirolamo, M. (1992). Habitual dietary intake and insulin sensitivity in lean and obese adults. *The American Journal of Clinical Nutrition, 55*(6), 1174–1179.

Luan de, C., Li, H., Li, S. J., Zhao, Z., Li, X., & Liu, Z. M. (2008). Body iron stores and dietary iron intake in relation to diabetes in adults in North China. *Diabetes Care, 31*(2), 285–286.

Luangtongkum, T., Jeon, B., Han, J., Plummer, P., Logue, C. M., & Zhang, Q. (2009). Antibiotic resistance in Campylobacter: Emergence, transmission and persistence. *Future Microbiology, 4*(2), 189–200. doi:10.2217/17460913.4.2.189 PMID:19257846

Ludwig, D. S., & Willett, W. C. (2013). Three daily servings of reduced-fat milk an evidence-based recommendation? *The Journal of the American Medical Association (JAMA). Pediatrics, 167*(9), 788–789.

Lundqvist, J., de Fraiture, C., & Molden, D. (2008). Saving water: from field to fork, curbing losses and wastage in the food chain. Huddinge, Sweden: Stockholm International Water Institute (SIWI). Retrieved September 8, 2015, from http://www.siwi.org/documents/Resources/Policy_Briefs/PB_From_Filed_to_Fork_2008.pdf

Lupton, D. (1996). *Food, the body and the self.* London, UK: Sage Publications.

Lu, S. C., Wu, W. H., Lee, C. A., Chou, H. F., Lee, H. R., & Huang, P. C. (2000). LDL of Taiwanese vegetarians are less oxidizable than those of omnivores. *The Journal of Nutrition, 130*(6), 1591–1596.

Macdiarmid, J. I., Kyle, J., Horgan, G. W., Loe, J., Fyfe, C., Johnstone, A., & McNeill, G. (2012). Sustainable diets for the future: Can we contribute to reducing greenhouse gas emissions by eating a healthy diet? *The American Journal of Clinical Nutrition, 96*(3), 632–639. doi:10.3945/ajcn.112.038729 PMID:22854399

Macey, R. (2006, June 22). Gas-friendly cows an emission possible to cut global warming. *The Sydney Morning Herald.* Retrieved from http://www.smh.com.au/news/national/emission-possible-gasfriendly-cows/2006/06/21/1150845247870.html

Mahony, P. (2014). Cowspiracy and the Australian red meaty industry. Retrieved from http://terrastendo.net/2014/11/09/cowspiracy-and-the-australian-red-meat-industry/

Marcuse, H. (1966). *One-dimensional man: studies in the ideology of advanced industrial societies.* Boston, MA: Beacon Press.

Margulis, S. (2004). Causes of deforestation of the Brazilian Amazon. World Bank Working Paper No. 22. Washington, DC: World Bank (WB). Retrieved from https://openknowledge.worldbank.org/bitstream/handle/10986/15060/277150PAPER0wbwp0no1022.pdf?sequence=1

Marinova, D., & Raphaely, T. (2015). Preface. In T. Raphaely & D. Marinova (Eds.), *Impacts of meat consumption on health and environmental sustainability.* Hershey, PA: IGI Global.

Marshall, A. (2014, August 18). White meat still Australia's choice. *The Land.* Retrieved from http://www.theland.com.au/news/agriculture/agribusiness/general-news/white-meat-still-australias-choice/2708428.aspx

Marshall, J. A., Bessesen, D. H., & Hamman, R. F. (1997). High saturated fat and low starch and fibre are associated with hyperinsulinaemia in a non-diabetic population: The San Luis Valley Diabetes Study. *Diabetologia, 40*(4), 430–438.

Marshall, J. A., Hamman, R. F., & Baxter, J. (1991). High-fat, low-carbohydrate diet and the etiology of non-insulin-dependent diabetes mellitus: The San Luis Valley Diabetes Study. *American Journal of Epidemiology, 134*(6), 590–603.

Marshall, J. A., Hoag, S., Shetterly, S., & Hamman, R. F. (1994). Dietary fat predicts conversion from impaired glucose tolerance to NIDDM. The San Luis Valley Diabetes Study. *Diabetes Care, 17*(1), 50–56.

Marsh, K., & Brand-Miller, J. (2011). Vegetarian Diets and Diabetes. *American Journal of Lifestyle Medicine, 5*(2), 135–143.

Marsh, K., & Bugusu, B. (2007). Food packaging – roles, materials and environmental issues. *Journal of Food Science, 72*(3), R39–R55. doi:10.1111/j.1750-3841.2007.00301.x PMID:17995809

Marsh, K., Saunders, A., & Zeuschne, C. (2015). Red meat and health: evidence regarding red meat, health and chronic disease risk. In T. Raphaely & D. Marinova (Eds.), *Impacts of meat consumption on health and environmental sustainability*. Hershey, PA: IGI Global.

Marsh, K., Saunders, A., & Zeuschner, C. (2015). Red meat and health: evidence regarding red meat, health and chronic disease risk. In D. Marinova & T. Raphaely (Eds.), *Impacts of Meat Consumption on Health and Enviuronmental Sustainability*. Hershey, PA: IGI Global.

Martella, V., Bányai, V., Matthijnssens, J., Bounavogolia, C., & Ciarlet, M. (2010). Zoonotic aspects of rotaviruses. *Veterinary Microbiology, 140*(3-4), 246–255.

Martin, D. S. (2011, August 18). From omnivore to vegan: the dietary education of Bill Clinton. *CNN*. Retrieved from http://edition.cnn.com/2011/HEALTH/08/18/bill.clinton.diet.vegan/

Martinez-Gonzalez, M. A., & Bes-Rastrollo, M. (2014). Dietary patterns, Mediterranean diet, and cardiovascular disease. *Current Opinion in Lipidology, 25*(1), 20–26.

Martinez, J. L. (2008). Antibiotics and antibiotic resistance genes in natural environments. *Science, 321*(5887), 365–367. doi:10.1126/science.1159483 PMID:18635792

Masson, J. F., & McCarthy, S. (1996). *When elephants weep: the emotional lives of animals*. New York, NY: Dell Publishing.

Mayer-Davis, E. J., Monaco, J. H., Hoen, H. M., Carmichael, S., Vitolins, M. Z., Rewers, M. J., & Karter, A. J. et al. (1997). Dietary fat and insulin sensitivity in a triethnic population: The role of obesity. The Insulin Resistance Atherosclerosis Study (IRAS). *The American Journal of Clinical Nutrition, 65*(1), 79–87.

Mayer, E. J., Newman, B., Quesenberry, C. P., & Selby, J. V. (1993). Usual dietary fat intake and insulin concentrations in healthy women twins. *Diabetes Care, 16*(11), 1459–1469.

McCullough, M. L., Feskanich, D., Stampfer, M. J., Giovannucci, E. L., Rimm, E. B., Hu, F. B., & Willett, W. C. et al. (2002). Diet quality and major chronic disease risk in men and women: Moving toward improved dietary guidance. *The American Journal of Clinical Nutrition, 76*(6), 1261–1271.

McDougall, J., Thomas, L. E., McDougall, C., Moloney, G., Saul, B., Finnell, J. S., & Petersen, K. M. et al. (2014). Effects of 7 days on ad libitum low-fat vegan diet: The McDougall Program cohort. *Nutrition Journal, 13*(1), 1–7. doi:10.1186/1475-2891-13-99 PMID:25311617

McEachran, A. D., Blackwell, B. R., Delton Hanson, J., Wooten, K. J., Mayer, G. D., Cox, S. B., & Smith, P. N. (2015). Antibiotics, bacteria and antibiotic resistance genes: Arial transport from cattle feed yards via particulate matter. *Environmental Health Perspectives*, *123*(4), 337–343. PMID:25633846

McEvoy, C. T., Temple, N., & Woodside, J. V. (2012). Vegetarian diets, low-meat diets and health: A review. *Public Health Nutrition*, *15*(12), 2287–2294.

McKenzie-Mohr, D., & Schultz, P. W. (2014). Choosing effective behaviour change tools. *Social Marketing Quarterly*, *20*(1), 35–46. doi:10.1177/1524500413519257

McLaren, D. S. (2000). The great protein fiasco revisited. *Nutrition (Burbank, Los Angeles County, Calif.)*, *16*(6), 464–465.

McLaughlin, A., & Mineau, P. (1995). The impact of agricultural practices on biodiversity. *Agriculture, Ecosystems & Environment*, *55*(3), 201–212.

McLaughlin, P., & Khawaja, M. (2000). The organizational dynamics of the U.S. environmental movement: Legitimation, resource mobilization and political opportunity. *Rural Sociology*, *65*(3), 422–439. doi:10.1111/j.1549-0831.2000.tb00037.x

McLeod, H. (2011). South Carolina scientist works to grow meat in the lab. *Reuters*. Retrieved from http://www.reuters.com/article/2011/01/30/us-food-meat-laboratory-feature-idUSTRE70T1WZ20110130

McMichael, A. J., Powles, J. W., Butler, C. D., & Uauy, R. (2007). Food, livestock production, energy, climate change and health. *Lancet*, *370*(9594), 1253–1263. doi:10.1016/S0140-6736(07)61256-2 PMID:17868818

McMichael, A. J., Powles, J. W., Butler, C. D., & Uauy, R. (2007). Food, livestock production, energy, climate change, and health. *Lancet*, *370*(9594), 1253–1263.

McTaggart, R. (1991). Principles for participatory action research. *Adult Education Quarterly*, *41*(3), 168–187. doi:10.1177/0001848191041003003

McWilliams, J. E. (2009). *Just Food: where locavores get it wrong and how we can truly eat responibly*. New York, NY: Little, Brown and Company.

Meadows, D. (1999). *Leverage points: places to intervene in a system*. Hartland, VT: The Sustainability Institute.

Meat & Poultry Staff (MPS). (2013). USDA plan for dark meat chicken. *Meat + Poultry*. Retrieved from http://www.meatpoultry.com/articles/news_home/Regulatory/2013/09/USDA_plan_for_dark_meat_chicke.aspx?ID=%7bCD131736-E7BE-4303-819B-7827E73C1332%7d&cck=1

Meat and Livestock Australia (MLA). (2014). China-Australia Free Trade Agreement announced. Retrieved from http://www.mla.com.au/News-and-resources/Industry-news/China-Australia-Free-Trade-Agreement-announced

Meat and Livestock Australia (MLA). (2014a). Community engagement. Retrieved from http://www.mla.com.au/Marketing-beef-and-lamb/Community-engagement

Meat and Livestock Australia (MLA). (2014b). Fast facts 2014: Australia's beef industry. Retrieved from http://www.mla.com.au/Cattle-sheep-and-goat-industries/Industry-overview/Cattle

Meat and Livestock Australia (MLA). (2014c). National livestock methane program. Retrieved from http://www.mla.com.au/Research-and-development/Environment-research/National-livestock-methane-program

Meat and Livestock Australia (MLA). (2015). Beef. Retrieved from http://www.mla.com.au/Prices-and-markets/Overseas-markets/South-Asia/Beef

Meatless Monday. (2014). New study confirms continued rise of the flexitarian diet. Meatless Monday. Retrieved November 17, 2014, from http://www.meatlessmonday.com/articles/new-study-confirms-continued-rise-flexitarian-diet/

Media, F. (2006). Fairfax annual report 2006. Retrieved from https://www.fairfaxmedia.com.au/ArticleDocuments/191/Fairfax_AR2006_2006.pdf.aspx?Embed=Y

Mee, L. (2006). Reviving dead zones. Scientific American. Retrieved December 1, 2014, from http://www.scientificamerican.com/article/reviving-dead-zones/

Meisterling, K., Samaras, C., & Schweizer, V. (2009). Decisions to reduce greenhouse gases from agriculture and product transport: LCA case study of organic and conventional wheat. *Journal of Cleaner Production*, *17*(2), 222–230. doi:10.1016/j.jclepro.2008.04.009

Mekonnen, M. M., & Hoekstra, A. Y. (2012). A global assessment of the water footprint of farm animal products. *Ecosystems (New York, N.Y.)*, *15*(3), 401–415. doi:10.1007/s10021-011-9517-8

Mellon, M., Benbrook, C., & Benbrook, K. L. (2001). *Hogging it: estimates of antimicrobial abuse in livestock*. Cambridge, MA: Union of Concerned Scientists.

Mensink, R. P., Zock, P. L., Katan, M. B., & Hornstra, G. (1992). Effect of dietary cis and trans fatty acids on serum lipoprotein[a] levels in humans. *Journal of Lipid Research*, *33*(10), 1493–1501.

Menzies, K., & Sheeshka, J. (2012). The process of exiting vegetarianism: An exploratory study. *Canadian Journal of Dietetic Practice and Research*, *73*(4), 163–168. doi:10.3148/73.4.2012.163 PMID:23217442

Metz, B., Davidson, O. R., Bosch, P., Dave, R., & Meyer, L. (2007). Climate change 2007: mitigation of climate change. Contribution of Working Group III to the Fourth Assessment Report of the Intergovernmental Panel on Climate Change (IPCC). New York, NY: IPCC, Working Group III. Retrieved September 8, 2015, from https://www.ipcc.ch/publications_and_data/publications_ipcc_fourth_assessment_report_wg3_report_mitigation_of_climate_change.htm

Micha, R., Michas, G., Lajous, M., & Mozaffarian, D. (2013). Processing of meats and cardiovascular risk: Time to focus on preservatives. *BioMed Central Medicine*, *11*, 136. doi:10.1186/1741-7015-11-136

Micha, R., Michas, G., & Mozaffarian, D. (2012). Unprocessed red and processed meats and risk of coronary artery disease and type 2 diabetes--an updated review of the evidence. *Current Atherosclerosis Reports*, *14*(6), 515–524. doi:10.1007/s11883-012-0282-8

Micha, R., Wallace, S. K., & Mozaffarian, D. (2010). Red and processed meat consumption and risk of incident coronary heart disease, stroke, and diabetes mellitus: A systematic review and meta-analysis. *Circulation*, *121*(21), 2271–2283.

Michigan State University. (2011). The environmental impact of imprudent antimicrobial use in animals: Open source teaching modules. Retrieved from http://amrls.cvm.msu.edu/veterinary-public-health-module/iii.-the-environmental-impact-of-imprudent-antimicrobial-use-in-animals

Miles, T. (2014). Official WHO Ebola toll near 5,000 with the true number nearer 15,000. Reuters. Retrieved October 23, 2014, from http://www.reuters.com/article/2014/10/22/us-health-ebola-who-idUSKCN0IB23220141022

Miller, R. V. (1998). Bacterial gene swapping in nature. *Scientific American*, *278*(1), 66–71. doi:10.1038/scientificamerican0198-66 PMID:9418300

Milman, O. (2015, January 16). Rate of environmental degredation puts life at risk, says scientist. *The Guardian*. Retrieved from http://www.theguardian.com/environment/2015/jan/15/rate-of-environmental-degradation-puts-life-on-earth-at-risk-say-scientists

Ministerio da Agricultura Pecuària e Abastecimento (MAPA). (2012). Exportação [Exports]. Retrieved from http://www. agricultura.gov.br/animal/exportacao

Mittal, G. S. (2004). Characterization of effluent wastewater from abattoir for land application. *Food Reviews International*, *20*(3), 229–256.

Mittal, G. S. (2006). Treatment of wastewater from abattoirs before land application: A review. *Bioresource Technology*, *97*(9), 1119–1135.

Moby. (2014). Save the humans. *The Huffington Post*. Retrieved from http://www.huffingtonpost.com/moby/moby-meat_b_5889850.html

Mombiot, G. (2015, May 19). Feces, bacteria, toxins: welcome to the chicken farm. The Guardian. Retrieved from http://www.theguardian.com/commentisfree/2015/may/19/chicken-welfare-human-health-meat

Monbiot, G. (2014, August 4). Eat more meat and save the world: the latest implausible farming miracle. *The Guardian*. Retrieved from http://www.theguardian.com/environment/georgemonbiot/2014/aug/04/eat-more-meat-and-save-the-world-the-latest-implausible-farming-miracle

Mood, A. (2010). Worst things happen at sea: the welfare of wild-caught fish. Fishcount.org.uk. Retrieved May 23, 2013, from http://www.fishcount.org.uk/published/standard/fishcountfullrptSR.pdf

Moodie, R. (2004). Introduction: getting your hands on. In R. Moodie & A. Hulme (Eds.), *Hands-on Health promotion (i-xix)*. Melbourne, Australia: IP Communications.

Morgan, D. J., Okeke, I. N., Laxminarayan, R., Perencevich, E. N., & Weisenberg, S. (2011). Non-perscription antimicrobial use worldwide: A systematic review. *The Lancet Infectious Diseases Commission*, *11*(9), 692–701. doi:10.1016/S1473-3099(11)70054-8 PMID:21659004

Morris, C., Kirwan, J., & Lally, R. (2014). Less meat initiatives: An initial exploration of a diet-focused social inovation in transitions to a more sustainable regime of meat provision. *International Journal of Sociology of Agriculture and Food*, *21*(2), 189–208.

Moss, L. (2014). The marketing of meat: why beef and pork producers are so focused on millennials. Retrieved from http://www.mnn.com/money/sustainable-business-practices/stories/the-marketing-of-meat-why-beef-and-pork-producers-are

Moss, M. (2015, January 19). U.S. research lab lets livestock suffer in quest for profit. *The New York Times*. Retrieved from https://www.organicconsumers.org/news/us-research-lab-lets-livestock-suffer-quest-profit

Motoki, C., Broggi, F., Falcão, S. N., Favoretto, T., & Casteli, T. (2013). Moendo gente: a situação do trabalho nos frigoríficos [Grinding people: The labour situation in refrigerators]. Sao Paulo, Brazil: ONG Repórter Brazil. Retrieved from http://escravonempensar.org.br/upfilesfolder/materiais/arquivos/moendo_gente_final.pdf

Mozaffarian, D., Aro, A., & Willett, W. C. (2009). Health effects of trans-fatty acids: Experimental and observational evidence. *European Journal of Clinical Nutrition*, *63*(2), S5–S21.

Mozaffarian, D., Hao, T., Rimm, E. B., Willett, W. C., & Hu, F. B. (2011). Changes in diet and lifestyle and long-term weight gain in women and men. *The New England Journal of Medicine*, *364*(25), 2392–2404.

Mozaffarian, D., Katan, M. B., Ascherio, A., Stampfer, M. J., & Willett, W. C. (2006). Trans fatty acids and cardiovascular disease. *The New England Journal of Medicine*, *354*(15), 1601–1613.

Mozaffarian, D., Micha, R., & Wallace, S. (2010). Effects on coronary heart disease of increasing polyunsaturated fat in place of saturated fat: A systematic review and meta-analysis of randomized controlled trials. *Public Library of Science (PLoS). Medicine, 7*(3), 1–10.

Mozaffarian, D., Pischon, T., Hankinson, S. E., Rifai, N., Joshipura, K., Willett, W. C., & Rimm, E. B. (2004). Dietary intake of trans fatty acids and systemic inflammation in women. *The American Journal of Clinical Nutrition, 79*(4), 606–612.

Mrazek, P., Biglan, A., Hawkins, J. D., & Cody, C. (2007). *Community monitoring systems: tracking and improving the well-being of America's children and adolescents.* Falls Church, VA: Society for Preventive Research.

Msangi, S., & Rosegrant, R. W. (2011). Feeding the future's changing diets: implications for agriculture markets, nutrition, and policy. In *Proceedings of 2020 Conference: Leveraging Agriculture for Improving Nutrition and Health.* New Delhi, India. Retrieved from http://cdm15738.contentdm.oclc.org/utils/getfile/collection/p15738coll2/id/124834/filename/124835.pdf

Muhirwa, D., Nhapi, I., Wali, U., Banadda, N., Kashaigili, J., & Kimwaga, R. (2010). Characterization of wastewater from an abattoir in Rwanda and the impact on downstream water quality. *International Journal of Ecology and Development, 16*(S10), 30–46.

Muirhead, S. (2011). Ag's go-to message not resonating. *FoodLink.* Retrieved from http://feedstuffsfoodlink.com/story-ags-goto-messages-not-resonating-71-66622

Munro, L. (2012). The animal rights movement in theory and practice: A review of the sociological literature. *Social Compass, 6*(2), 166–181. doi:10.1111/j.1751-9020.2011.00440.x

Murray, C. J. L., Lopez, A. D., Chin, B., Feehan, D., & Hill, K. H. (2006). Estimation of potential global pandemic influenza mortality on the basis of vital registry data from the 1918-20 pandemic: A quantitative analysis. *Lancet, 368*(9554), 2211–2218. doi:10.1016/S0140-6736(06)69895-4 PMID:17189032

Myers, N., & Kent, J. (2003). New Consumers: The influence of affluence on the environment. *Proceedings of the National Academy of Sciences of the United States of America of the United States of America, 100*(8), 4963–4968. doi:10.1073/pnas.0438061100 PMID:12672963

Myers, N., & Kent, J. (2004). *New consumers: the influence of affluence on the environment.* Washington, DC: Island Press.

Myhre, G. D., Shindell, D., Bréon, F.-M., Collins, W., Fuglestvedt, J., & Huang, J. … Zhang, H. (2013). Anthropogenic and natural radiative forcing. In T. F. Stocker, D. Qin, G.-K. Plattner, M. Tignor, S.K. Allen, J. Boschung, …, P. M. Midgley (Eds), Climate change 2013: the physical science basis. Contribution of Working Group 1 to the Fifth Assessment Report of the Intergovernmental Panel on Climate Change (IPCC) (pp. 659-740). Cambridge, UK and New York, NY: Cambridge University Press.

Nadalin, T. (2012, February 26). Fake meat tastes so good you can't tell the difference. *Herald Sun.* Retrieved from http://www.heraldsun.com.au/ipad/fake-meat-tastes-so-good-you-cant-tell-the-difference/story-fn6bfm6w-1226281498315

Nath, J. (2010). God is a vegetarian: The food, health and bio-spirituality of Hare Krishna, Buddhist and Seventh-Day Adventist devotees. *Health Sociology Review, 19*(3), 356–368. doi:10.5172/hesr.2010.19.3.356

Nath, J., & Prideaux, D. (2011). The civilised burger: Meat alternatives as a conversion aid and social instrument for Australian vegetarias and vegans. *Australian Humanities Review, 51*, 135–151.

National Antimicrobial Resistance Monitoring System (NARMS). (2010). 2010 Executive report. Retrieved from http://www.fda.gov/downloads/AnimalVeterinary/SafetyHealth/AntimicrobialResistance/NationalAntimicrobialResistanceMonitoringSystem/UCM312360.pdf

National Bureau of Statistics of China. (2014). *2014 China statistical yearbook.* Beijing, China: China Statistical Press.

National Cancer Institute (NCI). (2007). Greater than the sum: systems thinking in tobacco control. Tobacco control monograph No. 18. Bethesda, MD: US Department of Health and Human Services, National Institute of Health, National Cancer Institute. Retrieved SEptember 8, 2015, from http://cancercontrol.cancer.gov/brp/tcrb/monographs/18/m18_complete.pdf

National Cancer Institute (NCI). (2013). Sources of saturated fat in the diet of the U.S. population ages 2 years and older, NHANES 2005-2006. Retrieved from http://appliedresearch.cancer.gov/diet/foodsources/sat_fat.html

National Cancer Institute (NCI). (2013). Sources of saturated fat in the diets of the U.S. population ages 2 years and older, NHANES 2005-2006. Retrieved November 17, 2014, from http://appliedresearch.cancer.gov/diet/foodsources/sat_fat/sf.html

National Centre for Groundwater Research and Training (NCGRT). (2013). Future mining and dining booms depend on water. Retrieved from http://www.groundwater.com.au/news_items/media-release-future-mining-dining-booms-depend-on-water

National Chicken Council. (2014). The nutritional value of chicken. Retrieved December 17, 2014, from http://www.nationalchickencouncil.org/chicken-the-preferred-protein-for-your-health-and-budget/the-nutritional-value-of-chicken/

National Farmers' Federation (NFF). (2015). Major commodities: beef. Retrieved from http://www.nff.org.au/commodities-beef-cattle.html

National Health and Medical Research Council (NHMRC). (2003). Food for Health: Dietary Guidelines for Australian Adults. Canberra, Australia: Commonwealth of Australia, NHMRC. Retrieved September 8, 2015, from https://www.nhmrc.gov.au/_files_nhmrc/publications/attachments/n29.pdf

National Health and Medical Research Council (NHMRC). (2009). NHMRC levels of evidence and grades for recommendations for developers of guidelines. Canberra, Australia: Commonwealth of Australia, NHMRC. Retrieved September 8, 2015, from https://www.nhmrc.gov.au/guidelines-publications/information-guideline-developers/resources-guideline-developers

National Health and Medical Research Council (NHMRC). (2013). Australian dietary guidelines (2013). Canberra, Australia: Australian Government, NHMRC, Department of Health and Ageing. Retrieved September 8, 2015, from https://www.nhmrc.gov.au/guidelines-publications/n55

National Health and Medical Research Council (NHMRC). (2013). Australian dietary guidelines. Canberra, Australia: Australian Government, NHMRC, Department of Health and Ageing. Retrieved from https://www.nhmrc.gov.au/guidelines-publications/n55

National Health and Medical Research Council (NHMRC). (2013). Australian Dietary Guidelines. Canberra: Australian Government, NHMRC, Department of Health and Ageing. Retrieved from https://www.nhmrc.gov.au/guidelines-publications/n55

National Health and Medical Research Council (NHMRC). (2013). Australian Dietary Guidelines. Canberra: Australian Government, NHMRC, Department of Health and Ageing. Retrieved September 8, 2015, from https://www.nhmrc.gov.au/guidelines-publications/n55

National Institute of Food and Agriculture (NIFA). (2004). Decreasing production related bone diseases in poultry. Report from the Current Research Information System (CRIS), Accession No. 0404529. Washington, DC: United States Department of Agriculture, Research, Education and Economic Information System, NIFA. Retrieved from http://portal.nifa.usda.gov/web/crisprojectpages/0404529-decreasing-production-related-bone-diseases-in-poultry.html

National Institute of Food and Agriculture (NIFA). (2006). Analysis of sperm storage mechanisms in poultry. Report from the Current Research Information System (CRIS), Accession No. 0405837. Washington, DC: United States Department of Agriculture, Research, Education and Economic Information System, NIFA. Retrieved from https://portal.nifa.usda.gov/web/crisprojectpages/0405837-analysis-of-sperm-storage-mechanisms-in-poultry.html

National Institute of Food and Agriculture (NIFA). (2009). The effect of scrotal insulation on oxidative damage to sperm and testicular gene expression. Report from the Current Research Information Systems (CRIS), Accession No. 0208789. Washington DC, USA: United States Department of Agriculture, Research, Education and Economic Information System, NIFA. Retrieved from http://www.reeis.usda.gov/web/crisprojectpages/0208789-the-effect-of-scrotal-insulation-on-oxidative-damage-to-sperm-and-testicular-gene-expression.html

National Institute of Food and Agriculture (NIFA). (2010). The effect of scrotal insulation on testicular gene expression. Report from the Current Research information Systems (CRIS), Accession No. 0220407. Washington DC, USA: United States Department of Agriculture, Research Education and Economic Information System, NIFA. Retrieved from http://www.reeis.usda.gov/web/crisprojectpages/0220407-the-effect-of-scrotal-insulation-on-testicular-gene-expression.html

National Institute of Food and Agriculture (NIFA). (2013). Understanding and control of male fertility in swine. Report from the Current Research Information System (CRIS), Accession No. 0229466. Washington DC, USA: United States Department of Agriculture, Research, Education and Economic Information System, NIFA. Retrieved from http://reeis.usda.gov/web/crisprojectpages/0229466-understanding-and-control-of-male-fertility-in-swine.html

National Institute of Food and Agriculture (NIFA). (2014a). Molecular strategies to circumvent or mitigate seasonal infertility in boars. Report from the Current Research Information System (CRIS), Accession No. 1002384. Washington DC, USA: United States Department of Agriculture, Research, Education and Economic Information System, NIFA. Retrieved from http://portal.nifa.usda.gov/web/crisprojectpages/1002384-molecular-strategies-to-circumvent-or-mitigate-seasonal-infertility-in-boars.html

National Institute of Food and Agriculture (NIFA). (2014b). Poultry germplasm preservation: modulating membrane lipids for successful cryopreservation of semen from valuable genetic stocks. Report from the Current Research Information System (CRIS), Accession No. 1000438. Washington DC, USA: United States Department of Agriculture, Research, Education and Economic Information System, NIFA. Retrieved from http://www.reeis.usda.gov/web/crisprojectpages/1000438-poultry-germplasm-preservation-modulating-membrane-lipids-for-successful-cryopreservation-of-semen-from-valuable-genetic-stocks.html

National Institute of Health (NIH). (2014). Ferritin blood tests. Retrieved 17 November, 2014, from http://www.nlm.nih.gov/medlineplus/ency/article/003490.htm

National Institute of Health (NIH). (2014, July/August). Global leaders raise alarm on antibiotic resistance. Global Health Matters, 13. Retrieved from http://www.fic.nih.gov/News/GlobalHealthMatters/july-august-2014/Pages/antimicrobial-resistance.aspx

National Pork Board. (n.d.). Pork, the other white meat brand. Retrieved April 27, 2015, from http://www.porkbeinspired.com/about-the-national-pork-board/the-other-white-meat-brand/

National Renewable Energy Laboratory (NREL). (2012). Life cycle greenhouse gas emissions from solar photovoltaics. Retrieved June 10, 2015, from http://www.nrel.gov/docs/fy13osti/56487.pdf

National Research Council (NRC). (1999). *The use of drugs in food animals: Benefits and risks.* Washington, DC: National Academic Press.

Neales, S. (2014, November 7). China deal on cattle exports. *The Australian.* Retrieved from http://www.theaustralian.com.au/national-affairs/foreign-affairs/china-deal-on-cattle-exports/story-fn59nm2j-1227115289349

Neff, R. A., Chan, I. L., & Smith, K. C. (2008). Yesterday's dinner, tomorrow's weather, today's news? US newspaper coverage of food system contributions to climate health. *Public Health Nutrition, 12*(7), 1006–1014. doi:10.1017/S1368980008003480 PMID:18702838

Neff, R. A., Chan, I. L., & Smith, K. C. (2008). Yesterday's dinner, tommorrow's weather, today's news? US newspaper coverage of food system contributions to climate health. *Public Health Nutrition, 12*(7), 1006–1014.

Nellemann, C., MacDevette, M., Manders, T., Eickhout, B., Svihus, B., Prins, A. G., & Kaltenborn, B. P. (2009). The environmental food crisis: the environment's role in averting future food crises. A UNEP Rapid Response Assessment. Arendal, Norway: United Nations Environment Programme. Retrieved September 8, 2015, from http://www.grida.no/files/publications/FoodCrisis_lores.pdf

Nelson, M. L., Dinardo, A., Hochberg, J., & Armelagos, G. J. (2010). Brief communication: Mass spetroscopic characterization of tetracycline in the skeletal remains of an ancient population from Sudanese Nubia 350-550 CE. *American Journal of Physical Anthropology, 143*(1), 151–154. doi:10.1002/ajpa.21340 PMID:20564518

Nestle, M. (2002). *Food politics: how the food industry influences nutrition and health.* London, UK: University of California Press.

Neuendorf, K. A. (2002). *The content analysis guidebook.* Thousand Oaks, CA: Sage.

Neumann, C. G., Bwibo, N. O., Murphy, S. P., Sigman, M., Whaley, S., Allen, L. H., & Demment, M. W. et al. (2003). Animal source foods improve dietry quality, micronutrient status, growth and cognitive function in Kenyan school children: Background, study, design and baseline findings. *The Journal of Nutrition, 133*(11), 3941S–3949S.

Newport, F. (2012). In U.S., 5% consider themselves vegetarians: even smaller 2% say they are vegans. Retrieved from http://www.gallup.com/poll/156215/consider-themselves-vegetarians.aspx

Nicholson, A. S., Sklar, M., Barnard, N. D., Gore, S., Sullivan, R., & Browning, S. (1999). Toward improved management of NIDDM: A randomized, controlled, pilot intervention using a lowfat, vegetarian diet. *Preventive Medicine, 29*(2), 87–91.

Nieman, D. C., Sherman, K. M., Arabatzis, K., Underwood, B. C., Barbosa, J. C., Johnson, M., & Lee, J. et al. (1989). Hematological, anthropometric, and metabolic comparisons between vegetarian and nonvegetarian elderly women. *International Journal of Sports Medicine, 10*(4), 243–251.

Nijjar, R. (2011, June 4). From pro athletes to CEOs and donut cravers, the rise of the vegan diet. *CBC News Canada.* Retrieved from http://www.cbc.ca/news/canada/from-pro-athletes-to-ceos-and-doughnut-cravers-the-rise-of-the-vegan-diet-1.1049116

Noakes, M. (2013). *CSIRO total wellbeing diet recipes on a budget.* Melbourne, Australia: Penguin Books.

Noakes, M., & Clifton, P. (2005). *CSIRO total wellbeing diet.* Melbourne, Australia: Penguin Books.

Noakes, M., & Clifton, P. (2006). *CSIRO total wellbeing diet. Book 2.* Melbourne, Australia: Penguin Books.

Nodari, E. S. (2009). *Ethnicidades renegociadas: práticas socioculturais no oeste de Santa Catarina* [Ethnicities renegotiated: socio-cultural practices in Western Santa Catarina]. Florianópolis, Brazil: Editora da UFSC.

Nordgren, A. (2012). Ethical issues in mitigation of climate change: The option of reduced meat production and consumption. *Journal of Agricultural & Environmental Ethics, 25*, 563–584.

Nordhaus, T., & Shellenberger, M. (2009, November 17). Apocalypse fatigue: losing the public on climate change. *The Guardian*. Retrieved from http://www.theguardian.com/environment/2009/nov/17/apocalypse-public-climate-change

Norrby, R., Powell, M., Aronsson, B., Monnet, D. L., Lustar, I., & Bocsan, I. S. … Gyssens, I. C. (2009). The bacterial challenge time to react: A call to narrow the gap between multidrug-resistant bacteria in the EU and the development of new antibacterial agents. Technical Report. Stockholm, Sweden: European Centre for Disease Prevention Control (ECDC) and European Medicines Agency (EMEA).

Noss, R. F., Dobson, A. P., Baldwin, R., Beier, P., Davis, C. R., Dellasala, D. A., & Tabor, G. et al. (2012). Bolder thinking for conservation. *Conservation Biology, 26*(1), 1–4. doi:10.1111/j.1523-1739.2011.01738.x PMID:22280321

Noto, H., Goto, A., Tsujimoto, T., & Noda, M. (2013). Low-carbohydrate diets and all-cause mortality: A systematic review and meta-analysis of observational studies. *Public Library of Science (PLoS). One, 8*(1), 1–10.

O'Neill, S. J. (2013). Image matters: Climate change imagery in the US, UK and Australian mass media. *Geoforum, 49*, 10–19. doi:10.1016/j.geoforum.2013.04.030

O'Neill, S. J., Boykoff, M., Niemeyer, S., & Day, S. A. (2013). On the use of imagery for climate change engagement. *Global Environmental Change, 23*(2), 413–421. doi:10.1016/j.gloenvcha.2012.11.006

O'Brien, B. (1996). Animal welfare reform and the magic bullet: The use and abuse of subtherapeutic doses of antibiotics in livestock. *University of Colorado Law Review, 67*, 407–442.

O'Connor, A. (2011, August 18). Bill Clinton's vegan journey. *The New York Times*. Retrieved from http://well.blogs.nytimes.com/2011/08/18/bill-clintons-vegan-journey/

Oliveira, E. M., & Goldim, J. R. (2014). Legislação de proteção animal para fins cientificos e a não inclusão dos invertebrados - análise bioética[Animal protection legislation for scientific purposes and non-inclusion of invertebrates - bioethical analysis]. *Revista Bioética, 22*(1), 45–56. doi:10.1590/S1983-80422014000100006

Olson, L. (2013). Commentary by Linden Olson: words. *Pork Network*, Retrieved from http://www.porknetwork.com/pork-news/latest/Commentary-by-Linden-Olson-Words-204219211.html

Oomen, C. M., Ocke, M. C., Feskens, E. J., van Erp-Baart, M. A., Kok, F. J., & Kromhout, D. (2001). Association between trans fatty acid intake and 10-year risk of coronary heart disease in the Zutphen Elderly Study: A prospective population-based study. *Lancet, 357*(9258), 746–751.

Opara, E. C. (2004). Role of oxidative stress in the etiology of type 2 diabetes and the effect of antioxidant supplementation on glycemic control. *Journal of Investigative Medicine, 52*(1), 19–23.

Oreskes, N., & Conway, E. M. (2010). *Merchants of doubt*. New York, NY: Bloomsbury.

Organisation for Economic Co-operation and Development (OECD). (2014). Agricultural output. Retrieved from https://data.oecd.org/agroutput/meat-consumption.htm

Orlich, M. J., Singh, P. N., Sabate, J., Jaceldo-Siegl, K., Fan, J., Knutsen, S., & Fraser, G. E. et al. (2013). Vegetarian dietary patterns and mortality in Adventist Health Study 2. *Journal of the American Medical Association Internal Medicine, 173*(13), 1230–1238.

Packer, I. (2012). A engenharia legal e o papel do estado na transição para um 'capitalismo verde'[Legal engineering and the role of the state in the transition to 'green capitalism']. *Proposta, 36*(125), 26–31.

Paddenburg, T. (2014, July 12). Good times for WA agriculture as farmers ride a food boom fueled by two billion middle-class Asians. *Sunday Times: Perth Now*. Retrieved from http://www.perthnow.com.au/news/western-australia/good-times-for-wa-agriculture-as-farmers-ride-a-food-boom-fuelled-by-two-billion-middleclass-asians/story-fnhocxo3-1226986783946

Padler-Karavani, V., Yu, H., Cao, H., Chokhawala, H., Karp, F., Varki, N., & Varki, A. et al. (2008). Diversity in specificity, abundance, and composition of anti-Neu5Gc antibodies in normal humans: Potential implications for disease. *Glycobiology*, *18*(10), 818–830.

Pairottia, M. B., Ceruttib, A. K., Martinic, F., Vescea, E., Padovand, D., & Beltramoa, R. (2014). (in press). Energy consumption and GHG emission of the Mediterranean diet: A systemic assessment using a hybrid LCA-IO method. *Journal of Cleaner Production*.

Paluszkiewicz, P., Smolinska, K., Debinska, I., & Turski, W. A. (2012). Main dietary compounds and pancreatic cancer risk. The quantitative analysis of case-control and cohort studies. *Cancer Epidemiology*, *36*(1), 60–67.

Pan, A. (2012). Red meat consumption and mortality: Result from two prospective cohort studies. *The Journal of the American Medical Association (JAMA). Archives of Internal Medicine*, *172*(7), 555–563. doi:10.1001/archinternmed.2011.2287 PMID:22412075

Pan, A., Sun, Q., Bernstein, A. M., Manson, J. E., Willett, W. C., & Hu, F. B. (2013). Changes in red meat consumption and subsequent risk of type 2 diabetes mellitus: Three cohorts of US men and women. *Journal of the American Medical Association Internal Medicine*, *173*(14), 1328–1335.

Pan, A., Sun, Q., Bernstein, A. M., Schulze, M. B., Manson, J. E., Stampfer, M. J., & Hu, F. B. et al. (2012). Red meat consumption and mortality: Results from 2 prospective cohort studies.[formerly Archives of Internal Medicine]. *Journal of the American Medical Association Internal Medicine*, *172*(7), 555–563.

Pan, A., Sun, Q., Bernstein, A. M., Schulze, M. B., Manson, J. E., Willett, W. C., & Hu, F. B. (2011). Red meat consumption and risk of type 2 diabetes: 3 cohorts of US adults and an updated meta-analysis. *The American Journal of Clinical Nutrition*, *94*(4), 1088–1096.

Parillo, M., & Riccardi, G. (2004). Diet composition and the risk of type 2 diabetes: Epidemiological and clinical evidence. *The British Journal of Nutrition*, *92*(1), 7–19.

Parker, D. R., Weiss, S. T., Troisi, R., Cassano, P. A., Vokonas, P. S., & Landsberg, L. (1993). Relationship of dietary saturated fatty acids and body habitus to serum insulin concentrations: The Normative Aging Study. *The American Journal of Clinical Nutrition*, *58*(2), 129–136.

Parker-Pope, T. (2011, January 7). Nutrition advice from the China study. *The New York Times*. Retrieved from http://well.blogs.nytimes.com/2011/01/07/nutrition-advice-from-the-china-study/?_r=0

Parliament of the United Kingdom. (1969). Joint committee on the use of antibiotics in animal husbandry and veterinary medicine: Swann Committee Report. London, UK: Government of the United Kingdom, House of Commons, Her Majesty's Stationary Office (HMSO).

Pathak, H., Jain, N., Bhatia, A., Patel, J., & Aggarwal, P. K. (2010). Carbon footprints of Indian food items. *Agriculture, Ecosystems & Environment*, *139*(1-2), 66–73. doi:10.1016/j.agee.2010.07.002

Patterson, R. E., Kristal, A. R., & White, E. (1996). Do beliefs, knowledge and perceived norms about diet and cancer predict dietary change? *American Journal of Public Health*, *86*(10), 1394–1400. doi:10.2105/AJPH.86.10.1394 PMID:8876507

Payne, R. (2002). Animal welfare, animal rights and the path to social reform: One movements struggle for coherency in the quest for change. Virginia Journal of Soial. *Policy & the Law*, *9*(3), 587–633.

PDDR-Concórdia. (2014). *Plano diretor de desenvolvimento rural de Concórdia* [Plan for rural development of Concórdia]. Concórdia, Brazil: Secretaria de Agricultura. [Agriculture Department]

Pearson, M. (2011, January 4) Personal interview with Mark Pearson, Executive Director of Animal liberation in New South Wales (Interviewer N. Pendergrast).

Pearson, D., Friel, S., & Lawrence, D. (2014). Building environmentally sustainable food systems on informed citizen choices: Evidence from Australia. *Biological Agriculture and Horticulture, 30*(3), 183–197. doi:10.1080/01448765.2014.890542

Peatling, S. (2007). Farmers say $100m policy must go further. *The Sydney Morning Herald.* Retrieved from http://www.smh.com.au/news/federal-election-2007-news/farmers-say-100m-policy-must-go-further/2007/11/13/1194766675352.html

Pelaz, J., & Long, J. (2008). Characterizing the glycocalyx of poultry spermatoza: Ii low temperature storage of turkey semen and sperm mobility phenotype impact the cabohydrate component of membrane glycoconjugates. *Andrology, 29*, 431–439. doi:10.2164/jandrol.107.004259

Pelletier, N., & Tyedmers, P. (2010). Forecasting potential global environmental costs of livestock production 2000-2050. *Proceedings of the National Academy of Sciences of the United States of America (PNAS), 107*(43), 18371-18374.

Pelletier, N., Audsley, E., Brodt, S., Garnett, T., Henriksson, P., Kendall, A., & Troell, M. et al. (2011). Energy intensity of agriculture and food systems. *Annual Review of Environment and Resources, 36*(1), 223–246. doi:10.1146/annurev-environ-081710-161014

Pelletier, N., & Tyedmers, P. (2010). Forecasting potential global environmental costs of livestock production 2000-2050.[PNAS]. *Proceedings of the National Academy of Sciences of the United States of America, 104*(12), 4814–4819.

Pendergrast, N. (2010, November 10-11). Veganism, organisational considerations and animal advocacy campaigns. Paper presented at the Voicing the Unseen: Just Write It, Curtin University, Perth, Australia. Retrieved from http://hgsoconference.curtin.edu.au/local/pdf/Pendegrast.pdf

People for the Ethical Treatment of Animals (PETA). (2007). PETA to Gore: too chicken to go veg. Retrieved from http://www.peta.org/blog/peta-gore-chicken-go-veg/

People for the Ethical Treatment of Animals (PETA). (2011a). Meat and the Environment. Retrieved from http://www.peta.org/issues/animals-used-for-food/meat-environment/

People for the Ethical Treatment of Animals (PETA). (2011b). *PETA's vegetarian/vegan starter kit.* Sydney, Australia: PETA.

People for the Ethical Treatment of Animals 2 (PETA2). (2015). Meat's not green. Retrieved from http://features.peta2.com/meatsnotgreen/index.asp

People for the Ethical Treatment of Animals Asia-PAcific (PETA Asia-Pacific). (2008). Think you can be a meat-eating environmentalist? Think again! Retrieved from http://www.petaasiapacific.com/resources/literature/leaflets/Think_You_Can_Be_a_Meat-Eating_Environmentalist_Leaflet_GEN_ENG.pdf

Pereira, M. A., Jacobs, D. R. Jr, Van Horn, L., Slattery, M. L., Kartashov, A. I., & Ludwig, D. S. (2002). Dairy consumption, obesity, and the insulin resistance syndrome in young adults: The CARDIA Study. *Journal of the American Medical Association, 287*(16), 2081–2089.

Perez-Peña, R. (2008). Newspaper circulation continues to decline rapidly. *The New York Times.* Retrieved from http://www.nytimes.com/2008/10/28/business/media/28circ.html

Peters, H. P., & Heinrichs, H. (2004, June). Expertise for the public: the science-journalism interface in German discourse on global climate change. Paper presented at the 8th International Conference, Barcelona, Spain. Retrieved from http://juser.fz-juelich.de/record/42732

Peters, C. J., Wilkins, J. L., & Fick, G. W. (2007). Testing a complete-diet model for estimating the land resource requirements of food consumption and agricultural carrying capacity: The New York State example. *Renewable Agriculture and Food Systems*, *22*(2), 145–153. doi:10.1017/S1742170507001767

Petherick, A. (2012). Light is cast on a long shadow. *Nature Climate Change*, *2*(10), 705–706. doi:10.1038/nclimate1703

Petray, T. L. (2011). Protest 2.0: Online interactions and Aboriginal activists. *Media Culture & Society*, *33*(6), 923–940. doi:10.1177/0163443711411009

Pew Comission on Industrial Farm Animal Production (Pew Commission). (2008). Putting meat on the table: industrial farm animal production in America. Baltimore, MD: Pew Charitable Trusts and Johns Hopkins Bloomberg School of Public Health. Retrieved from http://www.ncifap.org/_images/pcifapsmry.pdf

Pham, N. M., Mizoue, T., Tanaka, K., Tsuji, I., Tamakoshi, A., Matsuo, K., & Sasazuki, S. et al. (2014). Meat consumption and colorectal cancer risk: An evaluation based on a systematic review of epidemiologic evidence among the Japanese population. *Japanese Journal of Clinical Oncology*, *44*(7), 641–650.

Phelps, N. (2007). *The longest struggle: animal advocacy from Pythagoras to PETA*. New York, NY: Lantern Books.

Philpott, T. (2011a). Meat industry still denying antibiotic resistance. Retrieved from http://www.motherjones.com/tom-philpott/2011/09/meat-industry-antibiotic-resistance

Philpott, T. (2011b). What the USDA doesn't want you to know about antibiotics and factory farms. Retrieved from http://www.motherjones.com/tom-philpott/2011/07/what-usda-doesnt-want-you-know-about-antibiotics-and-factory-farms

Pickett, H. (2012). Nutritional benefits of higher welfare animal products. Surrey, UK: Compassion in World Farming (CIWF).

Pietinen, P., Ascherio, A., Korhonen, P., Hartman, A. M., Willett, W. C., Albanes, D., & Virtamo, J. (1997). Intake of fatty acids and risk of coronary heart disease in a cohort of Finnish men. The Alpha-Tocopherol, Beta-Carotene Cancer Prevention Study. *American Journal of Epidemiology*, *145*(10), 876–887.

Pimental, D., Berger, B., Filiberto, D., Newton, M., Wolfe, B., Karabinakis, E., & Nandagopal, S. et al. (2004). Water resources: Agricultural and environmental issues. *Bioscience*, *54*(10), 909–918. doi:10.1641/0006-3568(2004)054[0909:WRAAEI]2.0.CO;2

Pimentel, D., & Pimentel, M. (1997). Sustainability of meat-based and plant-based diets and the environment. *Clinical Nutrition (Edinburgh, Lothian)*, *78*(3), 660S–663S. PMID:12936963

Pimentel, D., & Pimentel, M. (1997). Sustainability of meat-based and plant-based diets and the environment. *The American Journal of Clinical Nutrition*, *78*(3), 660S–663S.

Pinker, S. (2011). *The better angels of our nature*. New York, NY: Viking.

Planta, M. B. (2007). The role of poverty in antimicrobial resistance. *Journal of the American Board of Family Medicine*, *20*(6), 533–539. doi:10.3122/jabfm.2007.06.070019 PMID:17954860

Plumber, B. (2014). How the U.S. government spends millions to get people to eat more pizza. *The Washington Post*. Retrieved from http://www.washingtonpost.com/blogs/wonkblog/wp/2014/02/10/13-percent-of-americans-are-eating-pizza-on-any-given-day/

Plumwood, V. (1993). *Femenism and the mastery of nature*. New York, NY: Routledge.

Pohjolainen, P., Vinnari, M., & Jokinen, P. (2015). Consumers' perceived barriers to following a plant-based diet. *British Food Journal*, *117*(3), 1150–1167.

Pollan, M. (2007). *In defense of food, an eater's manifesto: eat food, not too much mostly plants*. London, UK: Penguin Books.

Pond, W. G., Bazer, F. W., & Rollin, B. E. (Eds.). (2011). *Animal welfare in animal agriculture: husbrandry, stewardship and sustainability in animal production*. Boca Raton, FL: CRC Press.

Popp, A., Lotze-Campen, H., & Bodirsky, B. (2010). Food consumption, diet shifts and associated non-CO2 greenhouse gases from agricultural production. *Global Environmental Change*, *20*(3), 451–462. doi:10.1016/j.gloenvcha.2010.02.001

Porcher, J. (2006). Well-being and suffering in livestock farming: Living conditions at work for people and animals. *Sociologie du Travail*, *48*, 56–70. doi:10.1016/j.soctra.2006.02.001

Portha, B., Giroix, M. H., Cros, J. C., & Picon, L. (1980). Diabetogenic effect of N-nitrosomethylurea and N-nitrosomethylurethane in the adult rat. *Annales de la Nutrition et de l'Alimentation*, *34*(5-6), 1143–1151.

Poultry Site. (2015a). Breeding and genetics products and services. Retrieved April 8, 2015, from http://www.thepoultrysite.com/products/2/breeding-and-genetics/

Poultry Site. (2015b). Influence of increasing slaughter age of chickens on meat quality, welfare and technical and economic results. Retrieved April 8, 2015, from http://www.thepoultrysite.com/articles/2540/influence-of-increasing-slaughter-age-of-chickens-on-meat-quality-welfare-and-technical-and-economic-results/

Pounis, G. D., Tyrovolas, S., Antonopoulou, M., Zeimbekis, A., Anastasiou, F., Bountztiouka, V., & Panagiotakos, D. B. et al. (2010). Long-term animal-protein consumption is associated with an increased prevalence of diabetes among the elderly: The Mediterranean islands (MEDIS) study. *Diabetes & Metabolism*, *36*(6), 484–490.

Prairie, Y. T., & Duarte, C. M. (2007). Direct and indirect CO$_2$ release by humanity. *Biogeosciences*, *4*(2), 215–217. doi:10.5194/bg-4-215-2007

Pratt, M. M., John, K., MacLean, A. B., Afework, S., Phillips, D. H., & Poirier, M. C. (2011). Polycyclic aromatic hydrocarbon (PAH) exposure and DNA adduct semi-quantitation in archived human tissues. *International Journal of Environmental Research and Public Health*, *8*(7), 2675–2691.

Preis, S. R., Stampfer, M. J., Spiegelman, D., Willett, W. C., & Rimm, E. B. (2010). Dietary protein and risk of ischemic heart disease in middle-aged men. *The American Journal of Clinical Nutrition*, *92*(5), 1265–1272.

PricewaterhouseCoopers (PwC). (2011). The Australian beef industry: the basics. Sydney, Australia: PwC. Retrieved from http://www.pwc.com.au/industry/agribusiness/assets/Australian-Beef-Industry-Nov11.pdf

PricewaterhouseCoopers (PwC). (2012). Too late for two degrees? London, UK: PwC. Retrieved from https://www.thepmr.org/system/files/documents/Low%20Carbon%20Economy%20Index%202012.pdf

Prochaska, J., & DiClemente, C. (1983). Stages and processes of self-change of smoking: Toward an integrative model of change. *Journal of Consulting and Clinical Psychology*, *51*(3), 390–395. doi:10.1037/0022-006X.51.3.390 PMID:6863699

Public Health Association Australia (PHAA). (2009). A future for food: addressing public health, sustainability and equity from paddock to plate. Retrieved September 8, 2015, from https://www.phaa.net.au/documents/item/563

Putt del Pino, S., Levinson, R., & Larsen, J. (2006). Hot climate, cool commerce: a service sector guide to greenhouse gas management. Washington, DC: World Resource Institute (WRI). Retrieved from http://www.ghgprotocol.org/files/ghgp/tools/HotClimateCoolCommerce_lowrez.pdf

Qi, L., van Dam, R. M., Rexrode, K. M., & Hu, F. B. (2007). Heme iron from diet as a risk factor for coronary heart disease in women with type 2 diabetes. *Diabetes Care*, *30*(1), 101–106.

Qin, L. Q., Xu, J. Y., Wang, P. Y., Tong, J., & Hoshi, K. (2007). Milk consumption is a risk factor for prostate cancer in Western countries: Evidence from cohort studies. *Asia Pacific Journal of Clinical Nutrition*, *16*(3), 467–476. PMID:17704029

Qiu, C., Zhang, C., Gelaye, B., Enquobahrie, D. A., Frederick, I. O., & Williams, M. A. (2011). Gestational diabetes mellitus in relation to maternal dietary heme iron and nonheme iron intake. *Diabetes Care*, *34*(7), 1564–1569.

Quiggin, J. (2010). Draught, climate change and food prices in Australia: Working Paper. Brisbane, Australia: University of Queensland, School of Economics and School of Political Science and International Studies.

Quinn, J. M., & McFarlane, P. N. (1989). Effect of slaughterhouse and dairy factory effluent on epilinium. *Water Resources*, *23*, 1267–1273.

Radford, T. (2013). Carbon dioxide levels now 61% higher than 1990. *Responding to Climate Change*. Retrieved from http://www.rtcc.org/2013/12/31/carbon-dioxide-levels-now-61-higher-than-1990/

Rajpathak, S. N., Crandall, J. P., Wylie-Rosett, J., Kabat, G. C., Rohan, T. E., & Hu, F. B. (2009). The role of iron in type 2 diabetes in humans. *Biochimica et Biophysica Acta*, *1790*(7), 671–681.

Rajpathak, S., Ma, J., Manson, J., Willett, W. C., & Hu, F. B. (2006). Iron intake and the risk of type 2 diabetes in women: A prospective cohort study. *Diabetes Care*, *29*(6), 1370–1376.

Raphaely, T., & Marinova, D. (2013). Sustainability humanistic education: A new pedagogy for a better world. *International Journal of Education. Economics and Development*, *4*(2), 170–189.

Raphaely, T., & Marinova, D. (2014). Flexataririanism: Decarbonising through flexible vegetarianism. *Renewable Energy*, *67*, 90–96.

Raphaely, T., & Marinova, D. (2014a). Flexitarianism: A more moral dietary option. *International Journal of Sustainable Society*, *6*(1-2), 189–211. doi:10.1504/IJSSOC.2014.057846

Raphaely, T., & Marinova, D. (2014b). Flexitarianism: Decarbonising through flexible vegetarianism. *Renewable Energy*, *67*, 90–96. doi:10.1016/j.renene.2013.11.030

Raphaely, T., Marinova, D., Crisp, G., & Panayotov, J. (2013). Flexitarianism (flexible or part-time vegetarianism): A user-based dietary choice for improving personal, population and planetary wellbeing. *International Journal of User-Driven Healthcare*, *3*(3), 34–58. doi:10.4018/ijudh.2013070104

Raphaely, T., Marinova, D., & Todorov, V. (2010). Sustainability education: What on earth are we doing? *Management and Sustainable Development*, *26*(2), 49–60.

Rayner, M., Clarke, D., Thomas, P., & Scarborough, P. (2010). Healthy planet eating: how lower meat diets can save lives and the planet. Oxford, UK: Friends of the Earth & Oxford University, British Heart Foundation Health Promotion Research Group. Retrieved September 8, 2015, from http://www.foe.co.uk/sites/default/files/downloads/healthy_planet_eating.pdf

Regan, T. (2001). *Defending animal rights*. Chicago, IL: University of Illinois Press.

Reid, M. A., Marsh, K. A., Zeuschner, C. L., Saunders, A. V., & Baines, S. K. (2013). Meeting the nutrient reference values on a vegetarian diet. *The Medical Journal of Australia*, *199*(4), S33–S40.

Reisch, L., Eberle, U., & Lorek, S. (2013). Sustainable food consumption: an overview of contemporary issues and policies. *Sustainability: Science, Practice, &. Policy*, *9*(2), 7–25.

Resnicow, K., & Page, S. E. (2008). Embracing chaos and complexity: A quantum change for public health. *American Journal of Public Health*, *98*(8), 1382–1389. doi:10.2105/AJPH.2007.129460 PMID:18556599

Responsible Use of Medicines in Agriculture Alliance (RUMA). (2012). RUMA comment on the Early Day Motion in the House of Commons on the use of antibiotics in intensive farming. Retrieved from http://www.ruma.org.uk/news/20121105.htm

Responsible Use of Medicines in Agriculture Alliance (RUMA). (2013). *Written evidence submitted to Science and Technology Committee by the Responsible Use of Medicines in Agriculture Alliance (RUMA) (AMR0045)*. London, UK: RUMA.

Rezai, A., Foley, D. K., & Taylor, L. (2009). Global warming and economic externalities. Working Paper 2009-3. New York, NY: The New School for Social Research, Department of Economics, Schwartz Center for Economic Policy Analysis.

Richards, E., Signal, T., & Taylor, N. (2013). A different cut? Comparing attitudes toward animals and propensity for aggression within two primary industry cohorts-farmers and meatworkers. *Society & Animals*, *21*(4), 395–413. doi:10.1163/15685306-12341284

Ricketson, M. (2007, February 16). Big jump in online readership. *The Age*. Retrieved September 9, 2015, from http://www.theage.com.au/news/business/big-jump-in-online-readership/2007/02/15/1171405370719.html

Riggs, D. (2012). *Meat free Monday cookbook*. London, UK: Kyle Books.

Ripple, W. J., Smith, P., Haberl, H., Montzka, S. A., McAlpine, C., & Boucher, D. H. (2014). Commentary: Ruminants, climate change and climate policy. *Nature Climate Change*, *4*(1), 2–5. doi:10.1038/nclimate2081

Ripple, W. J., Smith, P., Harberl, H., Montzka, S. A., McAlpine, C., & Boucher, D. H. (2014). Ruminants, climate change and climate policy. *Nature Climate Change*, *4*(1), 2–5.

Risku-Norja, H., Kurppa, S., & Helenius, J. (2009). Dietary choices and greenhouse gas emissions – assessment of impact of vegetarian and organic options at national scale. *Progress in Industrial Ecology*, *6*(4), 340–354. doi:10.1504/PIE.2009.032323

Rizzo, N. S., Jaceldo-Siegl, K., Sabate, J., & Fraser, G. E. (2013). Nutrient profiles of vegetarian and nonvegetarian dietary patterns. *Journal of the Academy of Nutrition and Dietetics*, *113*(12), 1610–1619.

Rizzo, N. S., Sabate, J., Jaceldo-Siegl, K., & Fraser, G. E. (2011). Vegetarian dietary patterns are associated with a lower risk of metabolic syndrome: The adventist health study 2. *Diabetes Care*, *34*(5), 1225–1227.

Robbins, J. (2010). What about Soy? *John Robbins: Tools, Resources, and Inspiration*. Retrieved from http://johnrobbins.info/blog/what-about-soy/

Robins, N., & Roberts, S. (2006). Making sense of sustainable consumption. In T. Jackson (Ed.), *The earthscan reader on sustainable consumption* (pp. 37–47). London, UK: Earthscan.

Robinson, T. N. (2010). Save the world, prevent obesity: Piggybacking on existing social and ideological movement. *Obesity (Silver Spring, Md.)*, *18*(1sSuppl 1), S17–S22. doi:10.1038/oby.2009.427 PMID:20107456

Rogelj, J., Schaeffer, M., Meinshausen, M., Shindell, D. T., Hare, W., & Kilmont, Z., ... Schellnhuber, H. J. (2014). Disentangeling the effects of CO2 and short-lived climate forcer mitigation. *Proceedings of the National Academy of Sciences (PNAS)*, *111*(46), 16325-16330.

Rogers, L. (2012). Pew comments on proposed antibiotics legislation. Retrieved from http://www.pewtrusts.org/en/about/news-room/press-releases/0001/01/01/pew-comments-on-proposed-antibiotics-legislation

Rogers, R. (2008). Beasts, burgers and hummers: meat and the crisis of masculinity in contemporary television advertising. *Environmental Communication: A Journal of Nature and Culture, 2*(3), 281-301.

Rogers, E. M. (1983). *Diffusion of innovations*. New York, NY: Free Press.

Rohrmann, S., Overvad, K., Bueno-de-Mesquita, H. B., Jakobsen, M. U., Egeberg, R., Tjonneland, A., & Linseisen, J. et al. (2013). Meat consumption and mortality--results from the European Prospective Investigation into Cancer and Nutrition. *BioMed Central Medicine, 11*(63), 1–12.

Romaguera, D., Norat, T., Vergnaud, A. C., Mouw, T., May, A. M., Agudo, A., & Peeters, P. H. et al. (2010). Mediterranean dietary patterns and prospective weight change in participants of the EPIC-PANACEA project. *The American Journal of Clinical Nutrition, 92*(4), 912–921.

Rooklidge, S. J. (2004). Environmental antimicrobial contamination from terracumulation and diffuse pollution pathways. *The Science of the Total Environment, 325*(1), 1–13. doi:10.1016/j.scitotenv.2003.11.007 PMID:15144773

Rosa, F. R. T. (2006). *Areas de pastagem versus agricultura: o que aconteceu em 2005* [Grazing areas versus agriculture: what happened in 2005]. São Paulo, Brazil: Scot Consultoria.

Rosell, M., Appleby, P., Spencer, E., & Key, T. (2006). Weight gain over 5 years in 21,966 meat-eating, fish-eating, vegetarian, and vegan men and women in EPIC-Oxford. *International Journal of Obesity, 30*(9), 1389–1396.

Rothgerber, H. (2013). Real men don't eat (vegetable) quiche: Masculinity and the justification of meat consumption. *Psychology of Men & Masculinity, 14*(4), 363–375. doi:10.1037/a0030379

Rouse, I. L., Armstrong, B. K., & Beilin, L. J. (1983). The relationship of blood pressure to diet and lifestyle in two religious populations. *Journal of Hypertension, 1*(1), 65–71.

Rowe, A. K. (2009). Potential of integrated continuous surveys and quality management to support monitoring, evaluation and the scale-up of health interventions in developing countries. *The American Journal of Tropical Medicine and Hygiene, 80*(6), 971–979. PMID:19478260

Roy, P., Nei, D., Orikasa, T., Xu, Q. Y., Okadome, H., Nakamura, N., & Shiina, T. (2009). A review of life cycle assessment (LCA) on some food products. *Journal of Food Engineering, 90*(1), 1–10. doi:10.1016/j.jfoodeng.2008.06.016

Rozin, P., Homes, J., Faith, M. S., & Wansink, B. (2012). Is meat male? a quantitative multimethod framework to establish metaphoric relationships. *The Journal of Consumer Research, 39*(3), 629–643. doi:10.1086/664970

Ruby, M. B., & Heine, S. J. (2011). Meat, morals and masculinity. *Appetite, 56*(2), 447–450. doi:10.1016/j.appet.2011.01.018 PMID:21256169

Rushton, J., Ferreira, J. P., & Stark, K. D. (2014). Antimocrobial resistance: the use of antimicrobial in the livestock sector (No. 68). Paris, France: Organisation for Economic Co-operation and Development (OECD).

Russell, G. (2010). Burning the biosphere, boverty blues (part 1). *Brave New Climate*. Retrieved from http://bravenewclimate.com/2010/01/05/boverty-blues-p1/

Russell, G. (2009). *CSIRO perfidy*. Fremantle, Australia: Vivid Publishing.

Russell, W. R., Gratz, S. W., Duncan, S. H., Holtrop, G., Ince, J., Scobbie, L., & Flint, H. J. et al. (2011). High-protein, reduced-carbohydrate weight-loss diets promote metabolite profiles likely to be detrimental to colonic health. *The American Journal of Clinical Nutrition, 93*(5), 1062–1072.

Ryder, R. (2005, August 6). All beings that feel pain deserve human rights. *The Guardian*. Retrieved September 6, 2015, from http://www.theguardian.com/uk/2005/aug/06/animalwelfare

Salas-Salvadó, J., Bulló, M., Babio, N., Martínez-González, M. A., Ibarrola-Jurado, N., Basora, J., & Ros, E. et al. (2011). Reduction in the Incidence of Type 2 Diabetes with the Mediterranean Diet. *Diabetes Care*, *34*(1), 14–19.

Salehi-Abargouei, A., Maghsoudi, Z., Shirani, F., & Azadbakht, L. (2013). Effects of Dietary Approaches to Stop Hypertension (DASH)-style diet on fatal or nonfatal cardiovascular diseases--incidence: A systematic review and meta-analysis on observational prospective studies. *Nutrition (Burbank, Los Angeles County, Calif.)*, *29*(4), 611–618.

Salehi, M., Moradi-Lakeh, M., Salehi, M. H., Nojomi, M., & Kolahdooz, F. (2013). Meat, fish, and esophageal cancer risk: A systematic review and dose-response meta-analysis. *Nutrition Reviews*, *71*(5), 257–267.

Salonen, A. O., & Helne, T. T. (2012). Vegetarian diets: A way towards a sustainable society. *Journal of Sustainable Development*, *5*(6), 10–24. doi:10.5539/jsd.v5n6p10

Salter, D. (2012, March 29). An hour to ponder why Fairfax bothers turning off the lights. *Crikey*. Retrieved from http://www.crikey.com.au/2012/03/29/history-of-earth-hour-at-fairax/?wpmp_switcher=mobile

Sampei, Y., & Aoyagi-Usui, M. (2009). Mass-media coverage, its influence on public awareness of climate-change issues and implications for Japan's national campaign to reduce greenhouse gas emissions. *Global Environmental Change*, *19*(2), 203–212. doi:10.1016/j.gloenvcha.2008.10.005

Sample, I., Harvey, F., & Campbell, D. (2013, June 11). UK raises alarm on deadly rise of superbugs. The Guardian. Retrieved from http://www.theguardian.com/society/2013/jun/11/uk-urge-global-clampdown-antibiotics-g8

Samraj, A. N., Laubli, H., Varki, N., & Varki, A. (2014). Involvement of a non-human sialic acid in human cancer. *Frontiers in Oncology*, *4*, 1–13.

Saunders, A. V., Craig, W. J., Baines, S. K., & Posen, J. S. (2012). Iron and vegetarian diets. *The Medical Journal of Australia*, *1*(Suppl 2), 11–16.

Scarborough, P., Allender, S., Clarke, D., Wickramasinghe, K., & Rayner, M. (2012). Modelling the health impact of environmentally sustainable dietary scenarios in the UK. *European Journal of Clinical Nutrition*, *66*(6), 710–715.

Scarborough, P., Appleby, P. N., Mizdrak, A., Briggs, A. D. M., Travis, R. C., Bradbury, K. E., & Key, T. J. (2014). Dietary greenhouse gas emissions of meat-eaters, fish-eaters, vegetarians and vegans in the UK. *Climatic Change*, *125*(2), 179–192. doi:10.1007/s10584-014-1169-1 PMID:25834298

Schlegelmilch, B. (1998). *Marketing ethics: an international perspective*. London, UK: International Thomson Business Press.

Schlesinger, S. (2010). Onde pastar? O gado bovino no Brasil [Where to graze? The cattle in Brazil] (1st ed.). Rio de Janeiro, Brazil: Federação de Órgãos para Assistência Social e Educacional (FASE).

Schlesinger, S. (2010). *Onde pastar? O gado bovino no Brasil* [Where to graze? The cattle in Brazil] (1st ed.). Rio de Janeiro, Brazil: Fase.

Schmidt, A., Ivanova, A., & Schäfer, M. S. (2013). Media attention for climate change around the world: A comparative analysis of newspaper coverage in 27 countries. *Global Environmental Change*, *23*(5), 1233–1248. doi:10.1016/j.gloenvcha.2013.07.020

Schneider, S., Nullmeier, F., & Hurrelmann, A. (2007). Exploring the communicative dimension of legitamacy: text analytical approaches. In A. Hurrelmann, S. Schneider, & J. Steffek (Eds.), *Legitamacy in an age of global politics* (pp. 126–155). Hampshire, UK: Palgrave Macmillan.

Schonfeld, V. (2010, January 18). Five fatal flams of animal activism. *The Guardian*. Retrieved from http://www.theguardian.com/commentisfree/2010/jan/18/five-fatal-flaws-animal-activism

Schösler, H., de Boer, J., & Boersema, J. J. (2012). Can we cut out the meat of the dish? Constructing consumer-orientated pathways towards meat substitution. *Appetite, 58*, 39–47.

Schreineir, D. F. (1994). *A formação de uma cultura do trabalho: cotidiano, trabalho e poder. Extremo Oeste do Paraná [The formation of a labour culture: daily life, work and power. Extreme West of Paraná] - 1970/1988.* [Masters in History Dissertation]. Universidade Federal de Santa Catarina (UFSC), Florianópolis, Brazil.

Schulz, M., Kroke, A., Liese, A. D., Hoffmann, K., Bergmann, M. M., & Boeing, H. (2002). Food groups as predictors for short-term weight changes in men and women of the EPIC-Potsdam cohort. *The Journal of Nutrition, 132*(6), 1335–1340.

Schwab, U. S., Ausman, L. M., Vogel, S., Li, Z., Lammi-Keefe, C. J., Goldin, B. R., & Lichtenstein, A. H. et al. (2000). Dietary cholesterol increases the susceptibility of low density lipoprotein to oxidative modification. *Atherosclerosis, 149*(1), 83–90.

Schwab, U., Lauritzen, L., Tholstrup, T., Haldorssoni, T., Riserus, U., Uusitupa, M., & Becker, W. (2014). Effect of the amount and type of dietary fat on cardiometabolic risk factors and risk of developing type 2 diabetes, cardiovascular diseases, and cancer: A systematic review. *Food & Nutrition Research, 58*, 1–26.

Science Codex. (2012). Climate change increases stress, need for restoration on grazed public lands. Retrieved from http://www.sciencecodex.com/climate_change_increases_stress_need_for_restoration_on_grazed_public_lands-102039

Scientist, N. (2005, December14). It's better to green your diet than your car. *New Scientist, 2530*. Retrieved from https://www.google.com.au/search?q=It%E2%80%99s+better+to+green+your+diet+than+your+car&oq=It%E2%80%99s+better+to+green+your+diet+than+your+car&aqs=chrome.69i57j0.461j0j8&sourceid=chrome&es_sm=91&ie=UTF-8

Scully, M. (2002). *Dominion: the power of man, the suffering of animals and the call to mercy.* New York, NY: St Martin's Press.

Secretariat of the Convention of Biological Diversity (CBD). (2010). Global biodiversity outlook 3. Montréal, Canada: CBD. Retrieved from http://www.cbd.int/doc/publications/gbo/gbo3-final-en.pdf

Sen, K. M. (1929). *Medieval mysticism of India.* London, UK: Luzac and Company.

Servico de Apoio às Micro e Pequenas Empresas (SEBRAE). (2013). Santa Catarina em números: Concórdia [Santa Catarina in numbers: Concórdia]. Florianópolis, Brazil: SEBRAE.

Shaban, R. Z., Cruickshank, M., & Christiansen, K.Antimicrobial Resistance Standing Committee. (2013). *National surveillance and reporting of antimicrobial resistance and antibiotic usage for human health in Australia.* Canberra, Australia: Government of Australia, Antimicrobial Resistance Standing Committee, Australian Health Protection Principal Committee.

Shah, R. (2011, August 22). Why Bill Clinton went vegan. *The Huffington Post*. Retrieved from http://www.huffingtonpost.com/2011/08/22/bill-clinton-diet_n_932439.html

Shapiro, P. (2004, Summer). On the killing floor. *AV Magazine*, Retrieved from http://aavs.org/our-work/av-magazine/

Shapiro, P. (2013). The Chicken Industry Loves Federal Handouts. *The Huffington Post*. Retrieved from http://www. huffingtonpost.com/paul-shapiro/the-chicken-industry_b_3947857.html

Shapiro, P. (2015). Feasting from the Federal Trough: how the meat, egg and dairy industries gorge on taxpayer dollars while fighting modest rules. In T. Raphaely & D. Marinova (Eds.), *Impacts of meat consumption on health and environmental sustainability*. Hershey, PA: IGI Global.

Sharma, S. (2015). Superbugs a grave global threat by 2030 due to industrial meat production. Retrieved from http://www.iatp.org/blog/201504/superbugs-a-grave-global-threat-by-2030-due-to-industrial-meat-productions

Shea, K., Florini, K., & Barlam, T. (2001). When wonder drugs don't work: how antibiotic resistance threatens children, seniors and the medically vaulnerable. Washington, DC: Environmental Defense Fund (EDF).

Shindell, D. T. (2013). Climate change: breaking the stalemate. *Milken Institute Review: A Journal of Economic Policy, 15*(1), 35-45.

Shindell, D. T., Faluvegi, G., Koch, D. M., Schmidt, G. A., Unger, N., & Bauer, S. E. (2009). Improved attribution of climate forcing to emissions. *Science, 326*(5953), 716–718.

Shindell, D. T., Kuylenstierna, J. C. I., Vignati, E., van Dingenen, R., Amann, M., Klimont, Z., & Fowler, D. et al. (2012). Simultaneously mitigating near-term climate change and improving human health and food security. *Science, 335*(6065), 183–189.

Shirani, F., Salehi-Abargouei, A., & Azadbakht, L. (2013). Effects of Dietary Approaches to Stop Hypertension (DASH) diet on some risk for developing type 2 diabetes: A systematic review and meta-analysis on controlled clinical trials. *Nutrition (Burbank, Los Angeles County, Calif.), 29*(7-8), 939–947.

Shneiderman, B., & Plaisant, C. (2014). Treemaps for space-constrained visualization of hierarchies: including the history of treemap research at the University of Maryland. Retrieved June 10, 2015, from http://www.cs.umd.edu/hcil/treemap-history/index.shtml

Shoemaker, J. K., Schrag, D. P., Molina, M. J., & Ramanathan, V. (2013). What role for short-lived climate pollutants in mitigation policy? *Science, 342*(6164), 1323–1324. doi:10.1126/science.1240162 PMID:24337280

Siekmann, J. H., Allen, L. H., Bwibo, N. O., Demment, M. W., Murphy, S. P., & Neumann, C. G. (2003). Kenyan school children have multiple micronutrient deficiencies, but increased plasma vitamin B-12 is the only detectable micronutrient response to meat of milk supplementation. *The Journal of Nutrition, 133*(11), 3972S–3980S.

Sievenpiper, J. L., & Dworatzek, P. D. (2013). Food and dietary pattern-based recommendations: An emerging approach to clinical practice guidelines for nutrition therapy in diabetes. *Canadian Journal of Diabetes, 37*(1), 51–57.

Silbergeld, E. K., Graham, J., & Price, L. B. (2008). Industrial food animal production, antimicrobial resistance and human health. *Annual Review of Public Health, 29*(1), 151–169. doi:10.1146/annurev.publhealth.29.020907.090904 PMID:18348709

Silverstone, A. (2011). The kind life with Alica Silverstone: vegan. Retrieved from http://thekindlife.com/?s=Vegan&submit=

Simon, D. (2013). *Meatonomics: how the rigged economics of meat and dairy make you consume too much - and how to eat better, live longer, and spend smarter*. San Francisco, CA: Conari Press.

Simons, M. (2012). *Journalism at the crossroads: crisis and opportunity for the press*. Brunswick, Australia: Scribe.

Simopoulous, A. P. (2008). The importance of the omega-6/omega-3 fatty acid ratio in cardiovascular disease and other chronic diseases. *Experimental Biology and Medicine, 233*(6), 674–688. doi:10.3181/0711-MR-311 PMID:18408140

Singer. (2010, September 14). Fish the forgotten victims of our plate. The Guardian. Retrieved from http://www.theguardian.com/commentisfree/cif-green/2010/sep/14/fish-forgotten-victims

Singer, P. (1991). *Animal liberation* (2nd ed.). London, UK: Random House.

Singer, P. (2002). *Ethica prática* [Practical ethics]. São Paulo, Brazil: Martins Fontes.

Singer, P. (2004). *Libertação animal* [Animal liberation]. Rio de Janeiro, Brazil: Lugano Editora.

Singer, P. (2012). *Personal interview with Peter Singer, Australian ethicist.* Interviewer N. Pendergrast.

Singer, P., & Mason, J. (2007). *The ethics of what we eat: why our food choices matter.* Emmaus, PA: Rodale.

Singh, P. N., Sabate, J., & Fraser, G. E. (2003). Does low meat consumption increase life expectancy in humans? *The American Journal of Clinical Nutrition, 78*(3), 526S–532S.

Sinha, R., Cross, A. J., Graubard, B. I., Leitzmann, M. F., & Schatzkin, A. (2009). Meat intake and mortality: A prospective study of over half a million people. *Archives of Internal Medicine, 169*(6), 562–571.

Sinha, R., Cross, A. J., Graubard, B. I., Leitzmann, M. F., & Schatzkin, A. (2009). Meat intake and mortality: A prospective study of over half a million people. *Journal of the American Medical Association Internal Medicine, 169*(6), 562–571.

Sinha, R., Peters, U., Cross, A. J., Kulldorff, M., Weissfeld, J. L., Pinsky, P. F., & Hayes, R. B. et al. (2005). Meat, meat cooking methods and preservation, and risk for colorectal adenoma. *Cancer Research, 65*(17), 8034–8041.

Sipetic, S. B., Vlajinac, H. D., Kocev, N. I., Marinkovic, J. M., Radmanovic, S. Z., & Bjekic, M. D. (2005). The Belgrade childhood diabetes study: A multivariate analysis of risk determinants for diabetes. *European Journal of Public Health, 15*(2), 117–122.

Sirri, F., Castellini, C., Bianchi, M., Petracci, M., Meluzzi, A., & Franchini, A. (2011). Effect of fast-, medium- and slow-growing strains on meat quality of chickens reared under the organic farming method. *Animal, 5*(2), 312–319. doi:10.1017/S175173111000176X PMID:22440776

Skolimowski, H. (1993). *A sacred place to dowell: living with reverence upon the Earth.* London, UK: Element Book Ltd.

Sluijs, I., & Beulens, J. W., van der A, D. L., Spijkerman, A. M., Grobbee, D. E., & van der Schouw, Y. T. (2010). Dietary intake of total, animal, and vegetable protein and risk of type 2 diabetes in the European Prospective Investigation into Cancer and Nutrition (EPIC)-NL study. *Diabetes Care, 33*(1), 43–48.

Smil, V. (2013b, July 19). Should humans eat meat [Excerpt]? *Scientific American,* Retrieved from http://www.scientificamerican.com/article/should-humans-eat-meat-excerpt/http://www.scientificamerican.com/article/should-humans-eat-meat-excerpt/

Smil, V. (2002). Eating meat: Evolution, patterns and consequences. *Population and Development Review, 28*(4), 599–639. doi:10.1111/j.1728-4457.2002.00599.x

Smil, V. (2013). *Should we eat meat? Evolution and consequence of modern carnivory.* Chichester, UK: Wiley-Blackwell.

Smil, V. (2013a). *Should we eat meat?: Evolution and consequence of modern carnivory.* Chichester, UK: Wiley-Blackwell. doi:10.1002/9781118278710

Smil, V. (2014). Eating meat: Constants and changes. *Global Food Security, 3*(2), 67–71.

Smith, L. C., & Haddad, L. J. (2000). Explaining child malnutrition in developing countries: a cross-country analysis. Research report. Retrieved June 10, 2015, from http://www.ifpri.org/sites/default/files/pubs/pubs/abstract/111/rr111.pdf

Smith, J. (2005). Dangerous news: Media decision-making about climate change risk. *Risk Analysis*, *25*(6), 1471–1482. doi:10.1111/j.1539-6924.2005.00693.x PMID:16506976

Smith, K. R., Desaia, M. A., Rogersa, J. V., & Houghton, R. A. (2013). Joint CO_2 and CH_4 accountability for global warming.[PNAS]. *Proceedings of the National Academy of Sciences of the United States of America*, *110*(31), E2865–E2874. doi:10.1073/pnas.1308004110 PMID:23847202

Smith, P., Martino, D., Cai, Z., Gwary, D., Janzen, H., Kumar, P., & Towprayoon, S. et al. (2007). Policy and technological constraints to implementation of greenhouse gas mitigation options in agriculture. *Agriculture, Ecosystems & Environment*, *118*(1), 6–28. doi:10.1016/j.agee.2006.06.006

Smith, R., & Coast, J. (2013). The true cost of antimicrobial resistance. *British Medical Journal*, *346*(1493), 1–5. PMID:23479660

Smolinska, K., & Paluszkiewicz, P. (2010). Risk of colorectal cancer in relation to frequency and total amount of red meat consumption. Systematic review and meta-analysis. *Archives of Medical Science*, *6*(4), 605–610.

Sneijder, P., & te Molder, H. (2009). Normalizing ideological food choice and eating practices: Identity work in online discussions on veganism. *Appetite*, *52*(3), 621–630. doi:10.1016/j.appet.2009.02.012 PMID:19501759

Snow, D. A. (2004). Framing processes, ideology and discursive fields. In D. A. Snow, S. A. Soule, & H. Kriesi (Eds.), *The Blackwell companion to social movements* (pp. 380–412). Malden, MA: Blackwell Publishing. doi:10.1002/9780470999103

Snowdon, D. A., & Phillips, R. L. (1985). Does a vegetarian diet reduce the occurrence of diabetes? *American Journal of Public Health*, *75*(5), 507–512.

Snowdon, D. A., Phillips, R. L., & Fraser, G. E. (1984). Meat consumption and fatal ischemic heart disease. *Preventive Medicine*, *13*(5), 490–500.

Snyder, C., Bruulsema, T., Jensen, T., & Fixen, P. (2009). Review of greenhouse gas emissions from crop production systems and fertilizer management effects. *Agriculture, Ecosystems & Environment*, *133*(3), 247–266. doi:10.1016/j.agee.2009.04.021

Sofi, F., Macchi, C., Abbate, R., Gensini, G. F., & Casini, A. (2014). Mediterranean diet and health status: An updated meta-analysis and a proposal for a literature-based adherence score. *Public Health Nutrition*, *17*(12), 2769–2782.

Soil Association. (2010). Telling porkies: the big fat lie about doubling food production. Bristol, UK: Soil Association. Retrieved from https://www.soilassociation.org/LinkClick.aspx?fileticket=qbavgJQPY%2Fc%3D&tabid=390

Song, P., Lu, M., Yin, Q., Wu, L., Zhang, D., Fu, B., & Zhao, Q. et al. (2014). Red meat consumption and stomach cancer risk: A meta-analysis. *Cancer Research and Clinical Oncology*, *140*(6), 979–992. doi:10.1007/s00432-014-1637-z PMID:24682372

Song, P., Lu, M., Yin, Q., Wu, L., Zhang, D., Fu, B., & Zhao, Q. et al. (2014). Red meat consumption and stomach cancer risk: A meta-analysis. *Journal of Cancer Research and Clinical Oncology*, *140*(6), 979–992.

Song, Y., Chavarro, J. E., Cao, Y., Qiu, W., Mucci, L., Sesso, H. D., & Ma, J. et al. (2013). Whole milk intake is associated with prostate cancer-specific mortality among U.S. male physicians. *The Journal of Nutrition*, *143*(2), 189–196. doi:10.3945/jn.112.168484 PMID:23256145

Soret, S., Mejia, A., Batech, M., Jaceldo-Siegel, K., Harwatt, H., & Sabaté, J. (2014). Climate change mitigation and health effects of varied dietary patterns in real-life settings throughout North America. *The American Journal of Clinical Nutrition*, *100*(Supplement_1), 490S–495S. doi:10.3945/ajcn.113.071589 PMID:24898230

Soto, E., de La Hoz, L., Ordonez, J., Herranz, B., Hierro, E., Lopez-Bote, C. J., & Cambero, M. (2009). The feeding and rearing systems of Iberian pigs affect the lipid composition and texture profile of dry-cured loin. *Journal of Animal and Feed Sciences*, *18*(1), 78–89.

Spencer, E. A., Appleby, P. N., Davey, G. K., & Key, T. J. (2003). Diet and body mass index in 38 000 EPIC-Oxford meat-eaters, fish-eaters, vegetarians and vegans. *International Journal of Obesity*, *27*, 728–734.

Spencer, E. A., Appleby, P. N., Davey, G. K., & Key, T. J. (2003). Diet and body mass index in 38000 EPIC-Oxford meat-eaters, fish-eaters, vegetarians and vegans. *International Journal of Obesity and Related Metabolic Disorders*, *27*(6), 728–734.

Spotts, P. (2012, March 28). Climate change report: time to start preparing for the worst. *Christian Science Monitor*. Retrieved from http://www.csmonitor.com/Environment/2012/0328/Climate-change-report-time-to-start-preparing-for-the-worst

Spretnak, C. (1999). *The resurgence of the real: body, nature and place in a hypermodern world*. New York, NY: Routledge.

Srinivasan, C. S. (2006, August 12-16). *WHO dietary norms: a quantitative evaluation of potential consumption impacts in the United States*. Paper presented at the 26th International Association of Agricultural Economists Conference, Gold Coast, Queensland, Australia. Retrieved from http://purl.umn.edu/25292

Stafford, J., Allsop, S., & Daube, M. (2014). From evidence to action: Health promotion and alcohol. *Health Promotion Journal of Australia*, *25*(1), 8–13. doi:10.1071/HE14001 PMID:24739773

Stahler, C. (n.d.). How often do americans eat vegetarian meals and how many adults in the U.S. are vegans? The Vegetarian Resource Group. Retrieved December 1, 2014, from http://www.vrg.org/journal/vj2011issue4/vj2011issue4poll.php

Stahler, C. (2011). How often do Americans eat vegetarian meals and how many adults in the U.S. are vegan? *Vegetarian Journal*, *30*(4), 10–11.

Stanton, R. (2012). A plant based diet - good for us and for the planet.[MJA]. *The Medical Journal of Australia*, *1*(Suppl2), 5–6.

Stanton, R. (2015). Meat and dietary guidelines. In D. Marinova & T. Raphaely (Eds.), *Impacts of meat consumption on health and environmental sustainability*. Hershey, PA: IGI Global.

Star, V. (2011). Alicia Silverstone starts vegan book club on "The kind life". Retrieved from http://vegetarianstar.com/2011/02/25/alicia-silverstone-starts-vegan-book-club-on-the-kind-life/

Steffek, J. (2009). Discursive legitimation in environmental governance: Discourse and expertise in forest and environmental governance. *Forest Policy and Economics*, *11*(5), 313–318. doi:10.1016/j.forpol.2009.04.003

Stehfest, E., Bouwman, L., van Vuuren, D., den Elzen, M., Eickhout, B., & Kabat, P. (2009). Climate benefits of changing diet. *Climatic Change*, *95*(1), 83–102. doi:10.1007/s10584-008-9534-6

Steiner, G. (2009, November 21). Animal, vegetable, miserable. Retrieved from http://www.nytimes.com/2009/11/22/opinion/22steiner.html?pagewanted=all

Steinfeld, H., Gerber, P., Wassenaar, T., Castel, V., Rosales, M., & de Hann, C. (2006). Livestock's long shadow: environmental issues and options. Rome, Italy: Food Agriculture Organisation (FAO). Retrieved from ftp://ftp.fao.org/docrep/fao/010/a0701e/a0701e00.pdf

Steinfeld, H., Gerber, P., Wassenaar, T., Castel, V., Rosales, M., & de Hann, C. (2006). Livestock's long shadow: environmental issues and options. Rome, Italy: Food Agriculture Organisation (FAO).

Steinfeld, H., Gerber, P., Wassenaar, T., Castel, V., Rosales, M., & de Hann, C. (2006). Livestock's long shadow: environmental issues and options. Rome, Italy: Food Agriculture Organisation (FAO). Retrieved from ftp://ftp.fao.org/docrep/fao/010/a0701e/a0701e00.pdf

Steinfeld, H., Gerber, P., Wassenaar, T., Castel, V., Rosales, M., & de Hann, C. (2006). Livestock's long shadow: environmental issues and options. Rome, Italy: Food and Agriculture Organisation (FAO).

Steinfeld, H., & Gerber, P. (2010). Livestock production and the global environment: Consume less or produce better? *Proceedings of the National Academy of Sciences of the United States of America*, *107*(43), 18237–18238. doi:10.1073/pnas.1012541107 PMID:20935253

Sterba, J. (2010). Kantians and utilitarians and the moral status of nonhuman life. In D. R. Keller (Ed.), *Environmental ethics: the big question* (pp. 182–192). Chichester, UK: Wiley Blackwell.

Stolba, A., & Wood-Gush, D. G. M. (1989). The behaviour of pigs in a semi-natural environment. *Animal Production*, *48*(2), 419–425. doi:10.1017/S0003356100040411

Stone, D. A. (1989). Casual stories and the formation of policy agendas. *Political Science Quarterly*, *104*(2), 281–300. doi:10.2307/2151585

Sulda, H., Coveney, J., & Bentley, M. (2010). An investigation of the ways in which public health nutrition policy and practices can address climate change. *Public Health Nutrition*, *13*(3), 304–313. doi:10.1017/S1368980009990334 PMID:19545472

Sullivan, N. (2007). *You can hear me now: how microloans and cell phones are connecting the world's poor to the global economy*. San Francisco, CA: Jossey-Bass.

Sweidan, M., Zhang, Y., Harvey, K., Yang, Y.-H., Shen, X., & Yao, K. (2005, November 28–30). Report on Proceedings of the 2nd National Workshop on Rational Use of Antibiotics in China. Paper presented at the 2nd National Workshop on Rational Use of Antibiotics in China, Beijing Children's Hospital, Beijing, China.

Tangvoranuntakul, P., Gagneux, P., Diaz, S., Bardor, M., Varki, N., Varki, A., & Muchmore, E. (2003). Human uptake and incorporation of an immunogenic nonhuman dietary sialic acid.[PNAS]. *Proceedings of the National Acadamy of Sciences of the United States*, *100*(21), 12045–12050.

Tantamango-Bartley, Y., Jaceldo-Siegl, K., Fan, J., & Fraser, G. (2013). Vegetarian diets and the incidence of cancer in a low-risk population. *Cancer Epidemiology, Biomarkers & Prevention*, *22*(2), 286–294.

Target 100. (2014a). About. Retrieved from http://www.target100.com.au/About

Target 100. (2014b). Cowspiracy. Retrieved from http://www.target100.com.au/Hungry-for-Info/Target-100-Responds/Cowspiracy

Tarré, R., Macedo, R., Cantarutti, R. B., de Rezende, C. P., Pereira, J. M., & Ferreira, E. et al. (2001). The effect of the presence of a forage legume on nitrogen and carbon levels in soils under Brachiaria pastures in the Atlantic forest region of the South of Bahia, Brazil. *Plant and Soil*, *234*(1), 15–26. doi:10.1023/A:1010533721740

Tassou, S. A., Ge, Y., Hadawey, A., & Marriott, D. (2011). Energy consumption and conservation in food retailing. *Applied Thermal Engineering*, *31*(2-3), 147–156. doi:10.1016/j.applthermaleng.2010.08.023

Taylor, J., Hafner, M., Yerushalmi, E., Smith, R., Bellasio, J. M., & Vardavas, R. … Rubin, J. (2014). Estimating the economic costs of antimicrobial resistance: model and results. Cambridge, UK: RAND Europe.

Taylor, N. (2012). Reversing meat-eating culture to combat climate change. Haslemere, UK: World Preservation Foundation (WPF).

Taylor, E. F., Burley, V. J., Greenwood, D. C., & Cade, J. E. (2007). Meat Consumption and risk of breast cancer in the UK Women's Cohort Study. *British Journal of Cancer, 96*(7), 1139–1146. doi:10.1038/sj.bjc.6603689 PMID:17406351

Taylor, N., & Twine, R. (2014). Introduction: Locating the 'critical' in critical animal studies. In N. Taylor & R. Twine (Eds.), *The rise of critical animal studies: from the margins to the centre* (pp. 1–16). New York, NY: Routledge. doi:10.5744/florida/9780813049205.003.0001

Taylor, P. W. (2011). *Respect to nature: a theory of environmental ethics.* Princeton, NJ: Princeton University Press.

ten Brink, P., Berghöfer, A., Schröter-Schlaack, C., Sukhdev, P., Vakrou, A., White, S., & Wittmer, H. (2009). The economics of ecosystems and biodiversity for national and international policy makers. Executive summary: responding to the value of nature. United Nations Environment Programme (UNEP). Retrieved from http://www.unep.org/pdf/TEEB_D1_Summary.pdf

Thai, K. V., Rahm, D., & Coggburn, J. D. (2007). *Handbook of globalisation and the environment.* London, UK: CRC Press. doi:10.1201/9781420016932

The Afro American. (1974, August 10). Tighten your belt: food prices to soar with world demand. *The Afro American.* Retrieved from https://news.google.com/newspapers?nid=UBnQDr5gPskC&dat=19740810&b_mode=2&hl=en

The Cambridge Declaration on Consciousness. (2012). Retrieved from http://yourbrainandyou.com/2012/08/24/the-cambridge-declaration-on-consciousness/

The Economist. (2014, June 14). Obesity. Chubby little emperors: why China is under- and over-nourished at the same time. *From the print edition: China.* Retrieved from http://www.economist.com/news/china/21604221-why-china-under-and-over-nourished-same-time-chubby-little-emperors

The Law Society of New South Wales Young Lawyers (LSNSWYL). (2014). Submission on the Agriculture Competitiveness Green Paper. Sydney, Australia: LSNSWYL, Animal Law Committee. Retrieved from https://www.lawsociety.com.au/cs/groups/public/documents/internetyounglawyers/920563.pdf

The Vegan Society. (2012). Donald Watson. Retrieved from http://www.foodsforlife.org.uk/people/Donald-Watson-Vegan/Donald-Watson.html

The Vegetarian/Vegan Society of Queensland Incorporated (VVSQ). (2010). A pound of flesh. Brisbane, Australia: VVSQ. Retrieved from https://www.voiceless.org.au/sites/default/files/PoundofFlesh220310.pdf

The Week Staff (WS). (2013). Farm subsidies: a welfare program for agribusiness. *The Week,* Retrieved from http://theweek.com/articles/461227/farm-subsidies-welfare-program-agribusiness

Theis, I. M. (2008). *Desenvolvimento-e-territótio: questões teóricas, evidências empíricas* [Development and territory: theoretical issues, empirical evidences]. Santa Cruz, Brazil: Edunsic.

Thomas, K. (1996). *O homem e o mundo natural: mudanças de atitude em relação as plantas e aos animais (1500-1800)* [Human and the natural world: attitude changes towards plants and animals]. Sao Paulo, Brazil: Companhia das Letras.

Thornton, P., & Herrero, M. (2010). The inter-linkages between rapid growth in livestock production, climate change and the impacts on water resources, land use and deforestation. Policy Research Working Paper 5178. Nairobi, Kenya: International Livestock Research Institute. Retrieved from https://openknowledge.worldbank.org/handle/10986/9223

Thornton, P., Herrero, M., & Ericksen, P. (2011). Livestock and climate change. Nairobi, Kenya: International Livestock Research Institute. Retrieved from https://cgspace.cgiar.org/bitstream/handle/10568/10601/IssueBrief3.pdf

Thornton, P., Herrero, M., & Ericksen, P. (2011). Livestock and climate change. Nairobi, Kenya: International Livestock Research Institute. Retrieved September 8, 2015, from https://cgspace.cgiar.org/bitstream/handle/10568/10601/IssueBrief3.pdf

Tilman, D., & Balzar, C., Hill, & Befort, B. L. (2011). Global food demand and the sustainable intensification of agriculture. Proceedings of the National Academy of Sciences of the United States of America of the Unites States of America, 108(50), 20260-20264. doi:10.1073/pnas.1116437108

Tilman, D., & Clark, M. (2014). Global diets link environmental sustainability and human health. *Nature, 515*(7528), 518–522. doi:10.1038/nature13959 PMID:25383533

Timmer, J. (2011). Lab-grown meats face long road to supermarket. *Wired*. Retrieved from http://www.wired.com/2011/11/artificial-meat-economics/

Ting, I. (2013, April 16). Hold the red, pass the white - meat that is. *The Sydney Morning Herald*. Retrieved from http://www.smh.com.au/national/hold-the-red-pass-the-white--meat-that-is-20130415-2hvv6.html?skin=text-only

Tobler, C., Visschers, V. H., & Siegrist, M. (2011). Eating green: Consumer's willingness to adopt ecological food consumption behaviours. *Appetite, 57*(3), 674–682.

Tobler, C., Visschers, V. H., & Siegrist, M. (2011). Eating green: Consumers' willingness to adopt ecological food consumption behaviours. *Appetite, 57*(3), 674–682. doi:10.1016/j.appet.2011.08.010 PMID:21896294

Todar, K. (2012). *The good, the bad, and the deadly.* Online Textbook of Bacteriology. Retrieved September 6, 2015, from http://textbookofbacteriology.net/resantimicrobial.html

Toledo, E., Hu, F. B., Estruch, R., Buil-Cosiales, P., Corella, D., Salas-Salvado, J., & Martinez-Gonzalez, M. A. et al. (2013). Effect of the Mediterranean diet on blood pressure in the PREDIMED trial: Results from a randomized controlled trial. *BioMed Central Medicine, 11*, 207.

Tomson, B. (2013). Study: livestock drugs, hardier bugs. Retrieved from http://www.politico.com/story/2013/09/livestock-antibiotic-study-proof-97078.html

Tonsor, G. T., & Olynk, N. J. (2011). U.S. meat demand: The influence of animal welfare on media coverage. *Journal of Agricultural Economics, 62*(1), 59–72. doi:10.1111/j.1477-9552.2010.00266.x

Tonstad, S., Butler, T., Yan, R., & Fraser, G. E. (2009). Type of vegetarian diet, body weight, and prevalence of type 2 diabetes. *Diabetes Care, 32*(5), 791–796.

Tonstad, S., Butler, T., Yan, R., & Fraser, G. E. (2009). Type of vegetarian diet, body weight, and prevalence of type 2 diabetes. *Diabetes Care, 32*(5), 791–796. doi:10.2337/dc08-1886 PMID:19351712

Tonstad, S., Stewart, K., Oda, K., Batech, M., Herring, R. P., & Fraser, G. E. (2013). Vegetarian diets and incidence of diabetes in the Adventist Health Study-2. *Nutrition, Metabolism, and Cardiovascular Diseases, 23*(4), 292–299.

Trichopoulou, A., & Lagiou, P. (1997). Healthy traditional Mediterranean diet: An expression of culture, history, and lifestyle. *Nutrition Reviews, 55*(11), 383–389.

Trostle, R. (2008). Global agricultural supply and demand: factors contributing to the recent increase in food commodity prices. Washington, DC: United States Department of Agriculture (USDA), Economic Research Service. Retrieved September 8, 2015, from http://www.ers.usda.gov/media/218027/wrs0801_1_.pdf

Trumbo, C. (1996). Constructing climate change: Claims and frames in US news coverage of an environmental issue. *Public Understanding of Science (Bristol, England)*, *5*(3), 269–283. doi:10.1088/0963-6625/5/3/006

Tsunehara, C. H., Leonetti, D. L., & Fujimoto, W. Y. (1990). Diet of second-generation Japanese-American men with and without non-insulin-dependent diabetes. *The American Journal of Clinical Nutrition*, *52*(4), 731–738.

Tucker, C. A. (2014). The significance of sensory appeal for reduced meat consumption. *Appetite*, *81*, 168–179.

Tukker, A., Huppes, G., Guinée, J. B., Heijungs, R., de Koning, A., & van Oers, L. … Nielsen, P. (2006). Environmental Impact of Products (EIPRO) Analysis of the life cycle environmental impacts related to the final consumption of the EU-25. Technical Report Series. Seville, Spain: European Commission, Joint Research Centre, Institute for Prospective Technological Studies. Retrieved September 8, 2015, from http://ec.europa.eu/environment/ipp/pdf/eipro_report.pdf

Tuomilehto, J., Lindstrom, J., Eriksson, J. G., Valle, T. T., Hamalainen, H., Ilanne-Parikka, P., & Uusitupa, M. et al. (2001). Prevention of type 2 diabetes mellitus by changes in lifestyle among subjects with impaired glucose tolerance. *The New England Journal of Medicine*, *344*(18), 1343–1350.

Tuomisto, H. L., & de Mattos, N. J. (2011). Life cycle assessment of cultured meat production. *Environmental Science & Technology*, *45*(14), 6117–6123. doi:10.1021/es200130u PMID:21682287

Turner-McGrievy, G. M., Barnard, N. D., Cohen, J., Jenkins, D. J., Gloede, L., & Green, A. A. (2008). Changes in nutrient intake and dietary quality among participants with type 2 diabetes following a low-fat vegan diet or a conventional diabetes diet for 22 weeks. *Journal of the American Dietetic Association*, *108*(10), 1636–1645.

Turner-McGrievy, G. M., Davidson, C. R., Wingard, E. E., & Billings, D. L. (2014). Low glycemic index vegan or low-calorie weight loss diets for women with polycystic ovary syndrome: A randomized controlled feasibility study. *Nutrition Research (New York, N.Y.)*, *34*(6), 552–558.

Turner-McGrievy, G. M., Davidson, C. R., Wingard, E. E., Wilcox, S., & Frongillo, E. A. (2015). Comparative effectiveness of plant-based diets for weight loss: A randomized controlled trial of five different diets. *Nutrition (Burbank, Los Angeles County, Calif.)*, *31*(2), 350–358.

Turnridge, J., Nimmo, G., & Francis, G., & Australian Group on Antimicrobial Resistance. (1996). Evolution of resistance in Staphylococcus aureus in Australian teaching hospitals. *The Medical Journal of Australia*, *164*(2), 68–71. PMID:8569574

Tuttle, W. (2005). *The world peace diet: eating for spiritual health and social harmony*. New York, NY: Lantern Books.

Twigg, J. (1983). Vegetarianism and the meanings of meat. In A. Murcott (Ed.), *The sociology of food and eating: essays on the sociological significance of food* (pp. 18–30). Aldershot, UK: Gower.

United Nations Department of Economic and Social Affairs (UNDESA). (2013). Population Division, Population Estimates and Projections Section. Retrieved from http://esa.un.org/wpp/unpp/p2k0data.asp

United Nations Educational Scientific and Cultural Organization (UNESCO). (2008). Baul songs. Retrieved from http://www.unesco.org/culture/ich/rL/00107

United Nations Environment Programme (UNEP). (2010a). Assessing the environmental impacts of consumption and production, priority products and materials. Paris, France: UNEP, International Panel for Sustainable Resource Management. Retrieved from http://www.greeningtheblue.org/sites/default/files/Assessing%20the%20environmental%20impacts%20of%20consumption%20and%20production.pdf

United Nations Environment Programme (UNEP). (2010b). State of biodiversity in Latin America and the Caribbean. Retrieved from http://www.unep.org/delc/Portals/119/LatinAmerica_StateofBiodiv.pdf

United Nations Environment Programme (UNEP). (2014). The emissions gap report 2014. A UNEP synthesis report. Nairobi, Kenya: UNEP. Retrieved from http://www.unep.org/publications/ebooks/emissionsgapreport2014/portals/50268/pdf/EGR2014_LOWRES.pdf

United Nations Framework Convention on Climate Change (UNFCC). (2014). National inventory submissions: Australia. Retrieved June 10, 2015, from http://unfccc.int/national_reports/annex_i_ghg_inventories/national_inventories_submissions/items/8108.php

United Nations Framework Convention on Climate Change (UNFCCC). (2014). Background on the UNFCCC: The international response to climate change. Retrieved from http://unfccc.int/essential_background/items/6031.php

United Nations. (2013). World population prospects: the 2012 revision. New York, NY: United Nations. Retrieved from http://esa.un.org/unpd/wpp/Publications/Files/WPP2012_HIGHLIGHTS.pdf

United States Department of Agriculture (USDA) and United States Department of Health and Human Services. (USDHHS). (2010). Dietary Guidelines for Americans, 2010. Washington DC: USDA and USDHHS. Retrieved September 9, 2015, from http://health.gov/dietaryguidelines/dga2010/dietaryguidelines2010.pdf

United States Department of Agriculture (USDA) and United States Department of Health and Human Services. (USDHHS). (2015). Scientific report of the 2015 Dietary Guidelines Advisory Committee. Retrieved September 7, 2015, from http://health.gov/dietaryguidelines/2015-scientific-report/pdfs/scientific-report-of-the-2015-dietary-guidelines-advisory-committee.pdf

United States Department of Agriculture (USDA) and United States Department of Health and Human Services. (USDHHS). (2015). Scientific Report of the 2015 Dietary Guidelines Advisory Committee. United States Department of Health & Human Services. Retrieved September 19, 2015, from http://health.gov/dietaryguidelines/2015-scientific-report/pdfs/scientific-report-of-the-2015-dietary-guidelines-advisory-committee.pdf

University of Maryland Medical Centre. (2013). Omega-6 fatty acids. Retrieved April 2, 2015, from http://umm.edu/health/medical/altmed/supplement/omega6-fatty-acids

Uribarri, J., Cai, W., Peppa, M., Goodman, S., Ferrucci, L., Striker, G., & Vlassara, H. (2007). Circulating glycotoxins and dietary advanced glycation endproducts: Two links to inflammatory response, oxidative stress, and aging. *The Journals of Gerontology. Series A, Biological Sciences and Medical Sciences*, *62*(4), 427–433.

Uribarri, J., Woodruff, S., Goodman, S., Cai, W., Chen, X., Pyzik, R., & Vlassara, H. et al. (2010). Advanced glycation end products in foods and a practical guide to their reduction in the diet. *Journal of the American Dietetic Association*, *110*(6), 911–916.

van Aelst, P., & Walgrave, S. (2011). Minimal or massive? the political agenda-setting power of the mass media according to different methods. *The International Journal of Press/Politics*, *16*(3), 295–313. doi:10.1177/1940161211406727

van Boeckel, T. P., Brower, C., Gilbert, M., Bryan, T., Grenfell, B. T., & Levina, S. A., … Laxminarayan, R. (2014). Global trends in antimicrobial use in food animals. Proceedings of the National Academy of Sciences (PNAS), 112(18), 5649-5654.

van der Werf, G. R., Morton, D. C., de Fries, R. S., Oliver, J. G., Kasibhatla, P. S., Jackson, R. B., & Randerson, J. T. et al. (2009). CO2 emissions from forest loss. *Nature Geoscience*, *2*(11), 737–738. doi:10.1038/ngeo671

van Dooren, C., Marinussen, M., Blonk, H., Aiking, H., & Vellinga, P. (2014). Exploring dietary guidelines based on ecological and nutritional values: A comparison of six dietary patterns. *Food Policy*, *44*, 36–46. doi:10.1016/j.foodpol.2013.11.002

van Puyvelde, K., Mets, T., Njemini, R., Beyer, I., & Bautmans, I. (2014). Effect of advanced glycation end product intake on inflammation and aging: A systematic review. *Nutrition Reviews*, *72*(10), 638–650.

van Sluijs, E. M., van Poppel, M. N., & van Mechelen, W. (2004). Stage-based lifestyle interventions in primary care: Are they effective? *American Journal of Preventive Medicine*, *26*(4), 330–343. doi:10.1016/j.amepre.2003.12.010 PMID:15110061

Vang, A., Singh, P. N., Lee, J. W., Haddad, E. H., & Brinegar, C. H. (2008). Meats, processed meats, obesity, weight gain and occurrence of diabetes among adults: Findings from Adventist Health Studies. *Annals of Nutrition & Metabolism*, *52*(2), 96–104.

Vanhonacker, F., van Loo, E. J., Gellynck, X., & Verbeke, W. (2013). Flemish consumer attitudes towards more sustainable food choices. *Appetite*, *62*, 7–16.

Varner, G. (2001). Sentientism. In D. Jamieson (Ed.), *A companion to environmental philosophy* (pp. 192–203). Malden, MA: Blackwell. doi:10.1002/9780470751664.ch13

Vegan Society NSW. (2012). Vegan dining campaign. Retrieved from http://www.vegansocietynsw.com/vs/html/campaigns_vegan_dining.html

Ventanas, S., Ventanas, J., Tovar, J., García, C., & Estévez, M. (2007). Extensive feeding versus oleic acid and tocopherol enriched mixed diets for the production of Iberian dry-cured hams: Effect on chemical composition, oxidative status and sensory traits. *Meat Science*, *77*(2), 246–256. doi:10.1016/j.meatsci.2007.03.010 PMID:22061597

Verain, M., Dagevos, H., & Antonides, G. (2015). Flexitarianism a range of sustainable food styles. In L. A. Reisch & J. Thøgersen (Eds.), *Handbook of research on sustainable consumption* (pp. 209–223). Cheltenham, UK: Edward Elgar Publishing. doi:10.4337/9781783471270.00023

Verain, M., Dagevos, H., & Antonides, G. (2015). Flexitarianism: a range of sustainable food styles. In L. A. Reisch & J. Thøgersen (Eds.), *Handbook of research on sustainable consumption* (pp. 209–223). Cheltenham, UK: Edward Elgar Publishing.

Vergnaud, A. C., Norat, T., Romaguera, D., Mouw, T., May, A. M., Travier, N., & Peeters, P. H. et al. (2010). Meat consumption and prospective weight change in participants of the EPIC-PANACEA study. *The American Journal of Clinical Nutrition*, *92*(2), 398–407.

Verheignjen, L. A. H. M., Weiersna, D., Hulshoff, L. W., & de Wit, J. (1996). Livestock and the environment. Finding a balance: management of waste for animal processing, Wageningen, the Netherlands: International Agriculture Centre. Retreived September 8, 2015, from http://www.fao.org/ag/againfo/programmes/en/lead/toolbox/Refer/IACwaste.PDF

Vermeulen, S. J., Campbell, B. M., & Ingram, J. S. (2012). Climate change and food systems. *Annual Review of Environment and Resources*, *37*(1), 195–222. doi:10.1146/annurev-environ-020411-130608

Vessby, B. (2000). Dietary fat and insulin action in humans. *The British Journal of Nutrition*, *83*(1), S91–S96.

Vieux, F., Darmon, N., Touazi, D., & Soler, L. G. (2012). Greenhouse gas emissions of self-selected individual diets in France: Changing the diet structure or consuming less? *Ecological Economics*, *75*, 91–101. doi:10.1016/j.ecolecon.2012.01.003

Vinnari, M., & Vinnari, E. (2014). A framework for sustainable transition: The case of plant-based diets. *Journal of Agricultural & Environmental Ethics*, *27*, 369–396.

Vranken, L., Avermaete, T., Petalios, D., & Mathijs, E. (2014). Curbing global meat consumption: Emerging evidence of a second nutrition transition. *Environmental Science & Policy*, *39*, 95–106.

Walsh, F. (2014, July 2). Antibiotic resistance: Cameron warns of medical 'dark ages'. BBC News. Retrieved from http://www.bbc.com/news/health-28098838

Wang, Z., Zhang, B., Wang, H., Zhang, J., Du, W., & Su, C. … Zhai, F. (2013). Multi-level and longitudinal association between red meat consumption and changes in BMI, body weight and risk of incident overweight among Chinese adults (In Chinese). Paper presented at the 11th National Nutritional Science Conference of Chinese Nutrition Society, Hangzhou, China. *Zhonghua Liu Xing Bing Xue Za Zhi*, 34(7):661-667. Retrieved from http://www.ncbi.nlm.nih.gov/pubmed/24257164

Wang, J., Joshi, A. D., Corral, R., Siegmund, K. D., Marchand, L. L., Martinez, M. E., & Stern, M. C. et al. (2012). Carcinogen metabolism genes, red meat and poultry intake, and colorectal cancer risk. *International Journal of Cancer*, 130(8), 1898–1907.

Wang, Y., & Beydoun, M. A. (2009). Meat consumption is associated with obesity and central obesity among US adults. *International Journal of Obesity*, 33(6), 621–628.

Wang, Y., Lehane, C., Ghebremeskel, K., & Crawford, M. A. (2009). Modern organic and broiler chickens sold for human consumption provide more energy from fat than protein. *Public Health Nutrition*, 13(3), 400–408. doi:10.1017/S1368980009991157 PMID:19728900

Wang, Z., Klipfell, E., Bennett, B. J., Koeth, R., Levison, B. S., Dugar, B., & Hazen, S. L. et al. (2011). Gut flora metabolism of phosphatidylcholine promotes cardiovascular disease. *Nature*, 472(7341), 57–63.

Warren, K. (2000). *Ecofemenist philosophy: a western perspective on what it is and why it matters*. Lanham, MD: Rowman and Littlefield.

Water Footprint Network. (2014a). Product gallery: Beef. Retrieved December 17, 2014, from http://waterfootprint.org/en/resources/interactive-tools/product-gallery/

Water Footprint Network. (2014b). What is water footprint? Retrieved December 17, 2014, from http://waterfootprint.org/en/water-footprint/what-is-water-footprint/

Waters, C. (2014). Accommodating the target: veganism, healthfulness, and hegemonic masculinities. Olympia, WA: Faunalytics (Formerly Humane Research Council).

Watson, R., & Pauly, D. (2001). Systematic distortions in world fisheries catch trends. *Nature*, 414, 534–536.

Watts, C. (2012). Methane and livestock: factoids help farmers least of all. Retrieved June 10, 2015, from http://www.climateinstitute.org.au/articles/opinion-pieces/methane-and-livestock-factoids-help-farmers-least-of-all.html

Weber, C. L., & Matthews, H. S. (2008). Food-miles and the relative climate impacts of food choices in the United States. *Environmental Science & Technology*, 42(10), 3508–3513. doi:10.1021/es702969f PMID:18546681

Wegener, H. C. (2003). Antibiotics in animal feed and their role in resistance development. *Current Opinion in Microbiology*, 6(5), 439–445. doi:10.1016/j.mib.2003.09.009 PMID:14572534

Weggemans, R. M., Zock, P. L., & Katan, M. B. (2001). Dietary cholesterol from eggs increases the ratio of total cholesterol to high-density lipoprotein cholesterol in humans: A meta-analysis. *The American Journal of Clinical Nutrition*, 73(5), 885–891.

Weidema, B. M., Wesnes, M., Hermansen, J., Kristensen, T., Halberg, N., Eder, P., & Delgado, L. (2008). Environmental improvement potentials of meat and dairy products: JRC Scientific and Technical Reports, No. EUR 23491. Seville, Spain: Institute for Prospective Technological Studies. Retrieved from http://ftp.jrc.es/EURdoc/JRC46650.pdf

Weobong, C. A.-A., & Adinyira, E. Y. (2011). Operational impacts of the tamale abattoir on the environment. *Journal of Public Health and Epidemiology, 3*(9), 386–393.

Westhoek, H., Rood, T., van den Berg, M., Janse, J., Nijdam, D., Reudink, M., & Stehfest, E. (2011). The protein puzzle: the consumption and production of meat, dairy and fish in the European Union. The Hague, the Netherlands: Netherlands Environmental Assessment Agency (NEAA). Retrieved September 7, 2015, from http://www.pbl.nl/en/publications/2011/meat-dairy-and-fish-options-for-changes-in-production-and-consumption

Westhoek, H., Lesschen, J. P., Rood, T., Wagner, S., De Marco, A., Murphy-Bokern, D., & Oenema, O. et al. (2014). Food choices, health and environment: Effects of cutting Europe's meat and dairy intake. *Global Environmental Change, 26*, 196–205.

Westhoek, H., Lesschen, J. P., Rood, T., Wagner, S., DeMarco, A., Murphy-Bokern, D., & Oenema, O. et al. (2014). Food choices, health and environment: Effects of cutting Europe's meat and dairy intake. *Global Environmental Change, 26*, 196–205. doi:10.1016/j.gloenvcha.2014.02.004

West, R. (2005). Time for a change: Putting the transtheoretical (stages of change) model to rest. *Addiction (Abingdon, England), 100*(8), 1036–1039. doi:10.1111/j.1360-0443.2005.01139.x PMID:16042624

West, R. O., & Hayes, O. B. (1968). Diet and serum cholersterol levels. A comparision between vegetarians and nonvegetarians in a Seventh-day Adventist group. *The American Journal of Clinical Nutrition, 21*(8), 853–862.

Wheat, D. (2015). Push for national chicken cage standards stalls. *Capital Press*. Retrieved from http://www.capitalpress.com/Nation_World/Nation/20150102/push-for-national-chicken-cage-standards-stalls

White, D. L., & Collinson, A. (2013). Red meat, dietary heme iron, and risk of type 2 diabetes: The involvement of advanced lipoxidation endproducts. *Advances in Nutrition, 4*(4), 403–411.

Whitman, W. B., Coleman, D. C., & Wiebe, W. J. (1998). Prokaryotes: The unseen majority.[PNAS]. *Proceedings of the National Academy of Sciences of the United States of America, 95*(12), 6578–6583. doi:10.1073/pnas.95.12.6578 PMID:9618454

Willett, W. C. (2012). Dietary fats and coronary heart disease. *Journal of Internal Medicine, 272*(1), 13–24.

Willett, W. C., & Stampfer, M. J. (2013). Current evidence on healthy eating. *Annual Review of Public Health, 34*, 77–95.

Willett, W. C., Stampfer, M. J., Manson, J. E., Colditz, G. A., Speizer, F. E., Rosner, B. A., & Hennekens, C. H. et al. (1993). Intake of trans fatty acids and risk of coronary heart disease among women. *Lancet, 341*(8845), 581–585.

Williams, C. (2012). *The framing of animal cruelty by animal advocacy organizations. (Sociology with Honors)*. ME, USA: The University of Maine.

Wilson, M. S., Weatherall, A., & Butler, C. (2004). A rhetorical approach to discussions about health and vegetarianism. *Journal of Health Psychology, 9*(4), 567–581. doi:10.1177/1359105304044040 PMID:15231057

Wilson, N., Nghiem, N., Ni Mhurchu, C., Eyles, H., Baker, M. G., & Blakely, T. (2013). Foods and dietary patterns that are healthy, low-cost and environmentally sustainable: A case study of optimization modelling for New Zealand. *PLoS ONE, 8*(3), 1–10. doi:10.1371/journal.pone.0059648

Winglee, M. (2014). China's meat, processed food consumption is rising: what does that mean for China? *Food Tank*. Retrieved from http://foodtank.com/news/2014/06/chinas-rising-appetite

Wirsenius, S., & Hedenus, F. (2010). Policy strategies for a sustainable food system: options for protecting the climate. In J. D'Silva & J. Webster (Eds.), *The meat crisis: developing more sustainable production and consumption* (pp. 237–253). London, UK: Earthscan.

Witte, W. (1998). Medical consequences of antibiotic use in agriculture. *Science*, *279*(5353), 996–997. doi:10.1126/science.279.5353.996 PMID:9490487

Wollenberger, L., Halling-Sorensen, B., & Kusk, K. O. (2000). Acute and chronic toxicity of veterinary antibiotics to Daphnia magna. *Chemosphere*, *40*(7), 723–730. doi:10.1016/S0045-6535(99)00443-9 PMID:10705550

Wong, L., Selvanathan, E. A., & Selvanathan, S. (2013, July 7-10). Changing patterns of meat consumption in Australia 2013. Paper presented at the Australian Conference of Economists, Murdoch University, Australia. Retrieved from http://www.murdoch.edu.au/School-of-Management-and-Governance/_document/Australian-Conference-of-Economists/Changing-pattern-of-meat-consumption-in-Australia.pdf

Wong, J. M. W. (2014). Gut microbiota and cardiometabolic outcomes: Influence of dietary patterns and their associated components. *The American Journal of Clinical Nutrition*, *100*(1), 369S–377S.

Wong, W. C., Pui, C. F., Tunung, R., Ubong, A., Noor Hidayah, M. S., Farinazleen, M. G., & Son, R. et al. (2012). Antibiogram patterm among cultures of listeria monocytogenes isolated from frozen burger patties in Malaysia. *Pertanika. Journal of Tropical Agricultural Science*, *35*(4), 793.

Woodford, J. (2007, March 23). How to live in a more sustainable way. *The Sydney Morning Herald*.

Woolhouse, M., Ward, M., van Bunnik, B., & Farrar, J. (2015). Antimicrobial resistance in humans, livestock and the wider environment. Biological Sciences, 370(1670), 1-7.

Woolhouse, M., & Farrar, J. (2014). Policy: An intergovernmental panel on antimicrobial resistance. *Nature*, *509*(7502), 555–557. doi:10.1038/509555a PMID:24877180

Working Party on Agricultural Policies and Markets (WPAPM). (2015). Global antimicrobial use in the livestock sector: final report. Paris, France: Organisation for Economic Co-operation and Development (OECD).

Works, M. (2009). The Guardian. *Media Works Asia Limited*. Retrieved from http://www.mediaworksasia.com/publication/detail/000000006/page1/index.html.bak

World Animal Protection (WAP). (2014). Fishing's phantom menace. Retrieved September 8, 2015, from https://www.worldanimalprotection.us.org/sites/default/files/us_files/summary_sea-change-tackling-ghost-fishing-gear-summary_us.pdf

World Bank. (2015) Population Estimates and Projections. Retrieved from http://data.worldbank.org/data-catalog/population-projection-tables

World Cancer Research Fund (WCRF), & American Institute for Cancer Research (AICR). (2007). Food, nutrition, physical activity and the prevention of cancer: a global perspective. Washington, DC: AICR.

World Cancer Research Fund (WCRF), & American Institute for Cancer Research (AICR). (2007). Food, nutrition, physical activity and the prevention of cancer: a global perspective. Washington, DC: AICR. Retrieved from http://www.aicr.org/assets/docs/pdf/reports/Second_Expert_Report.pdf

World Cancer Research Fund (WCRF), & American Institute for Cancer Research (AICR). (2011). Continious update project report. Colorectal cancer 2011: food, nutrition, physical activity and the prevention of colorectal cancer. London, UK and Washington, DC: WCRF and AICR. Retrieved September 8, 2015, from http://www.dietandcancerreport.org/cancer_resource_center/downloads/cu/Colorectal-Cancer-2011-Report.pdf

World Cancer Research Fund (WCRF). (2007). Food, nutrition and the prevention of cancer: a global perspective. Retrieved June 10, 2015, from http://www.wcrf.org/sites/default/files/Second-Expert-Report.pdf

World Cancer Research Fund (WCRF). (2007). Our Cancer Recommendations - Animal Foods. Retrieved February, 2015, from http://www.wcrf.org/int/research-we-fund/cancer-prevention-recommendations/animal-foods

World Cancer Research Fund. (2014). Our Cancer Prevention Recommendations. Retrieved November 17, 2014, from http://www.wcrf.org/int/research-we-fund/our-cancer-prevention-recommendations

World Commission on Environment and Development (WCED). (1987). *Our common future.* Oxford, UK: Oxford University Press.

World Health Organisation (WHO). (2001). WHO global strategy for containment of antimicrobial resistance. Geneva, Switzerland: WHO. Retrieved from http://www.who.int/drugresistance/WHO_Global_Strategy_English.pdf

World Health organisation (WHO). (2012). *Critically important antimicrobials for human medicine. 3rd Revision.* Geneva, Switzerland: WHO.

World Health Organisation (WHO). (2013). *WHO model list of essential medicines. 18th List.* Geneva, Switzerland: WHO.

World Health Organisation (WHO). (2014a). *Antimicrobial resistance: global report on surveillance 2014.* Geneva, Switzerland: WHO.

World Health Organisation (WHO). (2014b). WHO's first global report on antibiotic resistance reveals serious, worldwide threat to public health. Retrieved from http://www.who.int/mediacentre/news/releases/2014/amr-report/en/

World Health Organisation (WHO). (2015). Drug resistance: antimicrobial use. Retrieved from http://www.who.int/drugresistance/use/en/

World Health Organization (WHO). (2004). Global Strategy of Diet, Physical Activity and Health. Geneva, Switzerland: WHO. Retrieved September 10, 2015, from http://www.who.int/dietphysicalactivity/strategy/eb11344/strategy_english_web.pdf

World Health Organization (WHO). (2008). Obesity and overweight. Media centre. Retrieved September 8, 2015, from http://www.who.int/mediacentre/factsheets/fs311/en

World Health organization (WHO). (2008). *Protecting health from climate change, World Health Day 2008: Annex 1 Reducing your carbon footprint can be good for your health, a list of mitigating actions.* Geneva, Switzerland: WHO.

World Health Organization (WHO). (2009). Pandemic (H1N1) 2009 - update 100. Emergencies preparedness, response. Retrieved September 8, 2015, from http://www.who.int/csr/don/2010_05_14/en/

World Health Organization (WHO). (2011). Global status report on noncommunicable diseases 2010. Geneva, Switzerland: WHO. Retrieved Septemver 10, 2015, from http://www.who.int/nmh/publications/ncd_report2010/en/

World Health Organization (WHO). (2011). Guidelines for drinking-water quality (4th ed.). Geneva, Switzerland: WHO. Retrieved September 8, 2015, from http://www.who.int/water_sanitation_health/publications/2011/dwq_guidelines/en/

World Health Organization (WHO). (2011). *Tackling antibiotic resistance from a food safety perspective in Europe.* Copenhagen, Denmark: WHO.

World Health Organization (WHO). (2014). The top 10 causes of death. Media Centre. Retrieved March 27, 2015, from http://www.who.int/mediacentre/factsheets/fs310/en/index2.html

World Health Organization (WHO). (2014a). Cancer (Fact Sheet). Retrieved November 17, 2014, from http://www.who.int/mediacentre/factsheets/fs297/en/

World Health Organization (WHO). (2014a). Disability-adjusted life years (DALYs). Retrieved June 10, 2015, from http://www.who.int/gho/mortality_burden_disease/daly_rates/en/

World Health Organization (WHO). (2014b(). Nutritional landscape information system. Retrieved June 10, 2015, from http://apps.who.int/nutrition/landscape/report.aspx?iso=CHN&rid=1620&goButton=Go

World Health Organization (WHO). (2014b). Cardiovascular diseases (Fact sheet). Retrieved November 17, 2014, from http://www.who.int/mediacentre/factsheets/fs317/en

World Health Organization (WHO). (2014c). Diabetes (Fact sheet). Retrieved November 17, 2014, from http://www.who.int/mediacentre/factsheets/fs312/en/

World Health Organization (WHO). (2015a). Health topics: Diet. Retrieved April 2, 2015, from http://www.who.int/topics/diet/en/

World Health Organization (WHO). (2015b). Statistics: A(H5N1) reported human cases. Retrieved April 8, 2015, from http://www.who.int/influenza/human_animal_interface/EN_GIP_20150303cumulativeNumberH5N1cases.pdf?ua=1

World Health Organization (WHO). (2015c). H5N1 Influenza. Retrieved April 8, 2015, from http://www.who.int/influenza/human_animal_interface/avian_influenza/h5n1_research/faqs/en/

World Health Organization (WHO). (n.d.). Ten facts on food safety. Fact File. Retrieved March 2, 2015, from http://www.who.int/features/factfiles/food_safety/facts/en/index4.html

World Health Organization Regional Office for Europe (WHOEU). (2015). A healthy lifestyle. Retrieved March 1, 2015, from http://www.euro.who.int/en/health-topics/disease-prevention/nutrition/a-healthy-lifestyle

World Hunger Education Service. (2015). 2015 world hunger and poverty facts and statistics. Retrieved May 10, 2015, from http://www.worldhunger.org/articles/Learn/world hunger facts 2002.htm

World Wide Fund for Nature (WWF). (2014). Ecological footprint and sustainable consumption in China. Vaud/Beijing, Switzerland/China: WWF and China-ASEAN Environmental Cooperation Centre. Retrieved from http://www.footprint-network.org/images/article_uploads/China_EF_Sustainable_Consumption_2014_English.pdf

World Wildlife Fund (WWF). (2014). Living planet report 2014: summary. Gland, Switzerland: WWF. Retrieved from http://assets.wwf.org.uk/downloads/living_planet_report_2014_summary.pdf?_ga=1.38233059.1148581730.1413300010

Wrenn, C. L. (2012). Abolitionist animal rights: Critical comparisons and challenges within the animal rights movement. *Interface*, *4*(2), 438–458.

Wu, A. (2012, January 12). Livestock in China given too many antibiotics. The Epoch Times. Retrieved from https://kristinasaid.wordpress.com/2012/01/24/livestock-in-china-given-too-many-antibiotics/

Wyker, B. A., & Davison, K. K. (2010). Behavioural change theories can inform the prediction of young adult's adoption of a plant-based diet. *Journal of Nutrition Education and Behavior*, *42*(3), 168–167. doi:10.1016/j.jneb.2009.03.124 PMID:20138584

Xue, H., Liu, Y., Duan, R., Zhou, X., & Cheng, G. (2014). Trends in children's overweight and obesity in China and their influencing factors. *China School of Health*, *35*(8), 1258–1262.

Xue, X. J., Gao, Q., Qiao, J. H., Zhang, J., Xu, C. P., & Liu, J. (2014). Red and processed meat consumption and the risk of lung cancer: A dose-response meta-analysis of 33 published studies. *International Journal of Clinical and Experimental Medicine*, *7*(6), 1542–1553.

Xu, J., Yang, X. X., Wu, Y. G., Li, X. Y., & Bai, B. (2014). Meat consumption and risk of oral cavity and oropharynx cancer: A meta-analysis of observational studies. *Public Library of Science (PLoS). One*, *9*(4), 1–9.

Xu, X., Yu, E., Gao, X., Song, N., Liu, L., Wei, X., & Fu, C. et al. (2013). Red and processed meat intake and risk of colorectal adenomas: A meta-analysis of observational studies. *International Journal of Cancer*, *132*(2), 437–448.

Yang, W. S., Wong, M. Y., Vogtmann, E., Tang, R. Q., Xie, L., Yang, Y. S., & Xiang, Y. B. et al. (2012). Meat consumption and risk of lung cancer: Evidence from observational studies. *Annals of Oncology*, *23*(12), 3163–3170.

Yang, W., Li, B., Dong, X., Zhang, X. Q., Zeng, Y., Zhou, J. L., & Xu, J. J. et al. (2014). Is heme iron intake associated with risk of coronary heart disease? A meta-analysis of prospective studies. *European Journal of Nutrition*, *53*(2), 395–400.

Yates, R. (2015). *On human-nonhuman relations, a sociological exploration of speciesism*. Retrieved from http://onhumannonhumanrelations.tumblr.com

Yokoyama, Y., Nishimura, K., Barnard, N. D., Takegami, M., Watanabe, M., Sekikawa, A., & Miyamoto, Y. et al. (2014). Vegetarian diets and blood pressure: A meta-analysis. *Journal of the American Medical Association Internal Medicine*, *174*(4), 577–587.

Zarb, P., & Goossens, H. (2012). Human use of antimicrobial agents. *Revue Scientifique et Technique (International Office of Epizootics)*, *31*(1), 121–133. PMID:22849272

Zeuschner, C. L., Hokin, B. D., Marsh, K. A., Saunders, A. V., Reid, M. A., & Ramsay, M. R. (2013). Vitamin B12 and vegetarian diets. *The Medical Journal of Australia*, *199*(4), S27–S32.

Zhang, C., Schulze, M. B., Solomon, C. G., & Hu, F. B. (2006). A prospective study of dietary patterns, meat intake and the risk of gestational diabetes mellitus. *Diabetologia*, *49*(11), 2604–2613.

Zhao, Z., Li, S., Liu, G., Yan, F., Ma, X., Huang, Z., & Tian, H. (2012). Body iron stores and heme-iron intake in relation to risk of type 2 diabetes: A systematic review and meta-analysis. *Public Library of Science (PLoS). One*, *7*(7), 1–17.

Zheng, W., & Lee, S. A. (2009). Well-done meat intake, heterocyclic amine exposure, and cancer risk. *Nutrition and Cancer*, *61*(4), 437–446.

Zhu, H. C., Yang, X., Xu, L. P., Zhao, L. J., Tao, G. Z., Zhang, C., & Sun, X. C. et al. (2014). Meat consumption is associated with esophageal cancer risk in a meat- and cancer-histological-type dependent manner. *Digestive Diseases and Sciences*, *59*(3), 664–673.

Zhu, H., Yang, X., Zhang, C., Zhu, C., Tao, G., Zhao, L., & Sun, X. et al. (2013). Red and processed meat intake is associated with higher gastric cancer risk: A meta-analysis of epidemiological observational studies. *Public Library of Science (PLoS). One*, *8*(8), 1–10.

Zohar, D., & Marshall, I. (2001). *Spiritual intelligence: the ultimate intelligence*. London, UK: Bloomsbury Publishing.

Zuraw, L. (2014). UK review makes staggering antibiotic resistance predictions. Retrieved from http://www.foodsafetynews.com/2014/12/uk-review-releases-staggering-antibiotic-resistance-predictions/VcAwspOqqko

Zur, I., & Klöckner, C. A. (2014). Individual motivations for limiting meat consumption. *British Food Journal*, *116*(4), 629–642.

About the Contributors

Talia Raphaely Originally from Cape Town, South Africa, Talia has 30 years of international experience in behavioural and attitudinal change, communications and diverse media, sustainability awareness and consciousness and collaboration and partnership building for increasing sustainable outcomes. She has worked closely with multicultural and heterogeneous groups in a diverse array of organizational settings, including academia, media, research-based organisations, government bodies, non-government organisations, community-based organisations and industry. Talia works as an academic at the Curtin University Sustainability at Policy (CUSP) Institute in Perth, Western Australia and continues to undertake consultancy research relating to sustainability. She is the Editor-in-Chief (together with Dora Marinova) of the International Journal of Disease Control and Containment for Sustainability (published by IGI Global). Talia recognised for her work on flexitarianism (reducing meat consumption to within healthy levels as recommended by national and international guidelines), collaboration, empowerment and sustainability humanistic education.

Dora Marinova Originally from Sofia, Bulgaria, Dora Marinova moved to Perth, Western Australia in 1991. After being Head of School at the Institute for Sustainability and Technology Policy (ISTP) at Murdoch University, she is now Director of the Curtin University Sustainability Policy (CUSP) Institute. Dora has over 400 refereed publications and has supervised 50 PhD students to successful completion. She is a member of the National Health and Medical Research Council's Panel on Centres of Research Excellence in Population Health. Her research interests cover innovation models, including the evolving global green system of innovation and the emerging area of sustainometrics. Dora is the Editor-in-Chief (together with Talia Raphaely) of the International Journal of Disease Control and Containment for Sustainability (published by IGI Global) and Editorial Board member of the International Journal of Education Economics and Development (published by Inderscience, Switzerland) and Transformations: An Interdisciplinary Journal (published by EBSCO, USA). She is Elected Fellow of the prestigious Modelling and Simulation Society of Australia and New Zealand (MSSANZ) and International Environmental Modelling and Software Society (IEMSS).

* * *

Jonathan Balcombe has a PhD in ethology from the University of Tennessee, where he studied communication in bats. He has published over 50 scientific papers and book chapters on animal behavior and animal protection. He is the author of four books: *The Exultant Ark: A Pictorial Tour of Animal Pleasure* (University of California Press, 2011); *Second Nature: The Inner Lives of Animals* (Palgrave

Macmillan, 2010); *Pleasurable Kingdom: Animals and the Nature of Feeling Good* (Macmillan, 2006); and *The Use of Animals in Higher Education: Problems, Alternatives, and Recommendations* (Humane Society Press, 2000). His next book, *What a Fish Knows*, is scheduled for publication in 2016. Jonathan recently served as Department Chair for Animal Studies with Humane Society University, and is now Director for Animal Sentience with the Humane Society Institute for Science and Policy, in Washington, DC. In 2011 he was presented with the St. Francis of Assisi Award by the New Zealand Companion Animal Council.

Diana Bogueva has almost 20 years of international communication and media experience as a professional radio journalist working with several worldwide news media, including the Bulgarian National Radio, Bulgarian Section of BBC, London and Special Broadcasting Services (SBS), Sydney. She is currently working for SBS radio as a program producer and as a director of market research with Sydney Focus Groups. Diana is also completing her PhD studies at Curtin University focusing on meat consumption and sustainability. She is a member of the International Federation of Journalists and a Wood Badge member of the World Scout Organization.

Gremma Carey is a Research Fellow with the Regulatory Network at the Australian National University. She holds a PhD in social policy and population health from the University of Melbourne. Her research sit at the critical interface between public health, public administration and social policy. In particular, Dr Carey has investigated processes of 'joining up' within government and between government and non-government organisations. Carey has published a widely on different aspects of public administration and public health. She has a monograph on joined-up government and public administration with Melbourne University Press due for release in 2015 – 'Grassroots to Government: Joining-up in Australia', and is Chief Editor and contributor to the upcoming book 'Designing and Implementing Public Policy: Cross-sectoral Debates', to be published by Routledge in 2015.

Hans Dagevos (PhD) is senior researcher at the Agricultural Economics Research Institute (LEI Wageningen UR), and fellow at Economics of Consumers and Households, both at the Social Sciences Group of Wageningen University and Research Centre. Hans received his PhD with honours from Utrecht University in 1994. His research interests are in the field of the sociology of consumption with a special focus on food. His bibliography contains around 250 publications in de form of scientific papers, edited books, book chapters and book reviews, research reports, columns and weblogs.

Luciano Félix Florit Graduated in Sociology at the Universidad de Buenos Aires (1995), master's degree in Political Sociology at the Federal University of Santa Catarina (1998) and Ph.D. in Sociology at the Federal University of Rio Grande do Sul (2003). He is currently professor at the Regional University of Blumenau (Santa Catarina, Brazil) where he acts in the Postgraduate Program in Regional Development in Masters and PhD courses. His research fields are the Sociology of Development, Environmental Sociology and Environmental Ethics, currently focusing on Environmental Ethics and Regional Development with the following sub-themes: 1 Moral consideration of not human beings and development patterns; 2 - Environmental justice and development patterns; 3 - Development models, value systems and welfare.

Celia Green is a PhD researcher at the Australian National University, National Centre for Epidemiology and Population Health where she is currently investigating the worldwide lack of robust policies around environmentally sustainable diets. Her research interests include climate change and diet, public health and public policy, and social equity issues. Celia also has an honours degree in Veterinary Medicine and Surgery from Murdoch University and has extensive experience working as a veterinarian in both small and large animal practice.

Robert Goodland retired as lead environmental adviser at the World Bank Group after serving there for 23 years, and was the first ever winner of the World Conservation Union's Harold Jefferson Coolidge medal for lifetime achievement in the conservation of nature and natural resources.

Xiumei Guo is a research fellow at the Curtin University Sustainability Policy (CUSP) Institute in Australia. Her research interests are in the area of demography, economic development, energy efficiency and sustainability studies with a focus on China. Xiumei has conducted research for the Australian Research Council, has published widely and is an accomplished supervisor of research students.

Jonathan Hallett is a health promotion lecturer in the School of Public Health at Curtin University teaching public health practice, program evaluation and public health politics. His research with the Collaboration for Evidence, Research and Impact in Public Health (CERIPH) focuses on community-based intervention evaluation, evidence-based policy and advocacy methods and translational research in the areas of substance use, mental health, sexual health and dietary practices.

Amzad Hossain is a university associate at the Curtin University Sustainability Policy (CUSP) Institute, Australia and a guest teacher at the Institute of Environmental Science, Rajshahi University, Bangladesh. His interests cover research in indigenous culture, values education, self-reliance, appropriate technology, climate change and spirituality with relevance to individual, social, technological and environmental sustainability. He has published over 100 research papers, including 20 popular articles on the mystic Baul tradition of Bangladesh. Amzad has a MCom from Rajshahi University, MA (by research) in religious studies from Deakin University, Australia and two PhDs (in Anthropology and in Sustainability) from Murdoch University, Australia.

Andrew Joyce is a Research Fellow with the Centre for Social Impact Swinburne. Alongside his research on sustainable diets and intervention design, his research interests include the use of continuous quality improvement models and the role of evaluation within practice and policy settings. Andrew has a history of joint appointments between academia and practice, having previously worked for ten years as a public health lecturer at Curtin University and then Monash University together with industry related roles, coordinating and evaluating health and social policies and programs. More recently he has led the evaluation of state-wide policies in the area of public health and education.

John Kinsella is a Fellow of Churchill College, Cambridge University, Professor of Literature and Sustainability at Curtin University, and a Professorial Research Fellow at the University of Western Australia. He is the author of numerous prize-winning volumes of poetry, as well as volumes of short fiction, novels, plays, collaborative works, and critical volumes which include Disclosed Poetics, Activist Poetics, Contrary Rhetoric, and Spatial Relations.

Mira Marinova is a postgraduate student in medicine at Notre Dame University in Fremantle, Australia. She also works as a research assistant in studies related to chemotherapy and cancer treatment. With a background in physiotherapy, Mira previously worked for the How to Train Your Dragon Arena Spectacular show which toured in Australia, New Zealand, Canada and USA.

Kate Marsh is an Advanced Accredited Practicing Dietitian and Credentialled Diabetes Educator with a Masters of Nutrition and Dietetics and PhD in nutrition from the University of Sydney. She works in private practice in Sydney and has a particular interest in diabetes, polycystic ovary syndrome (PCOS), vegetarian nutrition and nutrition in pregnancy & pre-conception. Kate is convenor of the Dietitians Association of Australia (DAA) National Vegetarian Interest Group, and writes and speaks regularly on the topic of vegetarian and plant-based nutrition. She is co-author of The Low GI Vegetarian Cookbook and has published articles on vegetarian nutrition for a number of nutrition and medical journals including a recent supplement in the Medical Journal of Australia.

Xavier Mayes is a strategic communicator and postgraduate student studying sustainability and climate policy at Curtin University.

Never Mujere holds a Master of Philosophy in geography from the University of Zimbabwe. He is a lecturer at the Department of Geography and Environmental Science in the University of Zimbabwe where he is also reading for a Doctor of Philosophy in hydrology and water resources management. Never's research expertise covers geography, hydrology and environmental issues. He has conducted research for the Zimbabwe National Water Authority and published on water, water quality, climate change, river flow, flood frequency, rainflow variations and crop yields in Zimbabwe and in China, including two books. He is a founder of the non-governmental organization Environmental Management Trust (EMT), Zimbabwe.

Nick Pendergrast completed his PhD in Sociology at Curtin University in 2014. The research focused on the animal advocacy movement, primarily in Australia and the United States, and explored the range of ideologies, activism, organisations, and key actors that make up this movement. He also teaches Sociology and Anthropology at Curtin University, covering topics such as: human/non-human relations and the animal advocacy movement, intersectionality and social movements, gender and veiling, international trade and globalisation, the anti-neoliberal globalisation movement and anarchism, refugees and migration, nationalism and ethnicity, media and politics, and music and social change.

Ian Phau is a Professor in Marketing at Curtin University in Australia. His experience includes management consulting and market research in the luxury fashion industry in Europe and Asia. He has been involved in research and consulting projects, including portfolios for Hugo Boss, Araluen Botanic Park, Prada, Shiseido, AirAsia and BlackBerry. His research focuses on luxury branding, country image, and advertising appeals. Ian has extensive teaching experience in UK, Singapore, Hong Kong, Malaysia and Singapore. Ian is the Editor-in-Chief for the Asia Pacific Journal of Marketing and Logistics.

Geoff Russell is a computer programmer and author whose work has appeared in many magazines and newspapers in addition to some peer reviewed journals.

Angela Vince Saunders is an accredited practicing dietitian (APD) and accredited nutritionist (AN) with 30+ years' experience working in plant-based nutrition. She is a lifelong vegetarian and has raised her twin sons on a vegetarian diet, so she is both passionate about and experienced in vegetarian nutrition and its many health benefits. Angela obtained her degree in nutrition and dietetics from Loma Linda University in California, a university well known globally for its cutting edge research on vegetarian nutrition and longevity. She also has a MA in Leadership and Management which she obtained from Avondale College of Higher Education in 2001. Angela's dietetics career has been primarily in the clinical hospital setting, including Sydney Adventist Hospital, in charge of the Nutrition and Dietetics Department. Her current role as Senior Dietitian at Sanitarium Health and Wellbeing has given her opportunity to be a significant advocate for plant-based nutrition. Angela has published several scientific peer-reviewed papers on the health benefits of a vegetarian diet, including the 2012 Medical Journal of Australia 40-page supplement on vegetarian nutrition (www.mja.com.au/open/2012/1/2). She is an active spokesperson regarding the adequacy of a vegetarian diet for all age groups, busting the myths and controversies related to plant-based nutrition.

Cristiane Sbardelati graduated in history from the Centro Universitário de Brusque (2004). She specialized in education, society and culture at the Universidade de Blumenau (2008) and has a Master's degree in regional development from the Universidade de Blumenau (2015). Christiane is a post-graduate researcher at the Universidade de Blumenau focusing in regional development, including state, society and territory development and socioeconomic dynamics. She also teaches in extension programs.

Paul Shapiro serves as the vice president of farm animal protection for The Humane Society of the United States, the world's largest animal protection organization. He directs one of the organization's biggest advocacy teams, spearheading legislative initiatives to prevent farm animal abuse, engaging with major food corporations to help end the cruelest agribusiness practices from their supply chains, and leading a nationwide staff working in the field to increase demand for plant-based proteins by reducing consumption of animal products. In 1995, Shapiro founded Compassion Over Killing as a high school club, building it into a national organization over the next decade; he left in 2005 to start The HSUS's farm animal protection program. Today, COK is one of the world's leading organizations dedicated to combatting factory farming, with offices in Los Angeles and Washington, DC. Shapiro, an inductee into the Animal Rights Hall of Fame, has been interviewed by hundreds of print, broadcast and online news outlets as an authority on farm animal welfare and animal advocacy. He's also published dozens of articles about animal welfare in publications ranging from daily newspapers to academic journals.

Rosemary Stanton OAM has qualifications in science, nutrition & dietetics, and administration. She was awarded a doctorate for her many publications, and an Order of Australia for her contributions to public health through nutrition. Currently a Visiting Fellow in the Faculty of Medicine at the University of New South Wales, Rosemary is an invited member of many committees, including the NHMRC's Working Committee for the Dietary Guidelines and the Infant Feeding Guidelines for Health Workers, DOHA's Reference Group for the National Healthy Weight Guide, FSANZ Consumer and Public Health Dialogue, the University of Adelaide's Food Futures program, the Scientific Committee for Doctors for the Environment and the University of Technology's Institute for Sustainable Futures. She is an active member of the Public Health Association, as well as the Dietitians Association of Australia, the Nutrition Society, and the Australia New Zealand Obesity Society. Rosemary has authored many scientific papers,

33 books on food and nutrition (including several textbooks) and over 3,500 articles for magazines and newspapers. She maintains a close liaison with the Australian public and has been regarded by the media for almost 50 years as a source of reliable nutrition information.

Carol Zeuschner is the Director of Support Services at Sydney Adventist Hospital, Australia. She is an Accredited Practicing Dietitian with a Master of Science (Nutrition & Dietetics) from the University of Wollongong. She enjoys the challenges of working at the frontline in healthcare and understands the complex factors required to maximise food and nutrition for the health and recovery of patients. Carol has contributed to research on vegetarian nutrition, and has co-authored papers on vitamin B12, meeting nutrient reference values and adolescent nutrition, including publishing in the American Journal of Lifestyle Medicine and the Medical Journal of Australia.

Index

Become an IRMA Member

Members of the **Information Resources Management Association (IRMA)** understand the importance of community within their field of study. The Information Resources Management Association is an ideal venue through which professionals, students, and academicians can convene and share the latest industry innovations and scholarly research that is changing the field of information science and technology. Become a member today and enjoy the benefits of membership as well as the opportunity to collaborate and network with fellow experts in the field.

IRMA Membership Benefits:

- **One FREE Journal Subscription**

- **30% Off Additional Journal Subscriptions**

- **20% Off Book Purchases**

- Updates on the latest events and research on Information Resources Management through the IRMA-L listserv.

- Updates on new open access and downloadable content added to Research IRM.

- A copy of the Information Technology Management Newsletter twice a year.

- A certificate of membership.

IRMA Membership $195

Scan code to visit irma-international.org and begin by selecting your free journal subscription.

Membership is good for one full year.